Interdisciplinary Aspects of Climate Change

Environmental Education, Communication and Sustainability

edited by Walter Leal Filho

Vol. 29

PETER LANG
Frankfurt am Main · Berlin · Bern · Bruxelles · New York · Oxford · Wien

Walter Leal Filho
Franziska Mannke
(eds.)

Interdisciplinary Aspects of Climate Change

PETER LANG
Internationaler Verlag der Wissenschaften

Bibliographic Information published by the Deutsche Nationalbibliothek
The Deutsche Nationalbibliothek lists this publication in the Deutsche Nationalbibliografie; detailed bibliographic data is available in the internet at <http://www.d-nb.de>.

Copy editing and layout:
Kumpernatz + Bromann
www.kumpernatz-bromann.de

ISSN 1434-3819
ISBN 978-3-631-58153-7

www.peterlang.de

Table of Contents

Foreword

Climate change is a matter of great interest and worldwide concern. The latest evidences from scientific studies and officials documents produced by the International Panel on Climate Change (IPCC), a body established in 1988 as a joint effort of the World Meteorological Organization and the United Nations Environment Programme, show that the challenges posed by climate change need to be taken seriously if they are ever to be tackled properly. It is by no accident that the IPCC together with former US Vice-President Al Gore received the Peace Nobel Prize in 2007: climate change is one of the most important challenges of modern times and represents a field where integrated worldwide efforts are needed.

It is widely acknowledged that information, communication and education on climate change are important tools in the search for solutions to the social, economic or political problems climate change poses. But despite the fact that much has been written about climate change, most works to date tend to focus on specific issues (e.g. climate modeling, forecasts), as opposed to addressing the problem in an interdisciplinary way as it should be. This book tries to address this perceived gap by providing a wide range of perspectives on climate change, which goes over and above the traditional barriers seen among subjects. It is interdisciplinary in nature and comprehensive in scope.

This book was prepared in the context of "Climate 2008" (www.klima2008.net), the world's first scientific conference on climate change held on the internet. "Climate 2008" had the following aims:

- to introduce the latest findings on scientific research on climate change, including elements related to its environmental, social, economic and policy aspects;
- to introduce projects and other initiatives being undertaken in both industrialised and developing countries by universities and scientific institutions, government bodies, NGOs and other stakeholders;
- to discuss the problems, barriers, challenges and chances and potentials related to sustainability research projects in general.

Last but not least, the "Climate 2008" conference, which will repeated in 2009 under the heading "Climate 2009 / Klima 2009" (www.klima2009.net) was meant to encourage more networking and information exchange among participants and to catalyse new cooperation initiatives and possibly new projects.

As a follow-up to "Climate 2008", the Hamburg University of Applied Sciences has created the "International Climate Change Information Programme" (ICCIP). The aim of ICCIP is to analyse the problems which are connected with the communication of matters related to climate change and find ways to address them. In addition, ICCIP undertakes integrated communication and education projects, as well as organises events aimed at bringing about a broader understanding about climate change.

It is hoped this book, a follow-up to "Climate 2008" and one of the first outputs of ICCIP, will provide a long-needed contribution to a better understanding of the interdisciplinary nature of the subject matter of climate change, as well as offer an overview of some of the on-going interdisciplinary projects and initiatives in this field taking place in different parts of the world.

Walter Leal Filho, February 2009

Chapter 1

Impacts of Climate Change on Electricity Generation and Consumption

Solveig Mimler, Ulrike Müller, Stefanie Greis, Benno Rothstein

Abstract

The energy sector is highly dependent on the climate, as many climate parameters have an impact on energy demand, production and transport. The following chapter will concentrate on the impacts of air temperature changes on water temperatures concerning the electricity generation and consumption. On the one hand, thermal power plants using river water for cooling processes must meet statutory thresholds e.g. concerning mixed water temperature after waste heat discharge. On the other hand, electricity demand is influenced by temperature (heating demand in winter, cooling demand in summer). Both aspects have been further investigated for parts of Germany.

Regarding river water use, recent trends of river water temperatures and their extremes were analyzed in the first part. The trend analysis of water temperature showed significant negative trends for all rivers in spring and summer months whereas negative significant trends exist in winter and fall in some rivers. Also the frequencies of extreme values show a significant increase whereby the exceeding probability of extreme high water temperatures is inconsistent for the different rivers.

In the second part, the change in electricity consumption due to changes in mean air temperatures was examined for the state of Baden-Wuerttemberg, southwestern Germany. The analysis of mean changes in simulated future loads shows that between 1971 to 2000 and 2006 to 2035 significant positive trends exist for the months July, August and September and significant negative trends for January, May, June and December. However, these changes are marginal in absolute MW values.

Finally, for periods of extreme high air temperatures resulting in a possible bottleneck of electricity supply, existing adaptation options are presented. The electricity sector has to be aware of this context in order to adapt in the frame of existing options.

1. Introduction

Shifting in means and extremes of meteorological parameters due to climate change influences all areas of the electricity sector: generation, transportation/distribution and consumption. The following paper will concentrate on the impacts of air temperature changes on water temperatures concerning the electricity generation and consumption.

On the one hand, air temperature correlates with river water temperature. Thermal power plants in Germany mostly depend on usage of river water for cooling processes. Thereby they have to meet statutory thresholds concerning mixed water temperature after waste heat discharge for instance. Therefore, power plant operators are interested in how the water temperature will develop due to varying air temperatures if at all. In order to gain knowledge about future trends it is necessary to analyse recent water temperature trends and to identify the correlation between air and water temperatures for specific gauging stations. This issue will be the subject of the first section of this chapter.

On the other hand, electricity demand is influenced by temperature. Below a certain site-dependent temperature threshold, electricity consumption increases due to heating demand whereas above this threshold electricity consumption also increases but due to cooling. In order to estimate the impact of climate change on electricity demand, it is necessary to isolate the climate change effect from other load influencing variables. Section two of this chapter will describe the result of how regional climate model data is put into a load forecasting model for the exemplary region Baden-Wuerttemberg. With the help of this procedure, it is possible to simulate and to analyse future daily load curves (in terms of future climate).

Besides the above mentioned long-term impacts of climate change, the electricity sector is also confronted by extreme weather events like heat waves (such as the summer in 2003). During these periods, two critical situations can coincide and possibly cause a bottleneck in electricity supply: reduced electricity generation because statutory thresholds concerning warm water temperatures have to be kept and a higher electricity consumption due to rising cooling demands. The last section will propose possible adaptation options for this special situation.

2. The Electricity Sector and Water Use for the Cooling Process of Thermal Power Plants

In thermal power plants (e.g., fossil fuel-fired and nuclear power plants), heat is produced by combustion or by nuclear fission. In the case of gas turbine power plants, combustion gases directly drive a gas turbine whereas water is heated up to steam which drives a steam turbine in steam power stations (e.g., fossil fuel-fired and nuclear power plants). Finally, the mechanical energy is transformed into electrical energy using a connected generator.

The efficiency of a thermal power plant depends on the temperature interval upstream and downstream of the turbine. Since the temperature upstream turbines is limited by material conditions, it is energetically and economically preferable to cool down the steam (gas) after the turbine as far as possible. In order to achieve this either the steam can be drained off and used as local or district heat or the heat energy is dissipated to the cooling medium air or water (Kalide 2005). Due to the lower thermal absorption capacity of air, only power plants with a maximum thermal output of 900 megawatts thermal (MW_{th}) use air cooling systems. Water cooling systems are usually installed at bigger thermal power plant sites (European Commission 2001; Sauer 1984).

In water-cooled thermal power plants the steam downstream of the turbines is conveyed through a condenser where the remaining heat energy is dissipated to a second water cycle. After heating up the cooling water by about 10 Kelvin (K) it is either piped back to the receiving water (e.g. river) by means of a so-called once-through cooling system or the cooling water is cooled down in a cooling tower (Kalide 2005). Three alternative cooling tower operations exist that differ in terms of water demand, waste heat discharge into receiving water and evaporation loss; accordingly these are closed-circuit cooling, discharge cooling system and mixed operation (Sauer 1984).

The cooling tower operation reduces significantly the waste heat discharge into rivers (waste heat in MJ/s per 100 MW_{el}: once-through cooling system: 145-210, discharge cooling system: 84-125, closed-circuit cooling: 8-15). But this also results in evaporation losses up to 5% of cooling water volume depending on air temperature, humidity and pressure. Additional withdrawal of river water is then necessary to compensate the evaporation loss. Although linked to less water demand than a once-through cooling system the operation of a closed-circuit cooling system can intensify a low water level situation (Kobus & Bürkle 1996; Maniak 2005). Furthermore, an efficiency decrease of up to about 2-3% arises from cooling tower operation as opposed to once-through systems due to increasing station supply (e.g. water pumps and ventilators; Wagner 2003). Thermal power plants operating a cooling tower in Germany are able to switch between the above mentioned operating systems. Otherwise they are reliant on the once-through cooling system. Depending on power plant sites and on stream types, statutory thresholds exist concerning cooling water withdrawal and discharge (see table 1 and 2). These thresholds were set based upon the references of the "Working Group of Federal States on Water Issues" (Länderarbeitsgemeinschaft Wasser, LAWA). If these thresholds are reached, load reduction is triggered in order to reduce the evaporation loss and therefore the cooling water demand respectively the waste heat discharge into rivers (LAWA 1991; Maniak 2005).

Since this occurred, for instance in summer 2003 and 2006, the electricity sector is interested in the development of water temperatures in rivers to adjust production sites to possible changes (Müller et al. 2007).

Tab. 1: Thresholds regarding cooling water
(values in brackets in exceptional cases)

Cooling water	**Once-through cooling system**	**Discharge cooling system**	**Closed-circuit system**
Discharge temperature	30°C (33°C)	33°C	35°C
Temperature rise	10 K (15 K)	10 K (15 K)	15 K

Source: modified after Maniak 2005

Tab. 2: Thresholds regarding the receiving water
(values in brackets in exceptional cases)

River water	**Summer-warm streams**	**Summer-cool streams**	**Salmonid streams**
River water temperature (computed value)	28°C	25°C	18°C
Temperature rise	5 K (7 K)	3 K (5 K)	3 K

Source: modified after Maniak 2005

2.1 Water Temperature Analysis of German Rivers

Generally, a strong correlation exists between air and water temperature. Since a shifting in mean and extreme values of air temperatures has been observed all over Germany in recent decades, a change of water temperature means and extremes could be assumed (UBA 2007, Mohseni & Stefan 1999). This would affect the electricity sector more often in the future for the above mentioned reasons. Some studies in the United States and Austria, for instance, investigated the possible impacts of climate change on shifting mean values of water temperature in rivers (Cooter & Cooter 1990; Sinokrot et al. 1995; Webb & Nobilis 1994; 2007). However, neither the development of water temperatures in rivers nor the correlation between air and water temperature has recently been investigated in Germany until now. Therefore, the first objective of this study is to investigate whether and to what extent mean and extreme values of water temperature of several German rivers have changed during past decades. Secondly, a correlation analysis of air and water temperature for selected German rivers was conducted.

2.2 Data and Methodology

Publicly available water temperature time series were obtained from commissions in charge of rivers, watershed management departments or environmental agencies[1]. Daily mean values from over 100 gauging stations at several rivers

1 Arbeitsgemeinschaft für die Reinhaltung der Elbe (ARGE Elbe), Bayerisches Landesamt für Umwelt (LfU Bayern), Bundesanstalt für Gewässerkunde (BfG), Flussgebietsgemeinschaft Weser (FGG Weser), Internationale Kommission zum Schutz des Rheins (IKSR), Landesanstalt für Umwelt, Messungen und Naturschutz Baden-Württemberg (LUBW).

were collected. However, less than 50% of these time series exceed 20 years and are therefore appropriate for trend analysis. Moreover, most of them were not checked for homogeneity and plausibility by the agencies.

For a first analysis, 19 gauging stations along seven rivers (Weser, Elbe, Rhine, Main, Neckar, Danube and Isar) were utilised. Two criteria were used for their selection: the length of time series and the proximity of the gauging station to a thermal power plant site. The length of time series spans between 14 and 32 years and the data quality in terms of missing values within the time series varied between 0 and 12%. Missing data was interpolated using linear interpolation.

Furthermore, air temperature data was obtained from the German Weather Services (Deutscher Wetterdienst, DWD). Selected climate stations are located between 0 and 20 km away from the gauging stations. The air temperature time series match the length of water temperature time series. Missing values in air temperature time series (< 1%) were not interpolated since this is not necessary for a correlation analysis (Janssen & Laatz 2005).

A diagnostic trend analysis detecting a possible change in water temperatures was conducted. The existence, amplitude and statistical significance of trends in monthly and seasonal[2] mean values were calculated for each station using the Mann-Kendall trend-test, the Seasonal Mann-Kendall trend-test and the Sen's estimator of slope. Descriptive frequency methods were used to describe the appearance of positive extreme water temperatures. Furthermore, trends for the attainment of extreme values (90th-percentiles) were determined. Additionally, the empirical exceeding probabilities for different values were calculated using the Weibull formula. These values were chosen on the basis of statutory water temperature thresholds after waste heat discharge for the selected river taking into account the possible variation of daily water temperature up to 3 K. The procedure was applied to varying temporal scales: a) for the time period 1985 to 2005, b) for the first and c) for the second decade of that period. Finally, a correlation analysis was conducted in order to verify the quality of correlation between air and water temperatures (Hirsch et al. 1982; Gilbert 1987; Maniak 1997; Maidment 1993).

2.3 Results

The trend analyses result in positive significant trends for all rivers in spring and summer months (see figure 1). Winter and fall months show negative significant (> 90%) trends for some stations along the rivers Weser and Neckar.

Concerning the extreme value[3] analysis, positive, significant trends of extreme value frequencies are found for all rivers. The exceeding probability for the spe-

2 Meteorological seasons winter: Dec/Jan/Feb; spring: Mar/Apr/May; summer: Jun/Jul/Aug; fall: Sep/Oct/Nov.

3 Extreme values defined as 90th-percentile.

cific values 24, 25 and 26°C increased for four stations at the rivers Rhine, Main and Neckar, decreased for one station at the river Main and are inconsistent for four stations at the rivers Weser, Elbe, and Neckar (see figure 2).

The correlation analysis shows a good correlation between air and water temperatures. The coefficient of determination r^2 is between 0.74 and 0.86 for all stations with the exception of the station Hamburg-Seemanshöft/Elbe ($r^2 = 0.26$). This is probably due to the fact that this station is located inside the harbour of Hamburg where other anthropogenic factors (e.g. busy inland water transportation) influence the water temperature.

Fig. 1: Significant (> 90%) positive trends of water temperatures at all stations in summer months (Jun/Jul/Aug), calculated over the whole time series length, extrapolated to ten years (Greis 2007)

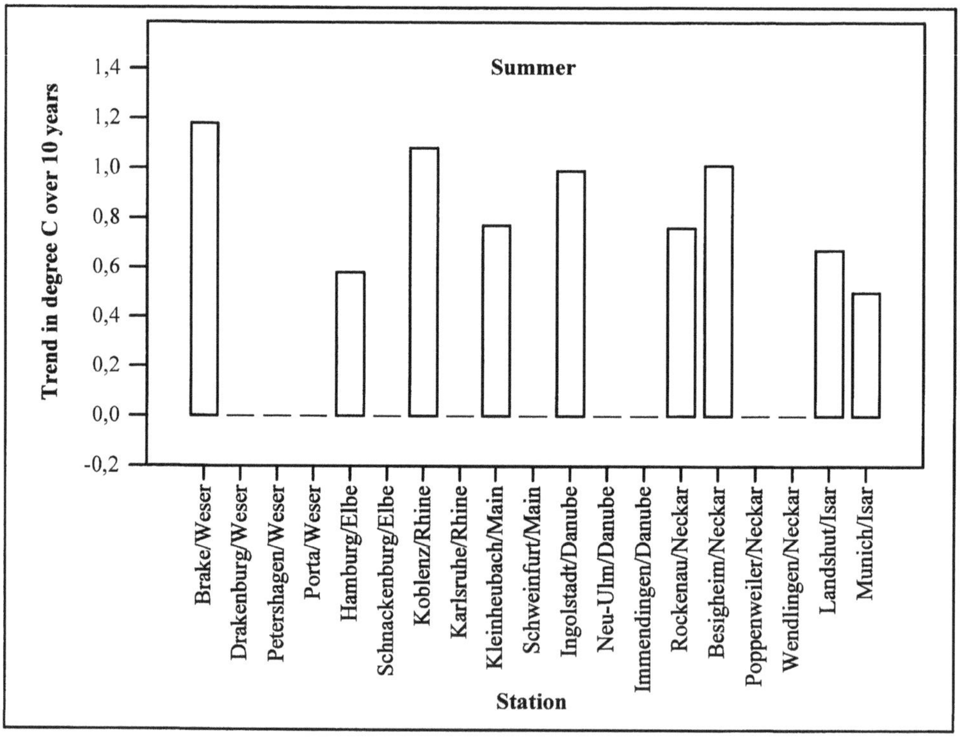

Shifting mean air temperatures do not only affect water temperatures but also the electricity demand as will be discussed in the next section.

Fig. 2: Change of the exceeding probability of different water temperature values for the rivers Weser, Elbe, Rhine, Main and Neckar comparing the decades 1985-1995 and 1995-2005 (Greis 2007)

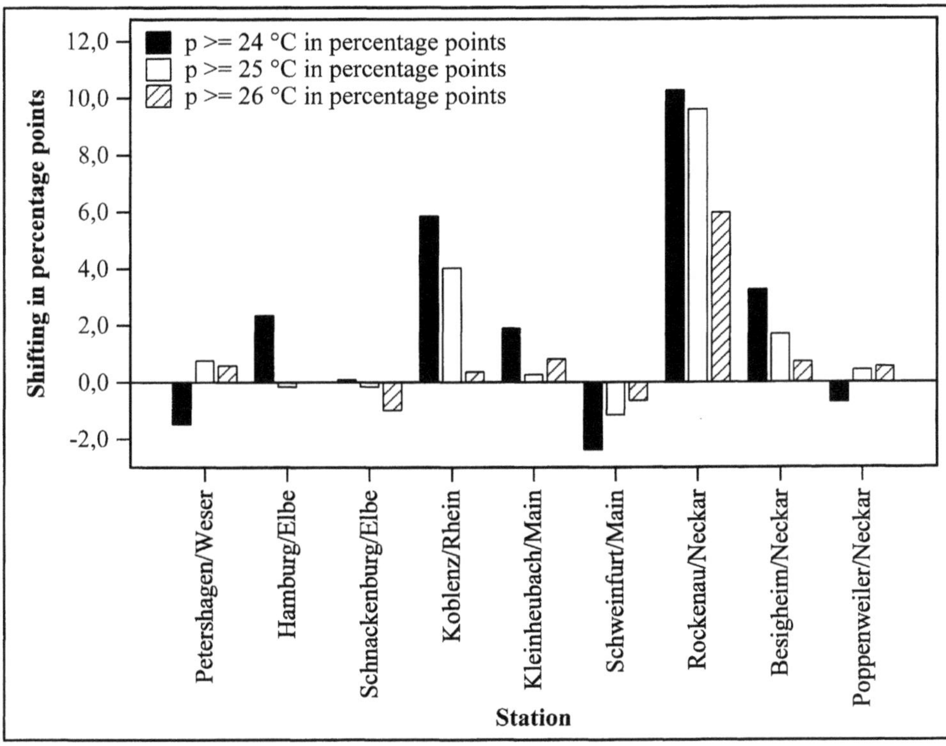

3. Impacts of Air Temperature Evolutions on the Electricity Demand in Baden-Wuerttemberg

Predicted air temperatures are a major parameter for electricity suppliers to estimate and forecast short-term electricity demand. There are numerous factors that influence electricity loads (e.g. economic activity). Among these, air temperature especially relates to heating demand in winter and cooling demand in summer (Mimler & Rothstein 2007).

3.1 Objective of the Study

The objective of the study on simulated electricity demand in Baden-Wuerttemberg is to quantify the impact of climate change on the consumption of electricity by freezing all other load influencing variables (like equipment of households with electric appliances, electricity prices, etc.) to the level of the year 2006. Thus, the results of the study represent the influence of a different climate on the consumption of electricity in Baden-Wuerttemberg. Besides air temperature, no other climate parameters (e.g. clouds or wind) were taken into account. This choice was made because temperature is the most important climate parameter that influences demand. However, no statements on underlying economic and human activity and their possible changes in the future are made.

To achieve this objective, simulated air temperature data was taken from the regional climate model REMO (developed by the Max-Planck-Institute for Meteorology, MPI-MET) and was used as input into a load forecasting model which had been estimated on the region of Baden-Wuerttemberg before.

The investigated time period was chosen to be 2006 to 2035 for two reasons:

1. The load simulation is based on climate model generated air temperature, and therefore the whole analysis sticks to climatologic research standards. Therefore, a 30-year period was chosen to be compared to an earlier 30-year reference time referring to the climatic normals (CLINO) periods defined by the World Meteorological Organisation (Schönwiese 1994) (in this case 1971-2000).
2. At the same time, the 30-year period from 2006 to 2035 is similar to the time horizon the International Energy Agency has chosen for its upcoming "World Energy Outlook 2008" [1]. In addition, an electricity supplier considers such a time scale for its current planning (e.g. the operational life time of a power plant is 40 years).

3.2 Data and Methodology

The study consists of five parts:

1. Estimation of the load forecasting model "Eventail" on the region of Baden-Wuerttemberg:

   The load forecasting model "Eventail" is based on mid-term regression models and is used for load forecasting and analyses in France (Bruhns et al. 2005). To estimate the model on the region of Baden-Wuerttemberg, two sets of data were used as input: historical load data of the supply region of the South-West German electricity supplier Energie Baden-Württemberg (EnBW) and averaged observed temperature data of the weather stations of six big cities[4] in the given region. The estimation of the load curves was conducted for the time frame May 2003-December 2007.
2. Model bias analysis of climate model regarding temperature data in Baden-Wuerttemberg:

[4] These cities are Freiburg, Karlsruhe, Mannheim, Stuttgart, Konstanz and Ulm.

Referring to the averaged temperature data taken for the forecast model estimation, climate model air temperature data (climate model REMO, emissions scenario A1B; see Jacob 2005a) was taken for six single grid points corresponding to the six chosen weather stations in the region and averaged to a single time series[5].

To analyse the climate model bias for the temperature, control run data from 1980-2000 (climate model REMO, control run, see Jacob 2005b) was compared to observed temperature data from the same time frame. The comparison took place for monthly means of daily means and daily quantiles, in order to evaluate if the climate model is able to reproduce means and variances. Furthermore, the non-parametric rank sum test after Wilcoxon, Mann and Whitney was applied to test for significant differences between the two time series for the different months.

3. Temperature changes between 1971-2000 and 2006-2035:

   To estimate changes in temperature from 1971-2035, a further pre-analysis was conducted with the two time series data sets that will be used for the load simulation later on: climate model control run temperature from 1971-2000 and climate model A1B-scenario temperature from 2006-2035.

   To state the changes estimated by REMO, monthly mean, minimum and maximum temperatures were computed and the yearly frequency of heat and frost days was examined. In regard to heat days, two different definitions were applied. One simple definition refers to a daily maximum temperature of at least 30°C (Schönwiese et al. 2003) and a more detailed definition to a daily maximum temperature of at least 30°C and a daily minimum temperature of at least 20°C to take into account tropical nights [2]. A frost day is defined by a daily minimum temperature below 0°C [2].

4. Impact of climate model bias on load curves: comparison of climate model temperature simulated load curves and observed temperature simulated load curves:

   Corresponding to the model bias of temperature data, the model bias of simulated load data was examined. Simulated loads of climate model control run temperature from 1980-2000 were compared to simulated loads based on observed temperature data for the same time.

---

5 A separate analysis before showed that this approach displays a lower model bias than taking the mean air temperature for the whole region of Baden-Württemberg.

Monthly and hourly means over the whole time period were computed and then compared by calculating the Root Mean Square Error (RMSE) and the Mean Absolute Percentage Error (MAPE) dependent on months and hours.

5. Load changes between 1971-2000 and 2006-2035:

   To examine the impact of climate change on the demand of electricity in Baden-Wuerttemberg, one past and one future load series were simulated on the parameters of the reference year 2006: Past daily load curves for 1971-2000 were obtained based on the climate model's control run temperature. Future daily load curves for 2006-2035 were simulated based on climate model temperatures of the scenario A1B.

   The electricity demand of single hours was averaged over months to display single mean curves for a chosen month or season. Furthermore, mean values in dependence of months were computed for each time series to calculate the load differences between 1971-2000 and 2006-2035. Finally, the Seasonal Mann Kendall test was applied to test trends in load changes and their significance.

3.3 Results

1. Estimation of the load forecasting model "Eventail" on the region of Baden-Wuerttemberg:

   The forecast model "Eventail" was estimated on the region of Baden-Wuerttemberg with an overall RMSE of 352.72 MW and a MAPE of 4.6%. This result is a bit higher than how the model regularly performs when fed with French electricity consumption input data[6], but still acceptable in the context of this study. As heating and cooling gradient the model output gives -59 respectively 84 MW which describes the variation in MW per temperature degree increase (below and above a certain temperature threshold). In relation to the load range (between 1712 and 9530 MW) and the mean load (6132 MW) for the whole estimation basis, these gradients are rather low. At the same time, they give information about the thermo-sensitive part of the input loads. For the heating, the low gradient can be explained by a lower share of electric heating in Germany respectively Baden-Wuerttemberg than in France[7].

---

6 Personal communication with Stéphanie Marcilly, EDF R&D, Commercial Innovation And Market Analysis, 31.8.2007.

7 In Germany, 4.1% of all households are equipped with electric heating (Frey et al. 2007), in France this share accounts for 40% (Grosmesnil 2002).

The cooling gradient indicates a low share of air conditionings in the region as well[8].

2. Model bias analysis of climate model control run and observed temperature data:

   Figure 3 shows the differences in quantiles for each month in absolute values. Mostly they stay below 1°C except in April and November. Both of these months display differences above 1°C and especially for April they rise up to 2.75°C for quantiles 75 and 100.

Fig. 3: Temperature differences in quantiles between climate model control run (REMO) and observed temperature 1980-2000 in absolute values (region Baden-Wuerttemberg, mean of 6 grid points)

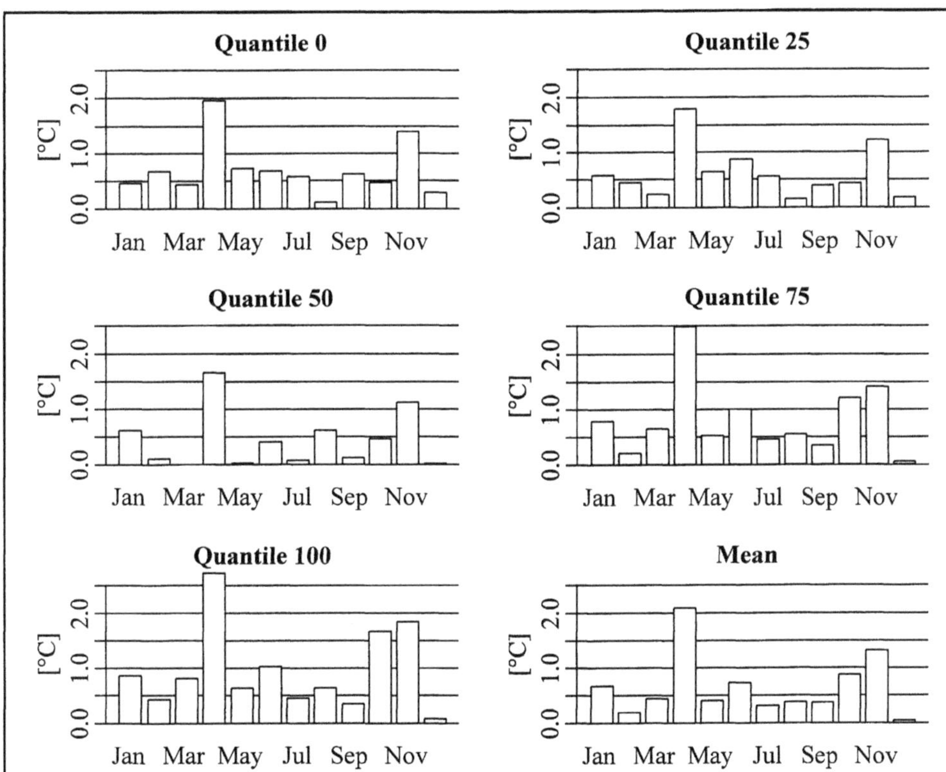

These observations are confirmed by the applied rank sum test after Wilcoxon, Mann and Whitney. This test proved significant differences between

8 In Germany, 1.4% of all households are equipped with air conditionings (Schlomann et al. 2004), in France, this share accounts for 10% (INSEE 2008).

both examined time series for April and November at a significance level of 0.99 and for October at a significance level of 0.95.

However, these results are based on the estimation of one single regional climate model. As there are still uncertainties in the simulations of climate models, data of different models should be analysed jointly.

3. Temperature changes between 1971-2000 and 2006-2035:

   The mean temperature increases by 0.35°C over the whole period (i.e. from 1971 to 2035). Figure 4 shows the changes in mean, minimum and maximum temperatures by months between 1971-2000 (REMO, control run) and 2006-2035 (REMO, emissions scenario A1B).

Fig. 4: Temperature changes between 1971-2000 and 2006-2035 (REMO climate model data, region Baden-Wuerttemberg, mean of 6 grid points)

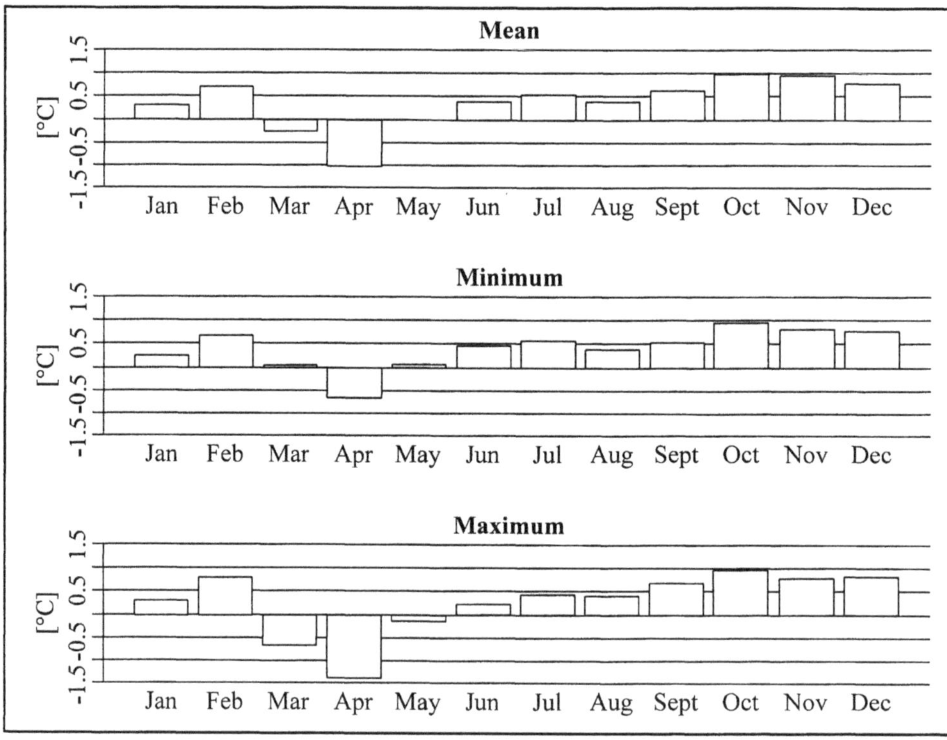

In the majority of all months, these statistical values rise, but this increase is below 1°C. The highest increases occur during October, November and December. Striking are the months April with a decrease of all values and May with nearly no changes at all. One possible explanation for the April results could be the described climate model bias (see above). In line with the estimated heating and cooling gradients given above these results imply that no major changes in mean loads can be expected after the simulation process.

Figure 5 shows the yearly frequencies of heat days corresponding to both definitions explained above. In both cases, the frequency varies along the years but shows an increasing tendency over the whole time period. In case of frost days there is a decreasing tendency.

Fig. 5: Number of heat days 1971-2035 (REMO climate model data, region Baden-Wuerttemberg, mean of 6 grid points)

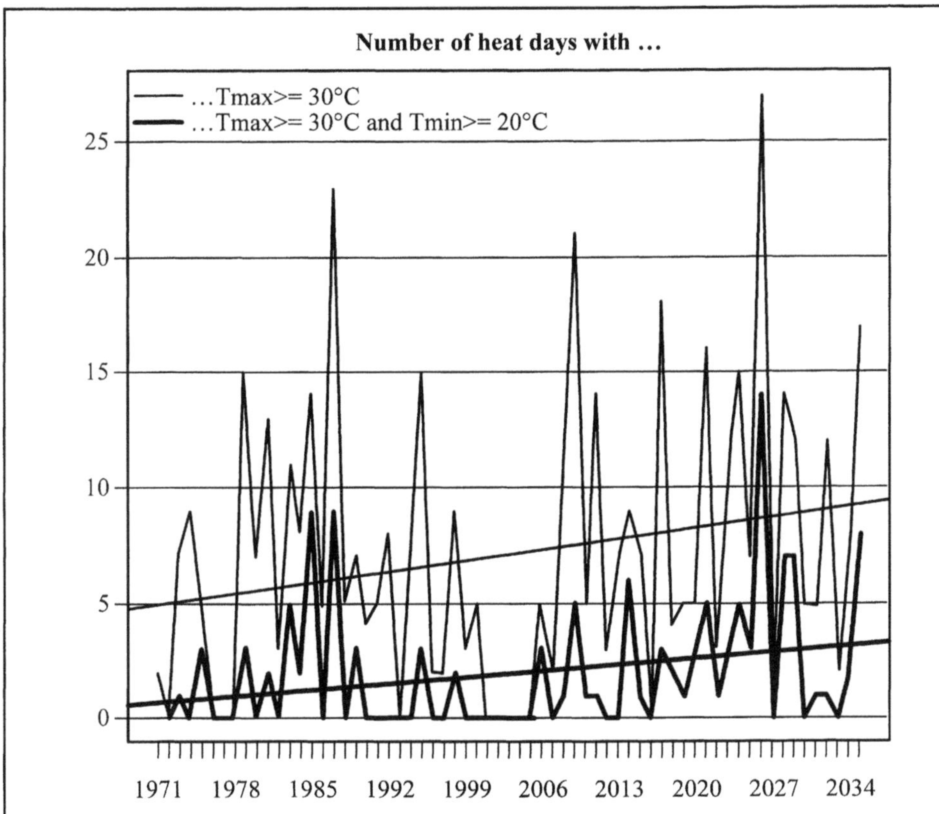

4. Impact of climate model bias on load curves: comparison of climate model temperature simulated load curves and observed temperature simulated load curves:

   To test the model bias between both load series, RMSE and MAPE were computed. At the top of figure 6, the MAPE for single months is displayed. It shows that April and November temperatures come with the highest errors. This is consistent with the results of the model bias analysis of climate model and observed temperatures (see above). The MAPE was computed for all days of the week jointly but also for work days and weekends separately. The higher MAPE during weekends is explained by the lower level of electricity consumption during these days. Thus, the same divergence of MW has a greater relative impact on weekends than on work days.

In the course of the day the MAPE varies between 0 and 0.8%. During nighttime, the error is larger while the mean load curve for 1980-2000 shows a low electricity demand (bottom of figure 6). This is due to the higher relative impact at lower consumption levels. But for the rest of the day – except for 12 a.m. – the MAPE corresponds to the shape of the load curve.

Fig. 6: MAPE for months and MAPE for hours in comparison to the daily mean load curve 1980-2000 (observed temperature and REMO climate model temperature, region Baden-Wuerttemberg, mean of 6 stations respectively grid points)

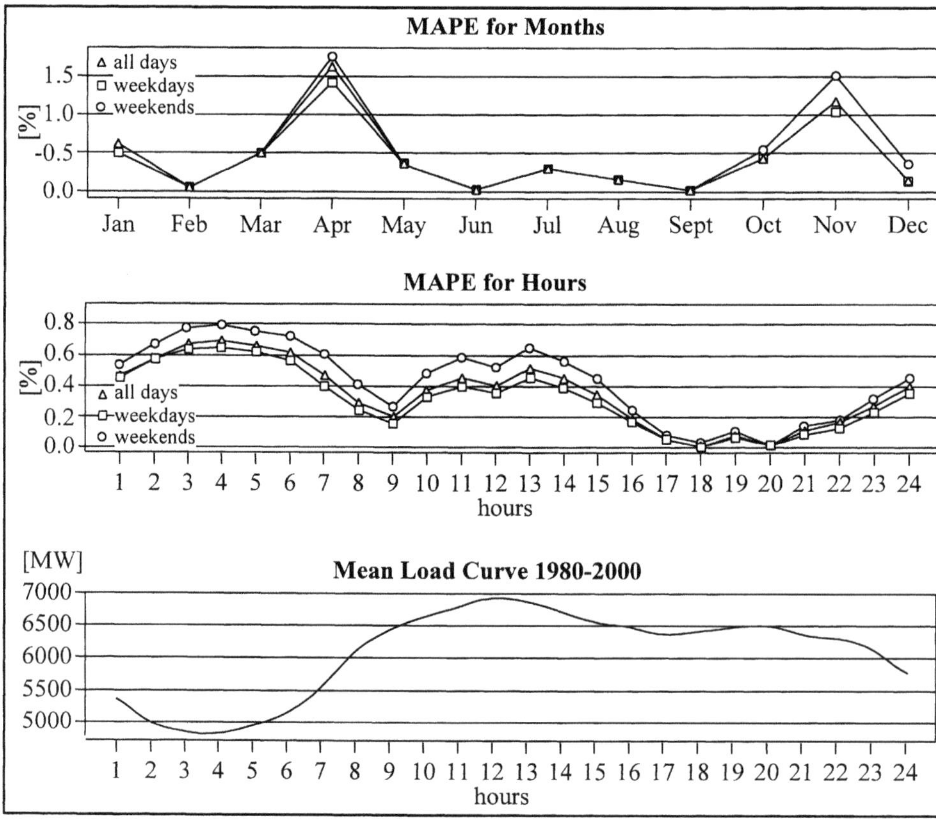

5. Load changes between 1971-2000 and 2006-2035:

   The mean load over the whole period decreases by 7 MW. This corresponds to 0.001% of the mean load of 6050 MW computed for 1971-2000. Consequently, the exemplarily presented mean curves do not show a remarkable difference in their shape neither for summer and winter (figure 7) nor for the more detailed mean load curves February and August (figure 8). A slight divergence is only observable in winter.

Fig. 7: Mean Load Curves for Summer and Winter (load simulation using REMO climate model data, region Baden-Wuerttemberg, mean of 6 grid points)

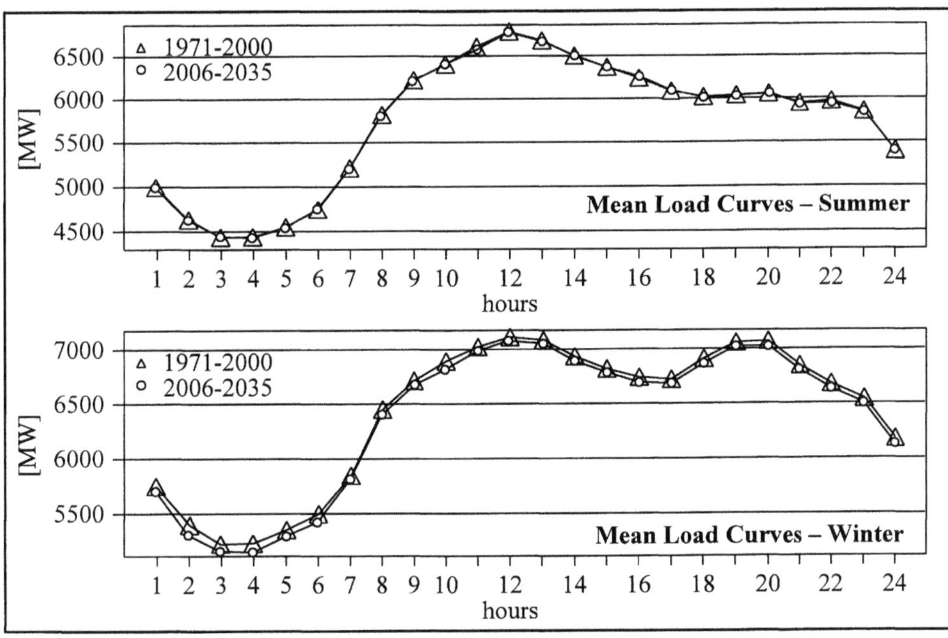

The Seasonal Mann Kendall test computed significant positive trends for July, August and September with little or no real evidence against and March with suggestive evidence against a trend. Significant negative trends were computed for January, May, June, December with little or no real evidence against and February, October with suggestive evidence against a trend (see table 3). Due to the results of the model bias analyses, no statements were made about April and November.

Fig. 8: Mean Load Curves for February and August (load simulation using REMO climate model data, region Baden-Wuerttemberg, mean of 6 grid points)

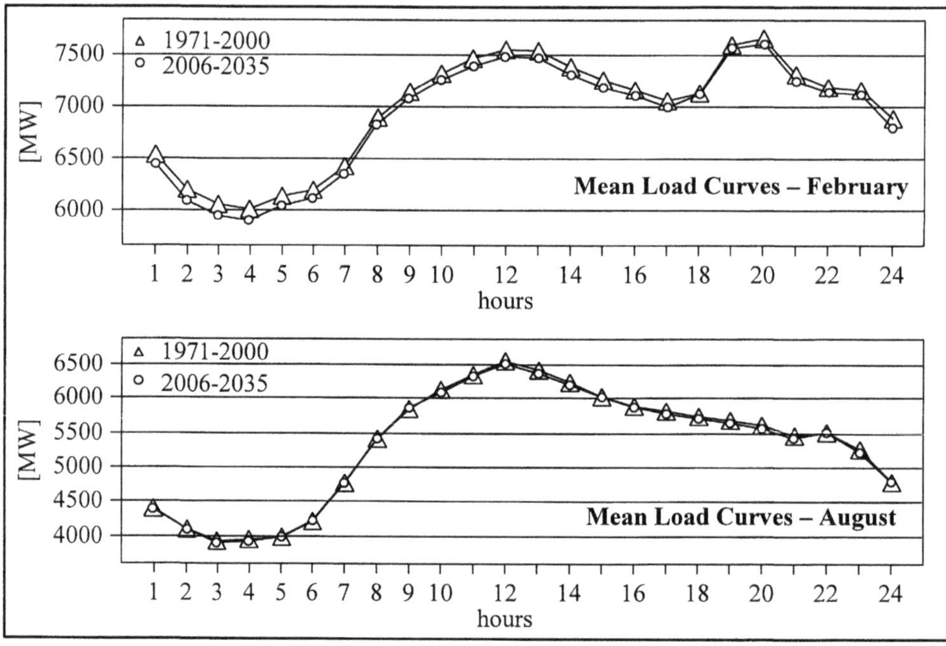

Tab. 3: Seasonal Mann Kendall test on load changes 1971-2035

	Jan	Feb	Mar	Apr	May	Jun	Jul	Aug	Sep	Oct	Nov	Dec
Trend	-	-	+	+	-	-	+	+	+	-	-	-
Code	0	1	1	1	0	0	0	0	0	1	3	0

Remarks:
Trend: + = positive trend, - = negative trend
Code: 3 = very strong evidence against H0 (= that there is a trend), 2 = moderate evidence against H0 (= trend), 1 = suggestive evidence against H0 (= trend), 0 = little or no real evidence against H0 (= trend)

But as Table 4 shows, the highest difference amounts to about 80 MW (November and December nights). This is equal to 0.015% and therefore marginal from the point of view of an electricity supplier. However, the obtained results are consistent with the results of the pre-analyses. The estimated heating and cooling gradients and the mean temperature changes 1971-2035 already implied low load changes.

Tab. 4: Changes in mean loads [MW] between 1971-2000 and 2006-2035

	Jan	Feb	Mar	Apr	May	Jun	Jul	Aug	Sep	Oct	Nov	Dec
Whole day	-18	-42	6	43	11	-3	18	17	5	-34	-50	-49
Morning	-11	-28	2	31	8	-5	14	22	12	-28	-34	-32
Forenoon	-11	-26	5	30	11	0	11	4	-2	-23	-31	-31
Noon	-14	-33	13	41	15	-1	18	7	-4	-25	-38	-40
Afternoon	-14	-34	18	43	13	-2	19	17	2	-30	-40	-40
Evening	-12	-26	8	35	10	-3	14	7	-4	-29	-33	-32
Night	-30	-72	-3	56	11	-4	26	23	9	-48	-83	-83

As a result, these changes lack any practical relevance in the frame of the daily business of an electricity supplier even though the Seasonal Mann Kendall test computes significant trends.

The presented results focused on mean changes in temperature over a long time period. Extreme events like for example the hot summer of 2003 were not subject in the frame of the presented analyses. Those events can result in a bottleneck of electricity supply requiring adaptation measures which will be discussed in the next section.

4. Adaptation Strategies

Section 2 und 3 dealt with impacts of climate change on water temperature of several German rivers in the context of electricity generation and with impacts on electricity consumption. In the following part, options for electricity companies to respond to long-term impacts of climate change in combination with extreme events (especially heat waves) are demonstrated. These adaptation options are mainly identified with the help of written and oral expert interviews. The polled specialists come from electricity companies mainly from Germany but also from France. Almost all the experts are working in key positions such as research and development, crisis-management, risk-management, transportation and distribution, production of electricity, technique and maintenance, environment, dispatching, coal logistic, and new construction of power plants. The options for adaptation of electricity companies to climate change are distinguished between *adaptation regarding higher water temperatures* and *adaptation regarding higher electricity demand during heat waves.*

4.1 Adaptation regarding higher water temperatures

As described in section 2, a once-through cooling system results in strong warming-up of the receiving water temperature. A thermal power plant operator can counteract with the following options.

Reducing the electricity output of a power plant

A simple adaptation option is decreasing the electricity output of a power plant. As a result, the waste heat discharge and therefore a further warming of the river can be minimised. This option can be used by an electricity supplier voluntarily (e.g. to avoid reducing the electricity output of a prioritised power plant which is situated downstream). In many cases though, the decrease is forced by authorities. Background of this measure is that the operation of a power plant is based on clearly defined authorisation guidelines concerning e.g. cooling water withdrawal and discharge. As electricity consumption during extreme hot weather periods is superior and a certain number of running power plants is needed to guarantee the grid stability, the use of this adaptation measure is limited (see section 4.2) (Rothstein 2007).

Installing a cooling tower / operating cooling towers more frequently

It is possible to install a cooling tower at a power plant site that was previously only equipped with a once-through cooling system. For instance the German electricity supplier EnBW did so at the power plant Altbach/Deizisau (EnBW 2003). As the retrofit is very expensive, electricity companies will certainly carry out a precise cost-benefit analysis before performing this kind of modification also taking into account efficiency losses due to increasing station supply.

If a power plant is already equipped with a cooling tower, a more frequent use of this kind of cooling system (instead of the once-through system) is certainly an important adaptation option. Nevertheless, cooling tower use reduces the efficiency of the electricity generation (see section 2) (Rothstein 2007).

Aeration measures at power plants and weirs

Another aspect is linked very closely to the aspect of rising water temperature: The absorbing capacity of water in terms of oxygen decreases with rising temperatures. This is amplified by the heat input through thermal power plants. Furthermore, the natural aeration of the water body is diminished as a consequence of the barrages built e.g. at the River Neckar in Germany. Especially in summer at low water events, critical oxygen rates can be observed (LfU 2004). In order to prevent the dying of fish, the oxygen content can be increased with aeration measures at power plants and weirs (UVM 2005).

In case of a drop below the permissible minimum oxygen content in the river the energy utilities commit themselves to implement voluntary measures in order to rise the oxygen content. For example, the company EnBW either puts special aeration equipment into operation or alters technical features so that aeration of water is made possible. Altogether turbines of six river power plants at the River Neckar are equipped with aeration facilities. In the meantime, these are required for new buildings as well as for replaced turbines (Pintér 2003; Rothstein 2007).

Adapted management of (human) resources

In addition to these more or less technical adaptation options, also adaptation measures in terms of management of (human) resources are beneficial: Resulting from the experiences of the hot summer 2003, the French electricity supplier EDF has re-enforced coordination from the field of production. Experts are assigned to act as contact persons for water related issues towards authorities. They forward information about the state of important rivers (Loire, Rhone, Garonne, Seine, Moselle and Meuse) and provide medium-term forecasts for temperature and discharge (EDF 2003).

Moreover, EDF re-enforces coordination for the water management of hydro and thermal power plants in cooperation with the national authority of waterways (Voies navigables de France, VNF). For instance a preventive damming of rivers helps manage low water level and high water temperatures. There are also questions of water sharing that need to be taken into account as water is used for energy and also for agriculture and irrigation. At last, even for the electricity sector there are sometimes different actors on the same river, as for the Rhône river where there is also the electricity producer CNR (Compagnie Nationale du Rhône) (Rothstein 2007).

Allowing exceedance of temperature thresholds

If the electricity supply is critical due to the risk of exceeding statutory thresholds, the regional authorities (such as the Ministry of Environment, of the Interior and of the Economics) will evaluate the individual case as to whether corresponding measures have to be taken. These authorities can possibly allow a selective, temporary exceedance of temperature threshold values under closely defined basic conditions. An example for these basic conditions is the commitment of the affected electricity supplier to take as much electricity as possible from the integrated network. Even if this is fulfilled, a further operation of local power plants is necessary to secure the network stability and the voltage level (idle power) respectively (Rothstein 2007).

Increasing some temperature levels for water rejects

In addition to the toleration of short-term exceedance of statutory thresholds, the LAWA in Germany is openly discussing if the permitted warming of the cooling water should be increased by 2 K. In France, water temperature levels are discussed to be adjusted in some specific situations where electricity is necessary (Hofmann 2006; Rothstein 2007).

Design of the future power plant mix

Last but not least the design of the future power plant mix plays a key role in creating a more resilient electricity generation. Within the next 20 years an expansion requirement of power plant capacities of 100,000 MW and a replacement requirement of 200,000 MW exists in the European Union (EU-15). This demand has increased further since the EU enlargement, due to old plants in the ten new EU member countries with an average efficiency factor of less than 35% (VGB 2003).

This special design comprises e.g. how explicit authorisation standards (e.g. the ratio between cooling water extraction and warming of cooling water) will be defined at water related regulations. Thus, the construction of new power plants opens up the opportunity to adapt to the already existing and expected climate changes (e.g. water temperature, discharge hydrographs) for the future generation of electricity. The research initiative "Power plants in the 21st century" (KW 21) analyses for example to what extent less or almost no cooling water is needed by power plants in future (abayfor 2006; Rothstein 2007).

4.2 Adaptation regarding higher electricity demand during heat waves

During extreme hot weather periods, high electricity consumption can coincide with periods of limited generation possibilities. In this section, only those adaptation options are regarded which can be implemented by electricity companies. Of course, many other business sectors (such as architecture, facility management, etc.) play an important role in minimising electricity demand.

For electricity companies, the most effective adaptation option to a higher electricity demand is load management. Load management describes goal-oriented options of action that aim at equalising the electricity supply and demand. It is therefore associated with the umbrella term Demand Side Management which tries to stimulate load reduction and long-term change in costumer behaviour. This includes all measures concerning energy efficiency (IEA 2000). Load management is important for the electricity sector because it facilitates the adjustment of loads to the momentary optimised generation of electricity. Furthermore, power plants are working according to their capacities as well as over and under voltage levels are avoided (ASE 2000). Eventually, reductions in generation of electricity can be realised. In the following, possible adaptation options are described.

Controllable consumers

In order to avoid an imminent power outage, the current load can be decreased by managing controllable consumers (especially industrial bulk consumers). They are able to minimise their electricity demand through advance planning or can totally delay it (EDF 2003). By contract, they agree to obtain less electricity in case of need and the energy utility can transfer loads and balance peaks. As service in return, they pay lower prices for the use of electricity. These contracts are re-negotiated every year. If there is a decrease of voltage despite the application of all measures regarding controllable consumers, some groups of consumers are disconnected automatically with the help of relays. The order of disconnection is defined and it is differentiated between three categories. Smaller towns belong to the first and hospitals to the third category (Rothstein 2007).

Calls for saving electricity

Not only industrial bulk consumers are able to decrease their electricity use in case of supply shortage but also private consumers. The population is then informed by media. In summer 2003, EnBW addressed the population with a press release and asked for saving electricity. Concrete advice was given, like a limited operation of air-conditioning and ventilation as well as the complete switch-off of computers and TVs (EnBW 2003). Especially between 11 a.m. and 3 p.m., the peak load time, a decrease of electricity use is most beneficial. This can also be realised by just transferring the use of household devices operation to other times of the day (EnBW 2003; Rothstein 2007).

Tariffs and contracts

Another possibility for the energy utilities to gain influence on the demand, especially of private customers, are variable tariffs. That means that the electricity price is higher in peak load times, (e.g. at noon), than during times with sufficient electricity supply (e.g. at night). Although the energy utility is primarily interested in a load balance because of optimised power plant efficiency, variable tariffs can help avoid power shortages, especially during heat events. For instance, EDF is offering the so called "L'option Heures Creuses". For eight hours a day (normally during the night), the reduced prices apply for the used electricity. In addition, EDF has established an adaptation measure on climate risks which is called EJP (Effacements Jours de Pointe). These contracts offer very low energy prices most of the years except some days in winter when electricity demand is high and the prices are high as well. Thus, the concerned consumers generally do not use energy, which helps to manage the energy balance. A development of even more specified contracts could be appropriate.

A relatively new measure relating to variable tariffs and contracts in the range of Demand Side Management is called Demand Side Bidding. It gives customers or groups of customers (industrial bulk customers or private customer associations) the possibility of participating in the electricity trading with short-term bilateral contracts. There are two forms of Demand Side Bidding. On the one hand, a contract about a defined amount of electricity for a fixed price can be negotiated. On the other hand, the customer can offer voluntarily to increase or reduce his electricity use to a certain limit. The benefit for the energy utility is a stabilisation of the load. The customer receives payment or reduced tariffs for his self-commitment (IEA 2002; Rothstein 2007).

Rolling blackouts

Furthermore, in extreme situations, rolling blackouts can be conducted when the produced electricity is too low and cannot be supplemented from the integrated network. Rolling blackouts are announced over the radio in case of need so that the people involved (e.g. bulk consumers, local authorities and regions) have enough time for preparation. At this procedure important facilities like hospitals, fire departments and police can be factored out. The underlying idea is that a great number of consumers are affected for a short period which means that the problem is pooled. Preliminary for the implementation of such a measure is that the disconnection plan is based on a practice which is specified and has to be approved by an inspecting authority (EDF 2003; Rothstein 2007).

5. Conclusions and Outlook

The main focus of this paper was to identify long-term impacts of shifting in air temperature means on water temperature of several German rivers and on the electricity demand. In general, the results of section 2 follow the assumption that a shift in mean values of air temperatures results in a shift of mean water temperatures. The water temperature time series, however, needs to be thoroughly checked for homogeneity and plausibility before further analyses are carried out. This could be done by means of comparison with neighbouring gauging stations in order to exclude anthropogenic impacts (e.g. hydro-engineering, change of location or method of measurement).

As shown in section 2, more than 19 water temperature time series are available in the meantime. Validation and further statistical analyses will allow comparing several time series at different gauging stations for one river as well as to compare the results for different rivers. Moreover, further correlation analyses between all air and water temperature time series will be the basis for estimating water temperature using air temperature data from a regional climate model. The objective of the whole study will be to connect the analysed and modelled data in order to estimate the possible impact on future power production. Shifting mean air temperatures do not only affect water temperatures but also the electricity demand as was discussed in section 3.

Here, the isolated impact of temperature evolutions on the demand of electricity in Baden-Wuerttemberg until 2035 is small in terms of mean changes. Although a trend test shows that there are significant tendencies in load changes the impact of these changes lacks practical relevance for electricity suppliers. However, for the interpretation of the results it has to be considered that only one climate model was used.

Furthermore, research focus should concentrate on the development of consumption patterns influencing the thermo-sensitive part of loads (i. e. equipment with electric heating and air conditioning), but also on other load influencing factors like economic and human activity. It could also be interesting to look at other regional scales and time horizons by using several regional climate models and different emissions scenarios.

Nevertheless, the presented study focused on mean changes in temperature over a long time period. Extreme events like for example the hot summer of 2003 were not subject in the frame of the presented analyses. Like as experienced in 2003, extreme temperatures can lead to an increase in electricity demand while electricity generation is reduced due to the same meteorological conditions (EnBW 2003) (see section 2).

How to adapt to these critical situations was shown in section 4. It became clear that most of the identified adaptation options already exist for other reasons (e.g. economical forces or political and environmental requirements). But climate change provides a further driving force to develop and adjust these instruments. The mentioned adaptation options addressed two major issues regarding climate change impact on the electricity sector. Further affected areas (e.g. grids affected by storms, transportation of coal affected by low water levels) demand adaptation measures as well.

It can be summarised, that on the one hand electricity companies are affected by climate change in many aspects. On the other hand a wide range of adaptation options can be used, in order to reduce the described vulnerabilities.

Beyond the results of this paper, the electricity sector is particularly interested in anticipating extreme weather situations like heat waves. To better understand such events characterised by reduced electricity generation and increased demand, data with higher spatial and temporal resolution is required.

Additionally, further investigations are necessary in order to meet the needs of electricity suppliers that benefit from the knowledge about the probability of extreme events. In a first study, this was already conducted for recent change in the probability of water temperature extreme values. This is a basis to estimate and analyse probabilities regarding the simultaneous appearance of reduced electricity generation and of higher electricity demand. This in turn can improve possibilities to formulate more sophisticated adaptation options and strategies including quantifications and feasibility studies.

References

Arbeitsgemeinschaft der bayerischen Forschungsverbünde (abayfor) (2006), "Forschungsinitiative Kraftwerke des 21. Jahrhunderts", pp. 1-12, www.abayfor.de/kw21/ projekte_liste.php.

Arbeitskreis Schulinformation Energie (ASE) (2000), "Energiewelten – Wie der Strom in die Steckdose kommt", CD-Rom, Frankfurt.

Bruhns, A., Deurveilher, G. and Roy, J.-S. (2005), "A Non-Linear Regression Model for Mid-Term Load Forecasting and Improvements in Seasonality", 15th PSCC, Liege, August 22nd-26th 2005, Session 17, Paper 2, p. 8, www.montefiore.ulg.ac.be/services/stochastic/pscc05/papers/ fp168.pdf.

Cooter, E.J. and Cooter, W.S. (1990), "Impacts of greenhouse warming on water temperature and water quality in the southern United States", Climate Research, volume 1, pp. 1-12.

EDF (2003), "Plan Aléas Climatiques", internal report, Paris, p. 47.

Energie Baden-Württemberg AG (EnBW) (2003), "Umwelt und Gesellschaft – Auf dem Weg zur Nachhaltigkeit", p. 89, www.enbw.com/content/de/ der_konzern/_media/pdf/umwelt_gesellschaft _jahresbericht_2003.pdf.

European Commission (2001), "Integrated Pollution Prevention and Control (IPPC). Reference Document on the application of Best Available Techniques to Industrial Cooling Systems", p. 313, http://eippcb.jrc.es/pages/FActivities.htm.

Frey, G., Schulz, W., Horst, J. and Leprich, U. (2007), "Studie zu den Energieeffizienzpotentialen durch Ersatz von elektrischem Strom im Raumwärmebereich", izes gGmbH, Saarbrücken, p. 118, www.bmu.bund.de/ files/ pdfs/ allgemein/application/pdf/studie_stromheizungen.pdf.

Gilbert, R.O. (1987), "Statistical methods for environmental pollution monitoring", Van Nostrand Reinhold, New York, p. 320.

Greis, S. (2007), "Gewässertemperaturentwicklung und industrieller Kühlwasserbedarf. Eine statistische Analyse der Wassertemperaturentwicklung ausgewählter deutscher Flüsse im Zuge rezenter Klimatrends und deren Bedeutung für den industriellen Kühlwasserbedarf", diploma thesis, University of Würzburg (unpublished), p. 134.

Grosmesnil, O. (2002), "La consommation d'énergie à usage domestique depuis quarante ans. L'électricité, numéro un dans les foyers", INSEE Première, issue 845, May 2002, p. 4, www.insee.fr/fr/ffc/docs_ffc/IP845.pdf.

Hirsch, R.M., Slack, J.R. and Smith, R.A. (1982), "Techniques of trend analysis for monthly water quality data", Water Resources Research vol. 18 issue 1, pp. 107-121.

Hofmann, M. (2006), "Der Fluss wird zur Badewanne", Heidenheimer Zeitung, July 28th 2006, p. 1.

Institut National de la Statistique et des Études Économiques (INSEE) (2008), "Le Tableau Economique de La Réunion édition 2007-2008", www.insee.fr/ fr/insee_regions/reunion/rfc/docs/ter2007_04_1_consommation_menages.pdf.

International Energy Agency (IEA) (2000), "International Energy Agency Demand-Side Management Programme Task VI – Developing Mechanisms for Promoting Demand-side Management and Energy Efficiency in Changing Electricity Businesses", research report no. 3, p. 309, www.ieadsm.org/ Files/ Tasks/ Task%20VI%20-%20Mechanisms%20for%20Promoting%20 DSM %20and% 20Energy% 20Efficiency% 20in%20Changing%20Electricity% 20Businesses/Publications/resrpt3_fin.PDF.

– (2002), "Market participants' views towards and experiences with Demand Side Bidding – Version 2, January 2002. Task VIII of the International Energy Agency Demand-Side Management Programme", p. 37, www.ieadsm.org/ Files/Tasks/Task%20VIII%20-%20Demand-side%20Bidding% 20in% 20a% 20Competitive%20Electricity%20Market/ Reports/Stage1ReportV2.pdf.

Jacob, D. (2005a), "REMO A1B Scenario run, UBA project, 0.088 degree resolution, run no. 006211, 1H data", World Data Center for Climate, CERA-DB „REMO_UBA_A1B_1_R006211_1H", http://cera-www.dkrz.de/WDCC/ui/ Compact.jsp?acronym=REMO_UBA_A1B_1_R006211_1H.

– (2005b), "REMO climate of the 20th century run, UBA project, 0.088 degree resolution, run no. 006210, 1H data", World Data Center for Climate, CERA-DB "REMO_UBA_C20_1_R006210_1H", http://cera-www.dkrz.de/ WDCC/ ui/ Cmpact.jsp?acronym=REMO_UBA_C20_1_R006210_1H.

Janssen, J. and Laatz, W. (2005), "Statistische Datenanalyse mit SPSS für Windows. Eine anwendungsorientierte Einführung in das Basissystem und das Modul Exakte Tests", Springer, Berlin, p. 755.

Kalide, W. (2005), "Energieumwandlung in Kraft- und Arbeitsmaschinen. Kolbenmaschinen – Strömungsmaschinen – Kraftwerke", Hanser, München, Wien, p. 421.

Kobus, H. and Bürkle, F. (1996), "Konkurrierende Ansprüche an ein Fließgewässer – das Beispiel Neckar." In: Lehn, H., Steiner, M. and Mohr, H. (ed.), "Wasser – die elementare Ressource", Springer, Berlin, Heidelberg, New York, p. 368.

Landesanstalt für Umweltschutz Baden-Württemberg (LfU) (2004), "Das Niedrigwasserjahr 2003. Oberirdische Gewässer", Gewässerökologie vol. 85, p. 46, www.hvz.Baden-Wuerttemberg.de/pdf/Niedrigwasserbericht.pdf.

Länderarbeitsgemeinschaft Wasser (LAWA) (ed.) (1991), "Grundlagen für die Beurteilung von Kühlwassereinleitungen in Gewässer", Erich Schmidt, Berlin, p. 109.

Maidment, D.R. (ed.) (1993), "Handbook of Hydrology", McGraw-Hill, New York, own counting.

Maniak, U. (2005), "Hydrologie und Wasserwirtschaft. Eine Einführung für Ingenieure", Springer, Berlin, Heidelberg, New York, p. 666.

Mimler, S. and Rothstein, B. (2007), "Klimawandel und Elektrizitätsnachfrage. Eine Untersuchung zu Änderungen in der Elektrizitätsnachfrage durch steigende Durchschnittstemperaturen und extreme Hitzeereignisse", poster publication, Deutscher Geographentag Bayreuth, September 29th-October 5th 2007.

Ministerium für Umwelt und Verkehr Baden-Württemberg (UVM) (2005), "Sauerstoffgehalt im Neckar stark abgesunken", press release, June 27th 2005.

Mohseni, O. and Stefan, H.G. (1999), "Stream temperature/air temperature relationship: a physical interpretation", Journal of Hydrology vol. 218, pp. 128-141.

Müller, U., Greis, S. and Rothstein, B. (2007), "Impacts on Water Temperatures of Selected German Rivers and on Electricity Production of Thermal Power Plants due to Climate Change", poster publication, 8. Forum DKKV/CEDIM: Disaster Reduction in Climate Change, October 15th-16th 2007, Karlsruhe, www.cedim.de/download/33_Mueller_etal.pdf.

– (2008), "Möglicher Einfluss des Klimawandels auf Flusswassertemperaturen und Elektrizitätserzeugung thermischer Kraftwerke", poster publication, Tag der Hydrologie 2008, March 27th-28th 2008, Hannover.

Pintér, I. (2003), "Der Neckar – ökologische Situation, wasserwirtschaftliche Probleme", Landesanstalt für Umweltschutz, p. 16, http://www2.lubw.baden-wuerttemberg.de/public/abt4/ fliessgewaesser/themen/neckar_probleme.pdf.

Rothstein, B. (2007), "Elektrizitätswirtschaft als Betroffene des Klimawandels – Eine Identifikation von Betroffenheiten und Ansätze zur Anpassung an den Klimawandel dargestellt am Beispiel der Energieunternehmen EnBW und EDF", postdoctoral lecture qualification (procedure ongoing) at University of Würzburg, unpublished, p. 367.

Sauer, E. (1984), "Abwärmetechnik. Kühlsysteme, Umweltprobleme, Abwärmenutzung", Verlag TÜV Rheinland, Köln, p. 188.

Schlomann, B., Gruber, E., Eichhammer, W., Kling, N., Diekmann, J., Ziesing, H.-J., Rieke, H., Wittke, F., Herzog, T., Barbosa, M., Lutz, S., Broeske, U., Merten, D., Falkenberg, D., Nill, M., Kaltschmitt, M., Geiger, B., Kleeberger, H. and Eckl, R. (2004), "Energieverbrauch der privaten Haushalte und des Sektors Gewerbe, Handel, Dienstleistungen (GHD)", final report to the Federal Ministry of Economics and Technology, project number 17/02, Karlsruhe, Berlin, Nürnberg, Leipzig, München, p. 27.

Schönwiese, C.-D. (1994), "Klimatologie", Ulmer, Stuttgart, p. 436.

Schönwiese, C.-D., Staeger, T., Trömel, S. and Jonas, M. (2003), "Statistisch-klimatologische Analyse des Hitzesommers 2003 in Deutschland", DWD (ed.), „Klimastatusbericht 2003“, Offenbach, pp. 123-132.

Sinokrot, B.A., Stefan, H.G., McCormick, J.H. and Eaton, J.G. (1995), "Modelling of climate change effects on stream temperatures and fish habitats below dams and near groundwater inputs", Climate Change vol. 30, pp. 181-200.

Umweltbundesamt (UBA) (2007), "Neue Ergebnisse zu regionalen Klimaänderungen. Das statistische Regionalisierungsmodell WETTREG", p. 27, www.wind-energie.de/ fileadmin/ dokumente/ Themen_A-Z/ Klimafolgen%20 und %20Klimaschutz/UBA-Regionale-Klimaaenderungen.pdf.

Verband der Großkessel-Besitzer VGB PowerTech E.V. (VGB) (2003), "Zahlen und Fakten zur Stromerzeugung", p. 14, www.vgb-power.de/ data/vgborg_/ DBsStatistiken/ VGB_ Stromerzeugung_Europa_D.pdf.

Webb, B.W., Nobilis, F. (1994), "Water temperature behaviour in the River Danube during the twentieth century", Hydrobiologia vol. 291, pp. 105-113.

– (2007), "Long-term changes in river temperature and the influence of climatic and hydrological factors", Hydrological Sciences Journal vol. 52, pp. 74-84.

[1] www.worldenergyoutlook.org/2008.asp

[2] www.dwd.de/lexikon

Chapter 2

The Role of Cross-Disciplinary Models in Reducing the Intensity of Environmental Release – An Example of the Irish Housing Sector[1]

Roy Manoj, Ellis Geraint

Abstract

The development of a more sustainable built environment is critical to meeting the challenge of climate change. This has a multitude of dimensions, one of which is the flow of materials and energy required to keep societies thriving. Most of these materials and energy eventually become an environmental burden as they are released as emission, dissipation and waste, the scale of which makes it a significant contributor to global climate change. No single discipline has been able to capture the appropriate conceptual understanding or propose relevant policy responses that are now required to make the built environment more sustainable. This paper takes an example, the case of construction material in the Irish housing sector, and discusses how the current material flow can be reduced and delayed through collective, cross-disciplinary actions involving inputs from the fields of spatial planning, waste management, building control, applied economics and interactive modelling. The methodology has two components: an analysis of material intensity and retention period of selected houses in Ireland during 1890's-2004 and a scenario analysis using contrasting sets of policy interventions (e.g. business-as-usual v/s integrated). Results show that in general Irish houses are becoming redundant (i.e. through abandonment or redevelopment) well before their expected retention period, thus intensifying current material flow and inflicting an avoidable environmental burden. However, based on the scenario analysis, there is a clear possibility of delaying and even reducing the quantity of this material flow. This is only possible if there is coordination between different disciplines and sectors, including civil society.

[1] This paper is based on a project funded under the Environmental Research Technological Development and Innovation (ERTDI) programme of the Republic of Ireland's Environmental Protection Agency (ref: 2005-FS-34-M1) and their support is gratefully acknowledged. The project has also benefitted from cooperation from Kildare County Council and their assistance is warmly appreciated. The authors also thankfully acknowledge the constructive comments of an anonymous reviewer.

1. Introduction

This paper focuses on the critical role the built environment plays in tackling climate change. There is a complex relationship between the built environment, greenhouse gas emissions and climate change that defies mono-disciplinary policy responses. Primarily drawing on debates from the UK and the Republic of Ireland, this paper briefly reviews some of the reasons for this and highlights why it demands a more effective cross-disciplinary response. In particular it takes the perspective that the sustainability of the built environment is best understood from a metabolic perspective (Fischer-Kowalski 1998; Fischer-Kowalski & Hüttler 1999), where energy and resource use is conceptualised as being flows passing through the socio-economic system, beginning as food, fuels, commodities etc. and ending as waste and emissions. The key drivers of this process are, of course, the societal demand, the economic process that meets the demand and the regulators that control/facilitate this society-economy transaction. The ability to understand this process from a spatial perspective offers insights on this issue, which have the potential to be effectively engaged through a range of existing regulatory systems such as those governing land use, waste and construction standards.

In developing these arguments, the paper discusses the development of the Spatial Allocation of Material Flow Analysis (SAMFA) model and its pilot run, based on the housing sector of Kildare County in the Republic of Ireland. The SAMFA model has been successful in combining the emerging concepts of Material Flow Analysis (MFA), spatial analysis, integrated planning, applied economics and scenario modelling. In so doing, it applies a cross-disciplinary approach to facilitate the development of integrated policy responses that can contribute to substantial progress in meeting agreed targets for green house gas emissions.

2. The built environment, climate change and cross-disciplinarity

The wider context for the development of the SAMFA model lies, of course, in the broader concerns over anthropogenic climate impacts. It is now beyond scientific doubt that this is becoming irreversible and deserves to be recognised as *the* leading global political and economic priority[2]. Scientists have been repeatedly endorsing the fact that human activity is to blame for continued global warming (IPCC 2007). Despite this, some of the world's richest countries are

2 For example, in 2004, Sir David King, the then Chief Scientific Adviser to the UK Government, noted that "in my view, climate change is the most severe problem we are facing today, more serious even than the threat of terrorism" (http://news.bbc.co.uk/1/hi/uk_politics/3584679.stm), while a report for the UK Treasury highlighted the fact that if urgent action was not taken, the costs and risks of climate change could cost up to 20% of global GDP per annum (Stern 2006).

still to commit to global treaties such as the 1992 Kyoto Treaty that aim to secure a reduction in CO_2 emissions. Even some countries that have committed to reducing CO_2 emissions have not been particularly successful in doing so, indicating the need for further, urgent action.

However, the challenge of climate change has resonance that ignores geographic, social and disciplinary boundaries and therefore has demanded responses that do not always fit neatly with traditional forms of organising institutions, public policy and even knowledge. As such, debates over the research required to support a transition into sustainable development have merged with epistemological enquires on the need for, and problems of, cross-disciplinary[3]and integrative research. Indeed, the very nature of the problem of how to address unsustainable patterns of development has demanded, and in turn stimulated, innovative, cross-disciplinary approaches to research and practice (Dovers 2005; Evans & Marvin 2006; Klein 2004; Petts et al. 2008; Russell et al. 2008).

A wide range of arguments have been presented for why sustainability and cross-disciplinarity are so entwined, primarily reflecting the fact that the inter-dependencies of human and natural systems requires an integration of social, economic and environmental systems under the paradigm of sustainability (Dovers 2005). In turn, effective responses to the problems generated by unsustainable patterns of development (such as climate change) require institutional frameworks for decision-making and implementation that have the capacity to integrate a broad spectrum of interests, knowledge and regulatory processes. To be effective, such institutional frameworks have to address the dynamics of society-economy interaction, and engage with environmental agenda through mechanisms that span multiple levels interventions, e.g. EU, national and local, with corresponding programmes of action.

Nowhere is this more applicable, or more important, than when considering sustainability of the built environment. The built environment encompasses all buildings, spaces and products that are created or modified by people. It comprises urban design, land use and transport system, and includes patterns of human activity within the physical environment (Handy et al. 2002; Srinivasan et al. 2003). The built environment is a key source of atmospheric emissions, being the locus of major energy consumption (e.g. in Europe around 41% of all energy is consumed in buildings, Míguez et al. 2006) and resource use. The environmental

---

[3] We note here the intricate discussions involving the important distinctions between inter-, multi-, pluri- and trans-disciplinarity (e.g. Petts et al. 2008) and the variable barriers and opportunities these present for sustainability researchers. However, we have chosen here to side-step the precise nature of integrative research in favour of the recognition that a variety of strategies are required to address the problems of the unsustainability of the built environment and for the purposes of this paper we will generalise on the precise nature of the integration by deploying the term "cross-disciplinarity", unless specified otherwise.

impact of built environment is clearly distributed in correlation with population density and it is in urban areas that its environmental impacts are most acutely felt. Indeed, cities cover only 2% of the earth's surface, yet consume 75% of all resources and produce 75% of all waste (Girardet 1999). The built environment is, therefore, clearly one of the key sources of greenhouse gases, with buildings (i.e. construction, occupation and use) amounting to 50% of the UK's total greenhouse gas emissions (RCEP 2007). While there are some encouraging conceptual and technical developments in assessing the environmental impacts of the built environment (e.g. Henderson et al. 2001; Ratti et al. 2005), it is suggested here that these are generally inadequate because they tend to be mono-disciplinary, address either operational or embodied energy (but not both) and fail to take a cumulative or holistic account of resource use in the built environment.

The degree to which the built environment contributes to climate change is clearly dependant on a huge number of factors ranging from issues such as built form, infrastructure aspects (e.g. the provision of public transport and types of building materials used), sources of energy supply, as well as individual lifestyle and consumer behaviour. Within this complexity, spatial planning can make a significant contribution to reducing energy demand by shaping the built environment (Bulkeley and Betsill 2003) – including design that promotes energy efficiency and maximizing location and density to reduce reliance on private transport. It can also create opportunities for commercially viable, socially affordable and environmentally better developments such as BedZED in London (Twinn 2003) and Vauban in Freiburg[4], as well as accommodate policy interventions such as London's Congestion Charge (Blow et al. 2003) and carbon taxing (Kasa 2000). Such issues have been central to the normative goals of planning for well over a decade, during which time, sustainable development has been adopted (rhetorically at least) as its primary *raison d'être* (Gunder 2006). There have however been major issues in translating the aspirations of climate change policy into realities on the ground (Bulkeley 2006; Haughton & Hounsell 2004; Owens & Cowell 2002). The reasons for this include the failure of politicians to agree on the difficult trade-offs required and particular relevance here is the conceptual and institutional limitations of planning. An inability to properly anticipate economic trends (e.g. mortgage lending and developer-led construction) into planning decision making is another grey area in this context.

An example from an Anglo-Irish context is that the current regulatory system of land use can specify some aspects of development, such as location and access, but is limited in what it can demand of developers in terms of issues such as detailed design, construction materials, pricing and the energy performance of proposed buildings. Indeed, an innovative planning policy introduced by the London Borough of Merton that required new commercial and industrial development to

4 See www.vauban.de.

generate at least 10% of its energy requirements from renewable sources was objected to by central government and other interests on the ground that it had no basis in national planning law and policy[5]. Furthermore, while most countries have adopted energy rating systems for buildings systems[6], these primarily relate to building operational energy efficiency. In order to enhance the operational energy needs of the built environment, a collective response is required from multiple sources, such as building design, efficiency of building maintenance, occupancy and occupant's behaviour (Ratti et al. 2005; Schlumpf et al. 1999). In addition, recent research suggests that in a building's lifetime, up to 50% of the total energy consumption may remain embodied (Treloar et al. 2000). Reduction in embodied energy requires other disciplines, such as building material engineers; construction management; resource management, production process, technological innovation and spatial planning. There is thus a lack of effective synergy between the various professions, socio-economic, regulatory systems and intellectual traditions that govern the development of the built environment, and this continues to act as a barrier to an appropriate response to climate change.

It is the complexity of tackling these issues that makes the built environment exactly the type of "problem" that demands a cross-disciplinary approach (Petts et al. 2008). A widespread persistence of these problems also highlights the limitations of knowledge and policy responses organised within traditional disciplines (Evans & Marvin 2006; Newell et al. 2005). A number of authors have also noted that research that attempts to integrate cross-disciplinary perspectives tends to be more intellectually and pragmatically challenging than research conducted within a single discipline (Dovers 2005), and has to overcome conceptual, social and institutional barriers (Horlick-Jones & Sime 2004). Indeed, as Evans and Marvin (2006) note, cross-disciplinary research is "hard work that requires significant investments on the part of those who undertake it" (p. 1010). Because of this, it has been noted that despite the high level of rhetorical commitment to cross-disciplinarity, the experience of researchers and policy-makers engaging in such urban research has been less encouraging, with many of the potential benefits proving to be elusive (ibid). This appears to be down to a range of complex structural, economic and cultural issues that include the fact that fields of enquiry still tend to be guided by epistemological structuring of disciplines rather than from a problem-solving perspective (Petts et al. 2008). In some

5 See www.themertonrule.org.

6 For example, within the EU the Danish system is considered to be one of the best energy rating systems as it has more comprehensive criteria, has different ratings for small, large and industrial buildings and proposes alternatives for improving the obtained score. In the UK the system is known as Standard Assessment Procedure (SAP) and in the Republic of Ireland, the Energy Rating Benchmark (ERBM) (Míguez et al. 2006).

cases, common ground between participating disciplines is also narrower than expected (Evans & Marvin 2006).

The way in which problems are framed and solving-strategies pursued appears to be central to realising the opportunities presented by cross-disciplinary approaches (Barnett et al. 2003; Horlick-Jones & Sime 2004). Indeed, Petts et al. (2008) draw specific attention to the way in which built environmental problems are often "pre-framed as physical or technical ones, with an associated failure to understand the nature and complexity of socio-technical systems in sectors like energy and waste" (p. 598). As noted above it is the issues of energy and waste that lay at the heart of the challenge of sustainability of the built environment and which critically calls into question the ability of spatial planning to effectively engage in cross-disciplinarity in a way that allows an effective conceptualisation and regulation of sustainable urban systems.

The consideration of the cross-disciplinary capacity of spatial planning, in an Anglo-Irish context at least[7], gives rise to a curious contradiction, as on the one hand it is characterised as being a "multi-disciplinary discipline", driven by the need to understand the full context and implications of urban development, while on the other it has failed to develop its own distinct disciplinary identity, which is a precondition for effective cross-disciplinary work (Pinson 2004). Indeed, in a UK context, the multi-disciplinary nature of planning is represented by the diverse variety of competencies the main professional body, the Royal Town Planning Institute (RTPI), expects accredited graduates to display (RTPI 2004), yet it is prone to protracted and recrudescent debates on the core values and distinguishing features of what it is to be a "planner" (e.g. Evans 1993; Eversley 1973; Healey 1985; Reade 1987; Tewdwr-Jones 1996). This debate is mirrored, within the Republic of Ireland, although not so publicly vociferously.

The professional insecurity that underpins the debate on the integrity of the planning discipline has been reified by regulatory systems in the UK and Ireland that have traditionally been characterised by; a narrow focus on physical land use rather than broader environmental impacts of development; a discretionary system that takes each application on its own merits, which then has had difficulties in considering cumulative impacts; and which justifies development decisions on the basis of an ill-defined concept of "the public interest", in which economic growth and the protection of property rights have been a dominant influence. Therefore, although planning has been seen as being central to the de-

---

7 It is acknowledged that the professional, disciplinary and regulatory aspects of spatial planning is closely related to institutional, cultural and political context (Sykes 2008), which also reflects country-specific histories of how the state and other stakeholders have intervened in the evolution of the built environment. *The Compendium of spatial planning systems and policies* provides an insight into the diversity in planning systems within a European context (EC 1997).

livery of sustainable development, in an Anglo-Irish context at least, the traditional concepts, ideology and models that support decision-making have not been well equipped to engage in true cross-disciplinary engagement related to the challenge of climate change.

We are particularly concerned here with how Anglo-Irish planning systems have been able to take into account local environmental limits and the broader patterns of energy use and waste management in the built environment. Indeed, these issues help understand how the discipline and regulatory system of spatial planning conceptualises and addresses *the* core issue of sustainable development; namely the ecological constraints of the bio-physical environment. We have therefore developed a model which seeks to assess the material flow of the built environment as witnessed by residential development. We have combined land and planning factors with socio-economic variables and some elements of fiscal policy and market forces.

3. Energy and resource use in the housing sector

It is undeniable that in climate change terms, housing is *the* most important element of the built environment, with the highest share of space, material and energy. Data from the UK shows that the energy requirements for buildings in the domestic and non-domestic sector exceed those for transport and industrial processes and the housing sector alone consumes almost 30% of total energy (Rylatt et al. 2003; Steemers 2003), while in Ireland this is 25% (Oh'Leary et al. 2008). Contemporary studies on Ecological and Carbon Footprint (EF and CF) have also highlighted the residential sector as a major contributor to environmental impacts – for example a recent study in Northern Ireland estimates the share of housing of the total EF and CF to be as high as 39% and 48% respectively (Maguire et al. 2008).

This impact, however, is not homogenous across the housing sector, with Ratti et al. (2005) observing that the variance in energy consumption in buildings with similar functions can be as much as 20-fold. They identify four broad factors to explain this variance: building design accounts for a 2.5x differences in energy consumption, urban geography for a 2x variation, system efficiency for a 2x variation, and occupant behaviour for a 2x variation. This variation clearly illustrates the potential of well focussed interventions on reducing the energy consumption in the housing sector, particularly when in a typical lifetime of 100 years, a residential building may consume between 80 and 100 GJ/m^2 of energy representing both initial and recurrent embodied, and operational energy demands (Grubert 2006; O'Leary et al. 2008; Yohanis & Norton 2002).

It also needs to be acknowledged that in addition to the quantifiable environmental impact, the housing sector is also underpinned by the symbolic cultural and personal significance of the home. In countries such as the UK and Ireland, characterized by high levels of homeownership, an individual's home is clearly not just seen as a vessel for environmental accounting, but usually represents a household's single most significant financial investment and a private refuge from the travails of modern life. Yet, all too often we see the housing sector being vulnerable to financial crisis such as credit crunch, social issues such as variable levels of availability of quality and affordable housing, and environmental problems such as flooding. As such, any environmental analysis of the housing sector cannot simply be viewed in techno-rational terms, but requires an understanding of the potential macro- and micro-economic impacts and an acknowledgement of the complex social factors involved.

Further complexity arises by taking into account the temporal dimension, where differences between embodied and operational energy becomes particularly significant. By definition, operational energy (OE) is that required for day-to-day running of the building – involving mainly heating and cooling, lighting and appliances. Although theoretically OE can be fixed, in practice this tends to increase exponentially (Yohanis & Norton 2002). For example, in RoI between 1990 and 2006 there has been an annual increase of 1.8% of primary energy use in the residential sector, mainly due to increased electricity consumption by domestic appliances (O'Leary et al. 2008). Commentators such as Schlumpf et al. (1999) and Ratti et al. (2005) have noted that factors such as housing type and space; household size; lifestyle choices; construction standards; and energy efficiency of appliances have a significant influence on OE. These factors represent an interface of personal, professional and commercial objectives, and socio-political norms.

The embodied energy (EE) is that which is expended up to the point that the building has been put into operation. EE is the energy required to provide a product through all upstream processes, traceable from the finished product to consideration of raw materials (Treloar et al. 2000). This includes both direct and indirect energy; the former is used directly during the main process and in association ancillary activities; the latter is required to manufacture inputs of goods and services during the main process and comprises 'energy paths'. An energy path comprises flows of goods or services and an associated amount of energy consumption. As buildings are a long-term investment, it is usual that EE periodically recurs through replacement of materials and components. Indeed, studies confirm that on average a recurrent EE of about 3 GJ/m^2 occurs every 25 years after construction and this tends to increase as the building ages (Atkinson et al. 1996; Yohanis & Norton 2002). The factors influencing EE are clearly different than those for OE and relate to: extraction of materials; manufacturing of components; transportation to building sites; and the construction process – all are related to economic activities.

It is noteworthy that while there are differences in the factors influencing OE and EE, many of the stakeholders that determine their magnitude are shared. For example, private households influence OE through their day-to-day energy consumption, while it is their changing lifestyle and demand for new material/homes and the way the market meets this demand which relates to EE. Similarly, construction of new houses involves policy makers, professionals and construction industries, who can influence OE by design and regulation of energy efficiency in buildings, while influencing EE through initiatives and legislation that influence the rates and spatial distribution of new dwellings and redevelopment of existing housing – for example through spatial planning and fiscal policy related to the housing market. Therefore, a range of stakeholders must act together to address the total energy use in buildings, supported by conceptual insights and evidence from a number of disciplinary perspectives (see table 1).

Tab. 1: Scope for cross-disciplinarity in decreasing energy use in housing sector

		Embodied Energy (EE)	
		Actors involved in decreasing EE use in housing sector	**Actions decreasing EE use in housing sector**
Operational Energy (OE)	Actors involved in decreasing OE in housing sector	a. Politicians b. Government bodies c. Analysts/professional bodies d. Industries/utility companies e. Private households f. Charity/non-government organisations	a. Setting up goals and objectives concerning resource and energy efficiency, with focus on housing provisions and standards b. Making and implementing the national/ regional/local level policies/ programmes/ plans; bench making and monitoring standards for construction industries/ utility services/households c. Education; research and development; policy analysis d. Promote uptake of energy efficient material and construction technology e. Energy-conscious decision making on the timing and material use while renovation/ demolition or building new homes f. Raise issues for the attention of all actors in general, and (a) and (b) in particular
	Actions decreasing OE use in housing sector	*Same as EE, with focus on:* a. Enhancing energy efficiency of existing housing stock b. Raising energy efficiency of existing and future housing stock c. Analyse people's lifestyle choice, and behaviour of utility providers d. Active promotion of energy-efficient house building e. Pursue a low-energy lifestyle f. Raising public awareness	*Cooperation between* • Policy making on housing provision and energy standards – all • Education; research and development – b) to e) • Energy efficient house building – b) to e) • Raising energy efficiency of existing housing stock – b. to e. • Pursue a low-energy lifestyle – c) and e) • Raising public awareness – all

First, it is the politicians who determine the policies, at various spatial levels, that influence building retention periods, energy efficiency and supply and quality of new building. Here, government bodies such as the Local Authority (LA) planners play a central role by facilitating the supply of land, imposing conditions on development and approving planning applications. The role of professionals is multifaceted: from developing guidelines on how, where and what type of housing should be constructed to when and how buildings are demolished and materials recovered and reused, as well as analyse both current and future policies. Industries also play a significant role by promoting technological innovation and its application in the housing sector, whilst conforming to existing policies. This includes development of energy efficient products and services, as well as innovative approach to material recovery and reuse. Individual household's knowledge and awareness of new products and services has a substantial influence through redevelopment/retrofitting decisions. Finally, environmental campaign groups and other non-government organisations may raise issues for the attention of all of these actors.

Having established the importance of the housing sector in terms of total energy demand and the complexity of the factors that can influence this, it becomes clear that at present there is a lack of a conceptual framework through which all these relationships can be synthesised. The next section describes the development and pilot of a model that aims to contribute to this goal.

4. The Spatial Allocation of Material Flow Analysis (SAMFA) model

4.1 Modelling framework

The SAMFA model has been developed with the aims of:

- Testing the spatial allocation of construction material[8] flows from the National Accounts to the local level;
- Highlighting the cumulative environmental impacts of existing policies and trends in the development of the built environment in general, and housing sector in particular;
- Generating policy scenarios that highlight the benefits of alternative policy options that incorporate both planning and socio-economic variables;
- Providing a cross-disciplinary, integrative medium for all relevant stakeholders to understand and prioritise policy approaches for reducing the environmental impact of the built environment, with emphasis on housing sector.

8 The construction material in this study refers to gravel, sand, clay and wood (Müller 2005; Tanikawa et al. 2005).

The model has been formulated by a cross-disciplinary team consisting of an architect-infrastructure planner, an industrial ecologist and a geographer-planner. It has been developed by first acknowledging the limitations of existing analytical models in this field and then drawing on a number of key papers that have indicated key energy-built environment relationships. These elements are reviewed briefly below, but critical to an understanding of the model is the fact that it attempts to spatially model Material Flow Analysis (MFA) at the most important policy level in determining energy use in the built environment – the LA, where (in an Anglo-Irish context) detailed policy determining spatial planning, building control, waste management and housing allocation are all determined.

MFA offers a methodological basis for accounting elements of EE other than the energy paths (Treloar et al. 2001). This approach follows the mass-balance principle, which implements the second law of thermodynamics (Eurostat 2001) and the concept of metabolism (Fischer-Kowalski 1998; Fischer-Kowalski & Hüttler 1999) in the context of socio-economic systems (Haberl et al. 2004). Put simply, it assumes that everything (i.e. materials/ products/ energy) that flows into and is extracted from a socio-economic system equals consumption (metabolically), stock and outflows (as materials/ products/ energy/ emissions/ dissipation/ waste).

In terms of the residential sector, this industrial ecology approach needs to be combined with an economic perspective, in order to fully take into account the effect of housing supply and demand (Müller 2005). In addition, it is suggested that the critical level of analysis is the single dwelling – so that supply is estimated in terms of units completed and waste in terms of units demolished. The main drivers of construction material flows in the housing sector are in fact the construction of new dwellings and demolition of old units. If these can be measured, the mass-balance principle can be satisfied. More importantly, we can address the critical issue of EE here by tracing the source of material and its throughput.

However, current MFA methodology works best at the national level (Hinterberger & Stefan 2004), and there has been a recent MFA study at that level in RoI (Maguire & Curry 2008). In contrast Bannett and Newborough (2001) argue that a city, town or urban district, i.e. the LA is the most effective level for energy auditing and emission estimation, and it is here where influential policy decisions are made. A critical function of a model that can influence climate change policies for the built environment is therefore an ability to allocate the national level material flow data to the LA level. A number of models have been developed in the field, but all are unable to satisfactorily analyse this data at the local spatial scale.

For example, the Resources and Energy Analysis Programme (REAP), developed by the Stockholm Environment Institute (Wiedmann et al. 2004) tries to solve this allocation problem using the household and PRODCOM (*PRODuct* of the European *COMmunity*) data, but is unable to spatially distribute these at the local level in a way that corresponds to the specific local level planning context. Its neglect of the local level planning factors in turn makes it inappropriate for the local spatial planning authorities. Another attempt to generate energy and CO_2 scenarios is the technique of 'regional downscaling' which allocates the national level data to the local or regional level, based on 34 parameters, such as population; households; mobility; employment; and floor space (Dürrenberger & Hartmann 2002). But again it does not incorporate the local level planning issues, such as land availability and planning constraints.

In developing a model that can overcome these constraints, there are a number of papers that suggest how spatial factors may influence the use of construction material in the housing sector. For example, Tanikawa et al. (2005) provides a growth curve to predict the year when a residential building may exceed its life expectancy. Elsewhere Schiller (2007) suggested that there are temporal as well as spatial variations in both material intensity and room volume. Based on these studies it is possible to estimate future levels of construction material waste, although they address neither recovery/recycle nor issues such as spatially disaggregated demand of construction material. Insight into these issues can be gained from Bramley (1993), who taking an economic perspective, highlights the influence of planning and land factors on the land availability for housing construction, and thus levels of housing completions. This study offers the possibility of incorporating elements of spatial planning framework and macro-economic and financial factors that determine the profitability of the construction sector into the spatial allocation of future housing supply. It has been argued that firms tend to combine both last year's data on land and planning factors and existing/future regulations to maximise their profit. Further, the economic and financial variables such as GDP and price of production factor (labour/ material) are most relevant at the regional and national level; their impact at the local level may not be that significant. This still leaves the spatial variability in demolition and associated waste unresolved.

The SAMFA model draws on the above research but offers a number of additional unique features. The initial development of the model is developed in the context of the Republic of Ireland and is based on the flow of construction material set out in the national level MFA (Maguire & Curry 2008) and allocates this to the LA level using a set of spatial attributes. The primary data are generated using a bottom-up approach, to which the national-level MFA studies offer a balancing factor. The model is a contribution to the growing number of tools aimed at assisting in the development and testing of various alternative development scenarios, helping to identify the spatial arrangements that minimise energy consumption and emissions. Its objective is to develop alternative energy use and emissions scenarios for individual Irish Local Authorities, which can then be used as a platform for engaging key stakeholders. It should at least involve policy-makers, professional planning officers, development industries and community representatives in scenario development, aligning itself with those needed to be involved in a city-wide energy audit (Barnett et al. 2003). In order to achieve this overall objective the SAMFA model has been designed with six interactive modules (M1-M6, figure 1) as follows:

- Module 1: Material intensity and building lifetime simulation
- Module 2: ED-level allocation of new housing units
- Module 3: Material-balance at different spatial scales
- Module 4: EE at different spatial scales
- Module 5: OE at different spatial scales
- Module 6: Total energy consumption of housing sector

Fig. 1: SAMFA modules and its relationship with housing sector and related actors

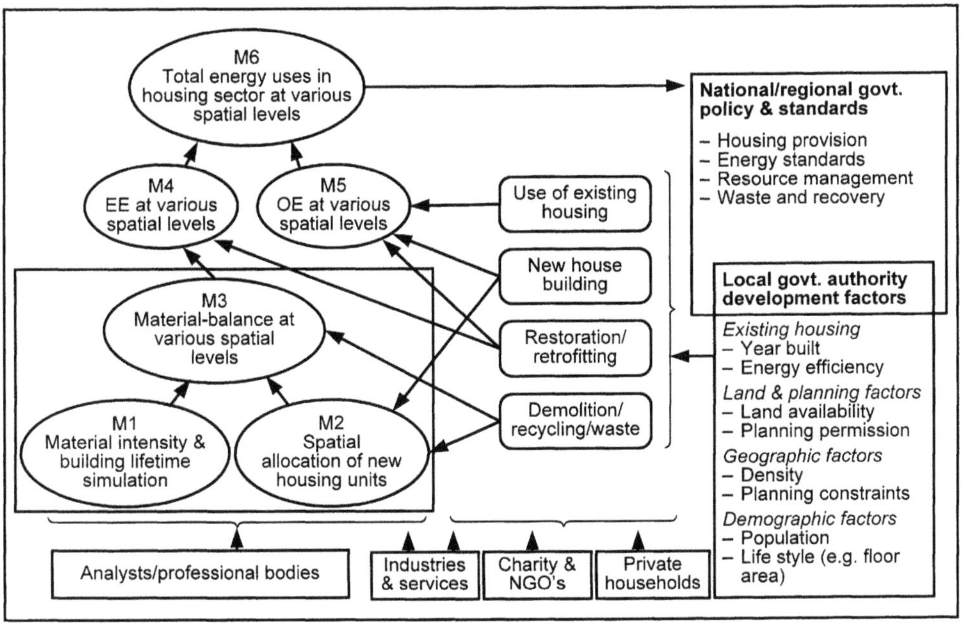

An elaborate account of the activities of individual modules has been presented in section 4.3, while how these are linked to each other and other stakeholders of the housing sector has been graphically elaborated in Figure 1. While the modelling components are closely related to the expert users, such as professionals and analysts, the background data are generated from, and the results delivered to all other stakeholders including private households. It is also noteworthy that Electoral Districts (EDs)[9] are the most appropriate level for the SAMFA model to maximise the policy impacts, allowing data assembly and disaggregation from, and aggregation of, results for, higher spatial levels, particularly at the County

[9] These are the smallest administrative units in the Irish Republic, with a total of 4334 in the country. They vary in population from 16 to 32,000 (see www.cso.ie/census/SAPs.htm).

level. In order to fully appreciate the data inputs and policy outputs of the model a brief account of the Irish context is presented below.

4.2 Overview of the Irish context

The Republic of Ireland (RoI) has a total population of 4.2 million, living at a density of around 60 persons/ km^2. Although Ireland has a strong rural tradition, the economy has been transformed from the mid-1980s onwards by a prolonged period of growth referred to as "The Celtic Tiger", with GDP per capita increasing by 6.3% during 1990–2001 (OECD 2001). Some of the consequences of this period of prosperity have been unparalleled demands for housing, major shifts in living patterns and increases in waste and greenhouse gas emissions. Historically Ireland has had a dispersed settlement pattern dominated by the rural. For example, in 1996 it had had the lowest share (about 35%) of population living in settlements of between 2,000 and 100,000 in Europe. Only 30% of its population live in cities with a population over 100,000, with the remaining 35% in dispersed rural settlements (Republic of Ireland 2003). Furthermore, about 40% of its total population is concentrated in Dublin and the Eastern Region, leaving a very low density in the rest of the country. Traditionally rural housing is in general 1-2 storeys high, detached or semi-detached, and in urban areas housing is mostly in low density suburbs (15-30 houses/hectare). The country has a relatively low level of housing availability, at 425 units/1000 population in 2005, which is 15th amongst 19 EU countries. However, in terms of house building completion rate (units/1000 population) it held the top position within 19 EU countries in 2005. Indeed house building has experienced significant growth in recent years – the annual number of units completed increased from under 20,000 in 1990 to more than 86,000 in 2005. In particular, since 2000 there has been a steady rise in the number of housing completions (CSO 2006). Despite this, housing construction is likely to remain at a high level due to ongoing low levels of housing availability and a relatively young demographic structure with approximately 37% aged under 25 years (Republic of Ireland 2003).

The recent and rapid economic growth has had substantial environmental consequences (McDonald & Nix 2005), particularly in terms of waste management (e.g. Davies 2003; Desmond 2006) and greenhouse gas emissions. Indeed, Ireland is one of the countries with a poor record of reducing its CO_2 emissions – by 1999 it had already overshot its agreed target for 2008-2012 by 16.3%, with only Spain performing more badly in relation to agreed targets within the EU (Míguez et al. 2006). The third Irish national communication to the UN Framework Convention on Climate Change noted that the main reason for failing to meet its Kyoto target was due to increased energy consumption by industries, businesses, households and transport – all key components of the built environment (Republic of Ireland 2003).

Fig. 2: The study area context

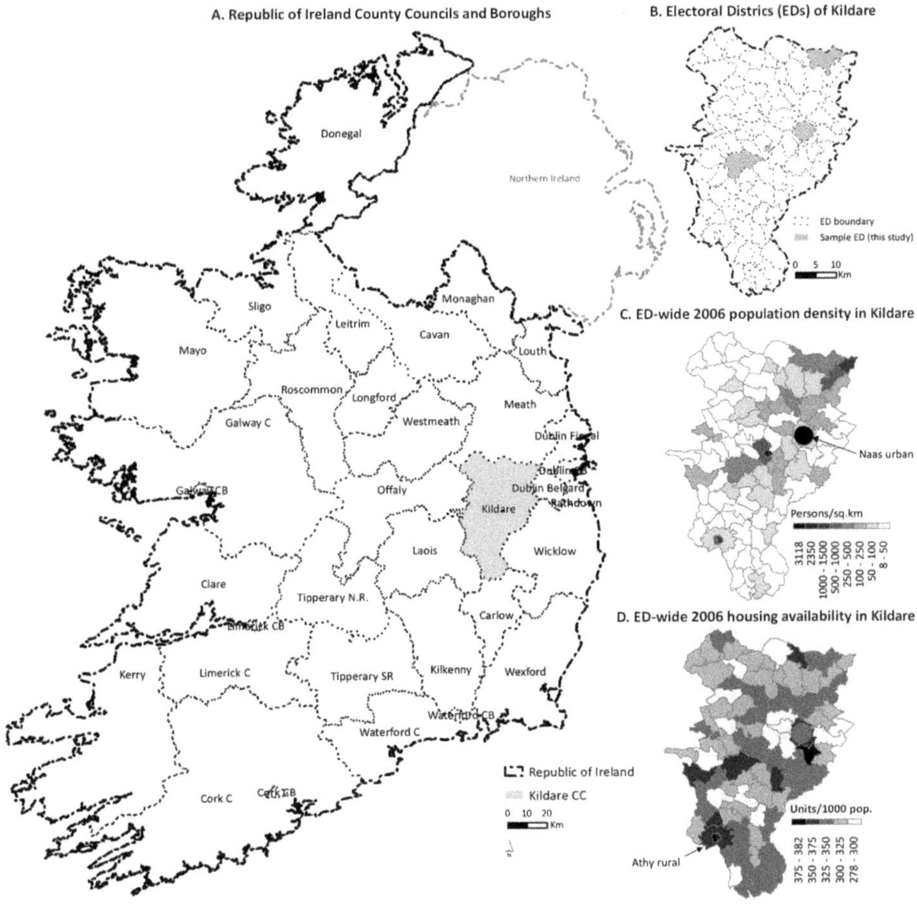

Source: Ordnance Survey Ireland (OSI); and Centreal Statistics Office Ireland (CSO)

The Government's response to this issue can be considered encouraging, but nevertheless, cautious and inadequate. For example, compared to some EU countries (e.g. UK, Denmark, France, the Netherlands and Luxembourg) where a system of energy rating for buildings is being widely used, RoI still deploys a voluntary system run by a commercial organisation, the National Irish Centre for Energy Rating (NICER) using its own labelling system know as the Energy Rating Benchmark, ERBM (Henderson et al. 2001). Furthermore, a recent national review on *Energy in the Residential Sector* in Ireland placed an exclusive focus on operational energy and neglected the impact of embodied energy (O'Leary et al. 2008).

There is therefore a strong case for more focussed policy initiatives for reducing the energy demands of the built environment as part of the broader strategy of tackling climate change. With this aim in mind, the SAMFA model has been initially piloted in one of the 29 Counties of Ireland, Kildare, which lies to the immediate west and south west of Dublin (figure 2A). Kildare is part of the Greater Dublin Regional Authority, which is part of the Eastern Regional Assembly (Ellis & Kim 2001) and has a total population of 186,000 people, in an area of 1692 km² (a density of 109 persons/ km²). There are 89 EDs in Kildare CC, of which four have been subject to this study (figure 2B). Due to the proximity of Dublin some parts of the county have been experiencing rapid urbanisation in recent times (e.g. ED Naas Urban which has a population density of over 3100 persons/ km^2, figure 2C), although other areas of the county are still characterised by traditional rural housing patterns. What is also noticeable is that the housing availability levels in general appear to be more in rural areas than in urban areas (figure 2D). In many respects therefore, Kildare represents many features of the current Irish built environment and provides a valuable laboratory for the testing of the SAMFA model.

4.3 Model application

A prerequisite for any modelling activity is the availability and accuracy of required data. The data needs should be based on the scope of the overall model as well as the individual modules. The modelling process, data requirement and availability of the individual modules have been presented below (table 2), followed by construction and comparison of two scenarios (table 3 and figure 3). To be clear, the model is being applied to the specific context of Kildare CC, and while most of the data refers to this specific context, some assumptions based on similar studies elsewhere were inevitable due to data limitation in Kildare CC/Ireland.

Module 1 (M1): Material intensity and building lifetime simulation

The objective of M1 is to provide M2 and M3 with two pieces of key information: construction material intensity (i.e. in tons/m^2 floor space); and probability of demolition at a given point in time (table 2). This is essentially a function of housing type, year built and spatial location. For the pilot application a simplified housing type (i.e. house and flat/apartment) and year built in Irish census periods[10] were applied as a basis for the assumptions found in Tanikawa et al. (2005) and Schiller (2007)[11]. The probability of demolition is derived at the ED

[10] I.e. pre-1919; 1919-1940; 1941-1960; 1961-1970; 1971-1980; 1981-1990; 1991-1995; 1996-2000; and 2001-2006.

[11] Schiller (2007) observed material intensity of Germany houses to decrease exponentially over time, while Tanikawa (2005) observed construction material intensity of UK houses to vary between traditional houses (0.78 ton/m^2) and flats/apartments (0.74 ton/m^2) constructed during the past decade. The construction material intensity of the Irish housing

level using the housing building records of the Irish Central Statistics Office (CSO). Taking the statistics for 2001 and 2006, it is possible to establish building life expectancy in Kildare, which varies between 38 and 65 years with a standard deviation (SD) of 15 to 20 years. This is, however, based only on those buildings that have already been demolished, which is only 5.7% of the 2000 stock (CSO 2002; 2007). This therefore potentially understates the life expectancy of all buildings, and we find, for example that 10% of the 2005 stock is still standing after 100 years and therefore not reflected in this data. This issue is overcome by following the approach of Müller (2005) where the following four types of information have been used to construct a distribution of life expectancy: minimum, maximum and average lifetime; and SD. In SAMFA we consider the average life expectancy found in the 2002 and 2006 statistics as the *minimum* life and assume the maximum life expectancy to be 120 years. The average of these two values is taken to be the average life expectancy of houses at the ED level, with the same SD calculated above. This enables us to calculate the probability of ED-level housing demolition in Kildare CC.

Tab. 2: SAMFA piloting: modelling components, data requirement and sources

Module	Modelling process	Data needs	Data sources
M1	Trend analysis	a) Year built, demolished or renovated b) Quantity estimation of construction material by housing type	a) CSO (2007; 2006; 2002; 1997); this study b) Tanikawa et al. (2005); this study
M2	Multiple regression	ED-level data on: a) Expected population b) Land availability (serviced land, brown field land, and additional land suitable for development) c) Distance from Dublin d) Local/regional development/planning factors	a) CSO (2007; 2006; 2002; 1997) b) Kildare CO (2005); Department of the Environment, Heritage & Local Government (DoEHLG)[12]; and Corine land cover 2000[13] c) Ordinance Survey Ireland (OSI); this study d) Kildare CO (2005)

sector is calculated in light of these two studies in this pilot. However, currently the authors are conducting a detailed survey of material intensity in four EDs in Kildare (see Figure 1). Following this survey it would be possible to add spatial dimension and more housing types in construction material intensity.

12 See www.npws.ie/en/MapsData/DownloadingandMetadata.

13 See http://dataservice.eea.europa.eu/dataservice/metadetails.asp?id=950.

Module	Modelling process	Data needs	Data sources
M3	Mass-balance	In addition to data items for M1 and M2, the existing and future standards of recycling and waste management	Collins et al. (2005)
M4	Trend analysis and assumptions	a) Existing standards of EE b) Trend of recurrent EE	a) Palmer et al. (2006) b) Yohanis and Norton (2002)
M5	Trend analysis and assumptions	Existing standards of OE use, and future trend	Intelligent Energy Europe[14]
M6	Cumulative of EE and OE	Summation of M 4 and M5; no additional data needed	This study

Module 2 (M2): ED-Level allocation of new housing units

The objective of module M2 is to provide module M3 with an ED-wide fraction of the County-level housing supply by type. The County-wide housing supply is a macro-level socio-economic indicator, which is assumed as a given in SAMFA. In M2, as many as four ED-level spatial criteria and three sub-criteria (for criterion land availability) have been selected and associated parameters calibrated to spatially allocate the County-level housing supply (see table 2). The criteria selected here follow Bramley (1993), but reflect the specific local/regional context. As such, economic and financial variables such as GDP and price have not been used here. Instead the land and planning factors that firms use to maximise their profit have been applied to simulate the ED-wide spatial distribution of a given LA level housing supply. The parameters have been calibrated with a goodness-of-fit of over 92%, which suggests that the criteria are highly relevant. The criteria and associated data sources are listed below:

- Expected population – The ED-level population trends have been estimated using the Small Area Population Statistics (SAPS) from the CSO. Using this and the targets of the Kildare County Development Plan 2005-2011, the ED-wide expected population has been calculated.
- Land availability for development – This has three sub-criteria: (i) supply of serviced land as in the County Development Plan 2005-2011; (ii) brown fields being available due to demolition of old houses; and (iii) additional land suitable for development. Land area for item (iii) has been calculated from ED areas excluding the land areas of items (i) and (ii), areas of nature conservation and existing built-up areas. Ideally item (iii) should also exclude areas of

14 See www.bsrec.bg/newbsrec/rosh/data/products/analysis/marketanalysis_dublin.pdf.

flood risks; however, spatial data on flood risk levels in Ireland is currently unavailable[15].

- Distance from Dublin – This is currently being measured in aerial distance from the geometric centroid of Dublin County Borough to that of individual EDs. GIS data from OSI have been used for this purpose.
- County development factor – There is no precise value to measure planning and development factors for individual EDs. Depending on the priority set in the County Development Plan 2005-2011, we have used an ordinal scale of 1 to 5 in order of low to high priority for housing development.

Module 3 (M3): Material-balance at different spatial scales

The outcome of modules M1 and M2 then become inputs for module M3, helping to compute the total material input. For example, the material intensity of a housing type at a given point in time (from M1) is multiplied with the total newly constructed floor space of that particular building type in individual ED (from M2) to calculate the corresponding material input. In parallel to this calculation, module M3 also calculates the material outflow using the ED-level housing demolition (from M1) and existing standards of recovery and reuse of construction and demolition waste. Data on waste management has been obtained from the national waste report of the EPA (Collins et al. 2005), which shows that around 90% of construction and demolition waste is currently recycled in Ireland. This does not indicate whether a proportion of the collected waste is reused in the house building industry, as studies such as Schiller (2007) observe that recovered construction and demolition waste is mostly used in sectors other than residential, such as road construction. However, it is assumed that there will be some reuse of demolition waste, particularly in redevelopment schemes and in the case of older buildings, whose fixtures may be of value to the architectural salvage sector. Such a provision has been made in the construction of the integrated scenario (table 3).

Modules 4 and 5 (M4 & M5): EE and OE at different spatial scales

Modules M4 to M5 are respectively designed for estimating the EE and OE use in the housing sector at the ED-level. We recall that EE can be both initial and recurred; the former through new construction and the latter through periodic renovations/ extensions. The amount of initial EE is calculated from the total material input, estimated in M3. No Irish-specific data is available for this, although a study by Environmental Change Institute (Palmer et al. 2006) notes that this can vary from 0.25 MWh/m^2 (Sustainable Homes – low estimate) to 0.8

15 The Office of Public Works is currently undertaking the Flood Hazard Mapping programme; the results are likely to be available sometime in 2008 (see www.floodmaps.ie/AboutTheSite.htm).

MWh/m^2 (traditional homes – high estimates). In this pilot study a similar range has been assumed. With regard to recurrent EE, two studies have been consulted – one shows that in Dublin the average frequency of renovation is once every 20 years[16]. The other (Yohanis & Norton 2002) shows that recurrent EE increases exponentially and periodically – 0.7 MWh/m^2 after the first, 1.73 MWh/m^2 after the second and 4 MWh/m^2 after the fourth 25 year period. Accordingly, the values observed in the second study have been used for Kildare, but at 20 year intervals. The amount of OE use also relates to both existing and new housing stock as different building standards apply to different periods[17]. Recently, O'Leary et al. (2008) observes a rising trend in household OE use in Ireland, mainly due to higher electricity consumption. In this pilot, the standards of the Dublin based study are used combined with the current rising trend. However the magnitude of the rising trend cannot be definitely determined, as such two levels have been chosen for two scenarios (table 3).

Module 6 (M6): Total energy consumption of housing sector

Modules M1-M5 provides all the information required to calculate the total energy use at the ED-level and above. It is important to highlight here that a number of the data elements discussed above are assumed, such as the rising trend of household OE. There are also a number of aspects where alternative policy options can be tested, such as increasing the minimum or maximum lifetime of housing stock, or changing values of the calibrated parameters. It is these uncertainties as well as opportunities for policy interventions that necessitate scenario analysis. It would be a societal benefit if a scenario can identify the optimum combination of activities that can lead to the required level of reduction in the total energy consumption. To this end, we have constructed and compared two alternative scenarios – a *business-as-usual* and an *integrated scenario* for the year 2011 which is the target year in the existing County Development Plan. In the former we try to replicate conditions that best describe the current trend, while in the latter we try to integrate a range of potential energy saving policy options. Further details of these scenarios are presented in Table 3, which also includes notes justifying why and how the variables/ parameters values have been set for the two scenarios.

---

16 See www.bsrec.bg/newbsrec/rosh/data/products/analysis/marketanalysis_dublin.pdf

17 Indeed, a Dublin based study notes the following variations in OE use in Irish housing: 295 kWh/m^2a (pre-1945); 300 kWh/m^2a (1945-1959); 308 kWh/m^2a (1960-1979); 247 kWh/m^2a (1980-1999); and 131 (since 2000). See footnote 13.

Tab. 3: Business-as-usual versus integrated scenario

Module	Variables/ data items (units)	Scenario assumptions and parameter values	
		Business-as-usual	**Integrated**
M1	a) Material intensity (ton/m^2) b) Maximum life expectancy (yrs) c) Demolition probability (%) d) Renovation frequency (yrs)	a) Existing trend b) 120 c) Calculated using existing lifetime and SD d) 20	a) 80% existing trend, while 20% at a 10% lower rate (i.e. 0.66 - house; 0.62 - flat)[1] b) Same as business-as-usual[2] c) Minimum lifetime: 60 yrs[3] d) 25[4]
M2	Criteria & parameters a) Expected population b) Land availability (hectare) c) Aerial distance from Dublin (km) d) Planning factor (none)	a) Current trend (power 1.025) b) Serviced land (power 0.4); brown fields (power 0.9); addition land (power 0.375) c) Negative exp. of distance multiplied by -0.001 d) Ordinal scale of 1 to 5 (power 0.005)	a) Same as business-as-usual[5] b) Same as business-as-usual except for brown fields land which doubled to give additional emphasis[6] c) Same as business-as-usual[7] d) Power is doubled, i.e. 0.01 to emphasis implementation of planning goals[8]
M3	Additional (to M1 & M2) criteria: a) Material reuse (%)	a) None	a) 20% reuse of waste generated from demolition of old buildings while constructing new housing at that site[9]
M4	a) Initial EE (MWh/m^2) b) Recurrent EE (MWh/m^2)	a) 0.8 (traditional homes – high estimates) b) Existing trend, at 20 years interval	a) 0.8 for 80% new units, and 0.25 (Sustainable Homes – low estimate) for the rest 20%[10] b) Existing trend, but at 25 years interval[11]
M5	OE (MWh/m^2)	Existing trend including current level of rise	Existing trend, but no further rise[12]
M6	Total energy use (MWh/cap*yr)	Kildare average 23.66, with minimum 17.5, maximum 66 and SD 7.4 (figure 3)	Kildare average 19.7, vol. with minimum 15.5, maximum 54.92, SD 5.7 (figure 3)

Notes: [1]The Irish Friends of the Earth (FoE 2006) calls for extension of "Greener Homes" grants and tightening of building regulations for new houses, including restricting construction of single dwellings. Thus some reduction of material intensity is expected. [2]There is no current policy on this issue, so current trend is maintained.

[3]Following FoE's (2006) calls, it is likely that existing building will continue to be used for longer than now. [4]Yohanis and Norton (2002). [5]In the Kildare County Development Plan (Kildare CC 2005) existing trend has been followed. [6]The Kildare County Development Plan (Kildare CC 2005) promotes densification, in turn emphasising development of brown fields land. The FoE (2006) also calls for tax relief on brown field sites. [7]There is no policy/ study to support otherwise. [8]Implementation of the objectives of the County Development Plan is a top priority of Kildare CC. [9]Waste reuse is the most favoured option of Irish Environment Protection Agency (EPA, Bolloch et al. 2007). The assumption used here reflects this preference. [10]Based on the target that 40% of new homes should be energy efficient by 2012 (FoE 2006; O'Leary et al. 2008). [11]Yohanis and Norton (2002). [12]As the recent trend shows a sign of flattening out (O'Leary et al. 2008).

Fig. 3: ED-wide energy use under the two scenarios

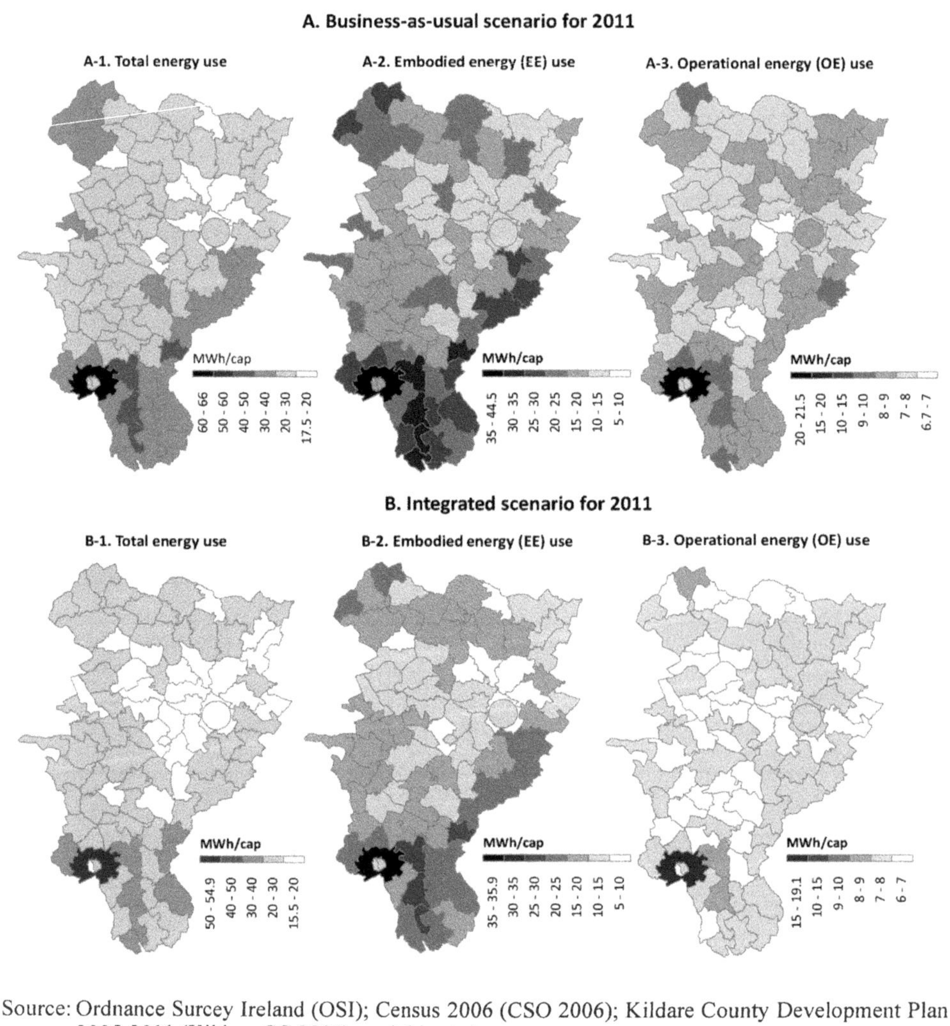

Source: Ordnance Surcey Ireland (OSI); Census 2006 (CSO 2006); Kildare County Development Plan 2005-2011 (Kildare CC 2005); and this study.

The results of the two scenarios have been mapped in Figure 3, which shows that the total energy consumption is significantly higher than a recent report produced by O'Leary et al. (2008) suggests. In the pilot study we find the average total energy use in Kildare in 2011 could be between 23.66 MWh/cap (business-as-usual scenario) and 19.7 MWh/cap (integrated scenario). Corresponding values for EE and OE are:

- Business-as-usual: average EE 19.7 MWh/cap (SD 6.4) – initial EE 3.8 MWh/cap (SD 0.8) and recurrent EE 15.9 MWh/cap (SD 6.5); and average OE 8.1 MWh/cap (SD 1.6)
- Integrated: average EE 16.1 MWh/cap (SD 5.04) – initial 3.3 MWh/cap (SD 0.71) and recurrent 13.1 MWh/cap (SD 5.2); and average OE 7.2 MWh/cap (SD 1.4)

There is a significant ED-wide variation in the total energy consumption for both scenarios (figure 3). There are many reasons for this, such as the age profile of the existing housing stock and the likelihood of new construction. For example, in Figure 1 we observed that one particular ED (named Athy Rural) has a lower population density but a higher housing availability. Most of its housing stock is quite old, and large in terms of floor area; as a result the ED appears to be using the highest amount of energy of all kinds under both scenarios. Unfortunately, current planning policies are unaware of this fact, and the ED did not get a higher ordinal value for policy factor in the integrated scenario.

The spatial difference notwithstanding, it is noticeable that the integrated scenario offers a substantial amount of reduction in all forms of energy use. In this scenario an average person uses about 17% less total energy than that in the business-as-usual scenario. The integrated scenario is therefore a 'clear winner' here, and may offer one of the best possible options to meet the national target. However, one must acknowledge the need for the collective effort that various disciplines must make to implement the scenario. The results also need to be discussed with caution – as a number of issues have been assumed based on studies conducted elsewhere.

5. SAMFA: Implications for policy

The pilot run is an initial testing of a model that is able to spatially allocate material flows. The initial application focuses on construction material in the housing sector, using data from County Kildare. This has provided insights for at least two key areas. First it provides information on the sustainability of the built environment. Second it suggests an approach that can be used to spatially allocate material flows from other sectors. The implications of the latter are explored in the conclusion, with this section focussing on findings directly related to the built environment.

It has been observed that about 60% of the housing stock in 2006 was constructed before the introduction of the Building Regulations for new construction in 1992 (Republic of Ireland 2003). These regulations significantly raised the energy standards for new housing, suggesting that a large majority of existing housing will have relatively high operational energy demands, particularly in relation to heating. Indeed, an initial use of SAMFA suggests that this is a contributing factor to the poor performance of the Irish Republic in relation to Kyoto targets. However, it also suggests that there is substantial potential for energy saving through relatively cheap retrofitting measures, such as double-glazing and draught proofing. Indeed, recent studies (Maguire & Curry 2008; Maguire et al. 2008) suggest that a staggering 39% reduction of Ecological Footprint (or 48% of Carbon Footprint) is possible through retrofitting in RoI. A high level of private ownership of houses (70% in 2005) means direct implementation of retrofitting through, for example, housing authorities, is inappropriate and that persuasive measures, such as market mechanisms and awareness raising of costs and benefits has to be relied upon. This also draws into question the effectiveness of the Irish system of voluntary energy rating, which is generally applied only by the most energy-conscious homeowners.

The model also underlines the influence of urban form and building design on reducing building energy use, both in embodied and operational terms. Indeed, SAMFA confirms that there is a negative correlation between building density and material use and thus, energy consumption. Put simply, the higher the building density the lower the material demand and energy use. While the existing planning guidelines in Ireland (and indeed in most developed nations) now promote higher density developments, SAMFA suggests that this has been poorly implemented. For example, the Guidelines for Planning Authorities on Residential Density recommends planning authorities to "*review and vary, if necessary, their Development Plans to promote higher residential densities*" (Republic of Ireland 2003, p. 7). Yet there is much anecdotal and research evidence that suggests the preference for country-side living and cost of houses in the main urban areas has resulted in a major problem of urban sprawl and long distance commuting (e.g. Hoggart 2003; McDonald & Nix 2005). This is translated into a changing mix of housing types – for example in Ireland between 2000 and 2005, the share of semi-detached houses has risen from 29.3 to 49.3% of the housing stock and flats/apartments increased from 18 to 21%. In contrast, the proportion of bungalows fell by nearly 10% (from 18.3 to 8.7%) and detached housing by nearly 13% (from 30.1 to 17.7%). In terms of residential densities a mixed picture is prevailing. In urban areas the average number of units per hectare has steadily increased from about 25 in 2000 to 31 in 2005, the average size of units has remained static at about 78 m^2/ unit. In contrast, in rural areas the size of units is growing (as evidenced in the national figure for housing floor space which rose from 144 m^2/unit in 2002 to 149 m^2/unit in 2005) despite a falling household size (from 3.34 in 1991 to 2.81 in 2006).

Thus it appears that the sustainability gains resulting from positive trends in urban development have been negated in excessive consumption on the rural front in Ireland. There is therefore a major issue in both the implementation of current planning guidelines and the differing supply of land for rural and urban based housing development. This particular issue has been specifically highlighted in SAMFA. The pilot run shows that there can be substantial gain in urban areas, such as ED Naas Urban, while rural areas, such as ED Athy Rural, continue to remain as 'big hitters' despite showing relative gains. In countries such as Ireland where the preference for rural living is deeply rooted within the country's history and where the rural lobby is strong while the planning restraint is weak, tackling energy consumption in rural areas becomes the major issue facing the development of the built environment. Faced with such a challenge, it is noteworthy that the integrated scenario shows a particular impact on the operational energy even within rural areas. Indeed, while the scenario shows an overall reduction of about 17% per capita in Kildare, in Athy Rural the corresponding reduction for OE is about 10%. However, it has to be acknowledged that the model operated with a given number of LA-level housing supply, which can vary significantly due to macro-economic factors. It is, of course, possible to use different figures for LA-level housing supply to construct other alternatives.

These findings can vastly enhance Local Authorities' capacity to develop alternative energy consumption and emissions scenario that should provide a basis for local development planning, which in turn may underpin the formulation of more informed and coherent regional/national plans. The process has benefited from interaction with the scientific and professional community, and the experience of developing the tool further contributes to the generation of evidence-based policy in this field. As noted in section 2, currently the land use planning systems of the UK and Ireland have a number of institutional and professional barriers to the development of a more holistic approach to sustainability planning. It is widely recognised that management of the built environment is faced with the complex task of coordinating between multiple organisations, each of which has its own agenda, priorities and setting (Bueren & Heuvelhof 2005). SAMFA has the potential to be adapted to a variety of complex settings, but will need a core team drawn from relevant stakeholder institutions (De V Borges & Sahay 2000). It is essential that the Local Authorities form the core participants, accompanied by a wider group drawn from the broader academic and professional communities. This includes professionals/organisations such as architects, construction companies, LA planning service, transport planners, waste managers and utility companies; and stakeholders such as members of strategic policy committee (SPC), county and city development board (CDB), environmental groups such as the Friends of the Earth; and LA community representatives. Finally, to address data-related issues, participation from the following organisations has also to be sought: CSO; OSI; EPA; DoEHLG; Department of Communications, Marine and Natural Resources; Construction Industry Federation; National Roads Authority; and the individual LAs.

Therefore, in terms of the implications for moving to a more sustainable regulation of the built environment, on the basis of this pilot run, it would appear that SAMFA does provide some additional insights for a more evidence-based approach to policy making. However, the value of the model lies not just in these terms, but in potential wider contributions. Indeed the adoption and use of the tolls and methods discussed has become more urgent and relevant when economies are going into recession or even depression, especially when the reduction in demand is led by reductions in construction investment. Policies aimed at decarbonising the building sector bring major economic advantages, in terms of employment and use of resources otherwise unemployed, in these circumstances.

6. Conclusion

The description of the SAMFA model provided in this paper is based on a number of premises: that the built environment is a priority for tacking the threat of climate change; that a critical arena for decision-making in this field lies at the LA level; and that a cross-disciplinary approach is required to make critical conceptual insights on which more progressive policy can be developed. Although the SAMFA model has initially been developed to map the changing material intensity of the residential element of the construction industry, and limited to one Irish County using some assumed data, the section above describes some of the insights the approach delivers. This points to the potential development of a much more robust and comprehensive model for guiding a wide range of policy decisions in support of sustainable development.

A first step here is to further develop the SAMFA model in terms of its accuracy, for example by validating the assumed built environment data (e.g. actual life expectancy of buildings through field surveys) and by incorporating data from each of the Irish County Councils and Boroughs to provide a complete spatial model of the national material flow for construction materials. An examination is also needed to test the validity of the current assumption that the macroeconomic and financial variables do not apply for a model like SAMFA. Further consideration should also be given in using the model to reproduce past results applying data-fitting and associated non-linear stochastic simulation techniques. Once complete, this will then provide a valuable predicting and monitoring tool that is capable of integrating national and local spatial planning objectives with actual environmental impacts, which is also fitting with both macro- and microeconomic variability and pattern.

Should the wider geographic development of the model prove to be similarly successful, there is then potential to transfer the modelling approach to other areas of the built environment, for example, commercial development and other key contributors of greenhouse emissions. Ultimately it may even be possible to develop a meta-model that includes a range of other economic sectors, so that those patterns of consumption that can have important environmental impacts can be allocated to a local scale. This would facilitate better links between behaviour, policy and implementation and provide important evidence to link environmental impacts (e.g. emissions, water quality etc.) with other spatial attributes, e.g. socio-economic categories, deprivation indices, voting patterns or access to different types of service. In so doing, the model offers the possibility of creating critical links between socio-technical systems with the most important challenge of all – the limitations of the bio-physical environment.

A final comment is reserved for the cross-disciplinary nature of the model presented here. The SAMFA model has been stimulated by recent developments in the field of industrial ecology, in particular successes in formulating mechanisms that can translate national trade accounts into material flows with the capability of illustrating national imperatives for resource planning. However, while effective global agreements and national strategies are essential elements of the governance for sustainable development, it is at the local level where the majority of required actions take place (Gilbert et al. 1996), hence the enduring sagacity of the phrase "thinking globally – acting locally". Thus while the national material flow accounts have some conceptual value, it is the ability to translate this to the local level where major breakthroughs can be achieved. The SAMFA model has attempted to do this using the concept of spatial allocation of the national material flows, based on an understanding of not just industrial ecology but also GIS and computer modeling. Furthermore, the model has been then adopted for maximum local policy impact by combining this technical expertise with an understanding of the local regulatory context for land use planning and waste management and the creation of a format which allows a wide range of local stakeholders to specify their own scenarios for the future development at the local scale. As such, we believe it highlights the synergistic value of cross-disciplinary working and begins to map out a future direction for a meta-discipline of sustainability planning.

References

Atkinson, C., Hobbs, S., West, J. and Edwards, S. (1996), "Life cycle embodied energy and carbon dioxide emissions in buildings", Industry and Environment 2, pp. 29-31.

Bannett, M. and Newborough, M. (2001), "Auditing energy use in cities." Energy Policy 29, pp. 125-134.

Barnett, J., Ellemor, H. and Dovers, S. (2003), "Interdisciplinarity and sustainability", pp. 53-76. In: Dovers, S., Stern, D. and Young, M. (eds.), New dimensions in ecological economics; integrative approaches to people and nature. Edward Elgar, Cheltenham.

Blow, L., Leicester, A. and Smith, Z. (2003), "London's Congestion Charge." Briefing Note no. 31. The Institute for Fiscal Studies, available at www.ifs.org.uk/bns/bn31.pdf, last accessed 22 September 2008.

Bolloch, O.L., Cope, J., Meaney, B. and Kurz, I. (2007), National Waste Report 2006. Environment Protection Agency (EPA), Dublin.

Bramley, G. (1993), "The impact of land use planning and tax subsidies on the supply and price of housing in Britain." Urban Studies 30, pp. 5-30.

Bueren, E. and Heuvelhof, E. (2005), "Improving governance arrangements in support of sustainable cities." Environment and Planning B: Planning and Design 32, pp. 47-66.

Bulkeley, H. (2006), "A changing climate for spatial planning." Planning Theory & Practice 7, pp. 203-214.

Bulkeley, H. and Betsill, M. (2003), Cities and climate change: Urban sustainability and global environmental governance. Routledge, London.

CSO (Central Statistics Office) (1997), Census 1996. The Stationary Office, Dublin.

– (2002), Census 2001. The Stationary Office, Dublin.

– (2006), Construction and housing in Ireland. The Stationary Office, Dublin.

– (2007), Census 2006, vol. – Housing. The Stationary Office, Dublin.

Collins, C., Bolloch, O.L. and Meaney, B. (2005). National waste report 2004. Environment Protection Agency (EPA), Wexford.

Davies, A. (2003), "Waste wars – public attitudes and the politics of place in waste management strategies." Irish Geography 36, pp. 77-92.

De V Borges, K.A. and Sahay, S. (2000), "GIS for public sector: Experiences from the city of Belo Horizonte, Brazil." Information Infrastructure and Policy 6, pp. 139-155.

Desmond, M. (2006), "Municipal solid waste management in Ireland: Assessing for sustainability." Geography 38, pp. 22-33.

Dovers, S. (2005), "Clarifying the imperative of integration research for sustainable environmental management." Journal of Research Practice 1, Article M1.

Dürrenberger, G. and Hartmann, C. (2002), "Regional energy and CO_2 scenarios: A decision support tool for policy makers." Environmental Impact Assessment Review 22, pp. 449-461.

Ellis, G. and Kim, J. (2001), "Dublin: City profile." Cities 18, pp. 355-364.

EC (European Commission) (1997), The EU compendium of spatial planning systems and policies. European Union (EU), Brussels.

Eurostat (Statistical Office of the European Communities) (ed.) (2001), Economy-wide material flow accounts and derived indicators: A methodological guide. Eurostat, Luxembourg.

Evans, B. (1993), "Why we no longer need a planning profession." Planning Practice and Research 8, pp. 9-15.

Evans, R. and Marvin, S. (2006), "Researching the sustainable city: Three modes of interdisciplinarity." Environment and Planning A 38, pp. 1009-1028.

Eversley, D. (1973), The planner in society. Faber and Faber, London.

Fischer-Kowalski, M. (1998), "Society's metabolism: The intellectual history of material flow analysis, part I: 1860-1970." Journal of Industrial Ecology 2, pp. 61-78.

Fischer-Kowalski, M. and Hüttler, W. (1999). "Society's metabolism: The intellectual history of material flow analysis, part II: 1970-1998." Journal of Industrial Ecology 2, pp. 107-136.

FoE (Friends of the Earth) (2006), A new national climate change strategy for Ireland. FoE, Dublin.

Gilbert, R., Stevenson, D., Giradet, H. & Stren, S. (eds.) (1996), Making cities work: The role of local authorities in an urban environment. Earthscan, London.

Girardet, H. (1999), Creating sustainable cities. Rep. Schumacher briefing no. 2. Green Books, Schumacher Society, Bristol.

Grubert, B. (2006), An environmental house analysis. The Unicorn Eco. Foundation Ireland.

Gunder, M. (2006), "Sustainability – planning's saving grace or road to perdition?" Journal of Planning Education and Research 26, pp. 208-221.

Haberl, H., Fischer-Kowalski, M., Krausmann, F., Weisz, H. and Winiwarter, V. (2004), "Progress towards sustainability? What the conceptual framework of material and energy flow accounting (MEFA) can offer." Land use Policy 21, pp. 199-213.

Handy, S.L., Boarnet, G.B., Ewing, R. and Killingsworth, R.E. (2002), "How built environment affects physical activity: Views from urban planning." American Journal of Preventive Medicine 23, pp. 64-73.

Haughton, G. and Hounsell, D. (2004), Regions, spatial strategies and sustainable development. Routledge, London.

Healey, P. (1985), "The professionalisation of planning in Britain." Town Planning Review 56, pp. 492-507.

Henderson, G., Tillerson, K. and Blaustein, E. (2001), Building energy labelling in existing buildings, pp. 97-106. In: Building energy labelling in existing buildings. The European Council for an Energy Efficient Economy (ECEEE) 2001 summary study, 2001. ECEEE.

Hinterberger, F. and Stefan, G. (2004), Modelling global resource use: MFA, land use and input-output models. In: Modelling global resource use: MFA, land use and input-output models. Input-output and general equilibrium: Data, modelling and policy analysis, Free University of Brussels. September 2-4 2004.

Hoggart, K. (2003), England, pp. 153-167. In: Gallent, N., Shucksmith, D.M. and Tewdwr-Jones, M. (eds.), Housing in the European countryside: Rural pressure and policy in western Europe. Routledge, London.

Horlick-Jones, T. and Sime, J. (2004), "Living on the border: Knowledge, risk and transdisciplinarity." Futures 36, pp. 441-445.

IPCC (Intergovernmental Panel on Climate Change) (2007). Climate change 2007: The physical science basis. IPCC, Geneva.

Kasa, S. (2000), "Policy networks as barriers to green tax reform: The case of CO_2-taxes in Norway." Environmental Politics 9(4), pp. 104-122.

Kildare CO (Kildare County Office) (2005), Kildare county development plan. Rep. vol. one 2005-2011. Kildare CO, Kildare.

Klein, J.T. (2004), "Prospects for transdisciplinarity." Futures 36, pp. 515-526.

Maguire, C. and Curry, R. (2008), Island limits – a material flow analysis and ecological footprint analysis of Ireland. Environment Protection Agency (EPA), Wexford.

Maguire, C., Curry, R. and McClenaghan, A. (2008), Northern visions – footpaths to sustainability. Sustainability and Resources Institute (SRI), Belfast.

McDonald, F. and Nix, J. (2005), Chaos at the crossroads. Gandon books, Kinsale.

Müller, D.B. (2005), "Stock dynamics for forecasting material flows: Case study for housing in the netherlands." Ecological Economics 59, pp. 142-156.

Míguez, J.L., Porteiro, J., López-González, L.M., Vicuña, J.E., Murillo, S., Morán, J.C. and Granada, E. (2006), "Review of the energy rating of dwellings in the european union as a mechanism for sustainable energy." Renewable and Sustainable Energy Reviews 10, pp. 24-45.

Newell, B., Crumley, C.L., Hassan, N., Lambin, E.F., Pahl-Wostl, C., Undral, A. and Wasson, R. (2005), "A conceptual template for integrative human-environment research." Futures 15, pp. 299-307.

O'Leary, F., Howley, M. and O'Gallachóir, B. (2008), Energy in residential sector. Rep. 2008 Report. Energy Policy Statistical Support Unit, Sustainable Energy Ireland (SEI), Cork.

Owens, S. and Cowell, R. (2002), Land and limits: Interpreting sustainability in the planning process Routledge, London.

Palmer, J., Boardman, B., Bottrill, C., Darby, S., Hinnells, M., Killip, G., Layberry, R. and Lovell, H. (2006), *Reducing the environmental impact of housing*. Rep. Final report. Environmental Change Institute, University of Oxford, Oxford.

Petts, J., Owens, S. and Bulkley, H. (2008), "Crossing boundaries: Interdisciplinarily in the context of urban environments." Geoforum 39, pp. 593-601.

Pinson, D. (2004), "Urban planning: An 'undisciplined' discipline?" Futures 36, pp. 503-513.

Ratti, C., Baker, N. and Steemers, K. (2005), "Energy consumption and urban texture." Energy and Buildings 37, pp. 762-776.

Reade, E. (1987), British town and country planning. Open University Press, Milton Keynes.

Republic of Ireland (2003). Third national communication under the united nations framework convention on climate change.

RCEP (Royal Commission on Environmental Pollution) (2007), The urban environment. The Stationary Office, London.

RTPI (Royal Town Planning Institute) (2004), Policy statement on initial planning education. RTPI, London.

Russell, A.W., Wickson, F. and Carew, A.L. (2008), "Transdisciplinarity: Context contradictions and capacity." Futures 40, pp. 460-472.

Rylatt, R.M., Gadsden, S.J. and Lomas, K.J. (2003), "Methods of predicting urban domestic energy demand with reduced datasets: A review and a new GIS-based approach." Building Service Engineering Research and Technology 24, pp. 93-102.

Schiller, G. (2007), "Urban infrastructure: Challenges for resource efficiency in the building stock." Building Research & Information 35, pp. 399-411.

Schlumpf, C., Behringer, J., Dürrenberger, G. and Pahl-Wostl, C. (1999), "The personal CO_2 calculator: A modelling tool for participatory integrated assessment methods." Environmental Modelling and Assessment 4, pp. 1-12.

Srinivasan, S., O'Fallon, L.R. and Dearry, A. (2003), "Creating healthy communities, healthy homes, healthy people: Initiating a research agenda on the built environment and public health." American Journal of Public Health 93, pp. 1446-1450.

Steemers, K. (2003), "Energy and the city: Density, buildings and transport." Energy and Buildings 35, pp. 3-14.

Stern, N. (2006), The stern review: The economics of climate change. HM Treasury, London.

Sykes, O. (2008), "The importance of context and comparison in the study of european spatial planning." European Planning Studies 16, pp. 537-555.

Tanikawa, N., Lawson, N., Hashimoto, S. and Moriguchi, Y. (2005), Spatial estimation and visualisation of regional MFA with GIS mapping. ISIE 2005, Stockholm. 13 June 2005.

Tewdwr-Jones, M. (1996), "Reflective planning theorising and professional protectionism." Town Planning Review 67, pp. 209-214.

Treloar, G.J., Love, P.E.D. and Holt, G.D. (2001), "Using input-output data for embodied energy analysis of individual residential buildings." Construction Management and Economics 19, pp. 49-61.

Treloar, G.J., Love, P.E.D., Faniran, O.O. and Iyer-Raniga, U. (2000), "A hybrid life cycle assessment method for construction." Construction Management and Economics 18, pp. 5-9.

Twinn, C. (2003), "BedZED." The Arup Journal 1, pp. 10-16.

Wiedmann, T., Barrett, J., Billing, M. and Birch, R. (2004), The REAP tool – creating physical accounts for the UK and evaluating policy scenarios. International workshop on sustainable consumption, University of Leeds.

Yohanis, Y.G. and Norton, B. (2002), "Life-cycle operational and embodied enegergy for generic single storey office building in the UK." Energy 27, pp. 77-92.

Chapter 3

Adapting the Modern City to New Challenges – Urban Agriculture as a Way out?

Frank Helten, Undine Giseke, Silvia Martin Han

Abstract

A globally effective potential for energy efficiency and climate protection presents itself especially in those areas where greenhouse gas emissions are concentrated and where the consequences of climate change will be increasingly felt: in fast growing urban growth centres and "megacities", particularly in developing and newly industrialising countries. The paper reflects on novel and pragmatic ideas, concepts and steps to overcome the new urban crisis of today's mega cities in newly developing countries caused by rapid urban and population growth and the man-made environmental degradations. Thus, a German funded research project on "Urban Agriculture as an integrative factor of climate-optimised urban development, Casablanca"[1], Morocco, will analyse and examine to what extent Urban Agriculture (UA) can make a relevant contribution to climate-optimised and sustainable urban development as an integrative factor in urban growth centres. Urban Agriculture is understood as every form of informal or formal agricultural production within a city, whereby "city" equates to the urban region. The reference area for the project is Casablanca, Morocco. In the first section (I) we will introduce the ongoing discussion on the urban crisis and especially the challenges of the Mega City against the background of climate change, energy problems and global food and nutrition shortages. The following section (II) will highlight aspects and projects of sustainable design efforts and examine whether these approaches are suitable to be applied to Casablanca. Next (III), the research and action framework of how the project is further dealing with the problems identified during the initial stage of the project will be described. The final section (IV) will briefly resume the activities over the first two years and point to the challenges ahead.

1 Federal Ministry of Education and Research. Funding Programme "Research for the Sustainable Development of the Megacities of Tomorrow".

1. Introduction

During the last hundred years or so, the shape and structure of urban growth has been characterised by an ever growing dependency on fossil fuel energy as the main driver of economic, technical, social and political development. During this period we witnessed the rise and plentiful diffusion of the automobile, the further spreading out of the urban fabric in terms of urban sprawl, the covering with concrete of ever more green and open space to serve the growing demands both of the highly mobile and of the spatially and functionally differentiated society[2]. Also, the blossoming of the consumer's and the event society promised the vision of a nearly problem free life where everybody could get a job and make a good living. The "Green Revolution" nourished the vision that hunger and food shortage even of the poorest of the Third World would no longer be an issue. However, since the early seventies, this seducing vision of our urban future, has been affected by less positive perspectives regarding pressing ecological and environmental problems. Since then, ever more bad news on environmental issues have shaped our everyday experience. The term "Megalogpolis", coined by Jean Gottman in 1961, clearly pointed to the problems of an uncontrolled urbanisation process thus raising the question whether the western model of modernisation would be appropriate any longer? Jane Jacobs writing on the Death and Life of Great American Cities (1961) as well as the outcomes of the innovative round table of W.H. Whyte on urban land use, resumed in his essay on "Urban Sprawl" (1957) were the most critical contributions in assessing the impact of western urbanisation and modernisation.

In addition to these critical accounts, the alarming results of the Club of Rome report on environmental issues on a global scale, submitted in 1972, urgently highlighted that more immediate action was needed to limit further degradation. Since then, some new ways of practically addressing the impact of rapid urbanisation and of thinking of new approaches on how to overcome the most pressing problems have led to a first step at dealing with a more ecological approach regarding urban development. As a result, some successful environmental politics and new planning tools and instruments have been implemented based on a modernised legal framework, communication and participation driven frameworks for action which opened up and broadened the way for the further involvement of civil society actors. Today we witness that within the developed world a lot has been achieved in limiting environmental problems by cleaning rivers, lakes and soils and in preserving rare species etc.

Nevertheless, as evidenced by the latest news and developments regarding environmental and food security problems, it is clear that all this is not sufficient to

---

2 Cf. for a recent account: Steinberger, Petra (2008): Rückzugsgefechte einer jungen Zivilisation. Zwischen Utopie und Apokalypse: Angesichts der Ölpreissprünge werden die Überlegungen über eine Welt ohne Öl immer konkreter. Süddeutsche Zeitung v. 9.6.2007 p. 13.

feel comfortable regarding our urban future[3]. In fact, the emerging megacities of today's newly developing countries overall show similar problems to those met during western urbanisation, however, on a quite different scale. Today, more than 50% or 3.3 billion of the world's population are living in cities. The number of slum dwellers has already surpassed 1 billion people. It is estimated that 93% of urban growth will occur in Asia and Africa. By 2050 over 8 billion people, two thirds of humanity, will be living in towns and cities[4]. In Morocco this number amounts to more than 60% or around 18 million inhabitants, of which nearly one third will be located in the Grand Casablanca region. According to the FAO, an estimated 850 million people in the world today suffer from hunger. Of those, about 820 million live in developing countries, the very countries expected to be most affected by climate change.

Thus, urban struggles are generating serious shortcoming in urban areas as, e.g. the widespread diffusion of diseases like cholera, tuberculosis, aids or bird flu, some of which appeared to be under control long ago. Moreover, it turns out that against the background of climatic challenges and high prices to be paid for energy and food, a new approach is necessary to limit and reduce the new urban crisis of the megacity of today. Solutions envisioned also have to consider that one of the most classical problems of the big city is to provide sufficient food at reasonable prices. This is a major challenge to be tackled within urban policies and not only in newly developing countries[5]. "The years 2007-2008 saw dramatic world food price rises, bringing a state of global crisis and causing political and economical instability and social unrest in both poor and developed nations"[6]. Thus, one could state that a new phantom is strolling around today's megacities, the phantom of hunger and food riots. Price rises particularly affected Asia and Africa. Egypt and Morocco were seeing protests and riots in late 2007 and early 2008 over the unavailability of basic food staples[7]. The New York Times reported on June 9th 2008 that

> "Today, Africa has less large-scale commercial agriculture than it had 50 years ago. Productivity has slowed to a crawl in India, Indonesia and China" and "According to the U.N. secretary general, Ban Ki-moon, world food production must rise 50 percent by 2030. This will require investments between $15 billion

---

3 Cf. „Nahrungmittelknappheit". Die Krise hat erst begonnen. Interview with FAO vice president Alexander Müller, Spiegel Online 1.6.2008. www.spiegel.de/wissenschaft/mensch/0,1518,556867,00.html.

4 UN Report 2008: World Urbanizations Prospects. The 2007 Revision, New York.

5 The German newspaper Süddeutsche Zeitung reported on June 14th 2008 that the number of inhabitants of New York suffering from hunger had increased to 3 Million people. In November 2007 this number amounted to 1.3 Million http://news.bbc.co.uk/2/hi/americas/7106726.stm.

6 An account of the recent impact of food shortages is given at: http://en.wikipedia.org/wiki/2007–2008_world_food_price_crisis.

7 Ibido.

to $20 billion a year in the farm economies of poor countries, including research into robust, high-yielding crops suited to poor regions like sub-Saharan Africa."

The new urban challenge has a multifaceted character. According to varying perspectives and research foci, one may emphasize the critical state of the housing conditions as outlined in Mike Davis "Planet of Slums" (2006). One may also point to the high degree of poverty struck people which amounts to 50% of the urban population in newly developing countries who spend between 40% and 60% of their income on food each year[8]. Or highlight the fact that clean water is in short supply, public transport is either too expensive or nonexistent, greenhouse gas emission are increasing, affecting the quality of clean air and rapidly damaging healthy environments and destroying the livelihood of many people. It would not be too difficult to extend this list by pointing to the lack in basic infrastructures such as waste water disposal, appropriate waste management, availability of recreational space etc. Eventually, it has to be stated that one of the major shortcomings in ensuring a sustainable urban re-development is caused by a lack of democratic structures and procedures which give way to more informal and often illegal dynamics of modelling and structuring the urbanisation process. Thus, we can characterise the actual stage of the new urban crisis as a highly uncertain and transformational stage, which, however, opens up chances to shape the nature of urbanisation in such a way as to install a more humane, just and a basically ecological balanced city. At the same time this situation can be described as being a borderline situation, as we still do not know whether we will be able to successfully cope with the known and still unknown challenges of climatic change, food shortages and environmental problems within the urban context.

2. How to address the problems?

Ways to go

There are only a limited number of ways out of this struggle available. In Europe as well as in the US, in some Arab and Maghreb countries, in Asia and especially in China a few approaches which strive to re-invent the city by drawing on environmental and ecological aspects in terms of a applying a *sustainable design* concept, have been (or are on the way to being) implemented. However, it appears that most of these efforts are still limited in size and scope and not really able to address the challenges of both the provision of adequate urban quality for everybody and of the limitation of climatic and environmental problems. Thus,

---

8 Cf. IDRC/UN-HABITAT. "Guidelines for Municipal Policymaking on Urban Agriculture" Urban Agriculture: Land Management and Physical Planning (2003) 1.3.

the most innovative Masdar City project[9] in Abu Dhabi, though quite impressive with its comprehensive and ecology driven master plan designed by Norman Foster, will house just 50000 people. The city will be C02 neutral and produce no waste. Fuel powered cars will not be allowed. The overall design of this project is highly inspired by local knowledge and classical Arabic practices of how to cool down the built environment without spending energy. A similar approach will be realised in the Chinese city of Dongtan, which will soon be set up on an island close to Shanghai. Here, a group of British planners (Ove Arup) has been asked to submit a master plan drawing on comprehensive and ecological issues[10] including the following services: sustainable energy management, cultural planning, waste management, renewable energy process implementation, economic and business planning, social development, sustainable building design, architecture, infrastructure, landscape design and ecology. Interestingly, some space for agricultural production has been foreseen in the master plan according to the requirements of the Chinese government in order to feed the future population and to avoid additional transport costs. However, it appears that the master plan of this project, even if it addresses the full scale of sustainability aspects, cannot serve as a model to be applied in the emerging megacity. Here urban policies have not only to negotiate diverging and often contradictory interests of many stakeholders, they must also address the basic question of how to overcome urban poverty, especially the high degrees of unemployment, housing shortages and food security. Even more so, they have to deal with the immediate impact of climatic change that to some extent cannot be avoided in the short term.

Looking for more pragmatic and alternative ways to fight poverty and environmental and ecological degradation in the emerging megacity, we decided to draw on solutions which could be implemented more integrative and process-oriented, and which were able to substantially contribute to alleviating the nutrition and environmental problems. *Urban Agriculture* (UA) appeared to be a promising alternative, because of its broad application potential suited to responding to the needs of the urban poor. Even more so, because of its overall affinity and impact on open and green space policies, urban planning, environmental politics and climatic change it was assumed that Urban Agriculture, when embedded into a more comprehensive perspective on urban development could be a key factor and a more or less silent driver allowing for an ongoing, pragmatic and sustainable improvement in the quality of urban life. As the first overview demonstrated, activities in this area have been getting more widespread since the nineties, especially in the aftermath of the IWF and World Bank promoted structural adjustment programs, which forced the urban poor or the

---

9 See www.spiegel.de/wirtschaft/0,1518,534205,00.html visited 17.6.2008 and www.fosterandpartners.com/Projects/1515/Default.aspx.

10 See (www.arup.com/eastasia/project.cfm?pageid=7047) visited 12/062008.

precarious classes to look for alternative employments and additional resources. Since then, more and more serious activities have been observed, which in the newly developing countries such as Tanzania led to successfully applied urban agriculture[11]. The relative success of these initiatives opened up the potential for researchers and policy makers to look further and thus, it was found that from a historical perspective urban agriculture has not totally disappeared from the landscape of urban activities. On the contrary, Chinese and Indian cities, as well as some Japanese and Taiwanese cities turned out to have a profound expertise in this area[12]. The most recent success story in urban agriculture comes from Havana and Cuba[13], where the shortage in primary energy and high costs for chemically processed fertilizers, since the refusal of Russia to deliver cheap petrol, led to a quite efficient agricultural and urban agricultural production, highly admired by ecologists from all over the world. The Cuban experience showed to some extent that alternative ways of urban agricultural production can be chosen and that urban agriculture can substantially contribute to achieving the objectives of feeding the urban poor and of reducing environmental and ecological degradation. Thus, being convinced that UA could be beneficial we also had to make sure that any implementation of UA would be coined by the urban situation itself. Reflecting the high degree of interrelatedness of Urban Agriculture to other urban functions it was becoming clear that UA had to be embedded and integrated into the local system of urban and regional planning and that solutions found needed to be tested, applied and re-examined to see whether they would deliver the expected outcomes or would fail.

Urban agriculture as a guideline for sustainable design

To strengthen urban agriculture as a promising and integrative factor of a climate-optimised approach of urban development of the mega city of today needs some further methodological considerations, which hopefully will contribute to sharpening the vision on how to strategically advance and make the concept more viable. "Urban Agriculture" seems to be an inconsistency, but the central questions posed concern on the one hand a possible contribution of Urban Agriculture to food provision and the fight against poverty (in line with the intentions formulated at the 1996 Habitat II Conference), and on the other hand the importance of this form of agriculture – regarded as production not only of food but also of landscape – to build up an urban open space system and ecological cycles in a city-region. A third and most important component is the climatic impact/interplay of urban and peri-urban agriculture on energy efficiency. Just to give an example: A metropolitan region – like that of Casablanca – will need around 1-1.3

11 Cf. The Worldwatch Institute (ed. 2007): State of the World. Our Urban Future. Washington, DC. www.worldwatch.org.

12 Cf. http://en.wikipedia.org/wiki/Urban_agriculture visited 17.6.2008.

13 Cf. www.ruaf.org/node/513 visited 17.6.2008.

Million tons of food to be imported per year. This amounts to around 3000 trucks transporting a weight of 45 tons each[14].

Limiting this amount would contribute to a reduction in not only the CO_2 emission but also the energy consumed in transporting food from rural to urban areas. It would also limit costs for storage and refrigeration and would reduce product losses. Additional direct and indirect benefits can be expected such as the availability of fresh food, creation of jobs, the better care of open spaces, the greening of the city, the reduction in pollution and a positive influence on the microclimate etc. Moreover, we have to consider that there are new technologies and practices available to produce vegetables and fruits not only on large but also on medium-, small and very small-sized plots of urban land, which opens up possibilities for the urban poor to make a living.

Thus, it appears that – strictu sensu – urban agriculture is not just agriculture in a traditional sense. The concept of Urban Agriculture goes beyond this and is within this project, based on the assumption that the city region of tomorrow, especially the peri-urban environment will grow discontinuously and not homogeneously. It will produce fragmented specific spatial patterns so that rural and urban environments will collide, exist alongside one another and interact. Within this context the concept of Urban Agriculture is conceived as a module for sustainable design of the city, as it saves and integrates a system of open space based on a productive urban landscape, a system which is evidently necessary for the ecological, social, economic, cultural, health and climatic requirements of a megacity. However, this module needs to be enlarged by considering three further dimensions to ensure a more realistic application. Thus, we have to consider the formal and informal dynamics of urban development, the role and impact of climate change and finally the underlying structure and activities related to governance issues and to the social production process of space as affected by various actors. Reflecting and researching on these dimensions led to a first account of the Casablanca situation which can serve as further input for a more sustainable design of the city.

Urban Laboratory Casablanca

Casablanca was once the leader in modern (European) architecture and city planning relying on the one hand on strict functional zoning[15] and on the other on the French colonial ideology of segregating the Moroccan Medina and the

---

14 What can be avoided shows the Havana experience: “Over 26000 popular gardens cover 2438,7 hectares in Havana and produce 25000 tons of food each year; a total of 299 square kilometres of urban agriculture produces 113525 tons/year” (Mario Gonzalvez Novo and Catherine Murphy in Bakker et al. 2000). Cited in: www.ruaf.org/node/513 visited 17.6.2008.

15 Cf. Cohen and Eleb (2002), Colonial Myths and Architectural Ventures. New York: Monacelli Press.

urban quarters of the Moroccan population from that of the French, due to so called hygienic concerns[16]. Since independence in 1956, things have changed a lot, making a metropolis of more than 5 Million out of a city with 700.000 inhabitants in the Grand Casablanca region and of 3.0 to 3.5 millions intra muros. Today, Casablanca does not differ much from the overall trends of rapid and accelerated urbanisation. There is disproportional (high) population growth[17] in the outskirts, whereas the inner city population remains quite unchanged.

Memories are made of this: Rick's (Fake) Café in Casablanca. Meeting point of Bogart fans, German researchers and business people

Due to positive economic perspectives, Casablanca's inner city, however, is speeding up its upgrading of its business district with tertiary and quaternary services and institutions and relocating former industrial zones, placed at the north eastern coastal line to the hinterland. Further densification of the existing built environment can be observed as well as ever more pressure on the few remaining open plots to be prepared for construction.

Periurban office space

Real estate business and speculation is a highly promising business with regards to the economic momentum of the actual conjuncture in Casablanca. New produc-

---

16 Institut National d'Aménagement et d'Urbanisme (2004): Intégraton à la ville et services urbains au Maroc. Rabat: INAU/ IRD. Ali Sedjari (ed., 2006): Les politiques de la ville: Intégration urbaine et cohésion sociale. L'Harmattan.

17 However, the newest studies in Moroccan reproduction rates point to a clear reduction in population growth. Cf. Courbage, Youssef & Todd, Emmanuel (2008): Die unaufhaltsame Revolution.Wie die Werte der Moderne die islamische Welt verändern. München: Piper.

tion lines for assemby and textile industries can be found at the southern and eastern edges of Casablanca as well as in newly developed locations and technology parks.

Casablanca Technology Park

Casablanca Technology Parkwithin the peri-urban fringe. Recent large scale housing projects, to fight the housing shortage and to limit the still ongoing process of tinkering slums, are built up in the suburban regions stretching out alongside the main axis of transportation, whereas an astonishing range of informal settlements ranging from expensive individual homes to traditional brick built wall enclosed settlements and poor "bidonvilles" to be found nearly everywhere, where some space is available and where the authorities either tolerate, ignore or do not sufficiently control the building activities. Casablanca is situated on a beautiful coast line suited to leisure, recreation and attracting tourists.

Coastline south of Casa

Thus, to some extent a quite dynamic gentrification process of the coast line and the adjacent countryside can be observed, accompanied by the development of individual houses, privately owned leisure, and recreational and tourist facilities in the form of country and golf clubs, resorts and hotels etc. which appear to be directly competing with agricultural activities not only in terms of land use but also with regards to consumption of water. Inner city parks and green and open space are in short supply and not always well maintained.

Small Urban Park near Medina

Agricultural activities are not very visible in the inner city. Urban Surfaces, where some fruits and vegetables have been grown in the past, are closed now or are retained to be "discovered" for more interesting investments.

Inner city agricultural surface, closed up in 2006.

A somehow different picture can be found in the peri-urban fringe, where agriculture is still ongoing, even if the soils are deteriorating sharply because of water shortages caused by droughts and salination of groundwater.

Small scale farming near Casablanca

But even here, real estate speculation and high rising financial expectations even of the small farmers are leading to further and an ongoing reduction in agricultural activities thus contributing to the further downgrade of some of the formerly green areas surrounding Casablanca. On the positive side, however, one can see a couple of new, remarkable, innovative but still limited activities of a few modern "*periurban agricultural pioneers*" who try to introduce modernized and less water intensive production concepts in agriculture or to revitalise and to re-open state owned but devastated farms.

New agricultural production concepts near Casablanca

This is especially interesting as the latter activity has been by a group of young academics[18] without employment, demonstrating that one of the potential drivers of revitalising the formerly well established peri-urban agriculture is the quest to get access to work, income and food. As a result of this short account, one can conclude that the opportunities for agricultural production in Casablanca and Casablanca region appear to be limited and it is not an easy task to achieve it. There are additional constraints which have to be considered and which will not be explicitly addressed in this paper: Constraints related to climatic change, neglect of application of the building code and disregard of legal plans for development, ownership of land, public and individual transport, to name but a few. However, regarding the observed transformational and open nature of the production of urban space in the Casablanca region, it appears that there are chances to limit the still ongoing process of further deterioration of the urban and peri-urban environment through accelerated and more or less uncontrolled urbanisation. Saying this it becomes clear that a novel and integrative effort is necessary to ensure that urban and peri-urban agricultural production on available plots can be beneficial in terms of:

- contributing to food production, and to feeding the city.
- producing healthy and fresh food.

---

[18] On site visit near Mediouna, Septembre 2006; UA Conference in Casablanca Septembre 2006.

- supporting small scale food production for affordable every day consumption, especially for the urban and peri-urban poor.
- creating job opportunities (e.g., using organic disposal for making compost).
- avoiding transport costs and expenses for primary energy to bring food to the end-consumer.
- allowing economic usage of waste water for irrigation.
- supporting the development of green and open spaces, thus creating a healthier and more aesthetic shaped environment.
- raising awareness and concern to care for open and green spaces.
- training and educating people in how to survive in urban environments with a small budget etc.

Regarding these prospects, considerations and insights into the actual condition of urban and peri-urban Casablanca four main research questions have to be answered:

1. To what extent can Urban Agriculture play a significant part in adapting to the consequences of climate change, in climate protection, and in energy efficiency?
2. To what extent is Urban Agriculture an innovative strategy for sustainable land conservation of urban green space in tomorrow's megacities?
3. To what extent can Urban Agriculture contribute to the struggle against poverty?
4. How can Urban Agriculture be integrated as a crucial element of urban development in accordance with local conditions?

3. Research Framework

As a result of our initial research effort on the urban and agricultural dynamics of Casablanca, the German-Moroccan project group decided in late 2007, to study the potential of urban agriculture in more detail and on different levels. The underlying spiralling – somehow iterative and recursive – research approach had to be followed up, to allow for ongoing feedback and a readjustment of results found so far. Furthermore, taking into account the latest suggestion of the main financial contributor, the German Federal Ministery of Education and Research, the project partners agreed to give more weight to climatic and energy related issues, already addressed within the original research framework of the project. As a result of these new suggestions and of the experience gained, it was felt necessary to enlarge the multidisciplinary and transdisciplinary composition of the consortium. To date the German Moroccan Research Group includes representatives from Urban and Landscape Planning, Climatology, Geography, Engineering, Agronomy, Urban Sociology, Geology, Political Economy as well as representatives from Moroccan NGOs and vocational associations. The institutions included are German and Moroccan Universities, Moroccan Schools, Urban Planning Agency Casablanca, Casablanca Region planning board, German and Moroccan private companies and organisation, Moroccan Regional Associations of agriculture, and Moroccan NGOs.

Reflecting the new challenges against the background of the experience gained during the initial stage, the project, then, decided to design a further detailed and differentiated research framework, drawing on three research layers to be integrated by a series of cross-cutting research and synthesising tasks.

The first layer on the *macro level* of all research activities has been conceived in order to develop a set of guiding principles as aims and visions of the role which urban agriculture could/should play within a climate-optimised urban development process. This includes the development of scenarios. The second layer on the *meso level* will provide climate-optimised modules of multifunctional spatial systems derived from research on resource efficient cycles in settlement structures and on settlement related landscape functions. These tasks will be achieved by taking into account the findings of action-research driven tasks on the third layer, the *micro level*, where four pilot projects will be implemented and continually observed, modified and adjusted if necessary during the next project stage. In addition, a series of comprehensive research tasks and accompanying activities have been identified to support and capitalise on the action research driven tasks. Thus, special work packages have been formed to focus on a) capacity building and communication to raise awareness for the concern of the project; b) to develop strategies, tool boxes and action plan to ensure the application and widespread dissemination/ exploitation of results on urban agriculture practices; c) to ensure scientific quality through scientific evaluation, prospects and transferability issues and finally d) to provide an mechanism to adjust organisational or internal shortcoming by project evaluation and audit.

The core activities of the next stage will be the further preparation and implementation of four pilot projects. These pilots each focus on a typified spatial situation of urban agriculture and are related to a central urban function such as: *production, housing, recreation or nutrition.* The pilot projects will be action research driven and conducted by bi-national and multidisciplinary teams. The following objectives will be addressed (table 1).

The pilots will serve as a starting point to figure out in more detail whether our assumptions concerning the potential of urban agricultural in real settings can be proved or not and to what extent they can contribute to achieving the objectives set. The project is aware of the fact that there are quite of lot of constraints that need further consideration and scientific investigation: How can we raise awareness regarding the new challenges of food and water shortages, reduced greenhouse gas emissions, a more climate optimised urban environment? How can waste water be recycled at reasonable costs? Are there cheap technologies available to cope with the problem of salinated water, how can we convince the urban poor to participate, to be more active, how can we handle the problem of ongoing real-estate speculation and artificial tightening of agricultural land? Thus, there are not only technical questions to be answered. As the results of the first stage have demonstrated, there are considerable governance-related and socially motivated problems which have to be solved. Results stemming from preparatory activities of the pilots show that the population of an informal settlement in the peri-urban fringe of Casablanca has witnessed considerable constraints in getting access to rentable land for agricultural purposes, even though they are willing to engage in agricultural activities to make a modest living and they are keen to spend less time and money to get access to fresh vegetables and fruits. Moreover it appeared that the few remaining farmers are more and more affected by the ever increasing costs of water, fertiliser and seed. Also, there are social problems as the growing crops have to be protected against vandalism or theft.

Tab. 1: Pilot projects on UA

Topic	**Urban/ Periurban Function**	**Primary Objective**	**Secondary Objective**	**Expected Climatic Benefit**
Industry & UA	Production	Re-use of wastewater for agricultural purposes	Improving the aesthetic dimension of industrial plants	Increasing green space, economic use of water
Informal Settlement & UA	Housing	Improvement of nutrition quality	Learning to grow food in urban settings	Using solar energy for agricultural and food production activities, reduction of transport costs to provide food, avoiding costs for cooling food, increasing green space
Peri-Urban Tourism & UA	Recreation	Using UA as an incentive for short distance tourism	Creating jobs and opportunities for attractive food markets in the peri- urban setting	Reduction of transport costs

Topic	Urban/ Periurban Function	Primary Objective	Secondary Objective	Expected Climatic Benefit
Healthy Food Production and UA	Nutrition	Establishing bio-production of food	Initiating customer relations between food production and gastronomy/touristic industry	Bringing the production of food close to the consumer, extending the peri-urban food production by adapting to specific nutrition modes, aesthetic landscapes

The outcomes of the pilots, will contribute to the adjusting and fine tuning of the respective micro objectives when being action-researched; they will also be fed back into the *meso level* of research activities where the main task will be to design climate-optimised models of multifunctional spatial systems, further relying on research on resource efficient cycles of agricultural production in settlement structures. Moreover, the landscape functions related to settlement needs in terms of food production, recreation, tourism and risk management against, e.g., unexpected flooding, water shortages and droughts, will be researched and evaluated according to the data collected in the pilot projects and by further investigating how the interplay and marriage/divorce between urban and rural takes place. It is clear that the climate-optimised modules will provide input for sustainable design issues to be explored within the pilots. At the same time they will serve to make the initial aims and visions, set as guiding principles at the macro level of the underlying research framework, more realistic and practical in terms of opening up chances for further dissemination, application and exploitation. A final and quite ambitious objective would be to insert the outcomes of this comprehensive approach as a module for a new master plan of sustainable design, badly needed to cope with the pressing challenges of Casablanca's urbanisation.

4. Conclusion: First Results and Challenges Ahead

The project on Urban Agriculture as an Integrative Factor of Urban Development, Casablanca achieved during the first two years a couple of interesting and promising results which need to be mentioned as they are the very precondition for the continuation and the potential success of the project.

Building up bi-national relations, networks and adequate communication modes

In the beginning, the project was quite effective in establishing a German-Moroccan research and action group composed of adequate competence and expertise and including a range of public and private institutions needed to successfully conduct a project of this size. During the initial stage of the project it has not been an easy task at all, to make the concerns and the objectives of the project clear to our Moroccan partners. This was mainly due to a different attitude and perspective on urban and peri-urban agriculture between the Moroccan and German actors involved. Thus, there was need for discussion, concerning the development of a common code of understanding of the research issues and to find a viable solution to conceptualise Urban Agriculture. A second and more basic difficulty was the language barrier. On the other hand, inside the German group some French speaking partners were available, however, to ensure a common understanding by all, an interpreter had been engaged. This setting influenced the communication process per se by being time and resource consuming; but, it appeared that this constellation also created some benefits as the issues discussed needed to be more carefully and clearly explained to ensure that all partners would be able to follow the ongoing presentations and discussions. Now, at the beginning of the second stage of research, the project stands on a broad network of relevant stakeholders who have committed themselves to contributing to the success of the project.

Raising awareness and fostering co-operation

Concerning the awareness raising activities of the project, it can be stated that the German project group has been especially supported by the Moroccan partners who were able to open doors and who facilitated access to political and vocational stakeholders in order to motivate them to participate in the project's activities. A conference held as early as 2006, a year after the start of the project at the University of Casablanca and at an agricultural farm, a further workshop with about 50 persons to deepen the issues of urban agriculture, contributed to demonstrating the potential of UA and confirmed that there was a considerable and growing interests among the Moroccan partners.

Producing applicable and practical knowledge for further research

A third activity, which produced tangible results was the effort of the German research group to become knowledgeable on the context of the ongoing urbanisation process in the Casablanca region and of the role and structure of agriculture. Thus a series of lightweight research tasks have been accomplished regarding issues like the state of urban agriculture in Casablanca region, urban planning issues, aspects of the Moroccan civil society, climate change, water shortage etc. The outcome of these analyses allowed for a further elaboration of the main research tasks and for a readjustment of the main problems to be addressed. Therefore, the project focuses on the urban and the peri-urban development of megacities, highlighting the rural-urban linkage as a new phenomenon of current urban growth centres.

Challenges ahead

The future work will start mid 2008 and will concentrate on the implementation of the pilots described above. The challenges ahead are the appropriate and timely implementation of the pilots and by deriving corresponding knowledge how to disseminate and further deploy the practices gained. But there is more. Regarding the rapid change of perspective of the recent past regarding food shortages, climatic changes and energy prices, we have to highlight two important issues that need adequate answers: First, the aggravation of the food problems of the big city needs to be addressed immediately. Morocco's urban poor are already suffering from food shortages. This number will grow and more effort will be necessary in the very near future to help those who cannot afford to pay the high-rising prices for food or for agricultural "raw material" like soil, water, fertiliser and seed. The project needs to discuss this unexpected acceleration of the food and agricultural problem during its first session. Second, the pressing problems of how to feed the city are rising to the background if climatic change and high prices to be paid for energy consumption need more attention. The project will address this problem by research and by providing a climate optimised module of multifunctional spatial systems. It appears to be of utmost importance to deepen and to try to accelerate these activities. Moreover there is considerable need to start as soon as possible with additional efforts in order to raise awareness for this novel threat and to ensure that the new insights into the interplay of climatic change, food security and energy consumption will be taken into account when the urban master plan of Casablanca will be developed further. This includes the potential reduction of distance that the consumed food has travelled ("food miles"). But this alone appears not to be sufficient to cope with the problems described above. There is an immediate need to change the overall dynamics of urbanisation which still relies on the classical separation of urban functions. The good news, however, is that the Moroccan King Mohammed VI, stated he wants to develop a comprehensive strategy "which tackles the root causes of the current situation and ... that, as far as Morocco is concerned, the government has adopted a "Green Plan" which seeks to make of Morocco's agriculture "one of the engines of economic development over the next 15 years."[19]

---

19 Cf. Maghreb Arabe Presse, 4.6.2008. www.map.ma/eng/sections/economy/morocco_calls_for_mu/view.

It is quite clear that the project would benefit from taking into consideration how to liaise with the intentions of the Green Plan in order to give more and sustainable weight to the project on Urban Agriculture.

References

Cohen and Eleb (2002), Colonial Myths and Architectural Ventures. New York: Monacelli Press.

Courbage, Youssef & Todd, Emmanuel (2008), Die unaufhaltsame Revolution. Wie die Werte der Moderne die islamische Welt verändern. München: Piper.

Daoud, Zakya (2005), Casablanca en mouvement. Des innovateurs dans la ville. Paris: autrement.

Davis, Mike (2006), Planet of Slums. New York: Verso.

Gottman, Jean (1961), Megalopolis. Cambridge, MA: M.I.T. Press.

IDRC/UN-HABITAT (2003), Guidelines for Municipal Policymaking on Urban Agriculture. Urban Agriculture: Land Management and Physical Planning 1.3.

Institut National d'Aménagement et d'Urbanisme (2004), Intégration à la ville et services urbains au Maroc. Rabat: INAU/IRD.

Jacobs, Jane (1961), Death and Life of Great American Cities. New York: random House Vintage.

Platt, Rutherford H. (ed.) (2006), The Humane Metropolis: People and Nature in the 21st Century City. Amherst: University of Massachusetts Press.

Revue Urbanisme no. 348, juin 2006: Dossier Éco-quartier. Paris. Url: www.urbanisme.fr.

Sedjari, Ali (ed.) (2006), Les politiques de la ville: Intégration urbaine et cohésion sociale. L'Harmattan.

Steinberger, Petra (2008), Rückzugsgefechte einer jungen Zivilisation. Zwischen Utopie und Apokalypse: Angesichts der Ölpreissprünge werden die Überlegungen über eine Welt ohne Öl immer konkreter. Süddeutsche Zeitung v. 9.6.2007, p. 13.

Technische Universität Berlin (2007), Urban Agriculture (UA) as an Integrative Factor of Cimate-Optimised Urban Development, Casablanca. Proposal description, Phase II, vol. I & vol. II. Berlin. Unpublished.

The Worldwatch Institute (ed.) (2007), State of the World. Our Urban Future. Washington, DC. www.worldwatch.org.

UN Report (2008), World Urbanization Prospects. The 2007 Revision. New York.

Wheeler, Stephen M. and Beatley, Timothy (eds.) (2004), The Sustainable Urban Development Reader. London: Routledge.

Whyte, W.H. (1957), Urban Sprawl in Editors of Fortune, the Exploding Metropolis. Garden City, NY: Doubleday Anchor.

Chapter 4

Policy-making in Climate Change – Experiences from the ASTRA Project

Walter Leal Filho, Franziska Mannke

Abstract

This chapter describes the work performed in the context of the project "Developing Policies and Adaptation Strategies to Climate Change in the Baltic Sea Region – ASTRA", co-financed by the European Regional Development Fund (ERDF) through the INTERREG IIIB Baltic Sea Programme. The ASTRA project undertook a regional assessment of the impacts of climate change in the Baltic Sea region and addressed various threats arising from climate change in the region, such as extreme temperatures, droughts, forest fires, storm surges, winter storms and floods. The final goal of the project was to develop a set of recommendations for strategies and policies for adapting to climate change and outline the role of information and communication in raising awareness on the adaptation strategies needed.

1. Introduction

Many natural systems are being affected by climate variation and climate changes, particularly in respect of temperature increases. Current scientific evidence indicates that global warming is progressing due to an increase in anthropogenic, i.e. man-made, greenhouse gas concentrations, mostly due to industrial activity. Still, uncertainties in the projection of future climate change remain – also due to the ambiguity of our social and economic systems. Although there are variations seen in the extent and scope of impacts of climate change, as well as regional (i.e. geographical) variations, the net annual costs of climate change are expected to increase over time as global warming continues. According to the International Panel on Climate Change (IPCC), "a portfolio of adaptation and mitigation measures can diminish the risks associated with climate change" (IPCC 2007: 8-9, 17, 20; Hilpert et al. 2007: 9).

Europe has warmed faster than the global average (Commission of the European Communities 2007: 4). According to the Green Paper on Climate Change Adaptation, the average temperature in Europe has increased within the last century by almost 1°C. Among the most vulnerable areas are coastal zones where sea level rise combined with increased risks of storms are among the major impacts. The Baltic Sea Region (BSR) for example, is a region affected by a rise in sea level to varying orders of magnitude, the reason being land uplift and subsidence (Meier et al. 2006: 41). In particular, Scandinavia will experience an increase in precipitation, a larger part of it in form of rain instead of snow. (Commission of the European Communities 2007: 4-5). Furthermore, climate change is expected to affect the biodiversity of the region. For example, more than fifty per cent of Europe's plant species could be vulnerable or threatened by 2080 (Commission of the European Communities 2007: 5).

It is a fact that climate change ignores national borders. Neighboring countries experience similar challenges, for example in the area of coastal protection, or when coping with extreme weather events. This suggests that concerted action is needed. This line of thinking has led the European Commission to take the decision to face the challenge of climate adaptation, working in partnership with its Member States and globally with many partner countries (Commission of the European Communities 2007: 3).

The above mentioned issues were taken up in the project "Developing Policies and Adaptation Strategies to Climate Change in the Baltic Sea Region – ASTRA", co-financed by the European Regional Development Fund (ERDF) through the INTERREG IIIB Baltic Sea Programme. The ASTRA project featured a regional impact assessment of the ongoing global change in climate trends, resulting in the development of recommendations for strategies and policies for adapting to climate change. The project addressed various threats arising from climate change in the Baltic Sea Region, such as extreme temperatures, droughts, forest fires, storm surges, winter storms and floods. The final goal of the project has been to ensure awareness and adaptation strategies for the socio economic system to prevent the adverse effects of climate change and to ensure a stable development of the BSR.

In the course of the project duration from July 2005 until the end of 2007, the project used scientific knowledge from state-of-the-art climate modelling and climate impact research as well as geomorphological and geological investigations to assess the effects of climate change impacts on both natural and socio-economic systems. As a first step, project partners studied the winter storm Gudrun of 8-9 January 2005 and related national, regional and local effects and responses. This served as an illustrative example of the various challenges climate change poses to the region. In addition, the development of adaptation strategies with reference to regional planning purposes has been fostered based on the results of internal working groups. Climate change impacts and vulnerability of regions have been studied in several regional and local case studies around the Baltic Sea. The analyses focused on single effects as well as on impacts on different socio-economic sectors. Entry points and integration to existing planning processes and methods have been identified. In some case study locations, e.g. in the Finnish cities Kokkola and Espoo, this has led to concrete implementation measures.

Moreover, a special attention to risk awareness has been paid and addressed through intensive dissemination exercises, such as a set of regional conferences with international participants as well as workshops in the countries around the Baltic Sea Region presenting the project partners' activities to local stakeholders.

Finally, policy recommendations for climate change adaptation have been delivered for different spatial scales from the local to the national and the BSR level.[1]

2. Policy-making in the Baltic Sea: Experiences from the ASTRA Project

The ASTRA project has its foundation in a successful previous European project titled SEAREG, which assessed the impacts of future sea level rise in the Baltic Sea Region. The positive response of planners that cooperated and/or were addressed there led to the idea to widen the scope of climate change impacts that affect the spatial development of the BSR. This approach has been new, as the focus did not only lie on coastal areas and problems of sea level rise. ASTRA aimed at taking all effects of climate change into account that affect spatial development in order to foster the joint, interdisciplinary elaboration of adaptation and mitigation strategies with spatial planners and further stakeholders.

2.1 Project set-up and management of work load aligned to overall objectives

Referring to the basic project design, the project wanted to involve its stakeholders at the municipal and regional level in a constructive way. It had as such a wide partner base from the very beginning on, involving as much as 36 partners. Among them were universities and research institutes which provided the scientific knowledge base and strategic recommendations. Furthermore, municipalities, city councils and local authorities took part as to directly benefit from the project-specific knowledge that was generated. Some of the latter served also as case studies where decision-makers started to take climate change issues into account, resulting in some reformed planning practices.

The large circle of partners consisted of so-called inner and outer circle partners. Those belonging to the outer circle were generally not financially involved in the project. However, they served as important dissemination channels receiving continuous project input and participating in the project's international conferences.

1 Hilpert, K., Mannke, F. and Schmidt-Thomé, P. (eds.): Towards Climate Change Adaptation in the Baltic Sea Region, Geological Survey of Finland, Espoo, Finland, 2007.

Concerning the scope of the project, the overall work load had been broken down into three so-called work packages, the first one comprising regional climate impact modeling and case study research, the second work package on communication and dissemination having a cross-cutting theme, and a third work package on the development of strategic recommendations for developing adaptation strategies. Each of the work packages was then broken down into a range of single tasks. Figure 1 illustrates how the single project results feed into the final project outcome, the so-called policy paper which provides guidance for decision-makers.[2]

Fig. 1: Feeding of single project results into final project outcome

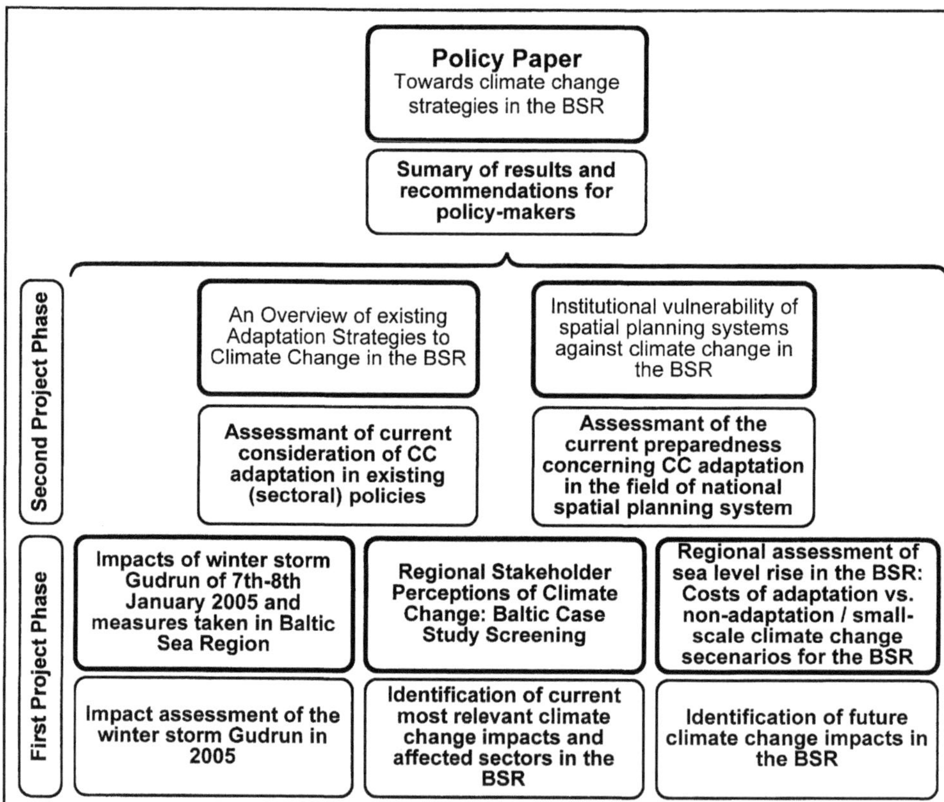

(adopted from Hilpert et al. 2007: 10)

Due to the interdisciplinary formation of the partner consortium, it was of utmost importance to create a common understanding of climate change. The underlying key concepts of climate change were elaborated in each of the interna-

---

2 The project results have been published in booklet "Towards Climate Change Adaptation Strategies in the Baltic Sea Region", ISBN 978-952-217-010-1, Espoo, Finland, 2007.

tional conferences and meetings in order to ensure that the relations between adaptation, mitigation, vulnerability and uncertainty were understood by anyone no matter which background one had. This refers back to the theory of organizational learning that is defined as the process of improving actions through better knowledge and understanding (Garvin, D.A. 2000: 77).

Fig. 2: Mitigation and adaptation as complementary approaches to climate change policy (adopted from Hilpert et al. 2007: 11 in: IPCC 2001: 4)

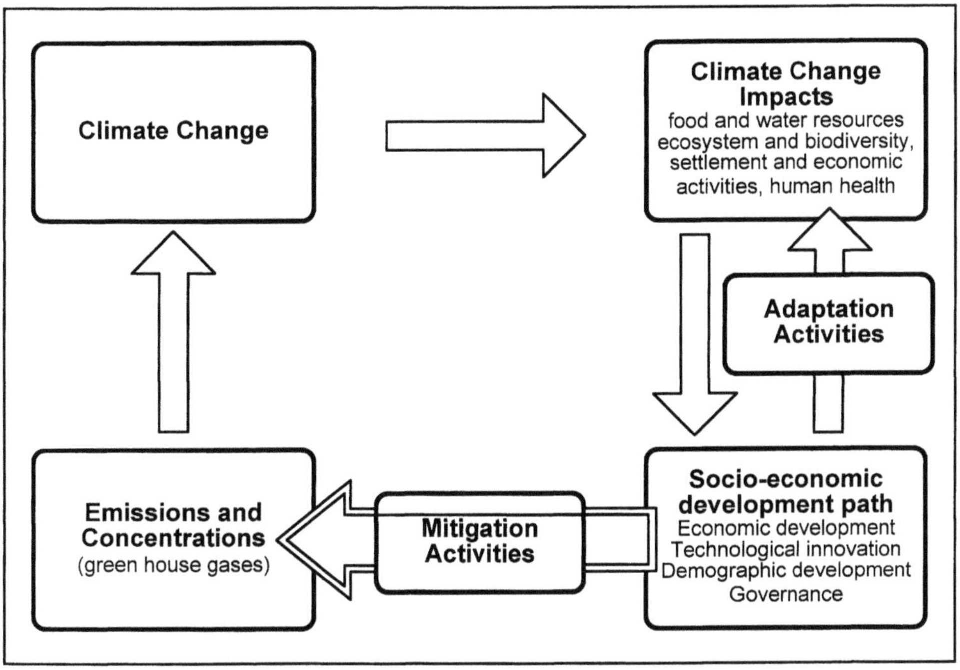

For example, the project visualized the complementary nature of the concepts of adaptation and mitigation which are intertwined by means of an easily understandable illustration. The message of the figure becomes clear: Even if the best mitigation strategies were implemented, further changes in climate and corresponding impacts cannot be avoided (Hilpert et al. 2007: 9). Thus, adaptation strategies are needed to cope with negative impacts, but also to seize beneficial opportunities that occur (IPCC 2007: 19; Hilpert et al. 2007: 4).

2.2 Information, Communication and Networking as integral part of project activities

Communication and dissemination events were an integral part of the project's activities. For example, all partners held meetings with local stakeholders to spread information about the project and its results, or conducted workshops to actively engage local stakeholders into joint collaboration on the issue (see box for Lithuanian best practice example). At the same time, the partners used this means of dissemination for networking purposes, in order to establish new stakeholder relations that did not exist before, which might prove to be valuable for future projects.

By means of three large international conferences, held in Klaipeda (Lithuania), Gdansk (Poland) and Riga (Latvia), a large group of stakeholders has been reached. Presentations as well as active involvement, e.g. in the course of a stakeholder workshop (see figure 3, for example), varied in order to not only spread latest climate change insights, but also to establish a dialogue with, and among, the international delegates.

Fig. 3: An example of an ASTRA activity

> **Regional workshops in Lithuania:** Lithuanian partners of the ASTRA project started a regional workshop session to raise public awareness of the issue of climate change adaptation. 55 regional stakeholders, representing local authorities, subordinate departments of the national Ministry of the Environment, research institutes, businesses, NGOs and the media, attended four workshop sessions. Workshops addressed climate change and its potential effects in general, later focusing on concrete climate change impacts in different sectors of the Klaipeda region and possibilities of adaptation. The final output of the communication process were guidelines for climate change adaptation that were presented to local stakeholders in Klaipeda as well as to other interested groups in other parts of Lithuania.

Each of the conferences had its own theme, often tailored to the specific needs of the hosting country in order to attract attention. For example, the event in Gdansk dealt with the theme flooding as this poses a major threat to the Polish city – a threat which might increase in future due to climate change impacts.

Fostering the transnational exchange of knowledge, experiences and best practice, the project raised awareness of the issue of climate change and adaptation in the Baltic Sea Region, and managed to build a network among the countries around the Baltic Sea.

2.3 Guidance for developing climate change adaptation strategies as main outcome

The final project outcome, the publication "Towards Climate Change Adaptation in the Baltic Sea Region", is dedicated to the specific needs of decision-makers, the main target group of the project. With its manageable scope and straightforward language, the book tries to convey latest climate impact know-how in a concise format to interested stakeholders. The publication is structured in three parts each building upon the findings of the previous ones and then, in the fourth chapter, resulting in specific recommendations for policy-makers.

Part I briefly introduces key concepts and relevant definitions, stressing the challenges when tackling climate change issues, and recalling the IPCC's opinion that early adaptation to climate change can help to reduce potential financial and humanitarian risks. Climate change impacts and corresponding adaptation needs will influence a range of sectors and regions in various ways. Furthermore, the chapter illustrates the need to solve the lingering conflict of interest among regions, sectors and administrative levels. Finally, it touches upon the crucial role of policy-making in addressing issues relating to climate change, enhancing adequate action and shaping a positive attitude towards climate change adaptation, not only among policy-makers, but also in public perception.

Part II recalled that summer seasons may become hotter and drier, and an increase in precipitation and milder climate may be expected for winter seasons, highlighting that expected impacts differ at regional scale. With reference to temperature rise, it was stated that the most affected areas during the summer months were the coastal areas Finland, Sweden and the Baltic States, whereas in winter it were the mountainous areas in Finland and Sweden. By means of concrete maps, it was shown that also precipitation patterns differ from region to region. The project partners came to the conclusion that, in the Baltic Sea Region, a range of specific sectors[3], i.e. urban and coastal areas, water, energy, transport, tourism and agriculture, is and will be influenced by climate change impacts. Two studies conducted in the course of the project, evaluating institutional preparedness in the BSR and analysing measures taken after winter storm Gudrun in 2005, stress once more the need to strengthen the awareness about adaptation to climate change, and to create adequate organisational frameworks. The chapter concludes that the formulation of strategic adaptation programmes at national level and the implementation of concrete measures at local level need to be enhanced. To achieve this, learning processes and cooperation among administrative levels were key aspects.

Part III displayed that, at European level, there are already some policies which take the aspects of a changing climate into consideration. With reference to the field of integrated coastal zone management it is claimed, however, that the national, regional and local actors have not yet implemented the recommendations to integrate climate change impacts sufficiently. Furthermore, it is shown that sectoral approaches at national level have begun to integrate the consideration of climate change impacts into first programmes and strategies. Still, identifying which policies take these impacts into account, has proven difficult for the pro-

3 In the framework of the project only a number of sectors could be assessed. A comprehensive investigation would have to analyse further sectors such as energy supply, water management, urban planning, transport and health as well as tourism, and deeper scientific research would be needed for the identified strategies.

ject partners, not because they lacked competencies, but due to partly missing distinctions between adaptation and mitigation strategies as well as the problem of obtaining more detailed information about the extent to which climate change impacts are considered. The project partners observed that most of the policies they identified considered only the mitigation aspect and did not refer to adaptation issues which form the complementary concept (see figure 2). In addition, it has been unclear whether identified policies referring to climate change impacts were implemented at regional or local level. An assessment conducted by ASTRA project partners could hardly identify local policies that integrate climate change impacts and adaptation needs. Thus, it had to be concluded that climate change adaptation has not been taken up at regional or local level as yet and that much action was still needed.

3. Recommendations for policy-makers

Resulting from the project work, the project partners designed recommendations for policy-makers, revealing a range of starting points on how adaptation strategies to climate change could be developed. Generally, adequate adaptation strategies can only be developed when existing threats and bottlenecks are recognized and a joint baseline target state is defined – answering the questions to what degree risk and vulnerability are generally accepted.

An analysis of current, i.e. sectoral, policies in the beginning of this development process can provide valuable insights on identifying where adaptation to climate change is already being considered, and where it should be incorporated. To achieve this, three different approaches were followed:

1. the hazard-based approach where conclusions are based upon the analysis of a specific hazard;
2. the vulnerability-based approach where the set of questions "who is vulnerable to what, in what way and where?" provides the starting point of discussion; and
3. the policy-based approach which investigates the efficacy of existing or planned policies concerning the changing climate.

For all of the approaches, concrete examples were provided by the project partners (Hilpert et al. 2007: 36pp).

The next step is about developing adequate strategies. The ASTRA project highlighted to this purpose appropriate frameworks that have been developed for policy-makers, specifically addressing their demand for a user-friendly guidance. In this context, the United Nation's "Adaptation Policy Framework for Climate Change (APF)" has been mentioned, as well as the UK Climate Impacts Programme (UKCIP) Wizard, a web-based step-by-step guide. Another example given relates specifically to a bottom-up, participatory process: The United Nations Framework Convention on Climate Change (UNFCCC) developed in its "National Programme on Adaptation (NAPA)" a guide specifically taking into account the local coping capacity (Hilpert et al. 2007: 38).

Figure 4 illustrates the general steps that should be considered closely when developing adaptation strategies. The illustration implies three basic requirements, namely that

1. an agenda setting process has already taken place;
2. decision-makers are not only aware of the need for adaptation, but also willing to take action; and
3. that appropriate collaboration structures have been established.

Fig. 4: General framework for the development of adaptation strategies

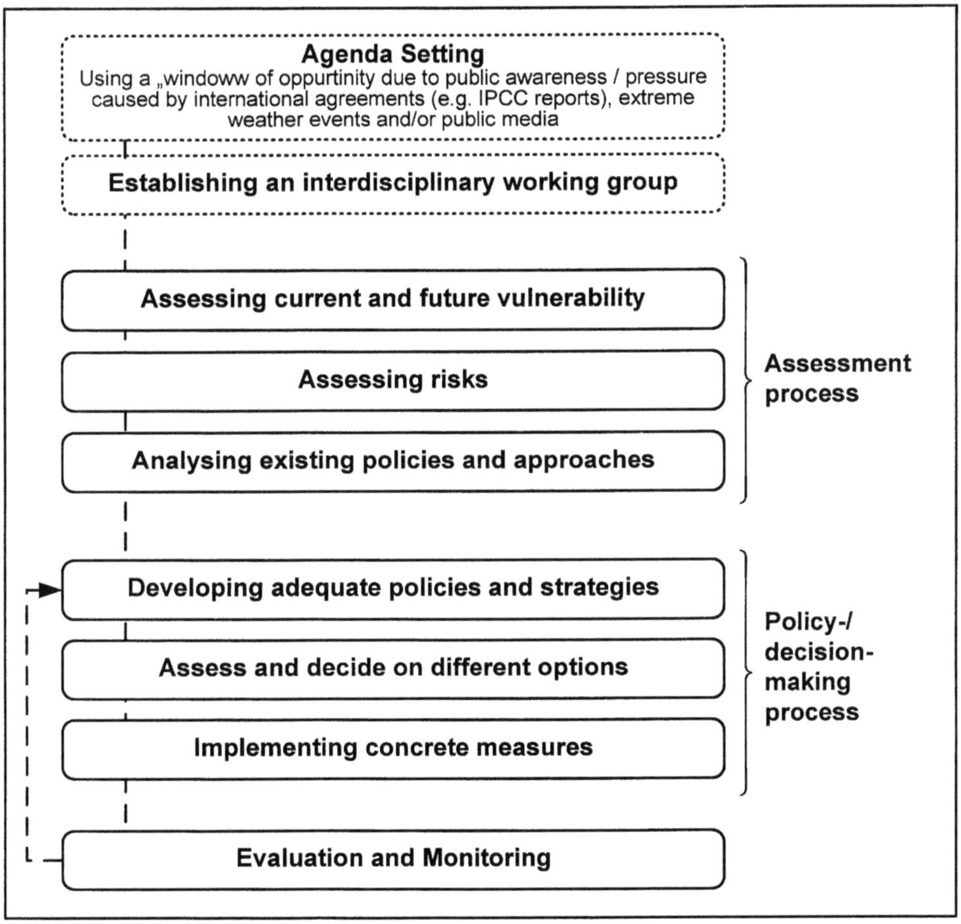

(adopted from Hilpert et al. 2007: 38)

According to the Green Paper on Adaptation, information-sharing on adaptive response measures could to a large extent lead to reduced learning costs across Member States, regions, municipalities, and communities (Commission of the European Communities 2007: 11).

The findings of the ASTRA project underlined that future policies should integrate the issue of climate change impacts not only to enhance the resilience of societies and economies, but also as to maintain their competitiveness. As such, policy-makers should aim at designing "climate-proof" policies, i.e. integrating climate change adaptation as well as mitigation strategies. No-regret and win-win options for both adaptation and mitigation should be developed. Hereby it is advisable to take more than one perspective into account: the technical as well as institutional and societal issues (Hilpert et al. 2007: 40).

As climate change is a cross-cutting issue, adaptation to climate change impacts should be mainstreamed into policies and planning, thus being a vital contribution to sustainable development. Spatial planning can be seen as adequate field to meet the requirements of steering the complex issue of climate change adaptation (Fleischhauer 2006: 162). Decisions on land use and building regulations need to be based on forecasts, e.g. in terms of demographic change or economic development, and on climate change scenarios. Here, spatial planning uses a range of instruments, ranging from hard regulations to soft supporting instruments. These instruments represent a structured way of assessing vulnerability and risk, e.g. by means of a strategic environmental assessment, the implementation of adaptation needs, e.g. by means of laws or construction codes, or spatial concepts incorporating climate change impacts, e.g. through the integration of flood protection maps into spatial plans.

With reference to the above mentioned, a change of behavior within societies and communities is said to be dependant to a large extent on awareness of the problem (Commission of the European Communities 2007: 12). The ASTRA project conferences confirmed that the need to adapt to climate change impacts is not yet perceived as being urgent. Also, it has been disclosed that the concept of adaptation and related strategies were not sufficiently known (Hilpert et al. 2007: 42). With reference to the discipline of spatial planning, delivering education and professional training can serve as means to mainstreaming climate change into daily business. A comprehensive effort could comprise a broad range of actors, e.g. students, practicing planners, architects, construction engineers, environmental specialists, public servants, insurance specialists and politicians. Furthermore, minimum requirements for spatial planning, land use and land use change might have a key role not only for awareness raising among the public, decision-makers and professionals, but also foster proactive approaches at all levels (Commission of the European Communities 2007: 12).

As climate change impacts can affect society as a whole, a holistic approach is needed to find adequate solutions. Today's already perceived impacts affect businesses, e.g. farmers or insurances, as well as private households when people live in vulnerable places. Over and above, climate change impacts can have negative impacts on public welfare, for example in sectors such as water or energy supply. Therefore, adaptation will need to be taken into account by many actors on all levels – internationally, nationally, regionally and locally. A key role is hereby assigned to public authorities.

With reference to the concept of governance, policy networks advance into the centre of attention, and it is acknowledged that actors external to the political arena become increasingly involved in a joint process by developing new partnerships and building up capacity for self-organization (Hilpert et al. 2007: 44).

Last, but not least, it is noted that having recognized climate change adaptation as a new market, this might stimulate further economic activities that result in innovative products and processes, e.g. in the construction or insurance industry, design of flood protection measures and so on. This is in line with the EU's Green Paper on Adaptation which claims that early adaptation will bring economic benefits by anticipating potential damages and threats. It also points out that competitive advantage could be gained for those European enterprises which are at the forefront of adaptation strategies and technologies (Commission of the European Communities 2007: 9).

4. Conclusions

According to the International Panel of Climate Change, the warming of the climate system is unmistakable as is now evident from observations of increases in global average air and ocean temperatures, the widespread melting of snow and ice, and rising global average sea level.

To shoulder the complex task of adapting to climate change, a joint effort by the entire society is needed. Here, public bodies as well as private actors are challenged to find solutions: The market could influence public attitude, e.g. by adjusting insurance rates for housing areas or economic activities in vulnerable areas, and could lead to technological innovation. Public authorities could stimulate private action towards climate change adaptation, e.g. through establishing a positive attitude towards the issue by raising awareness through information and communication.

In particular in the European Union and on a national level, incentives could be given in the form of financial assistance for further research and pilot projects. With reference to complementary policies, the existence of adaptation policy guidelines on a national level would support action on local level.

Tasks concerning climate change adaptation on regional and local level are adjusting general guidelines to local needs and implementing adaptation measures.

Tools and approaches to initiate local processes and develop adaptation strategies are already at hand. Policy-makers are advised to design adaptation strategies in a participatory approach, involving all affected and interested stakeholders from the very beginning on.

A major emphasis is thus placed on mainstreaming adaptation to climate change into different fields of policy by means of integrating it into every relevant policy, as outlined in Table 1.

Tab. 1: Conclusions from the ASTRA project for climate change adaptation in the Baltic Sea Region (adopted from Hilpert et al. 2007: 45).

• Awareness-raising activities in relation to climate change adaptation still have to be continued; communication, education and professional training are key aspects.
• Vulnerability assessments regarding different sectors should be conducted at national, regional and local level in order to identify needs of adaptation.
• Cooperation across sectors, administrative levels and regions is required when developing adaptation strategies by involving both public and private stakeholders. Climate change should be mainstreamed into existing holistic cross-sector approaches like Integrated Coastal Zone Management (ICZM).
• Organisational structures should encourage learning processes and enhance the transfer of knowledge and information about climate change. These processes as well as cooperation among administrative levels are key aspects.
• The implementation of adaptation strategies and recommendations has to be secured, and monitoring structures should be initiated. New technical options should be developed and pilot projects should be launched.
• Climate change adaptation should not only be regarded from the technical point of view but also in societal, economical and political terms. Adaptation and mitigation strategies should become integrated in everyday life (political and individual).

In this context, a special role may be played by targeted information, education and communication strategies (Leal Filho et al. 2007) which may catalyse further participation and engaging more people in taking up the challenge of dealing with climate change. Communication and dissemination among policy-makers on the one hand and the public on the other hand will help to make climate change a more manageable issue that can be related to other policy objectives. This also provides the opportunity to bring together *individual responses* to climate change by encouraging individual action, an effective complement to measures at the policy level.

5. References

Commission of the European Communities (2007), Green Paper from the Commission to the Council, the European Parliament, the European Economic and Social Committee and the Committee of the Regions. Adapting to climate change in Europe – options for EU actions 29 June 2007 (Brussels, Belgium: Commission of the European Communities).

Fleischhauer, M. and Bornefeld, B. (2006), "Klimawandel und Raumordnung", Raumforschung und Raumordnung no. 3, pp. 161-171.

Garvin, D.A. (1993), "Building a Learning Organization." In: Harvard Business Review on Knowledge Management, Harvard Business School Publishing, (July-August Boston 1993), pp. 78-91.

Hilpert, K., Mannke, F., and Schmidt-Thomé, P. (2007), Towards Climate Change Adaptation in the Baltic Sea Region. Espoo, Finland: Geological Survey of Finland.

Intergovernmental Panel on Climate Change (2001), Summary for Policymakers. In: McCarthy, J.J., Canziani, O.F., Leary, N.A., Dokken, D.J. and White, K.S. (eds.), Climate Change 2001: Impacts, Adaptation, and Vulnerability. Contribution of Working Group II to the Third Assessment Report of the Intergovernmental Panel on Climate Change. Cambridge, UK: Cambridge University Press, pp. 2-34.

– (2007), Summary for Policymakers. In: Parry, M.L., Canziani, O.F., Palutikof, J.P., van der Linden, P.J. and Hanson, C.E. (eds.), Climate Change 2007: Impacts, Adaptation and Vulnerability. Contribution of Working Group II to the Fourth Assessment Report of the Intergovernmental Panel on Climate Change. Cambridge, UK: Cambridge University Press, pp. 7-22.

Leal Filho, W., Mannke, F., Schmidt-Thome, P. (2007), Information, Communication and Education on Climate Change – European Perspectives. Frankfurt, Germany: Peter Lang Scientific Publishers.

Meier, M., Broman, B., Kallio, H., Kjellström, E. (2006), "Projections of future surface winds, sea levels, and wind waves in the late 21st century and their application for impact studies of flood prone areas in the Baltic Sea Region." In: Schmidt-Thomé, P. (ed.), Sea Level Change Affecting the Spatial Development of the Baltic Sea Region. Espoo, Finland, pp. 23-44.

Chapter 5

The Climate Stabilization Fund

–

Global Auctioning of Emission Allowances to help Forests and People

Michael Dutschke

Abstract

The initial research interest in the layout of this paper was to secure sufficient finance for reducing emissions from deforestation and degradation in developing countries under the UN Framework Convention on Climate Change. After analyzing the interest groups involved in the REDD discussion, the author proposes a Climate Stabilization Fund, based on proceeds of an international allowance auction. The auction-based model proposed fulfils a whole variety of goals besides providing financial proceeds, including a gradual long-term integration of developing countries into shared responsibility for confronting climate change.

1. Scope of an REDD Mechanism

During the actual first commitment period of the Kyoto Protocol, reducing emissions from the land-use sector in developing countries has not played a role, even though roughly 20 percent of all anthropogenic greenhouse gas emissions are estimated to originate in deforestation (Nabuurs & Masera et al. 2007). For the period after 2012, the parties to the U.N. Framework Convention on Climate Change (UNFCCC) plan to incentivize policies and activities to reduce emission from deforestation and degradation (REDD) in developing countries. Decision 1/CP.13 of the Bali Conference of the Parties (COP 13), the "Bali Roadmap" gives a prominent role and a tight timeframe to REDD. Together with the worldwide emission reduction targets, the REDD mechanism shall be decided in December 2009 during the Copenhagen COP 15. However, the context between international target setting and REDD is not only a temporal one. With its contribution to greenhouse gas concentrations in the same range as the total emissions of the United States; forestry plays a pivotal role in achieving the UNFCCC's ultimate objective to stabilize greenhouse gas concentrations in the atmosphere (Dutschke 2007).

Today, international negotiations actually circle around methodological issues for REDD, like international leakage, permanence and the determination of national baselines. The author argues that these are second-order questions, because the basic incentive structure for REDD remains unclear. The first-order question is a distributional one. A future REDD can be either participative or non-participative. The degree of participation will have repercussions on scope and size of a future REDD Mechanism. Only once these can be assessed, second-order methodological questions can be answered.

Participative REDD Mechanism: An all-in approach will be the first-best option. It will be most effective, because emissions and uptakes will be accounted for across different country situations. A participative mechanism will also be most carbon efficient, because it will offer most supply transparency. Under this option, cross-border carbon leakage will not occur, because emissions are accounted for wherever they occur. As long as any form of compensation for carbon services are involved, baselines will remain an issue. Monitoring will be most cost-efficient on a large scale, because the increased demand for satellite-based imagery will drive remote sensing technology development. Three levels of permanence need to be distinguished for REDD: (1) Permanence within country borders will be provided for, as long as the particular country party remains within the system. (2) Permanence on a global scale can be expressed as the risk that above certain temperatures increase thresholds, terrestrial carbon stocks may release more carbon than they sequester, an effect that has been known under the keyword "savannization". Little is known about these thresholds for the forest dieback to occur, and latest studies suggest that tropical forests are not as sensitive as initially expected (Gullison & Frumhoff et al. 2007). Should dieback occur, losses may eventually be temporary, due to the extension of vegetation zones. From the atmospheric point of view, a savannization in parts of the tropics might be compensated by increased cover and density of boreal forests in tundra and taiga ecosystems, at least over the longer run. (3) For tropical countries where savannization potentially occurs, global warming may lead to imbalances or even debits in any given REDD compensation system. To cater for this concern, monitoring methodologies for factoring out non-direct human influences and / or an international climate insurance mechanism could be helpful.

Partial REDD Mechanism: A non-participative approach will be much smaller in scope. It will create fewer incentives, only include some developing countries, and raise specific methodological issues, like international leakage, time consistency of reporting, and transitional problems for countries that graduate to a higher responsibility in the climate regime, among others. In the case of the afforestation & reforestation Clean Development Mechanism (A/R CDM), a partial flexible mechanism led to failure, because the precautions taken in its design are so prohibitive that hardly any projects get off the ground.

2. Interest groups in REDD negotiations

We will review the interest groups involved in the negotiations, and contrast their individual interest with the ultimate objective of the Climate Convention to stabilize greenhouse gases (GHGs) in the atmosphere "at a level that would prevent dangerous anthropogenic interference with the climate system". They all suffer from the classical prisoners' dilemma in environmental politics: *If I do not know for certain, whether my neighbour will really protect the climate, why should I take the pain to do so?* Fully acknowledging that each single country's circumstances are different, the following paragraphs will identify the stakes and interests of "archetypical" country categories in REDD. The last paragraph summarizes the preconditions for broad participation.

2.1 Annex I – industrialized country parties

Annex I represents the group of countries that were part of the Organization of OECD in 1992 plus the former Soviet Union. The UNFCCC and the Kyoto Protocol both have cemented this static division between "rich" and "poor" emitters. Meanwhile however, Hungary, South Korea and Mexico have joined OECD, and several others are set to follow. Annex I representatives readily admit that their countries have been responsible for the bulk of the greenhouse effect. This is why the Kyoto Protocol during the first commitment period set quantitative emission limitations for Annex I only. When the Kyoto Protocol was negotiated, differentiated commitments and the means of compliance were conflictive, especially the inclusion of forest-related mitigation activities (Jung & Michaelowa et al. 2004). The European Union has widely seen itself as a clean energy technology provider, which is why it largely opposed the inclusion of sinks and sources from land use in any compensation mechanism. In spite of its more constructive approach on REDD today, apprehension subsists that emission reductions in the energy and transport sectors could be bypassed by assumingly "cheap" emission reductions from land use, land-use change and forestry (LULUCF). Instead of giving too much mention to the different Annex I country groupings that were influential during the negotiations around the first commitment period, we will concentrate on Annex-I Parties' interests with respect to REDD after 2012. Since 2002, the issue of tropical deforestation has become fashionable in public opinion. Environmental NGOs are most influential in European decision making; to a lesser extend so in the US. NGO agenda setting focuses on biodiversity and socioeconomic issues, which could already be observed during the discussion around the A/R CDM. In spite of the fact that the idea of offsetting emission targets by mitigation activities abroad was first brought up by Norway, among European negotiators still some skepticism towards markets can be felt. This explains the tendency to impose safeguards that run counter carbon efficiency. Canada, the US and Australia, on the other hand, believe in robust market mechanisms. Nevertheless, with full market integration of credits from REDD, neither one of these sides can be too sure: Should emission targets be too lenient for future commitment periods, it would be quite likely that in the mid term, credits from REDD could flood the market. With stringent binding targets, Annex I countries might end up seeing themselves in the hands of some corrupt and inefficient tropical governments to provide the much-needed offsets from REDD. It is however in the interest of Annex I countries to get developing countries to take their appropriate share in the responsibility for global climate change mitigation. Practical proposals from Annex I go into the direction of "demonstration activities", in order to gain trust in administrative and technical capacities of developing country parties to really bring down emissions in their land use sector. In Bali, influential Annex I countries committed to investment in REDD activities and capacity building in tropical countries via the World Bank's Forest Carbon Partnership Facility (FCPF) and direct transfers. Nevertheless, no regulation has been found so far, whether these early activities will be accounted under a post-2012 climate regime. This will limit private-sector willingness to take high stakes in REDD business. Advanced representatives of Annex I countries today see the necessity of a "global deal" involving REDD (Stern 2008). Yet no leadership initiative comes from Annex I on REDD.

2.2 Coalition for Rainforest Nations – CfRN

The CfRN gathered around a proposal by Papua New Guinea supported by Costa Rica, and it has taken on board a variety of tropical and sub-tropical countries. It is a sub-grouping among G77 countries, but it has developed remarkable leadership. The initiative is welcome and carefully supported by most Annex I Parties. Nevertheless, inside the Coalition there are centrifugal forces with respect to the different country circumstances. These will become virulent once any substantial funds will have to be shared. There are countries on all different stages of the forest transition curve (Rudel & Coomes et al. 2005); the Congo Basin countries, where deforestation is no major issue yet, but which are threatened by deforestation and degradation and want support for sustainable resource use, countries where deforestation is a day-to-day reality, like PNG or Indonesia, and countries where deforestation is declining or where forest cover even increases, like Costa Rica. Besides this baseline issue, institutional capacities and legal instruments to implement REDD differ widely among partners to the Coalition. There are also different interests regarding national or subnational implementation of REDD activities. So far, the CfRN has been very prudent in representing a constructive common negotiation position. There is a risk in subdividing this group by allocating public funding to a selected subsection of this group only. In order to build up trust, and in the absence of enough finance to support all rainforest nations, at least all different approaches proposed within the Coalition should be tested in selected countries. Apart from the REDD issue, the CfRN does not represent any common standpoint towards a future climate regime.

2.3 Alliance of Small Island States – AOSIS

The small island states can be assumed to represent the conscience of humanity, because climate change is an existential threat for them, and most of them do not have the means to protect themselves. It is in their best interest not to disregard terrestrial carbon stocks. In UNFCCC negotiations, they are represented by the state of Tuvalu, who opposes any trade-off between industrial emissions and REDD credits. Were the latter allowed to offset emissions in other sectors, leakage (or "displacement", in REDD jargon) would be a main concern.

2.4 Least Developed Countries – LLDC

At the lowest range of development, there are countries that are practically excluded from the benefits of globalization, or they participate in international trade with only one or few assets. Countries like the Ivory Coast have nothing to lose. They hardly participate in international economy, and they are busy managing their own internal disorder caused by a vacuum of power in situations where mobility has increased and different ethnics mix. They are used to receiving the remainders from rich men's tables. Their economic hope is the exploitation of their natural resources, because it requires little investment in capital stocks. The hope however has not materialized so far, because of investment uncertainty and the lack of credibility in contracting. Property rights cannot be enforced in these countries, because they lack most features of nation states. If at all, deforestation occurs illegally or semi-legally.

Climate change has an extremely low priority for LLDC representatives, because living conditions of their populations hardly translate into trade receipts or state budgets (and vice-versa). Capacity building and institution building however have the potential to open up income sources for LLDCs, because they have the potential to create buyers' and investors' faith in the country, while at the same time creating income opportunities within the administration sector.

These countries' preference in the negotiations will be an ODA-like fund to direct investment allocation in REDD according to equity principles. They are the ones to insist that REDD funds be "new and additional" to existing obligations of industrialized countries, which makes it likely that a new debate over funding additionality will arise.[1] As the exploitation of natural forests is only just begin-

---

1 The precursor is the provision on the CDM within the Marrakech Accords, according to which "such funding does not result in a diversion of official development assistance and is separate from and is not counted towards the financial obligations of those Parties". This

ning, at least in the Congo basin, a growth baseline is promoted by these negotiators.

2.5 Developing countries – Group of 77

In contrast to industrialized nations, developing countries (G77) came late to the party. Most industrialized countries in their history have profited massively from deforestation. In European history, timber was used for mining, iron smelter, and for ship building in the colonial phase. Today's deforestation in developing countries mainly serves export purposes to cover Annex I demand for tropical timber, meat, soybeans, rubber or palm oil. Long-term maintenance of the forest resource only becomes interesting when timber increases in value and the resource starts to become scarce. However, as long as there is unprotected natural forest left in the world, the logging companies will continue plundering them.

Underprivileged rural poor benefit from unclear tenancy and often invade areas that were previously opened by logging companies. Marginal charcoal production or slash-and-burn agriculture offer a living. Stopping deforestation and its social environment may thus lead to unrest.

REDD offers an alternative income source for maintaining and protecting forests stocks. This however requires a complete shift in forest governance. A monitoring system has to be built up, property rights have to be enforced. Is it really worth all this effort? How much is in for the individual negotiator's country and how reliable is this source of income? Will Annex I countries really commit to cuts so deep they rely on complying them with the help of REDD credits? After all, the initial experience was not so good. In the run-up to the Milan decision on A/R CDM, the EU promoted temporary crediting as well as social and environmental safeguards, but when it came to implementing those projects, EU backed up and blocked any A/R credits from being imported into its emissions trading system.

This time, DC negotiators want to see real commitment, before they become engaged. They ask for deep targets, before they can agree on participation in an REDD mechanism. Nevertheless, future targets and REDD will be jointly negotiated until the Copenhagen Conference of the Parties. This is why Brazil as a leader among G77 promotes a voluntary fund from Annex II countries. Once they know how much money is in, they can better assess risks and chances of participation.

On the other hand, there is competition between DCs for REDD funds A strict denial can therefore lead to isolation, as the benefits, in case there will be, will go to other developing countries. For countries that benefit from timber sales,

so-called "financial additionality" clause under the CDM can hardly be operationalized [Dutschke & Michaelowa 2006].

ideally REDD funds come in at a time when forest resources already become scarce and structural change is required anyway.

2.6 Emerging economies

Representatives of emerging economies see themselves in a delicate position: Structural change is underway in their countries. There is a need to liberate rural workforce and to deepen the integration of broad layers of the society into the market economy. A modernization and intensification of the primary sector is overdue in the interest of economic development anyway. These countries are approaching the lower turning point of the forest transition curve. Nevertheless, rural elites are still powerful. As land prices are low, producers of agricultural goods benefit from low factor costs. In order for rural modernization to occur, economic incentives are needed, and a compensation for land-management based mitigation would come in handy.

On the other side, representatives of emerging economies are being incited to take over a higher degree of responsibility in the climate regime. The 4th IPCC Assessment Report leaves no doubt that emission reductions in industrialized countries only will fail to reaching the ultimate goal of the Climate Convention, which is the stabilization of the level of greenhouse gases in the atmosphere. Agreeing to a land-use target may be seen as a first step in this direction. There is a risk that other sectors will be expected to follow suit. In the phase of rapid industrialization however, this may deteriorate the country's competitiveness towards non-restricted developing countries. There are two possible solutions to this dilemma:

1. Postpone participation as long as possible and offer contribution in the REDD Mechanism as a last-minute compromise (the actual Brazilian position).
2. Promote differentiated targets for all countries. This latter option offers a long-term solution, but it will weaken the Group of 77 as a negotiation group. In order for such a proposal to be acceptable, it will need to be linked to reliable and long-term north-south transfers.

2.7 OPEC countries

The most vocal country representatives for the OPEC group are the ones of Saudi Arabia. Compensating developing countries for REDD bears some similarities with the OPEC approach of "response measures". The argument is that emission reduction policies will in the long term lead to a decline in petrol demand, and countries mainly depending on oil export will have costs to restructure their economies (Barnett 2001). REDD may be considered a mute acceptance of this principle. Additionally, REDD does not lead to a decrease in oil demand. We may therefore expect low-level support from OPEC countries for REDD policies. Solidarity with the group of developing countries may also be seen as a strategy to avoid being singled out as countries that given their high emissions per capita and per-unit GDP would be among the first to graduate to Annex-I countries and take over climate targets.

2.8 Criteria for compromise

Negotiating Parties are actually navigating on eye-sight. There is a remarkable lack of trust. As seen above, for REDD to be successful, it needs to cater for a variety of concerns and provide incentives for the most diverging country situations. A viable compromise needs to square the circle according to the principles of (1) effectiveness, (2) efficiency, and (3) equity (Stern 2008).

1. Effectiveness: REDD shall not distract attention from the need to reduce emissions from fossil sources, which are the main drivers of the greenhouse effect;
2. Efficiency: REDD funds shall be concentrated where most emissions can be avoided;
3. Equity: Transfers need to be reliable and accessible to all countries, where they are needed.

First and utmost, any compromise needs to create trust. Therefore, it shall involve all country parties. It shall allocate emission allowances where they are most effective for the creation of wealth, while observing the principle of equity. The system to be created should be compatible with UNFCCC rules and procedures so far, and it should be open to future modifications without creating inconsistencies.

3. A proposal for action

The actual Kyoto Protocol system of tradable emission allowances (Assigned Amount Units – AUUs) has produced several imbalances, most of all the so-called "windfall profits". This term describes profits that do not result from the operative performance of an enterprise, but which result from the possibility to sell allowances. A good example are the EU power utilities that in the first EU trading period received EU Emission Allowances for free, but still factor opportunity costs from not selling these allowances into actual electricity prices. As electricity is no internationally tradable good, there was no possibility for competitors to undercut these electricity prices.

On the other hand, the fight against climate change and its effects can only be won with the help of massive North-South transfers in pursuit of the ultimate objective of the Climate Convention of stabilizing the greenhouse gas (GHG) level in the atmosphere.

During the last three years, the inclusion of Reducing Emissions for Deforestation and forest Degradation (REDD) in the international climate regime has been high on the agenda of the Climate Convention. The IPCC estimates that at CO_2 prices below 100 US$, the mitigation potential from forestry until 2030 is nearly 14 Gt CO_2e, half of which can be found in developing countries (Nabuurs & Masera et al. 2007). The opponents of including these emission reductions and carbon uptakes as credits under an international allowance trading system fear that these would disrupt the market. Nevertheless, the actual allowance market is far from being perfect (as exemplified above), and it is not an end in itself. What is true is that the low-cost fraction of these emissions would compete with emission reductions in the energy and transport sectors. In case of open competition, REDD credits could well be the preferred option, and emission reductions in other sectors might be neglected for some years. It may be argued that technological progress will make industrial emission reductions cheaper in the future. Still following this line of arguments, mitigation potentials from deforestation and forest degradation are finite[2] and should therefore be used first. A recent study finds that, when banking is allowed, an unregulated inclusion of REDD into the carbon market would bring compliance costs down, but would definitively not lead to market flooding (Piris Cabezas & Keohane 2008). In order to balance the diverging goals of increased energy efficiency and at the same time mobilize GHG benefits in the land-use sector, some authors propose different means to limit market access for land-use related carbon credits (Hare & Macey 2007; Ogonowski & Helme et al. 2007). Limiting market liquidity will in any case lead to a lower price for REDD credits and a lower investment in REDD policies than would be efficient from a global climate change mitigation perspective.

The actual paper presents a simple model that provides sufficient finance for REDD and adaptation, without competition towards other targets under the Climate Convention. In line with the "triple e" (effectiveness, efficiency, equity) criteria above, it shall further fulfil the following requirements:

- Achieve early REDD
- Secure long-term finance for non-market goals
- Harness private-sector contributions for REDD
- Create incentives for industrial GHG reductions
- Make REDD and other emission reductions work in parallel.

---

2 Natural forest systems with their old-growth carbon density and biodiversity take centuries to recover, and can thus be seen as non-renewable resources in the policy scope.

4. The main elements

The overwhelming majority of the states of the world have committed to the ultimate objective to stabilize greenhouse gases in the atmosphere "at a level that would prevent dangerous anthropogenic interference with the climate system". One and a half decades after this decision, there is growing agreement over the significance of this phrase: two degrees C above pre-industrial levels is an increase in temperature that may still be acceptable during this century. The widely agreed time horizon for achieving stabilization is the year 2050. By then, the GHG concentration level should be down to 50% today's level, equivalent to a target GHG level of 450 ppm CO_2 equivalents.[3] This vision would have been turned down as utopia few years ago. Just recently, the parties at the G8 summit in Hokkaido, including the US, have agreed on this long-term policy goal. Instead of talking of emission reductions needed, we should thus consider the amount of greenhouse gases allowable until reaching the concentration level to be our worldwide GHG budget. This idea was first expressed under the concept of "Greenhouse Development Rights" (Athanasiou & Kartha et al. 2006). In the present paper however, this approach is carried further and operationalized. In 2000, 48 percent of all greenhouse gas (GHG) emissions came from developing countries (Baumert & Herzog et al. 2005). With the steep increase in China's emissions, we can safely assume that half of all GHG emissions today come from developing nations not included in Annex I of the UNFCCC.

Departing from the budget approach, the core elements of the proposal for the Climate Stabilization Fund are the following:

1. COP 15 in December 2009 agrees on a global emission budget until 2050 in tons of CO_2 equivalents, derived from a "safe" target temperature level (e.g. 2 degrees C above pre-industrial levels);
2. For the forthcoming 2nd commitment period (CP2), a global emissions target is derived as a fraction of the budget. This fraction is determined in a fashion that global emissions are set to decline after the year 2020.
3. Initially 50% of the allowances under the CP2 target are allocated free of charge to developing countries. These free Assigned Amount Units (f-AAUs) are non-tradable.
4. Until 2050, the share of f-AAUs is gradually decreased to zero.
5. The UNFCCC Secretariat will start auctioning in 2010 of the 50% of AAUs that have not been allocated to developing countries. In a five-year CP, each year, one-fifth of the tradable AAUs would be auctioned.

---

3 Scientific discussions actually question, whether 450 is a "safe" level for reaching the 2-degrees target [Athanasiou & Kartha et al. 2006; Baer & Athanasiou et al. 2007].

6. UNFCCC Parties would automatically qualify as bidders, but transnational enterprises and sector organizations may also register. Under Kyoto, enterprises are only trading on a secondary market. It may however be attractive to them to participate in the primary market too, thereby increasing market liquidity.
7. Banking: Unused AAUs can always be carried over to the subsequent commitment period, independently whether these are tradable, or f-AAUs.
8. A share of the auction benefits realized covers administrative expenses and the cost of the UNFCCC compliance branch.
9. The proceeds of these auctions are used for purposes that are not directly producing GHG benefits or for GHG benefits or that the international community wants to keep outside the market for emission reductions. In the first category, there is capacity building for national inventories, improvement of governance structures, and adaptation to inevitable climate change. In the second category, there are emission reductions from land use, commonly referred to as REDD.

The Climate Stabilization Fund can offer a common roof for all the existing funds under the Kyoto Protocol, including the Special Climate Change Fund, and the Adaptation Fund.

5. Calculation

In order to get an idea about the quantities to divide, we resort to IPCC estimates. There is growing consensus on a stabilization level at 2 degrees C above pre-industrial levels to be reached by the middle of this century, with reductions between 50 and 85 percent of current emissions.

Current annual emissions are roughly 20 Gt CO_2e (Stern 2008). Cumulative allowable emissions for the lowest IPCC stabilization scenario amount to 1,100 Gt CO_2e over the whole century (Fisher & Nakicenovic et al. 2007). During the first 12 years of this century, the world will have emitted roughly 20 percent of the century budget. Linearly bringing down the emissions by half until 2050 requires worldwide emission cuts of 0.26 Gt CO_2e annually. The stabilization budget for a 50-percent pathway between 2012 and 2050 is in the order of 600 Gt CO_2e of GHG emissions. On average, the available quantity is 15 Gt $CO_2e\ y^{-1}$. Assuming a low price level of 10 EUR per ton CO_2e and an initial 50% free allocation to developing countries, the proceeds will be in the range of 100 billion EUR per year. In the same pace that the commitment period budgets decline, the share of AAUs auctioned increases, thus stabilizing the auction receipts. The opportunity costs for halving emissions from deforestation have been estimated in the order of 2-22 billion EUR annually (Stern 2008). Even adding the same amount for readiness building and transaction costs, the remaining amount for adaptation surpasses the actually available funding for this purpose by orders of magnitude.

6. Institutional design

Theoretically, one auction per commitment period could be sufficient for the primary allocation. At least in an initial phase, it is recommended to opt for higher market transparency and auction a pro-rata share on an annual basis. This auction occurs on an electronic platform, in order to keep transaction costs low. The auctioneer should not be an interested party, which is why the UNFCCC Secretariat is the natural choice. It keeps an AAU registry account for every registered bidder. Country Parties are represented by their national focal point. These may need to deposit confidentially an authorized budget item allocation by their respective treasury ministry for the maximum allowance purchase bid.

For cashing in payments and the distribution of revenues, several creative solutions can be found. The World Bank is one option, but not necessarily the preferred one. Actually, the Bank is preparing as a trustee for the Forest Carbon Partnership Facility (FCPF), a much smaller exercise aimed at gaining experience on future allocation systems for REDD. It consists of the Readiness Fund on the one hand and the Carbon Fund on the other. REDD country readiness is an upfront investment in the infrastructure needed. The Carbon Fund will tender emission reductions from REDD according to price and minimum quality. It will acquire these credits from states and from subnational activities, as appropriate. For the management of REDD finance, the FCPF could be scaled up or at least be used as a blueprint. In order to achieve an equitable distribution, the Climate Stabilization Fund could be subdivided in continental branches, according to the continents' share in total terrestrial carbon stocks, but independent from their actual variation. Based on country and activity baselines, each branch would seek for the most cost-efficient investment within its respective continent.

As national governments are generally unwilling to leave the decision over billions of dollars to a supra-national institution, the second-best solution could be to channel the revenues into a dedicated budget line within national budgets, which is documented to the Climate Secretariat and used according to REDD principles, under the discretion of the country's legislative body. Revenues from direct private-sector bidders could still enter a supra-national fund.

7. Discussion

Allocating emission allowances via an auction is not a new idea (e.g. EU Commission 2008; Fuentes 2008). Most proposals however suggest national auctioning to private bidders after an emissions target has been fixed. In contrast, under the current proposal, no national target is needed. This bears the risk that economically potent states may acquire and withhold allowances. In the existing allowance trading systems, no monopolistic behaviour has been observed so far. As the long-term budget is commonly known and prices are transparent, there is little risk for asymmetric information at the scale needed to provide sufficient incentive for the formation of a carbon monopoly. Additionally, annual auctions will allow for fine-tuning by the market regulator (the Conference of the Parties) if needed.

Without the needed market narrowness, auctions will not produce the desired results. The scarcity needed is determined by the global commitment period target. The correct market signals are provided by the long-term emission budget, but a functioning allowance auction does not depend on it.

Politically, the proposed model has good chances to be implemented. This is so, because to a national electorate, global emission limits are easier to communicate than national ones, and a long-term stabilization goal is easily accessible to common sense. The politicians who actually decide on the long-term target are unlikely to be punished by their voters. From a political economy point of view, this is the explanation why the US is backing the Japanese and European proposal for a global reduction target of 50% by the year 2050 (Black 2008; G8 2008).

The same effect occurs with the integration of non-Annex I countries into the climate regime. Since the IPCC Fourth Assessment Report, there is no doubt that Annex I emission reductions will not stop global warming, if developing countries will not take over responsibilities. Allocating most of their emissions freely during the first years is an acceptable solution under the prevailing short-term perspective of political decision-making. A differentiation within non-Annex I countries is due anyway. Index-based models like "Graduation and Deepening" (Michaelowa & Butzengeiger et al. 2005) can be applied in a fair sharing of the reduction burden among developing countries. Initially, it is important to avoid perverse incentives while non-Annex I countries are not fully integrated. As a base year, the year 2010 could be chosen, which is the year that emissions from Annex I and non-Annex Parties are expected to be equivalent. The individual allocation to non-Annex I Parties could be allocated according to two principles: (1) A per-capita survival emission could be determined as a share of the 2050 estimated per-capita emission. For low-emitting countries, this will be equivalent to a growth cap. (2) For the first commitment period, stabilization on the estimated 2010 level could be envisaged. For fast-growing economies above the survival emissions, this will require participation in the auctioning system to the extend they overshoot their stabilization level. In any case, the preservation of terrestrial carbon stocks will help developing countries offset their own emissions and save on f-AAUs. CDM and Joint Implementation (JI) activities will converge, because the system is now closed. This means that the host country will have an eye on the additionality of mitigation activities, because non-additional activities will be financed out of the host country's f-AAU allocation. Seen from another angle; project-based activities will allow developing countries to convert f-AAUs into freely tradable emission permits.

Sharing the same budget however means that common accounting procedures are needed. An end to the split treatment of land use sinks and sources for Annex I and non-Annex I countries is overdue. Adapting Annex I land use accounting (like foreseen in Kyoto Protocol Article 3 paragraphs 3 and 4) for developing country Parties (Dutschke & Pistorius submitted) may be a common ground.

The proposal of the Climate Stabilization Fund is both simple and commonsense. It does not disrupt existing flexible compliance mechanisms, closes the divide between industrialized and developing countries, reduces international leakage, and makes bargaining around national targets for Annex I obsolete.

Allowance banking is a central feature of the Kyoto Protocol compliance system that needs to be preserved in order for the Climate Stabilization Fund to work. It offers an incentive for developing countries to bank their f-AAUs towards compliance with tighter future targets. The value of f-AAUs will be the one of AAUs, discounted by the number of years, before they will be used for compliance. REDD activities undertaken on a national or subnational level will increase the number of bankable f-AAUs. This means that on the long run, carbon benefits from forestry will enter the emissions trading system.

8. Variations and further research

The proposed model offers flexibility for possible adaptations and variations. For its functioning, it is unimportant whether the long-term budget is binding or non-binding. In order to secure environmental integrity however, it is recommended to change the long-term budget only in case new scientific evidence is detected by the IPCC, similarly to the treatment of the global warming potential of different GHGs, which is now beyond reach of political negotiation.

The feature of non-tradability of f-AAUs is intended to safeguard survival emissions and to avoiding industrialized countries from buying up freely allocated amounts without any underlying mitigation activity. This regulation could be loosened in the sense that trading is allowed between developing country Parties, as long as these are entitled to receive f-AAUs. The advantage would be that the present value of f-AAUs would increase and so would the pressure for carbon-efficiency.

It is conceivable that after one or two commitment periods of financing REDD through the Climate Stabilization Fund, the compliance market is opened up to REDD credits. In this case, more finance would become available for adaptation and other common tasks that are non-quantifiable in carbon metrics.

Starting AAU auctions three years before the commitment period for which they are issued is an elegant solution to the need of upfront finance in REDD, besides assisting country parties in mid-term carbon planning. Again, the Climate Stabilization Fund does not depend from the moment in time when allocation takes place, as long as it is *before* the respective commitment period.

As the actual debate around REDD is highly polarized between market and non-market instruments, the Climate Stabilization Fund should not be confused with a non-market solution. The scope of the present article was on how to collect the amount needed for REDD. Spending should be based on a competitive tendering system, in order to identify the low-cost opportunities first. It is conceivable that for equity considerations, the distribution of funds among world regions could be politically predetermined.

Further research should look into indicators for gradual phase-out of f-AAUs for advanced developing countries and the distribution of the Climate Stabilization Fund. There should be limits to arbitrariness which goals to finance with it, and how to divide the cake.

With the focus of this paper being on forestry, the author encourages further debate on potential co-benefits of the Climate Stabilization Fund model.

9. Acknowledgements

The author wants to express his gratitude for comments and suggestions received by Axel Michaelowa and Bill Hare.

References

Athanasiou, T., Kartha, S. et al. (2006), Greenhouse Development Rights. An approach to the global climate regime that takes climate protection seriously while also preserving the right to human development. United States, EcoEquity, San Francisco, CA (United States); EcoEquity, Earth Island Institute, San Francisco, CA (United States); Christian Aid, London (United Kingdom); Stockholm Environment Institute SEI, Stockholm (Sweden), p. 10.

Baer, P., Athanasiou, T. et al. (2007), The right to development in a climate constrained world. Berlin/DE, Heinrich Böll Foundation, ChrstianAid, Stockholm Environment Institute, p. 100.

Barnett, J. (2001), The Issue of 'Adverse Effects and the Impacts of Response Measures' in the UNFCCC. Tyndall Centre for Climate Change Resarch Working Paper 5. Norwich/UK, p. 29.

Baumert, K.A., Herzog, T. et al. (2005), Navigating the Numbers: Greenhouse Gas Data and International Climate Policy. Washington D.C., World Resource Institute, p. 132.

Black, R. (2008), US to set 'binding' climate goals. B. N. Website. London, UK, BBC News http://news.bbc.co.uk/go/pr/fr/-/2/hi/science/nature/7263225.stm.

Dutschke, M. (2007), "CDM Forestry and the Ultimate Objective of the Climate Convention." Mitigation and Adaptation Strategies for Global Change 12(2).

Dutschke, M. and Michaelowa, A. (2006), "Development assistance and the CDM – how to interpret "financial additionality." Environment and Development Economics 11(2), pp. 235-246.

Dutschke, M. and Pistorius, T. (submitted). "Will the future be REDD? Consistent carbon accounting for land use." International Forestry Review.

EU Commission (2008), Proposal for a Directive of the European Parliament and of the Council amending Directive 2003/87/EC so as to improve and extend the greenhouse gas emission allowance trading system of the Community (presented by the Commission). 2008/0013 (COD) COM(2008) 16 final.

Fisher, B.S., Nakicenovic, N. et al. (2007), Issues related to mitigation in the long term context. Climate Change 2007: Mitigation. Contribution of Working Group III to the Forth Assessment Report of the Inter-governmental Panel on Climate Change. In: Metz, B., Davidson, O.R., Bosch, P.R., Dave, R., Meyer, L.A. Cambridge, UK and New York, US, Cambridge University Press. III, pp. 169-250.

Fuentes, U. (2008), Innovative Finance Mechanisms – The German International Climate Initiative. A. I. E.G. Seminar. Paris, German Ministry for the Environment, Nature Conservation and Nuclear Safety.

G8 (2008), Environment and Climate Change. G8 Summit 2008. Hokkaido, JP, G8: 6 www.g8.utoronto.ca/summit/2008hokkaido/2008-climate.html.

Gullison, R.E., Frumhoff, P.C. et al. (2007), "Tropical Forests and Climate Policy." Science 316, pp. 985-986.

Hare, B. and Macey, K. (2007), Tropical Deforestation Emission Reduction Mechanism. A Discussion Paper, Greenpeace International: 52 www.greenpeace.org/raw/content/international/press/reports/TDERM-Exec.pdf.

Jung, M., Michaelowa, A. et al. (2004), Common Policy on Climate Change: Carbon Sinks, Domestic Stakeholders, and E.U. Foreign Policy. Global Warming and European Foreign Policy. P.G. Harris. New York, St. Martin's Press.

Michaelowa, A., Butzengeiger, S. et al. (2005), "Graduation and Deepening: An Ambitious Post-2012 Climate Policy Scenario." International Environmental Agreements 5(1), pp. 25-46(22).

Nabuurs, G.J., Masera, O. et al. (2007), Forestry In Climate Change 2007: Mitigation. Contribution of Working Group III. Forth Assessment Report of the Intergovernmental Panel on Climate Change. In: Metz, B., Davidson, O.R., Bosch, P.R., Dave, R. and Meyer, L.A. Cambridge, UK and New York, US, Cambridge University Press. III, pp. 541-584.

Ogonowski, M., Helme, N. et al. (2007), Reducing Emissions from Deforestation and Degradation: The Dual Markets Approach. International Future Action Dialogue Washington D.C., p. 20.

Piris Cabezas, P. and Keohane, N. (2008), Reducing Emissions from Deforestation and Degradation in Developing Countries (REDD): Implications for the Carbon Market. Washington D.C., Environmental Defense Fund, p. 13.

Rudel, T.K., Coomes, O.T. et al. (2005), "Forest transitions: towards a global understanding of land use change." Global Environmental Change (15), pp. 23-31.

Stern, N. (2008), Key Elements of a Global Deal on Climate Change. London/UK, London School of Economics, p. 56.

Chapter 6

Social Learning on Climate Change among Regional Agents – Insights from China, Eastern Europe and Iberia

J. David Tàbara, Francesc Cots, Xingang Dai, Maria Falaleeva, Zsuzsanna Flachner, Darryn McEvoy, Saskia Werners

Abstract

The building of individual capacities and institutional arrangements to adapt and mitigate climate change can be understood as a process of social learning. From empirical evidence gathered within the EU project ADAM (www.adamproject.eu) we look at climate impacts and responses in form of new social-ecological arrangements, capacities, and agents' transformations derived from increasing climate awareness, risks and opportunities in three regions of the world: 1. The Alxa league, Inner Mongolia, China 2. The Tizsa flood plain, Hungary and 3. The Guadiana river basin, Iberia. Several key factors appear to constrain or enhance *climate social learning* among agents at the regional level, and in particular: (a) the type and resources for multi-level and cross-sector interaction with empowered agents (b) the institutional and individual capacity of becoming aware, anticipate, and control the negative personal and system's effects of continuing with the existing patterns of social-ecological interaction and development (c) the available degrees of freedom to modify ones' practices and social-ecological interactions while creating a better-off situation from such transformations. In the three regions, climate has become an overarching rationalising discourse of many disparate biophysical changes and economic opportunities which triggers to new climate related actions in a reframing and social learning mode.

1. Introduction

Human societies are increasingly becoming aware of the need to adapt and to prevent the negative side effects of changes occurring to a new environment created by themselves (Folke et al. 2005; Murphy 2002). The making of robust long-term climate policies at local and regional levels requires the development of integrated processes capable to mainstream climate concerns into the implementation of policies in multiple domains and scales. While it is true that mitigation and adaptation respond to different concern, trade-offs and problem structures, the making of long term global policies which are coherent at the local and regional level demand certain degree of integration and coordination between the two. It requires new institutional designs that contribute to the reframing of the current global political situation and development priorities in a way that are able to empower agents and organisations at the very local level. This entails improving the understanding of the factors that condition the opportunities for agent and institutional collaboration and engagement to support social-ecological transformation and/or in some cases, to help protecting existing institutional arrangements. In this chapter, we refer to such processes of agent and institutional reconfiguration derived from a conscious awareness and willingness to act and deal with the common problem of global warming as contributing to *climate learning*. In order to specify the concept of climate learning we use the general approach to social learning as conceptualised in Palh-Wostl et al. (2007) and Mostert et al. (2007). For *regional agents* we understand those individuals, organisations and networks of actors capable of modifying existing systems of rules of interaction and behaviour, known as institutions, which mostly operate at the meso level but which also maintain important links at the local and also at the global context. In particular, our regional agents are empowered actors and stakeholders able respond and influence the social and political structure and the environmental context in which they act regarding adaptation and mitigation to climate change[1].

The increasing penetration of the climate discourses in many development and sectoral policies both in Europe and beyond does not always respond to explicit and orchestrated public climate strategies but is often the result of a combination of multiple and sometimes autonomous factors, many of them relating to the actions taken by non-state agents. How people get involved in climate related actions in the first place, how they learn to cope with climate change in their own local or regional contexts and what drives their learning processes is still poorly understood. As we gain deeper insights about the nature of the impacts, the vulnerability of the elements at risk, the causes and possible options to deal with climate change, our representation of the system of concern and of our relationships with it also changes. The development and implementation of feasible, effective and equitable climate options in particular places is dependent on our capacity to integrate the insights of multidisciplinary science, albeit other experiences and non-expert sources of knowledge need to be taken into account. Thus, learning, in this domain can be understood as continuous process of awareness raising, stakeholder collaboration and agent transformation in a mode that is able to put in place new ways of working between science, policy and the wider stakeholder communities and create more robust institutional designs to adapt to and mitigate the risks of climate change.

In this research, in order to know what learning means in the context of local and regional climate change we look at the social and ecological commonalities and

[1] For further discussions on agency (the more dynamic part of society) and structure (the more permanent one) in policy sciences see Dessler (1989), and on the debate on the multi-level governance of climate change see Gupta (2007a; 2007b).

contrasts encountered in the making of climate related policies and strategies in three distinct regions of the world: the Alxa league, located in Inner Mongolia (China), the Tisza floodplain (Hungary), and the Guadiana river basin (Iberian Peninsula). Populations living in these areas share in common that they are increasingly vulnerable to climate related pressures, including mounting desertification, water stress, and risks of occurrence of extreme events such as floods and heat waves. However, the institutional contexts and the governance traditions upon which societal responses and long-term policies are developed differ greatly in these three cases. Our research examines the impacts and the triggers of adaptive and mitigation responses, including the making of climate appraisals at the regional level from a comparative, empirical and analytical perspective. Our work is part of the EU project ADAM (2006-2009) on Adaptation and Mitigation Strategies. A methodology, called *Policy Appraisal Framework* (PAF; Weaver et al. 2006), was developed to support and assess in an integrated and social-ecological robust way the appraisal processes in each region. This methodology is used as a heuristic devise to examine the factors that influence the way regional climate appraising processes are carried out and how they can be improved in a social learning and reframing mode. Such a methodology is intended to constitute a general and flexible guidance – and not a rigid yardstick – for climate researchers and practitioners who want to evaluate and reflect onto what extent the existing climate appraising and policy processes yield socially and ecologically robust climate options in specific contexts of action. For 'socially and ecologically robust' we understand that the processes and outcomes of the appraisals deem and actually correspond to the particularities and conditions of the social and ecological systems in which the assessment takes place. In the quest for producing 'climate assessments in context', our research looks at issues such as how multiple sources of knowledge are integrated for what purposes in particular regions and to what extent the existing processes climate appraisal contribute to agent engagement, collaboration and transformation. We argue that in the long-term, enhancing, protecting and integrating cultural and institutional diversities and the corresponding biological diversity of the social-ecological systems of reference of the agents involved, may be one of the most socially and ecologically robust strategies to mitigate and adapt to the risks of global warming and thus may be as one of the most critical components of social learning on climate change.

2. Social learning on climate change

In order to address the issues of whether and how learning occurs in the context of climate change at the regional and local levels we use the general concept and approach devised within the EU project Harmonicop (2002-2005; www.harmonicop.info) and specified in Pahl-Wostl et al. (2007) and Mostert et al. (2007). According to this approach, social learning can be defined as the process by which agents and organisations continuously frame and reframe the issues at stake and develop enhanced content and relational capabilities to deal with common problems which individuals cannot resolve on their own. Social learning goes beyond the ideas of learning technical capabilities and entails the development of capacities to change – or in some cases, protect – particular social and ecological *relationships* and institutional arrangements. Thus social learning entails not only increased abilities to deploy more means – mainly to do the same, but being able to question, modify and attain new collective goals to do something different. As shown in figure one, social learning can simply be depicted as occurring at the interplay between: 1. A *context,* formed by a given governance structure and a biophysical system 2. A *process* of social interaction, which develops from a set of relational practices, and 3. A series of *outcomes* which feed back again into and affect the original context.

Fig. 1: The social learning concept as specified in the EU project Harmonicop

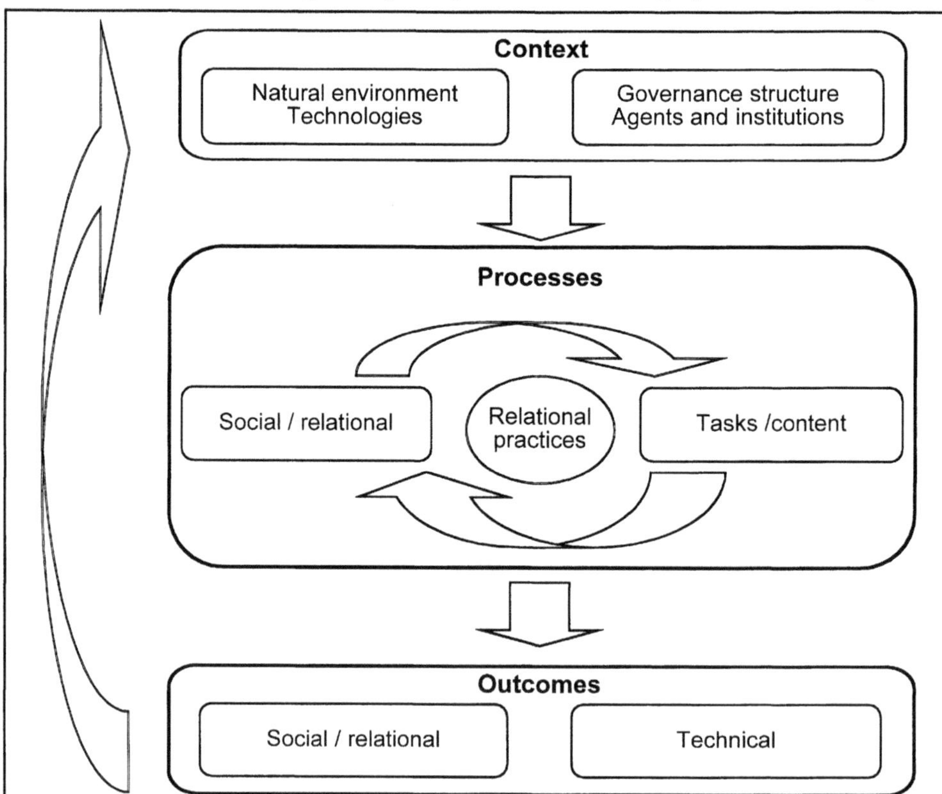

Within the Harmonicop project, is was emphasised that social learning is affected by: (a) how the *framing and reframing* of the problems at stake takes place in a particular context, (b) how the management of boundaries affects who

participates (or not) in the processes carried out, (c) the type of *negotiation strategies* involved, (d) the kind of *ground rules* established in the processes of interaction, (e) the type and strength of the leadership that exists in the coordination of collective action and (f) how resources are distributed among different agents. Therefore, a first step in any process of social learning is to be aware that a common problem, interest, or opportunity exists between a given group of people that needs collective action. This can be enhanced by processes of system representation and integration of plural sources of knowledge (including the use of computer and expert tools and other integrated deliberative procedures). But while collaboration between agents constitutes a necessary condition for social learning, it alone may not lead to the transformation of the original social-ecological context. For the eventual success of social learning – in terms of improving capacities to deal with such common problem – will always depend on to the extent agents are willing to transform themselves and the relationships that they maintain between each other and between the biophysical system of reference accordingly (figure 2; Tàbara 2008).

Fig. 2: Social learning as a continuous process of perception and system awareness, of agents' collaboration, and agents' transformation.

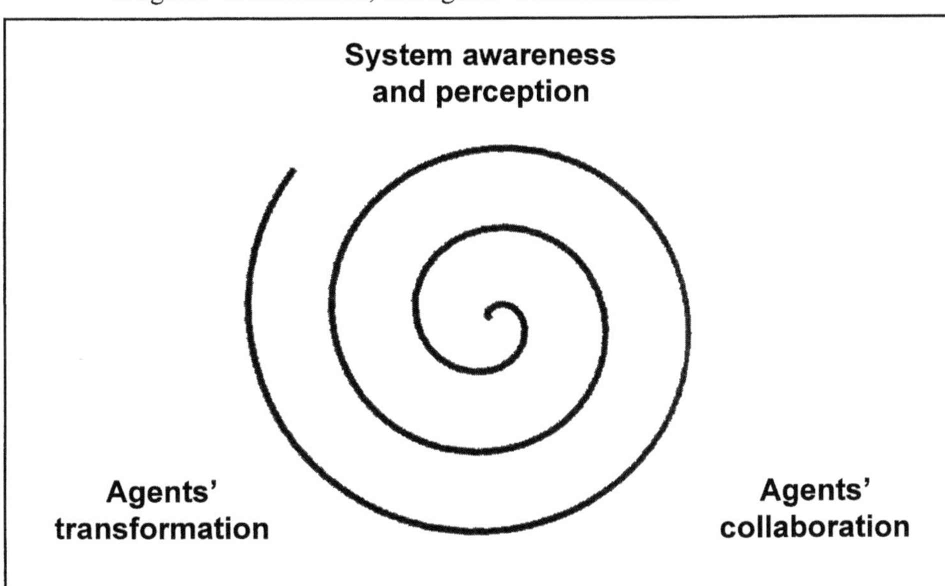

In the present chapter we start from these insights and look at the actual processes of learning within the climate domain. Our focus is not so much at the global scale in the form of how new policy architectures are being implemented (Siebenhüner 2006) but rather at how local and regional agents organise and interact with other agents operating at different scales and domains to cope with or take advantage of the new climatic conditions (see Pelling et al. 2008). Our empirical work was carried in the form of workshops, in-depth interviews and documentary research and was carried in three selected regions: 1. The Alxa league, located in Inner Mongolia, China, 2. The Tisza floodplain in Hungary, and 3. the Guadiana river basin, south of the Iberian Peninsula, between Spain and Portugal. Specially, our comparative analysis intends to address the following questions:

1. To what extent is climate affecting both the environmental and institutional *contexts* in the three regions, and to what extent those changes are the result of conscious collective *processes* aimed at mitigating and adapting to climate change?
2. What are the main triggers and driving forces that have led to the development of such processes and to the implementation of regional climate policies and actions? To what extent are these actions actually related to a conscious and explicit regional and national strategy that aims at dealing with climate change? What are the barriers to change?
3. What are the main impacts of climate change, what are the key vulnerabilities in each of the case studies, and what opportunities are available to respond?
4. What new capacities and transformations can be observed on specific agents and regional institutional arrangements to cope with climate change in a fashion that can be referred to as processes of *social learning*?

With these questions and framework in mind, we now proceed to examine the particular impacts, responses and triggers of existing regional policies and climate-related actions and the process of reframing previous non-climate related issues.

2. The Alxa and the Inner Mongolia, China

2.1 Context

Our research area, the Alxa sub-region, constitutes a vast expanse of arid and semi-arid land of 270.000 km^2 situated in the Autonomous Region of Inner Mongolia, Northern China. With a population of only 0,2 million people, it is crossed by three of the largest deserts within Western Inner Mongolia. This area is one of the most important origins of China sandstorms, the intensity and scale of which can even be felt in Beijing and cities in Japan and South Korea (there are even accounts of sands having reached to North America). Desertification is a growing concern in China, where one third of its territory is already affected by land degradation. Being deserts fragile and very sensitive to perturbations, climate has become an additional stress factor to these already degraded environments, which already have suffered from decades of overgrazing and terracing (Shapiro 2001).

China is increasingly taking part of global climate politics despite originally having been being part of the non-annex parties of the Convention of Climate Change (PRC 2004). An important aspect of this international collaboration is related to it increasing participation in the Clean Development Mechanisms. Of particular note in the Chinese climate policy context was the launch of the first national climate change plan in June 2007 (National Development and Reform Commission, People's Republic of China, 2007). This new legislation puts forward three major targets to be achieved by 2010 – reducing energy consumption by 20%, increasing renewable energy to 10% of primary energy supply, and increasing reforestation by 20%. It also highlights some of the key sectors and thematic areas for adaptation (agriculture, forestry and water in particular), with a stated goal of restoring 52 million hectares of grassland previously subject to desertification by 2010. The National Coordination Committee on Climate Change will oversee activity, formulating policies and measures and providing guidance for local responses to climate change[2]. However, it is still uncertain what is going to be the eventual contribution of China in the reduction of CO_2 emissions, although whatever it will be it will affect the global environment (Congjie & Dongping 2007; Day 2004; Economy 2004; Ho 2006).

To gain experience of some of the issues, and to gain an understanding of how local perceive the likely impacts of climate change, workshops and a field trip were organised by the Chinese Academy of Sciences in May-June 2007. These provided valuable opportunities for EU researchers working on the ADAM project to consult with local stakeholders and politicians and to identify the main options regarding climate adaptation and mitigation in this region (ADAM 2007a).

2.2 Processes

During the last few decades, a series of actions and policy measures have been carried out in the Inner Mongolia region which claim to be taken to cope with the new climate risks. These include: fencing off forest and desert grazing areas (Rogers et al. 2006; Yonghuan et al. 2006; Jiang et al. 2006), aerial seeding and afforestation, changing in the composition of livestock, rainwater harvesting (by means of collecting water in underground pools), and developing new forms of sand tourism to compensate for the reduced income from traditional farming activities. In the unique, forested and isolated mountain range of the Hellan Mountains, surrounded by three deserts and threatened by reduced precipitation, besides enclosure of grazing areas, artificial rainwater has been tested and argued as one of the best options to save this vulnerable ecosystem. Given the mounting *aridification* of the area, there are growing conflicts for the use of water resources, mainly along the Heine river. In this regard, some of the 'adaptation measures' are now directed to

2 For a summary, see: http://mt.china-embassy.org/eng/xwdt/t340287.htm.

extending the building of water infrastructures, such as reservoirs and water transfers, in an attempt to bring water and new development opportunities to areas where water resources are becoming increasingly precious and scarce (ADAM 2007b).

From the mitigation side, and more generally within the whole Inner Mongolia region, the potential for reducing GHG emissions is enormous, given the number of large coal power stations now in operation and planned (Aden 2006). Large international energy investors, mostly in the field of the development of Carbon Capture and Storage (CCS) and Integrated Gasification Combined Cycle (IGCC) technologies are now turning their attention into China. However, although local and regional agents can play an important role in mitigation strategies in this sector, the goals and the means to achieve these goals are not so much dependent of local purposes or actions, but much more related to the wider national and international strategies on development and climate change.

2.3 Results

Looking at the biophysical effects, and as part of a massive reforestation programme, ongoing activities include the creation of a 'Great Green Wall' of trees and plants stretching over Northern China including Inner Mongolia (Three North Forest Shelterbelt Program), and in Inner Mongolia itself there is a grain for green program subsidising tree and grass planting. Tree planting is seen as a major focus of addressing the desertification problem and can even involve other neighbouring countries (for instance, Japan sends volunteers each year to help with this restoration activity). Haloxylon woodlands in particular have been identified as one of the best windbreak and sand-fixing types of vegetation (Dai et al. 2007), however, evidence shows that the coverage of this tree line area in the Alxa League has shrunk from 1,133,000 hectares to 384,500 hectares since the 1960s. This said, the effectiveness of ongoing tree planting campaigns is not clear, as evidence from the ADAM fieldtrip suggested that in some cases trees were dying after being planted too close together (and hence competing for limited water resource). This showing that some of selected species are not the appropriate for the increasing and extreme arid conditions, which have revealed, among others, a drop in the groundwater table levels from 2 meters to 4 meters during the last 50 years (ADAM 2007).

The rising awareness of climate impacts and of its opportunities for development reform, as well as new internal and external political and civic pressures are beginning to produce some changes both in the Chinese environmental science and policy contexts (Heggelund & Backer 2007; Teng & Alun 2007). China counts with a notable tradition of atmospheric, terrestrial and climate monitoring, and the participation of Chinese scientists has been crucial in many international research programmes regarding global change (NRC 1992). In the present situation, scientists seem to be benefiting from the current wave of interest in the development of new climate international research networks and cooperation. In particular, the role played by transnational epistemic communities, more than NGOs, has been underlined as a key factor giving salience to the climate politics in China (Schroeder 2007). New scientific concepts and insights coming from environmental science and knowledge from ecology seem to start informing policy reforms in this domain. It is likely that concepts such as desert management following an integrated ecosystems approach and others, will gradually be used to reorient current land use and natural resource practices, in a way some authors on ecological modernisation, not without criticism, argue to have begun to penetrate in China in other domains (Huan 2007; Carter & Mol 2007; Zhang et al. 2007).

Nevertheless, the implementation of measures such as the grassland enclosure and resettlement have not free of questioning both on social and ecological grounds. For instance, some authors argue that some of these policies do not necessary guarantee that all resettled people are now better off than before (Jiang 2006; Rogers & Wang 2006; Webber & McDonald 2004). Herdsmen who have been moved to different areas now follow modernisation processes that entail adopting lifestyles radically different from their original or traditional practices and identities and that there are risks of social disarticulation; and where grazing has been almost totally eradicated, new environmental problems may emerge such grass overgrowth which may lead to greater fire risks. In this line, Williams (1996, 2002) has argued that in Inner Mongolia fencing has been an expression of the current move towards increasing land privatisation and more a catalyst of land degradation than its solution. For him, in some cases such measures have not increased the trust of local communities or reduced the political fear to existing political institutions; nor local people have not been better consulted and engaged in this move towards privatisation than when former collectivization processes were imposed. However, this focus on technological fixes aimed at expanding water supply instead on promoting agent-based social learning, as it has been widely recognised from the findings on integrated water resource management tend to increase demand and in the long term aggravate the persistence of the problems to be tackle and the non-adaptability of the agents.

Therefore, and in line to those strands of thought that distinguish between development as a outcome and development as process (Dickinson & Weber 2007), it is clear that the main political climate challenge at regional and local levels go beyond determining what particular measures need to be implemented that accommodate to the short term goals of property regime reform and economic growth. Rather, the main questions to be tackled relate to deciding what appropriate policy appraising and decision-making *processes* can be developed as to produce equitable, and socially and ecologically robust adaptation and mitigation results compatible with the long-term quest of sustainable development (Lemos et al. 2007). This necessarily entails processes that explicitly aim at protecting not only grasslands, but also enhancing the whole social conditions and traditional knowledges and communities that can potentially contribute to a sustainable use of resources, much of them embedded and expressed in cultural and institutional diversity (Becker & Ostrom 1995; Ostrom 2005; Tàbara et al. 2004). While many current policies aimed at alleviating poverty and restoring grassland translate in an augment in the number of populations being dependent on or participating in larger economic markets – thus with a higher contribution GHG emissions –, it is also important to consider to what extent the (often more adapted) traditional cultures and lifestyles can also contribute to mitigate or reduce the need of adopting new carbon-intensive lifestyles worldwide.

3. The Tisza flood plain, Hungary

3.1 Context

The Tisza River is the largest tributary of the Danube, receiving water from the Carpathian Mountains in Romania, Slovakia and Ukraine. Within Europe, the Tisza is unique in terms of wetlands and conservation areas, albeit its much regulated river bed and frequent floods. Until the 18th century the Tisza floodplain area was utilized in a complex way providing secure income for the communities along the river (Bellon 2004). Activities were mainly organized around the operation of a system of small streams and channels between the river main bed and the floodplain, cutting through the natural levee (Balogh et al. 2000). Since then the Tisza River has been heavily modified. River regulation has reduced river length by 40% and the floodplain area by 80% (Szlávik 2003). To cater for large-scale mono-agriculture and river transport, the river was canalised and straightened and the floodplains drained. The major changes of the Tisza River were introduced by the Vásárhelyi Plan initiated in the 19th century. These changes meant the destruction of the traditional water management system and the social and ecological decline of the region. Next to the flooding there are other problems that need to be managed such as draught, water stagnation and the degradation of peat lands and wetlands. The area also suffers from socio-economic deprivation, with high unemployment rate, ageing, migration and minority rights problems (Sendzimir et al. 2004; Linnerooth-Bayer et al. 2006; Flachner et al. 2004). At the same time the region has a large undeveloped potential for recreation and nature conservation (Vári et al. 2003).

The main observed signs of climate change are variation of precipitation and a slight increase in temperature. In the Tisza region, these result in a higher frequency and intensity of extreme floods and droughts. Between 1998 and 2006 there has been at least one severe flood each year, of a magnitude that previously happened once per 100 years (Timár & Rácz 2002). At the same time, the plains between Danube and Tisza are especially drought prone. This condition has been a research interest of several Hungarian scientists, investigating whether climate change can explain the droughts (Footitt & McKenzie Herder 2007). The predictions suggest more irregular rainfall and a warmer climate in the Carpathian basin (Lang 2006). Climate change is connected to the three main water related problems of the Tisza regions 1. floods; 2. in-land water stagnation; 3. droughts.

In 2004 Hungary was one of the former socialist countries in Eastern Europe to enter the European Union. During socialist rule a strong central government was established. European Union accession and the accelerated implementation of European directives and legislation (e.g. the common agricultural policy, water framework directive and Natura 2000) have added to the complexity of governance responsibilities and financial flows. Divergence in the objectives and mandates of organisations slows down the implementation of cross-sectoral initiatives. The severe budget deficit forces the Hungarian government to cut expenses. At the same time, European standards and practices for agriculture and water management offer an opportunity for climate learning and to make existing governance and planning practises more effective, sustainable (Commission of the European Communities 2001).

3.2 Processes

The recurring flooding events and the recent predictions of an increased incidence of floods and droughts led to the need of designing a new water management plan for the Hungarian Upper Tisza River: the New Vásárhelyi Plan (VTT plan). This plan has bee regarded as the main mechanism for mainstreaming climate adaptation in the Hungarian Tisza region. The plan's main aims are to reduce flood risk in combination with nature conservation and rural development. It identified micro regions for water retention (flood polders and formerly used small water bodies) and enacts the revitalization of floodplains as the preferred adaptation strategy in integrated floodplain management and rural development (Adam 2007c; Matczak et al. 2007; Werners et al. 2007). The first version of the plan in 2000 focussed on flood defence and had mainly been the work of the water authorities. In autumn 2002 a new NGO Bokartisz Public Utility (www.bokartisz.hu) in collaboration with other civic organizations along the Tisza asked for a review of the plan. At the same time they promoted their own on shallow flooding based floodplain rehabilitation concept to replace or complement the classical flood protection measures in the first version of the plan (Bokartisz 2002). This more integrated approach promised more long term benefits for the Tisza region (Flachner et al. 2004). In February 2003 the national government ordered by decree (1022/2003) that in the further development of the water plan flood retention reservoirs were to be combined with floodplain rehabilitation, rural development and nature conservation (VITUKI 2004). A large group of regional and national actors collaboratively prepared the implementation of the new water plan. Floodplain areas and smaller water bodies were identified for water retention and rehabilitation together with rural development. Major decisions were taken at the inter-ministerial committee led by the Ministry of Environment and Water where representatives of the region and of civic organizations were invited. However, the implementation of the plan does not comply with the initial timing and has been partially delayed (National Development Agency 2005).

A series of interviews conducted in the framework of the ADAM project focussed on existing challenges facing implementation of adaptive floodplain management in the Tisza area and the possibilities for enhancing capacities for climate adaptation in the region. The results of this study were discussed by stakeholders in a series of visioning workshops. As a result, formal relationships, institutions and participation has been progressively being redefined. Actors called for clarifying the roles of non-state actors and expressed a particularly strong interest in establishing the so-called agencies beyond the state (Matczak et al. 2007). In particular, the Tisza regional case identified various responses to increasing climate pressures: 1. Re-evaluation and modification of the institutional and legislative framework to support adaptation and mitigation through climate proof floodplain management as proposed in the new water management plan, 2. Supporting land use and water management that increases the buffering capacity of the region and the adaptive capacity of its inhabitants 3.Encouraging local markets, investment and awareness raising.

3.3 Results

Of particular note have been the international research and regional development projects which have increasingly offered the necessary space to reflect on the processes and problems of implementation of climate adaptation strategies in the region. For instance, in 2007 the VAHAVA project generated significant interest in coping with the impacts of climate change at the national level. This was followed by the KLIMAKKT national research project 'Preparation for climate change: environment – risk – society'. The latter project aimed to develop tools for a more efficient adaptation to climate change in the field of environment, agriculture, hydrology and hazard management. The first results were obtained in crop production, a most vulnerable field of Hungarian economy. In addition the VAHAVA project resulted in a breakthrough at governmental level: the coordinator was invited to initiate the development of the climate change strategy. Awareness raising and dissemination activities on the importance of adaptation policies and measures are key tasks in Hungary to support the development of the national climate change strategy. Whereas the VAHAVA and KLIMAKKT project focus primarily on adaptation, the national climate change strategy covers both adaptation and mitigation.

The national process of developing the climate strategy (NCCS) in Hungary is supported scientifically by the VAHAVA and KLIMAKKT national projects. The first draft of the NCCS was analysed and reviewed by NGOs and scientists. To prepare for climate change, actors suggested an informed multi-stakeholder approach (e.g.

dialogues) at national, basin and regional level to prepare action plans for adaptation and to strengthen regional and local representation (e.g. after the Bokartisz Public Utility). Nation wide an "Hungarian Water and Climate Alliance" could be established as a national umbrella to continue building bridges between the climate and water sector, encourage capacity development to better cope with climate impacts and facilitate obtaining financial support for national, basin and regional level adaptation plans. Consensus was been reached on the improvement of the national climate change strategy. In February 2008 the government approved the strategy, which opened further possibilities to mainstream climate change into relevant sectoral policies, such as the new water management plan. In particular, the main elements of the implementation of the national climate change strategy are:

- two-yearly national program for climate change strategies
- coordinated by the ministry responsible for the environment
- new advisory body: the committee on climate change
- new inter-sectoral coordinating body: working group on climate changes (Láng 2006).

In short, the progressive penetration of the climate change framing and discourse into regional development and the VTT management plan led to the following results:

- Establishment and operation of a regional platform for harmonization of knowledge, actions and policies at different levels (Élő Tisza Association – Tisza Alliance) to develop the institutional capacities to implement complex program at Tisza valley;
- Establishment of micro grant scheme to support small-scale innovations (e.g. local oil press to reduce energy cost for farmers and increase biofuel use);
- Support the national climate strategy approval (February 2008) which provided the framework for an improved harmonization among sectoral policies, regional floodplain management and other Tisza Alliance activities;
- Modelling of land use and water system development – defining alternatives for flood protection and land use changes (ARES project) with special attention to combined intensive and extensive agriculture;
- Recent initiatives by the Tisza Alliance to put together a catalogue of small scale adaptation options for the region;
- An offer by the Hungarian Ministry of Regional Development to prepare a handbook for climate change adaptation and mitigation
- In line with the Tisza valley long term vision program (2008-2009) climate change mitigation and adaptation options are now discussed by local, regional and national stakeholders. The process provides an opportunity to link the different decision making levels and increase political and public awareness.

4. Guadiana river basin

4.1 Context

The Guadiana river basin lies between Spain and Portugal and constitutes one of the three main drainage units of the Iberian Peninsula, with a total drainage area of 66,800 km^2. Its climate is semi-arid with low irregular precipitation (440mm/year; CHG 2000). Strong modification of the natural hydrological regime, in the form of dams, illegal dwells and increasing urbanization pressures in the last four decades have caused increased water scarcity along the Portuguese-Spanish border. The main problems within the river basin are the overexploitation of the aquifers through the large extractions for agricultural use, agricultural pollution, and river ecosystem fragmentation by dams (Cosme & Sousa et al. 2003; WWF 2003b). The Portuguese part of the basin shows a less a structured economy and includes some of the poorest municipalities in the country, with an ageing population and unemployment. Portuguese authorities often associate the deprived economic indicators to the lack of water availability (WWF 2003; Costejà & Maestu 2005). Other processes altering the river basin include the building of large urban and leisure developments for tourism, which originally used only to be concentrated near coastal areas, now also spread inland. Climate change potentially increases the pressure on the available water resources and thus the potential conflicts between agriculture/irrigation, tourism, the construction of large urbanisations and the environment. Current groundwater reserves are already threatened by salinisation and overexploitation, due to an increase in the demand of water and a decrease of the supply.

The Guadiana river basin constitutes a natural unit but not an administrative one. Administratively, in Spain, the Guadiana crosses various Autonomous Communities – Extremadura, Castilla-La Mancha and Andalucía – being specially represented the provinces of Badajoz, Ciudad Real and Huelva. In Portugal, the Plano de Bacia Hidrográfica do Rio Guadiana (PBHG) is administratively divided between the Alentejo region and the Algarve region. Given the growing water shortages, summer droughts and spreading desertification both regional and national political institutions in the two countries are beginning to realise the need to mainstream climate change into development policies and to carry out a more integrated management. An ADAM workshop carried out with stakeholders in the Lower Guadiana, in Portugal in December 2006 helped to identify some of the main impacts, responses and drivers of climate related actions (ADAM 2007d; 2007e; Moreno Rodríguez 2005; Santos et al. 2002; Tàbara 2007).

4.2 Processes

Portugal and Spain being part of the EU, the main processes regarding climate change in the Guadiana river basin are mostly driven or highly conditioned by the implementation of the European policies. The European Climate Change

Programme (ECCP) I and II together with the EU commitments taken before the UNFCCC after Kioto have become the overarching framework in which the following national instruments have been developed: 1. The Portuguese National Climate Change Programme (2001), 2. The Portuguese Strategy on Climate Change (2001) 3. The Spanish Strategy on Climate Change and New Energies (2007), and 4. The Spanish National Adaptation Plan (2006). During this period, several national bodies have been created in both countries to implement climate change policies, including Portuguese Climate Change Commission, the Spanish Inter-ministerial Group in Climate Change, the Climate Change Policies Coordination Commission, the Spanish Bureau of Climate Change and the National Council on Climate Change. Among their duties are to promote and facilitate the climate change policy across the Government bodies with relevant competences and ensure that these issues are considered in the full range of sectoral policies. Changes regarding climate institutional arrangements capacities in both countries and in particular at river basin scale can be observed although still in a rather underdeveloped stage (McEvoy et al. 2007).

The EU policy on Structural Funds have also impacted on the governance structure of the Guadiana river Basin. The European Commission has used successive reforms of Structural Funds to condition the activity of sub-estate actors and they have responded taken on a more active role. By means of initiatives such as INTERREG, the European Commission has promoted the setting up of transnational networks focusing on the definition of common interests and the intensification of horizontal co-operation among the substate actors belonging to different Member-States. Thanks to this type of initiatives, the regions have gradually developed a more important role both in the European arena and in national decision-making and policy-implementing spheres (Jones & Keating 1995). Today, regions interact directly with the EU bodies to obtain resources from the distributive policies and further their interests. Due to the transboundary nature of the Guadiana river Basin, important parts of it constitute eligible zones for cross border initiatives and projects under INTERREG. In order to promote cross border cooperation in the area, the following institutional agreements have been adopted:

1. Agreement between CCR Alentejo and Junta de Extremadura (1992);
2. Agreement between CCR Algarbe and Junta de Andalucia (1995);
3. Agreement between CCR Algarbe and Junta de Extremadura (1995).

These agreements have created three Working Communities for their respective geographical scopes, which are new structures of governance that are not a new tier of local or regional government, but a space for exchanges and cooperation between public- and private-sector actors. Institutions operating across boundaries can have responsibility for a portfolio of topics (for instance, economic and regional development, energy, environmental protection, tourism and leisure, agriculture, to name a few). Decisions in these policy arenas are likely to have a significant impact on climate change. The Operational Programme on Cross Border Cooperation Portugal-Spain 2007-2013, which was formally approved by the European Commission on October 2007, seeks to foster cross border cooperation on four main aspects: cooperation and joint management for the promotion of competitiveness and employment support; cooperation and joint management for the conservation of the environment, patrimonial and natural resources and risks management; cooperation and joint management in land use planning and accessibilities; promotion of cooperation and institutional and social integration.

During the ADAM workshop carried out in December 2006, among the main possible responses to adapt to climate pressures the following were mentioned: increase agricultural and economic diversification as well as the scale of farm organisations (e.g. by concentrating on certain more productive varieties, although this has a negative effect on agricultural biodiversity and perhaps on the long-term system adaptability), reuse of grey waters, promotion of rural tourism, and local produce, protecting but also modernising the 'dehesa' multifunctional ecosystem, which is formed by well adapted extensive farming tree varieties such as acorn and cork trees, the use new varieties of draught and heat resistant crops and the implementation of new management systems. Reforestation measures, mainly with pine (pyrhofic) species have been also implemented, not without criticism.

4.3 Results

In many regions of Portugal and Spain, important changes in the current institutional designs are motivated by the influence of European policies. However, it is also true that an important factor relates to the growing involvement of agents that traditionally were excluded from the regional policy-making processes. Now, thanks to the new climate change policy and problem-framing situation, a new space and venue for participation and collaboration has been created. While it is difficult to say to what extent such new spaces and opportunities are actually transforming both agents and institutional arrangements in the Guadiana river basin, it is true that new the new cross-boundary development institutions cannot ignore the issue of climate any more. Rural and EU development policies have created new areas for collaboration between adjacent regions of different countries. Due to the wide range of policies and actors that directly or indirectly influence climate change mitigation and adaptation policies in the Guadiana River Basin, the need for an adequate coordination at different levels of government and between different programmes and sectoral policies is a matter of concern. In this regard, several transboundary institutional arrangements for the use of shared water resources exist between Spain and Portugal (Maia 2000). Furthermore, the *Convention on Cooperation for the Protection and Sustainable Use of the Portuguese-Spanish River Basins*, formed the Commission for the Convention Development and Application (CDAC). The Convention defines the framework for cooperation between the two countries to protect inland waters (surface waters and groundwater) and their dependent ecosystems and to use the shared basins waters in a sustainable fashion. However, these institutional developments are very little known to the general public and even to the local administrative bodies in the Guadiana basin (Newater 2005) and climate change is still often not mentioned.

Some movements towards implementing actions regarding climate change mitigation and adaptation in the Guadiana river basin are beginning to materialise. First, in the elaboration and implementation of national and regional climate change strategies; second, as a result of the requirements established by the Water Framework Directive (WFD); third, as an effect of the formulation of national and regional rural development programmes; and fourth, because of the transformation provoked on the socio-economic structure and environmental conditions of the river by the EU Structural Funds. Although with limited success, these regulations do promote the participation of all public and private actors in the policy making process at all scales and the inclusion of "good governance" criteria and principles in the daily management of public institutions.

In short, within the Guadiana river basin, and characteristically to other regions within the EU, some regions are acquiring greater decisional powers in the recent years and new regions are emerging (Cots et al. 2007). However, the 'new regions' do not longer correspond to the divisions that once were useful to the particular goals of the old nation-states, but are now created upon other new cooperation goals, including environmental quality procurement and sustainability, that transcend country borders. The mainstreaming of climate change into regional development goals and processes translates into new ways of cooperation and collaboration, some capable to erode existing (and dysfunctional) regional or national divisions while at the same time developing new ones. This forms of 're-regionalisation' embraces the adoption of new forms of governance and resource management related to adaptive management of shared resources such as transboundary river basins. Therefore, climate change poses specific risks to geographical regions which do not fit with the existing administrative and political arrangements. New procedures of knowledge integration, coordination and agent engagement, are necessary to know how the 'new regions' can be best rearranged and what new political institutions can best be compatible with the setting of long-term climate goals. The new situation is likely to demand some flexibility and innovation of the existing national settings, given that originally they were developed in completely social-ecological conditions and for completely different purposes than mitigating and adapting to global environmental change.

5. Discussion and conclusion

The present research illustrates that climate change discourses are increasingly affecting the way many development, energy, water and land use policies are being *framed* in these very different regions of the world. The globalisation of the climate framing in regional policymaking discourses however, is not always the result of an conscious, explicit and orchestrated process, but responds to the way multiple stakeholders and agents interpret a number of different types of risks, take advantage of new opportunities, make sense of and rationalise a multiplicity of signals about biophysical change as 'climate change'. Indeed, climate serves as a logical justification of many goals and measures which existing institutions and policy makers wanted to pursue anyway. For instance, in Spain, the building of new dams has been argued in the official communications to the UNFCCC as contributing to climate mitigation in the same guise that the same has been claimed in China together with policies about controlling population growth. But most importantly, once such 'climate rationalisation' and reframing have occurred, the new situation creates new opportunities for collective action in line with climate change demands, which in turn modify the original biophysical and institutional contexts where agents interaction occurred in the first place. Hence, social learning on climate change is a result of a series of loops of framing-reframing, the first ones relating to connecting many social and ecological processes to climate change while the next ones regard to actually developing explicit strategies, in a purposeful and structured way, to actually deal with climate mitigation and adaptation. This process has been observed and is actually happening in the three regions.

Therefore, an important aspect in the definition and characterisation of social learning processes relate to that of *reframing*. In our regions we can observe that reframing takes place in many ways, both in the way the processes affecting the ecological systems are perceived as in the way the roles of existing agents in the changing governance structures are understood. For instance, while traditionally in the Tisza floods were perceived unquestionably as 'bad' – due to the temporal harm provoked to local most vulnerable populations-, in the same ways as in Iberia forest fires are perceived also 'bad', current perceptions are changing. New more socially and ecologically strategies are being designed to 'learnt to live with floods' – in the same way that in the Mediterranean region people are beginning to use and managed fires to improve the quality of their forests. Indeed, floods or fires are neither necessarily 'good' nor 'bad' but what can be regarded as part of social learning is how agents learnt to take advantage of these natural and ecological processes and to adapt their individual behaviours and social-ecological interactions and institutions accordingly. We can say that this is also happening in a differentiated manner in all three regions, with a particular degree in the Tizsa, and to a less extent in the Guadiana and the Alxa – Inner Mongolia regions.

Tab. 1: Climate trends, impacts, responses and drivers in Alxa, the Tizsa floodplain, and the Guadiana river basin.

Region	Climate change		Adaptation domain		Mitigation domain	
	Recent trends and signals	**Main impacts**	**Some responses**	**Triggers and driving forces**	**Some responses**	**Triggers and driving forces**
Alxa league, Inner Mongolia, China	Increased temperature: about 1°C from 1970s to 1990s. Increased aridification (desertification). General environmental deterioration and biodiversity loss.	Reduced agriculture and livestock productivity and water availability; Impact on human settlements leading to migration.	Desert enclosure and fencing; changes in livestock; State-organised ecological emigration; conversion of farmland into forestland or grassland, aerial seeding, artificial precipitation, Heihe River water redistribution; promotion of sustainable tourism activities in the desert. Water harvesting (underground pools). Artificial rainwater in the Hellan mountains. Reforestation to stop the expansion of the desert ('the green wall'),	Sand storms and increased water stress. Political reorganisation. Institutional changes regarding land privatisation and land ownership regimes. Poverty alleviation policies linked to new development opportunities elsewhere. New scientific findings and concepts now informing local and regional policies. Opportunities for international cooperation in the form of research and development projects. Increased participation of local and private agents.	Still not developed at the regional level, but huge potential within the coal power generation and electricity sector. Post-Bali climate commitments may impact Alxa's coal industry.	New opportunities and potential for large international investments, e.g. in Carbon Capture and Storage (CCS) and Integrated Gasification Combined Cycle (IGCC) technologies. Increased participation of local and private agents. Increasing role of transnational epistemic communities (Schroeder).

Tab. 1: Climate trends, impacts, responses and drivers in Alxa, the Tizsa floodplain, and the Guadiana river basin – *continued*

Region	Recent trends and signals	Climate change	Adaptation domain		Mitigation domain	
		Main impacts	Some responses	Triggers and driving forces	Some responses	Triggers and driving forces
Tisza floodplain, Hungary	Increased temperature; changes in precipitations regime e.g. heavy rainfalls; increased risks of droughts; increasing frequency and magnitude of floods and in-land water stagnation; biodiversity loss.	Physical damage for agriculture and infrastructure; water availability and water quality problems; (indirect) migration of the population from the declining areas. Highly transformed ecosystem has lower buffering capacity; unstable social and economic conditions increase the risk for population and economy.	Complex program of land rehabilitation based on traditional techniques of shallow flooding, water retention areas, diversification of land use (VTT plan); Traditional adaptation measures – dike construction, relocation of settlements,	Disastrous floods; increasing social awareness and pressure (NGO, etc.); scientific research at national level (VAHAVA plan); Policy debates at the national level (Hungarian National Climate Policy). EU accession: need to comply with EU standards for water management and rural development (WFD, CAP, Natura 2000, etc.); EU funding (Regional Development Fund, Cohesion Fund, CAP).	Development of for energy crop production (scientific research and interest from commercial companies); promotion of integrating the idea of biodiversity conservation (wetlands) as carbon storage; promoting local market and small-scale production to reduce energy and fuel consumption.	EU and national climate change programmes and strategies; opportunity for investments in bio-energy production; cross-compliance of the adaptation and mitigation goals – biodiversity conservation might also support adaptation and economy (tourism) in the region.

Tab. 1: Climate trends, impacts, responses and drivers in Alxa, the Tizsa floodplain, and the Guadiana river basin – *continued*

Region	Recent trends and signals	Climate change	Adaptation domain		Mitigation domain	
		Main impacts	Some responses	Triggers and driving forces	Some responses	Triggers and driving forces
Guadiana river basin, Iberian Peninsula	Increased temperature. Increased desertification. Erratic rainfall. Impact on local biodiversity. Impact on traditional landscapes such as the 'dehesa' (formed out of cork oaks, and other adapted species). Depopulation. Increase of forest fires.	Agriculture; Tourism; Water availability; water quality problems: oak tree disease, increased risk of forest fires, migration of endemic species, sea level rise (impacts on tourism, erosion of the coast, etc.)	Agricultural and economic diversification,; increase scale of farm markets, by concentrating on certain products. Dry crop irrigation and water reuse; promotion of rural tourism, local handcrafts and products; promoting and modernising the traditional dehesa ecosystem. Identify and plant drought and heat tolerant crops & management systems ; New transboundary institutional arrangements, linked to regional and rural development policies.	Economic globalisation of local farm products (e.g. olive oil). European, national and regional climate change, water management and rural development policies, European funds. Pressure of NGO s and civil society movements. Some new transboundary institutional regional arrangements triggered and supported by EU regional and rural development policies, such as INTERREG, leading to 're-regionalisation'.	Production of renewable energies, mainly sun, and others like bioenergy crops – although now questioned on other environmental grounds. Conservation and promotion of Dehesa, reforestation and soil conservation.	European, national and regional climate change programmes and strategies, local mobility plans. New opportunities for technological innovation in renewable (sun) energy. Beginning of the implementation of national climate and clean energy policy.

In the case of China, and with regard to key drivers of change, it is important to remark that beyond direct exposure to climate effects non-biophysical triggers seem to be playing a specially important role in mobilising resources towards climate adaptation and mitigation. In a society explicitly committed to alleviating poverty (Wang & Fang 2007) and dominated by the quest of economic growth (Harris 2004), this new cultural mood can not be neglected. Indeed, it is likely that the most important drivers for mitigation and adaptation actions, at least in the short and mid term in China, will be motivated more for the prospects of economic development and growth than the direct experience to the effects of climate change. At present, Chinese authorities see that any strategy towards climate adaptation and mitigation must be combined and create synergies, rather than be in conflict, with real improvements in the quality of life of local people. In some of areas of the Inner Mongolia, what is seen as a first priority is to alleviate poverty and improve their standard of living and then, consider other issues like climate change. In the Northwest China, the current worldwide increase in the price of food is making these trade-offs between some of the longer-term climate adaptation and mitigation goals and the more urgent demands of supplying the population with enough basic resources of more evident. Therefore, the major challenge is how to coordinate these conflicts between different scales and policy domains in an integrated and coherent strategy, taking also into account the growing pressures exerted by developing countries.

Table 1 summarises the main impacts, responses as well as triggers and driving forces that have led to adaptation and mitigation actions as identified in the three regions.

Certain transformation within existing governance structures appear to be also occurring in the three regions, although it is very difficult to discriminate to what extent these changes are the sole result of climate change impacts and discourses. For instance, actors in the Tisza called for clarifying the roles of non-state actors. To prepare for climate change, actors suggested an informed multi-stakeholder approach (e.g. dialogues) at national, basin and regional level to prepare action plans for adaptation and to strengthen regional and local representation (e.g. after the Bokartisz Public Utility). It was proposed that an "Hungarian Water and Climate Alliance" could be established as a national umbrella to continue building bridges between the climate and water sector, to encourage capacity development to better cope with climate impacts and facilitate obtaining financial support for national, basin and regional level adaptation plans. An alternative to national coordination of the new integral water plan (the New Vásárhelyi Plan) is to establish a multi-stakeholder implementing agency with half political representation, half other stakeholders. The Tisza Alliance (www.elotisza.hu) that was established in June 2006 could play a role in representing regional and local actors and play an active role in *climate learning* to increase adaptive capacity at local, regional level. In the case of the Guadiana, one of the main drivers of the transformations observed within regional governance structures of the Guadiana basin is to be found on the need to adapt the European directives to national and regional legislations. In this manner, Pasquier (2005) argues that the introduction of concepts and ideas from the European level to the national and regional ones constitutes a learning process based on the transmission of cognitive resources in terms of public policies, administrative agreements and institutional designs. These cognitive resources provide to national and regional actors new elements which can be applied to their territorial policies. Transformation at the level of agents also occur in terms of changes in the use of time – e.g. diversifying economic activities – but are dependent both of the capacities of agents to shift to a better-off situation or activity which is compatible with climate policy goals and of the degrees of freedom of agents to do so. Given the very different political contexts studied such capacities and degrees of freedom for agent collaboration and transformation largely vary across the three cases.

More schematically, Table 2 syntheses the similarities and difference in the dimensions of context, processes, results and feedbacks that characterise our approach to social learning in the three regions.

With the work carried out so far these regions, and acknowledging the fact that further research is needed, we can extract the following conclusions:

- Several factors appear to constrain or enhance social learning on climate change regional agents and these include: (a) the type and resources for multilevel and cross-sector interaction with empowered agents (b) the capacity of becoming aware, anticipate, and control the negative personal and system's effects of continuing with the existing patterns of social-ecological interaction and development and (c) the available degrees of freedom to modify ones' practices and social-ecological configuration while creating a better-off situation from such transformations.
- In the selected regions, the identified actions and measures which apparently have been considered to mitigate or adapt to climate change originally were triggered by direct exposure to climatic risks or as part of an explicit, conscious and structured climate policy but for other reasons. Such non-direct triggers include policies related to land privatisation and political reform, consolidation of new agricultural markets, new opportunities for regional development and transboundary cooperation, or possibilities for supporting new scientific projects and epistemic communities. This does not mean that these *other reasons* have nothing to do with climate change. On the contrary, a careful consideration and integration of these institutional, economic and social factors is fundamental for the success of long-term climate strategies, because is part of the framing-reframing processes that characterise social learning in this domain.

Tab. 2: Contexts, processes, results and feedbacks regarding climate learning in the three regions

Region	Context	
	Biophysical	**Institutional**
Inner Mongolia, China	Very arid and worsening biophysical conditions in terms of diminishing rainfall, temperature rise and loss of biodiversity.	*National*: Strong (communist) interventionist state but with increasing penetration of privatisation and market forces. *Regional:* With well defined but limited regional autonomy.
Tisza, Hungary	Greater risk of floods and droughts; problem of water quality; declining biodiversity and agriculture production. Land abandonment	*National:* transition economy. Recent incorporation into the EU and need to comply with EU regulations (CAP, WFD, Natura 2000), also possibilities to fund regional initiatives through EU funds *Regional*: economically deprived region; low social capital; however growing cooperation between actors at the local and regional level.
Guadiana, Iberia	One of the most important hotspots in Western Europe regarding current and forecasted increase in temperatures and lowering precipitation due to climate change.	*National:* Portugal and Spain fully incorporated in the EU governance structure and policy. Strong participation in EU and global markets. *Regional:* regional competences clearly defined within Spain and Portugal but not so at cross boundary regional level.
Region	**Processes**	
	Relational	**Tasks/content**
Inner Mongolia, China	Poorly developed. Agent transformation are due to resettlement of of large amounts of of people from the increasing desertified areas to urban areas.	Strongly developed, in the form of large investments and work force employment.
Tisza, Hungary	Participatory processes poorly developed at the "official" level. At the same time, actors show the tendency for establishing platforms for cooperation beyond the official structures.	Flexible adaptive measures are still poorly developed; large investments are made in traditional flood protection measures.
Guadiana, Iberia	New institutional partnerships developed as a result of implementation of EU policies, mainly INTERREG agreements.	Mostly oriented to support regional development strategies and economic growth and to reduce dependency on agriculture by increasing labour and economic diversification (e.g. tourism). Forestation is seen as a way to deal both with climate mitigation and adaptation.
Region	**Results**	
	Relational qualities & social capital	**Technical qualities**
Inner Mongolia, China	Arguably destructive to the conservation and develop-ment of social capital at the regional level due to migrations.	Mainly focused on developing technical and economic solutions and capacities to deal to climate change, e.g., fencing and extending water transfers to increase the standard of living.
Tisza, Hungary	Development of social capital in a form of inter-action and new networks between actors, primarily at the local and regional levels; ini-tiatives for new networks; attempt to influence decision-making process. Delayed and limited implementation undermines trust building.	At present technical solutions are largely oriented towards investments in business as usual flood protection measures.
Guadiana, Iberia	A notable milieu of new social transboundary networks appearing within the region, but still unable to influence core development policies.	New technical capacities relating to climate change appear to be strongly developed at the level of basic research, but still lack application and specific translation in development of measures at the local level.

Tab 2: Contexts, processes, results and feedbacks regarding climate learning in the three regions – *continued*

Region	Feedbacks	
	On biophysical environment	**On governance structure**
Inner Mongolia, China	Notable impact on reversing some impacts in the short term but also leading to new problems	Limited by a rigid hierarchical structure, but enhanced by participation of epistemic communities.
Tisza, Hungary	Protection measures are under implementation. There is a risk however, that traditional measures can not support long-term adaptation strategies.	Still limited by traditional system of governance and power distribution, but enhanced by growing social capital, experience in participation and, also, EU standards for deliberativedecision-making.
Guadiana, Iberia	Individual strategies and policies aimed at adapting to climate change include agricultural production diversification, conservation of dehesa landscape, and increase of water retention capacity.	New regional climate change strategies (e.g. Andalusia) being developed and implemented. However, still needs a transboundary river basin approach and further integration with national and local strategies.

- Hence, this does not mean either that climate change has not played a role in the eventual implementation or justification of current regional measures to deal with climate change. In the case of the Tizsa flood plain, for instance, it was the recurring risks of floods and droughts that triggered the attendant preventative actions to cope with floods, in a similar guise that in Inner Mongolia the increasing desertification also motivated a number of policy measures to deal with aridification. However, those impacts originally were not framed as climate change problems, given that such peculiar science and policy framing is very recent at the regional level, a fact which is specially clear in the case of the Guadiana river basin. Climate discourse operates as a *rationalising* devise that helps to make sense of and justify many disparate signals of biophysical change but also development opportunities and institutional reforms that would not appear otherwise. This explains the appeal that climate change exerts on corporate and private non-state actors and their increasing support that this discourse receives in many international and also local development policy strategies.

Thus, and more generally, and taking into account a long-term and global perspective, we could state as working hypotheses that need further research and validation that:

- The main challenge remains not in producing end-of-pipe portfolio of possible 'fit-all' measures to abate climate changes but to develop long-term durable *processes* that are able to integrate local and regional concerns, experiences and conditions, as well as long and short term goals and mitigation and adaptation responses and action in way that is consistent with the total social-ecological resilience of the planet. This process, in order to support social learning, needs to facilitate the emergence of new opportunities for social-ecological system awareness, and agents' transformation by allowing greater degrees of freedom and multi-level, multi-domain collaboration.

- The conservation and promotion of cultural, institutional and biological diversity may be one of the most critical aspects in any integrated strategy to radically mitigate and adapt to climate change both globally and at regional level. This is so because in many cases, the preservation of such diversities entails the conservation lifestyles, social-ecological practices and distinct communities which already require less greenhouse gases to define and practice their happiness and identities. The preservation of knowledges and cultural practices embedded in those communities appear to be absolutely necessary to successfully adapt and cope to accelerated climatic changes both locally and globally in the long term.

References

ADAM (2007a), Report of the Inner Mongolia stakeholder workshop. ADAM project report.

– (2007b), Scoping document P3d Inner Mongolia case study. ADAM project report.

– (2007c), Scoping document P3d Tisza case study. ADAM project report.

– (2007d), Scoping document P3d Guadiana case study. ADAM project report.

– (2007e), Scoping stakeholder workshop and interviews for Guadiana region. ADAM project report.

Aden, N.T. and Sinton, J.E. (2006), "Environmental implications of energy policy in China." Environmental Politics 15(2), pp. 248-270.

Balogh P., Sendzimir, J. and Vári, A. (2005), Modelling Biocomplexity in the Tisza River Basin within a Participatory Adaptive Framework. Proceedings of the International Environmental Modelling and Software Society iEMSs 2004. Osnabrück, Germany [URL]: www.iemss.org/iemss2004/.

Becker, C.D. and Ostrom, E. (1995), "Human ecology and resource sustainability: The importance of institutional diversity." Annual review of ecological systems 26, pp. 113-133.

Bellon, T. (2004), Living Together with Nature. Farming on the River Flats in the Valley of the Tisza. Acta Ethnographica Hungarica 49, pp. 243-256.

Carter, N.T. and Mol, A.P.J. (2007), Ecological Governance in China. London: Routledge.

Commission of the European Communities (CEC) (2001), European Governance – A White Paper, COM(2001) 428 final. Brussels.

Confederación Hidrográfica del Guadiana (CHG) (2000), Memoria del Plan Especial del Alto Guadiana. Confederación Hidrográfica del Guadiana.

Congjie, L. and Dongping, J. (eds.) (2007), The China's Environmental yearbook 2005: crisis and breakthrough of China's environment. Chinese Academy of Sciences: Brill.

Cosme, N., Sousa, S., Estrela, M.A., Álvarez, R. et al. (2003), Environmental data on a case study form the transboundary catchment of Guadiana River. Report of the Transcat project.

Costejà, M. and Maestu, J. (2005), The special plan of the Upper Guadiana Basin. Moving from traditional towards participatory decision making? Harmonicop Project.

Cots, F.J., Tàbara, D., Werners, S. & McEvoy, D. (2007), Climate change and water adaptive management through transboundary cooperation. The Case Of The Guadiana River Basin. CAIWA Conference Proceedings. Basel Switzerland.

Day, K. (2004), China's Environment and the Challenge of Sustainable Development. Arkmond, NY: M.E. Sharpe.

Dessler, D. (1989), "What's at stake in the agent-structure debate?" International Organisation 43 (3), pp. 441-473.

Dickinson, D. and Weber, M. (2007), "Environmental resettlement and development, on the steppes of Inner Mongolia, PRC." Journal of Development Studies 43(3), pp. 537-561.

Economy, E.C. (2004), The river runs black. The environmental Challenge to China's future. Ithaca, NY: Cornell University Press.

Flachner, Z., Molnár, G. and Kajner, P. (2004), Case study on the Cigánd-Tiszakarád water retention system: integrated nature oriented floodplain management, flood management, rural development and landscape rehabilitation (Bodrogköz esettanulmány, VTT. 1 kutatás keretében). Budapest.

Folke, C., Hahn, T., Olsson, P. and Norberg, J. (2005), Adaptive Governance of Social-Ecological Systems, Annual Review of Environment and Resources 30, pp. 441-473.

Footitt, A. and McKenzie Hedger, M. (2007), Climate change and water adaptation issues. no. 2/2007. EEA Technical report. Agency, E.E., Copenhagen, DK.

Gupta, J. (2007a), "The multi-level governance challenge of climate change." Environmental Sciences 4 (3), pp. 1-7.

– (2007b), Climate change: a 'glocal' problem requiring 'glocal' action, Environmental Sciences 4(3), pp. 139-148.

Harris, P.G. (2004), "'Getting Rich is Glorious': Environmental values in the People's Republic of China." Environmental Values 13, pp. 145-165.

Heggelund and Backer (2007), "China and UN environmental policy: institutional growth learning and implementation." International Environmental Agreements: Politics, Law and Economics 7(4), pp. 415-439.

Ho, P. (2006), China's Limits to Growth: Greening State and Society. Oxford, UK: Blackwell.

Huan, Q. (2007), "Ecological modernisation: a realistic green road for China?" Environmental Politics 16(4), pp. 683-687.

Jiang, H. (2006), "Decentralisation, ecological construction, and the environment in post-reform China: a case study from Uxin Banner, Inner Mongolia." World Development 34 (11), pp. 1907-1921.

Jiang, G., Han, X. and Wu, J. (2006), "Restoration and management of the Inner Mongolia grassland require a Sustainable Strategy." Ambio 35(5), pp. 269-270.

Jones, B. and Keating, M. (eds.) (1995), The European Union and the Regions. Oxford University Press. Oxford.

Láng, I. (2006), The project "VAHAVA", Executive summary, Ministry for the Environment and Water Management (KvVM) and the Hungarian Academy of Sciences (MTA), Budapest, May 2006.

Lemos, M.C., Boyd, E., Tompkins, E.L., Osbahr, H. and Liverman, D. (2007), Developing adaptation and adapting development. Ecology and Society 12(2): 26. [online] URL: www.ecologyandsociety.org/vol12/iss2/art26/.

Linnerooth-Bayer, J. and Vari, A. (2006), Floods and Fairness in Hungary. In: Verweij, M. and Thompson, M. (2006), Clumsy Solutions for a Complex World. Palgrave Macmillan.

Maia, R. (2000), Portuguese-Spanish river basins: bilateral agreements' evolution and context. Water Science and Technology 42, pp. 227-233.

Matczak, P., Flachner, Z. and Werners, S.E. (2007), Institutions for adapting to climate change in the Tisza river basin. In: Proceedings Earth System Governance 2007, Amsterdam.

Mostert, E., Pahl-Wostl, C., Rees, Y., Searle, B., Tàbara, J.D. and Tippett, J. (2007), Social learning in European river-basin management: barriers and fostering mechanisms from 10 river basins. Ecology and Society 12(1): p. 19. [online] URL: www.ecologyandsociety.org/viewissue.php?sf=28.

McEvoy, D., Cots, F., Lonsdale, K., Tàbara, J.D. and Werners, S. (2007), The Role of institutional capacity in enabling climate change adaptation: The case of the Guadiana river basin. Proceedings of the Symposium on Transborder Environmental and Natural Resource Management, Kioto, Japan, 5-7 December.

Moreno, J.M. (Coord.) (2005), Main conclusions of the preliminary general assessment of the impacts in Spain due to the effects of Climate Change. Madrid: ECCE Project, OECC and UCLM.

Murhpy, R. (2002), The Internalisation of autonomous nature into society. The Sociological Review 50, pp. 313-333.

National Development Agency (2005), Progress reports of Environmental Protection and Infrastructure Operational Programme (EIOP) in the first quarter of 2005. National Development Agency, Budapest, HU.

National Development and Reform Commission, People's Republic of China (2007), China's National Climate Change Programme, available at: www.pew-climate.org/docUploads/ChinaNationalClimateChangeProgramme%20 June% 2007.pdf.

National Research Council (1992), China and global change. Opportunities for collaboration. Washington: National Academy Press.

Newater (2005), Transboundary river basin management: State of the art review on transboundary regimes and information management in the context of adaptive management, Newater Report Series no. 10.

Ostrom, E. (2005), Understanding institutional diversity. Princenton, USA: Princenton University Press.

Pahl-Wostl, C., Craps, M., Dewulf, A., Mostert, E., Tàbara, J.D. and Taillieu, T. (2007), Social learning and water resources Management. Ecology and Society, 12(2):5 [Online] URL: www.ecologyandsociety.org/ viewissue. php? sf=28.

Pasquier, R. (2005), "Cognitive europeanization and the territorial effects of multilevel policy transfer: local development in French and Spanish regions." Regional and Federal Studies 15(3), pp. 295-310.

Pelling, M., High, C., Dearing, J. and Smith, D. (2008), Shadow spaces for social learning: a relational understanding of adaptive capacity to climate change within organisations. Environment and Planning A, 40 (4), pp. 867-884.

Rogers, S. and Wang, M. (2006), "Environmental resettlement and social dis/re-articulation in Inner Mongolia, China." Population and Environment 28, pp. 41-68.

Santos, F.D., Forbes, K. and Moita, R. (eds.) (2002), Climate Change in Portugal. Scenarios, Impacts and Adaptation Measures – SIAM Project. Portugal: Gradiva, Lisbon.

Schroeder, M. (2007), Transnational NGO Cooperation in Chinese Climate Politics. Proceedings of the IHDP Earth System Governance 2007 Amsterdam www.2007amsterdamconference.org/.

Sendzimir, J., Balogh, P. and Vari, A. (2004), "Experimental design framework to restore ecological services and economic resilience in the Tisza river basin." Forstliche Schriftenreihe, Universität fur Bodenkultur, Wien, Bd. 18, pp. 1-33.

Shapiro, J. (2001), Mao's war against Nature. Cambridge, UK: Cambridge University Press.

Sienbenhüner, B. (2006), Social learning in the field of climate change. GELEN research group at University of Oldenburg. Draft manuscript, available at http://biogov.cpdr.ucl.ac.be/bioinstit/papers/SIEBENHUENER_paper.pdf.

Szlávik, L. and Ijjas, I. (2003), Action Plan on Flood Prevention for Tisza River. In: Precautionary Flood Protection in Europe, International Workshop, 5-6 February 2003, Bonn.

Tàbara, J.D. (2005), Sustainability Learning for River Basin Management and Planning in Europe. HarmoniCOP integration report. www.harmonicop.info.

– (2007), "A new climate for Spain: A Late Accommodation of Environmental Foreign Policy in a Federal State." In: Harris, Paul (ed.), Europe and Global Climate Change: Politics, Foreign Policy, and Regional Cooperation, pp. 161-184. Cheltelham, UK, Northhamptom, MA, USA: Edward Elgar.

Tàbara, J.D. and Giner, S. (2004), "Diversity, civic virtues and ecological austerity." International Review of Sociology 14(2), pp. 262-283.

Tàbara, J.D., Haxeltine, A., Valkering, P., Offermans, A., Schilperoord, M. and Weaver, P. (2008), The Water Transitions Play©. A participatory tool for exploring opportunities and constraints of sustainability transitions. Internal Matisse Working Paper (www.matisse-project.net).

Teng, F. and G. Alun (2007), "Climate change: national and local policy opportunities in China." Environmental Sciences 4(3), pp. 183-194.

The People's Republic of China (PRC) (2004), Initial National Communication on Climate change, *available* at: http://maindb.unfccc.int/public/country.pl?country=CN.

Timár, G. and Rácz, T. (2002), The effects of neotectonic and hydrological processes on the flood hazard of the Tisza region (East Hungary). EGU Stephan Mueller Special Publication Series 3.

VITUKI (2004), Rebirth of the River Tisza – The new Vásárhelyi Plan. Vásárhelyi Plan Intersectorial Committee Bulletin. Water Resources Research Institute, Budapest.

Wang, D. and Fang, C. (2007), Migration and poverty alleviation in China. United Nations Publications: International Organisation for Migration (IOM), IOM research series no. 27.

Weaver, P. M., Haxeltine, A., van de Kerkhof, M. and Tàbara, J.D. (2006), "Mainstreaming action on climate change through participatory appraisal." International Journal on Innovation and Sustainable Development 1(3), pp. 238-259.

Werners, S.E., Flachner, Z., Matczak, P. and Falaleeva, M. (2007), Drivers, Barriers and Opportunities for Adaptive Floodplain Governance in the Hungarian Tisza River basin. Proceedings of the IHDP Earth System Governance 2007 Amsterdam www.2007amsterdamconference.org/.

Williams, D.M. (1996), "Grassland enclosures: catalyst of land degradation in Inner Mongolia." Human Organisation 55(3), pp. 307-313.

– (2002), Beyond the Great Walls: Environment, Identity, and Development on the Chinese Grasslands of Inner Mongolia. Standford, CA: Standford University Press.

World Wide Fund For Nature (WWF) (2003), Results overview for the Guadiana river basin Portugal). WWF Water and Wetland Index – Critical issues in water policy across Europe.

Yonghuan, M. and Shengyne, F. (2006), "The protection Policy of Eco-environment in Desertification Areas in Northern China: Contradictions and Countermeasures." Ambio 35(3), pp. 133-134.

Zhang, L., Mol, A.P.J. and Sonnenfeld, D.A. (2007), "The interpretation of Ecological Modernisation in China." Environmental Politics 16(4), pp. 659-668.

Chapter 7

Climate Change Mitigation Policies: Costs and Effectiveness

Dalia Streimikiene, Remigijus Ciegis

Abstract

This article analyses the main problems and challenges of GHG emission reduction in the energy sector. The aim of the article is to assess climate change mitigation costs in various GHG emission sectors in Lithuania and to compare them with external costs of GHG emissions evaluated during several projects conducted in the European Union (EU). An analysis of EU energy and climate change mitigation policies is presented. The greenhouse gas (GHG) emission reduction measures are briefly discussed and their costs are assessed based on recent scientific publications. GHG emission reduction costs and potential of GHG emission reduction was evaluated in Lithuania. External costs of GHG emission are presented based on results of the EU funded CASES project. The GHG emission damage costs were compared with GHG abatement costs in different time periods. The conclusion and recommendations for GHG emission reduction policies are developed. The main finding of the article is that energy efficiency improvement should be considered as the most efficient GHG emission reduction measure. The performed analysis in Lithuania indicated that costs of GHG emission abatement expressed in Lt/tCO_2 are the lowest for energy efficiency improvement measures. These GHG emission abatement costs are lower than external costs of GHG emissions evaluated in several projects.

Introduction

The last three decades of research, investment and deployment of energy efficiency technologies, practices, management strategies and market mechanisms have led to the conclusion that energy efficiency improvement is the most efficient GHG emission reduction policy. However several other cost efficient GHG emission reduction options exist. These include renewable energy options, carbon capture and storage and nuclear energy. EU energy policy promotes the use of renewable and carbon capture and storage options, however first of all the cheapest GHG mitigation options should be employed.

The costs of GHG emission abatement should be compared with the damage costs of GHG emissions (external costs of GHG emissions) seeking to implement first of all GHG emission abatement measures in sectors where costs of abatement are lower than GHG emission damage costs.

The aim of the article is to analyse climate change mitigation options in the energy sector and compare their costs and effectiveness seeking to define the most efficient GHG abatement options. The main targets of the article:

- Analysis of EU energy and climate change mitigation policies
- Analysis of GHG emission reduction measures and costs based on new scientific literature;
- Analysis of GHG emission reduction costs and GHG emission reduction potentials in Lithuania based on newest studies performed;
- Review studies on external costs of GHG emissions;
- Development of recommendations for GHG emission reduction measures.

EU energy and climate change mitigation policies

At the centre of the new energy policy is the EU's commitment to reduce its greenhouse gas (GHG) emissions by at least 20% by 2020 compared with 1990 levels, not least because CO_2 emissions from energy make up 80% of EU GHG emissions. By using less energy and using cleaner, locally produced energy, the EU aims to increase energy security by limiting its growing exposure to increasingly volatile prices for oil and gas, while stimulating competitiveness in the European energy market.

The recent Energy Policy for Europe (EPE) pursues three main objectives:

- Increasing security of supply,
- Ensuring the competitiveness of European economies and the availability of affordable energy, and
- Promoting environmental sustainability and combating climate change.

In general terms, the EPE is based on five pillars. First, the EU aims to increase its energy efficiency by saving 20% of its energy by 2020. This will save about 780 million tones of CO_2 from being emitted into the atmosphere. Second, the share of renewable energy sources in the total energy mix is intended to triple to 20% by 2020, while aiming for a 10% biofuel component in vehicle fuel by 2020.The third pillar focuses on reducing the carbon emissions from hydrocarbons. The fourth and fifth pillars of the EPE are the EU's carbon market and an open and competitive internal energy market (Behrens & Egenhofer 2007).

The White paper on an energy policy White Paper for a Community Strategy and Action Plan COM (97)599 final sets the broad principles of design of energy policy in the EU. Energy policy must form part of the general aims of the Union's economic policy based on market integration, deregulation, limiting public intervention to what is strictly necessary in order to safeguard the public interests and welfare, sustainable development, consumer protection and economic and social cohesion. The EU Green paper on European Strategy for Sustainable, Competitive and Secure Energy (SEC (2006) 317) sets the main priorities for EU energy strategy. The general EU policy objectives considered most relevant to the design of energy policy are: competitiveness of the EU economy, security of supply and environmental protection. These objectives should help to address central policy concerns such as job creation, boosting overall productivity of the EU economy, protection of the environment and climate change. Overall competition of economy is pursued by liberalizing the EU electricity and gas markets and restructuring of energy sector. For fostering competitiveness of the EU economy and concomitant income and added value creation, the promotion of one internal market at Union levels is considered essential. Cross-border trade on level playing-field terms would foster competition (Behrens & Egenhofer 2007).

Security of supply is the priority concern of EU energy policy. The Green Paper on energy supply security (COM (2000) 769 final) states, that the EU will become increasingly dependent on external energy sources. It was stressed in this paper that the EU has very limited scope to influence energy supply conditions but it can intervene on the demand side mainly by promoting energy saving in buildings and the transport sector. The EU is not in a position to respond to the challenges of climate change and to meet its Kyoto protocol commitments. The Green Paper identifies two main policy priorities: controlling the growth of demand and managing supply dependency.

The Commission's new Green Paper on energy efficiency COM (2005) 265 stresses the importance of energy efficiency improvement for the control of demand growth and security of supply. According to estimates, the economic potential for improving energy efficiency by 2010 for all sectors combined is 20% of the total annual primary energy consumption of the current level.

The Green Paper on energy security emphasizes the role of technology development. Technology may contribute to the increase of energy efficiency, to achieving security of supply and reduction of GHG emissions, in particular by improving access to indigenous energy resources in particular renewable energy resources. The state aid is foreseen for promoting the use of RES and combined heat and power production by way of tax exemptions or reductions. The White Paper on renewable sources White Paper for a Community Strategy and Action Plan on renewable energy sources COM (97) 599 final states that member states should formulate indicative targets contribute to the ambitious indicative target of doubling the overall share of RES in the EU by 2010. It sets an indicative target of 12% for the contribution by RES to the total primary energy consumption within the EU by 2010 and contains a strategy and action plan to achieve this target. Pursuant to the White paper on Renewables the Directive 2001/77/EC on the promotion of electricity produced from renewable energy sources in the internal electricity market (hereafter called the *RES-E Directive*) was passed in 2001. It adds the indicative target contribution of 22.1% by renewables-based electricity to total EU electricity consumption in year 2010. 2003/30/EC Directive on the promotion of the use of biofuels or other renewable fuels in transport (*RF Directive*) sets that Member States must ensure by the end of 2005 a 2% minimum proportion of biofuels, gasoline and diesel fuels sold on their market. In the longer term the target is to achieve a share of 5.75% of biofuels for transport in the total amount of fuels in Europe by 2010 and 20% by 2020.

The recent 2006/32/EC Directive on energy end-use efficiency and energy services sets the targets for EU member states to reduce final energy consumption by 9% during the nine year period to 2015.

2002/91/EC Directive on the energy performance of buildings sets a target of realizing a savings potential of around 22% by 2010 for energy used in heating, air – conditioning, hot water and lighting.

2004/8/EC Directive on the promotion of cogeneration based on a useful heat demand in the internal energy market aims to increase energy efficiency and improve security of supply by creating a framework for promotion and development of high efficiency cogeneration of heat and power based on useful heat demand and primary energy savings taking into account the specific national circumstances especially climate and economic conditions. The strategic goal of EU-15 is to double the share of electricity produced by CHP by 2010.

All these directives have a positive impact on GHG emission reduction and achieving the Kyoto target. The EU has ratified the Kyoto Protocol committing itself to 8% GHG emission reduction in the period 2008-2012 from 1990. Equally the New Member States are determined to meet their individual targets under the Kyoto Protocol. Therefore the GHG emission reduction in the energy sector is the priority issue in the EU energy policy. Therefore the main targets of the EU energy policy which can be addressed by selecting the appropriate indicators are to increase security of the energy supply, opening of energy markets, promotion of renewables and cogeneration, increase of energy efficiency and reduction of impact on the environment. All these policies have a positive impact on energy intensity decrease and reduction of GHG emission.

GHG emission reduction policies and measures

There are a range of measures according to the type of policy instrument employed, following the categorisation used by the United Nations Convention on Climate Change (UNFCCC). This involves distinguishing between: Economic and fiscal instruments; Regulations and standards; Voluntary agreements; Information and awareness; Research and development.

The first category in the UNFCCC classification is that of economic and fiscal instruments and in turn this covers three main classes of instrument: subsidies, taxes and emissions trading. These instruments are the most commonly used (along with regulation) in Annex 1 countries and are found in all sectors, accounting for over one third of the reported policy instruments in the energy and transport sectors. They are less commonly reported by developing countries though, as discussed below, subsidy removal is an important tool in some countries. In many ways, the first and most effective step in reducing greenhouse gas emissions in a cost-effective manner is the removal of subsidies on fossil fuels. But the main forms of subsidy used in a climate change context are those designed directly either to lower energy intensity or to reduce carbon intensity. Thus many measures aim directly at subsidising non-fossil fuels, particularly renewables (to reduce carbon intensity), or energy efficiency measures (to reduce energy intensity).

Fuel taxes have always been common in OECD countries, as noted above, particularly in relation to motor fuels. In Europe and Japan in particular, motor fuel taxes are very high indeed, amounting in many cases for around 70-80% of the final consumer price. Increasingly, OECD countries are also introducing wider fuel or carbon taxes – for instance, Japan has imposed an energy tax and is exploring carbon tax options. The European Union has set a framework for the taxation of energy products which sets out minimum levels of taxation for energy products and electricity but allows for exemptions or reductions to promote renewable sources. In practice, many EU countries have gone beyond this and have imposed higher levels of taxation, or specific carbon taxes (Malaman & Pavan 2002).

The EU GHG emission trading scheme implemented in 2005 is the first international emission trading scheme implemented all over the world. There are 3 flexible GHG mitigation tools under Kyoto protocol: emission trading, Joint Implementation (JI) and Clean Development Mechanisms (CDM). Under the terms of the Kyoto protocol, Annex 1 countries can invest in either JI or CDM Projects and can also host JI projects. They cannot, however, host CDM projects, which must be implemented in non Annex 1 countries. JI is a relatively straightforward way of transferring the credits from emissions reducing projects between countries which are subject to caps. CDM is more complicated, both conceptually and administratively. First, it has two objectives: not just to enable one party to meet its emissions reduction objectives, but also to promote sustainable development in the other (non Annex 1) country. Second, because the non Annex 1 country by definition has no emissions cap, strict criteria are needed to ensure that the emissions reductions are additional – i.e. that they would not have occurred in the absence of the CDM measure concerned. The key difference, as compared with the ETS, is that these mechanisms are project-based (Streimikiene & Bubniene 2004).

Regulations and standards are a very common approach, particularly in the OECD, where they account for about one quarter of the policy instruments reported. It is difficult to generalise about regulatory approaches since they apply across so many sectors and in such a variety of ways. However, common approaches include: building standards, appliance standards, vehicles regulations and other environmental regulations. Regulatory approaches are widespread in

OECD countries, all of which report a range of such measures. They undoubtedly have an important place in promoting efficiency and in overcoming market failures – not only the failure to internalise environmental costs but also some of the well recognised barriers to greater energy efficiency, such as consumer indifference, incomplete incentives in landlord/tenant relationships etc.

Voluntary agreements are often more effective when combined with other measures (eg taxes) or backed up by sanctions. Examples include the UK Climate Change Agreements, which are voluntary agreements but offer the carrot of relief from the Climate Change Levy. Voluntary agreements have also been used extensively in the Netherlands, for many years. These agreements are long term and integrated into the environmental permit-setting process. They are also backed up by sanctions – those who do not join up to the agreements incur direct obligations to undertake cost-effective voluntary measures. However they do not apply directly to energy intensive industrial sectors. A major example of a voluntary agreement, designed to improve vehicle standards, is the agreement the EU has reached with a number of car manufacturing associations in Europe, Japan and Korea.

Many studies have identified a huge range of cost-effective opportunities available across the economy but which are not currently being implemented. One of the barriers to exploiting these opportunities is often seen to be a lack of consumer information and awareness. If this can be overcome by the provision of better information, it may be possible for consumers to make their own decisions in a way which promotes improved energy efficiency and lower emissions without the need for further intervention. Most governments, in both developed and developing world, have therefore introduced information measures; education and awareness campaigns and the promotion of consumer-relevant energy efficiency information are particularly common. Measures such as energy labelling of appliances or buildings are also fairly widespread. Sometimes such measures are combined with subsidies (for instance, subsidies for energy efficiency surveys in industry or for the installation of energy monitoring and targeting systems) or regulation (as with some labelling schemes, or the requirement in some countries for the documentation for a house sale to include an energy audit).

Research and development is the fifth of the UNFCCC's categories. In principle, it is probably better to take the category widely to include research, development, demonstration, technology choice and deployment and technology transfer, because of the importance of all these aspects of the technology issue to the longer term emissions picture.

All these policies and measures can achieve GHG emission reduction by changing the structure of fuels used towards the fuels with lower carbon content (nuclear) or renewable energy sources, by implementing energy efficiency measures in supply and demand side and by implementing carbon capture and storage (Kuriyan & Kammen 2007). These main three GHG emission reduction options in the energy sector imply different economic costs. Based on Kammen (2007) the projected costs for 2006 and 2016 of the main GHG emission reduction options in the US electricity sector are presented in Table 1.

Tab. 1: The costs of GHG emission reduction alternatives in electricity generation (Kammen 2007).

GHG mitigation	Cost 2006	Projected costs 2016	Explanation
$30-per-ton carbon-emissions tax	About 3 cents per kWh	About 3 cents per kWh	Taxing carbon emissions and using the revenues to develop clean energy sources, discourages GHG emissions and encourages the investments in alternative energy sources. It could be balanced by a reduction in income taxes or other taxes.
Energy efficiency	Savings 1 cent/kWh to costs of 3 cents/kWh	Savings of 3 cents/kWh to costs of 3 cents/kWh	The technologies are readily available. This is the most efficient GHG emission reduction option.
Solar power	At least 20 cents/kWh	10 to 14 cents/kWh	Solar energy systems, including solar thermal systems, could be installed in all homes and business. Many of them could produce and export surplus power. The cost of solar electricity systems has fallen by 5% per year for the last decade. Solar is still the most expensive of clean technologies but integrating it directly into buildings could accelerate cost declines.
Wind power	5 to 7 cents/kWh	3 to 6 cents/kWh	Already cheaper than natural gas in the right locations. Germany is installing wind generators that will supply 10% of its total electricity. Wind power requires land on which to put turbines, but the land can be used concurrently for crops or grazing. The latest designs protect birds. The costs are declining.
Biomass fuel	7 to 9 cents/kWh	3 to 6 cents/kWh	Biomass fuel made from agricultural waste or "energy crops" can be basified and blended with natural gas, or made into liquid fuel (ethanol) that could replace one-third or more of gasoline use.
Hybrid and ethanol vehicles	$2.000 to $5.000 more than conventional cars	$1.000 cents/kWh	Hybrid gas vehicles or better still plug-in electric hybrids (especially those that use ethanol instead of gasoline) could reduce gas consumption by 80% or more in USA. Consumers may need incentives and education. A distribution network for ethanol is needed.
Nuclear power	At least 3 cents/kWh	At least 3 cents/kWh	The nuclear industry has learned to open existing reactors for more efficiently than the past and safety has improved. However, waste management is with political and technical problems. Proponents cite nuclear powers exceptionally low generation cost and zero emissions. Detractors cite its and tremendous federal subsidies. Trying to create a carbon-fine world without nuclear power would be difficult.
Carbon capture and storage	3 cents/kWh	3 cents/kWh	Storing carbon dioxide, especially from coal-fired power plants, underground or in deep ocean has received generous industry funding. The technologies becoming cheaper however it is temporary measure because of the risk of future leakage and the cost may not fall much because of the uniqueness of each geologic respiratory

Carbon tax is an economic tool of GHG emission reduction which can also stimulate the changes in fuel supply structure towards lower carbon content fuels, promote energy savings and implementation of energy efficient technologies and carbon capture and storage technologies. As one can see from the figures presented in Table 1 energy efficiency improvements are the most efficient GHG emission reduction measures as the increase of electricity price by 3 Cents/kWh provides for a cost saving of about 1 cent/kWh. In 2016 even higher savings equal to GHG mitigation costs can be achieved by implementing energy efficiency measures. The second cheapest options are nuclear energy and carbon storage however because of political reasons extension of nuclear is still limited in many countries and carbon capture and storage technologies needs special geological conditions for installations and therefore the costs of such technologies depends on the uniqueness of each sites geological conditions. Solar options are the most expensive. The wind power technologies are becoming cheaper and by 2016 their costs are expected to approach the nuclear option costs.

Fig. 1: Estimated carbon reduction possibilities in Euros per ton of carbon dioxide equivalent in 2030

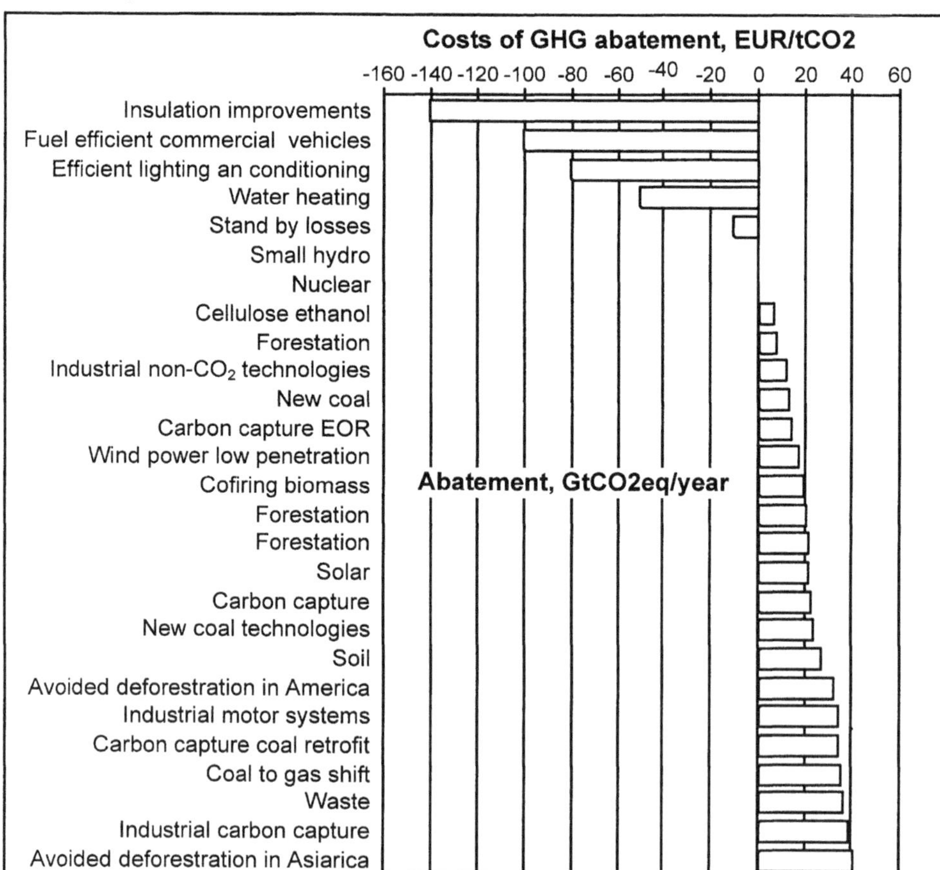

A particularly useful view of low carbon energy options has been developed by the Vattenfall Utility Cooperative in Sweden (Kammen 2007). In Figure 1 estimated carbon reduction possibilities in Euros (€) per ton of carbon dioxide equivalent are presented.

In Figure 1 estimated costs (some of which are negative, implying overall savings to the economy through the adoption of clean energy options and the amount of carbon they are able to offset or avoid by 2030) are presented. The vertical axis shows the costs or benefits of the specific climate change mitigation option, while the horizontal axis shows the estimated amount of savings. In Figure 1 27 options for GHG emission abatement are presented starting from the most efficient one. Each abatement measure corresponds to 1 Gt of GHG emissions reduction. As one can see from data presented in Figure 1 there are about 7 Gt of negative and zero cost opportunities. About 27 Gt of GHG emissions can be abated at costs bellow 40 EUR/ton (Nemet & Kammen 2007).

The key insight from this analysis is that a portfolio of low carbon opportunities exists, and that state and local governments can enact a wide range of cost-effective GHG emission reduction measures. As one can notice from analysis conducted above the most efficient GHG emission reduction measures are: insulation improvements in buildings, fuel efficient commercial vehicles, efficient lighting and air conditioning, reduction of stand by losses. Small hydro and nuclear are also cheap GHG emission reduction options. Carbon capture, solar energy, wind, new coal are GHG emission reduction options implying lower than 20 EUR/ton costs for GHG emission abatement (Kammen & Pacca 2004).

In the next section we will deal with GHG emission abatement costs in Lithuania seeking to compare them with GHG emission reduction costs presented in other studies. Further these costs will be compared with external costs of GHG emissions in order to compare costs of GHG reduction with achieved benefits of GHG emission abatements.

GHG emission reduction costs in Lithuania

Based on information provided in the recently conducted Phare project "Institutional capacities strengthening for the implementation of Kyoto protocol requirements in Lithuania" (Ministry of Environment 2008) the costs of GHG emission reduction and GHG emission reduction potential in various GHG emission sectors (fuel combustion, industrial processes, agriculture and waste sector) are presented in Table 2. GHG emission reduction costs in the project were evaluated based on analysis of various information sources about implemented programmes, projects etc. and their effects on GHG emission reduction in various sectors and costs related with implementation of these projects and programmes.

Seeking to evaluate GHG emission reduction costs and potential in fuel combustion sector in Lithuania in 2008-2012 the indicator of carbon intensity of total primary energy supply in Lithuania was developed based on GHG inventory submitted to United Nations Framework Convention on Climate Change (UNFCCC) in 2007. This indicator is in 2006 amounted to 62 kgCO_2 eq/GJ. For conversion of saved energy in the final sector the conversion coefficients to save primary energy were applied. These conversion factors account for losses in energy production, transportation and distribution losses.

After evaluation of saved primary energy it is converted to GHG emissions avoided by applying coefficient – 62 kgCO_2 eq/GJ.

GHG emissions avoided because of switching from fossil fuel to renewables are evaluated based on renewable's potential in primary energy supply – 12% or 11,4 TWh in 2010. Seeking to avoid double counting in Table 2 from the primary energy necessary for electricity production and biofuels used in transport are subtracted.

Potential GHG emission reduction potential because of use of renewables in 2008-2012 is evaluated based on the energy policy targets set in the national energy efficiency programme adopted in 2006 and the National energy strategy adopted in 2007.

Tab. 2: GHG emission reduction potential and costs in Lithuania in 2008-2012

GHG emission reduction costs	Costs of GHG emission reduction, Lt/tCO_2eq	Potential in 2010, GJ	Average GHG emission reduction potential in 2008-2012 m., Mt CO_2eq/year (62 CO_2eq kg/GJ)
Fuel combustion sector: GHG reduction potential: 18,3 Mt Average costs: 2-170 Lt/tCO_2eq			
Energy saving	2-20	0,8 TWh	1,8
Use of waste energy resources	32,4	2 TWh	4,5
Use of biofuels in transport	35,4	0,7 TWh	1,6
Renewable energy resources in electricity production (the conversion coefficient from electricity to primary energy -2,707)	170	0,9 TWh	5,4
Cogeneration extension	125	3,2 TWh	4,15
Renewable energy resources except listed above		2,26 TWh	5
Agriculture: GHG reduction potential – 0,1 Mt ; Average reduction costs: 1125 Lt/tCO_2eq			
State programme for emissions fro agriculture sources to water reduction	1125		0,1
Waste: GHG reduction potential 0,1 Mt; Average costs: 1370 Lt/tCO_2eq			
State waste management plan	1370		0,1
Industrial processes: GHG reduction potential: 2,4 Mt, average costs : 315-560 Lt/tCO_2eq			
Conversion of cement production from wet to dry technologies	560		0,5
Modernization of chemical industry for	315		1,9
Total:			**20,9**

(Ministry of Environment 2008)

As one can see from the information presented in Table 2 the most efficient way to reduce GHG emissions in Lithuania is to implement GHG abatement measures in the fuel combustion sector. The GHG emission reduction potential in this sector makes 18,3 Mt. GHG emission reduction costs in the range from 2 to 170 Lt/tCO_2 eq or 0,6-34 EUR/tCO_2 eq in this sector. In other GHG emission sectors GHG emission reduction potential in 2008-2012 is only 2,6 Mt/year and GHG emission reduction costs are significantly higher. Also we see from Table 2 the most efficient GHG emission reduction measure in Lithuania is implementation of energy saving measures in all sectors of the economy. The cost of GHG emission abatement by implementing energy efficiency improvements is 2-20 Lt/tCO_2 eq or 0,6-6 EUR/t CO_2 eq.

In the following section analysis of external GHG emission costs will be provided seeking to compare the costs of GHG emission abatement in Lithuania with external costs of GHG emissions.

External costs of climate change

At present, many estimates of the marginal external costs of the emissions of greenhouse gases (particularly CO_2) have been made. The estimates range from slightly negative (< 0) to over 400 USD per ton CO_2 currently emitted. Tol (Tol 1999; 2005; Tol & Dowlatabadi 2001) constructed a probability density function of published estimates. The function is highly skewed to the left, with a long right tail of sparse but high estimates. The mean value of the published estimates is 25 USD per ton of CO_2, but 50% of the studies report costs of less than 4 USD/ton (this is the median value). On the other extreme, 5% of the studies report costs of over 95 USD/ton. If only peer-reviewed studies are taken into account, the mean estimate drops to 12 USD/ton with a standard deviation of 23 USD/ton.

Most researchers agree that the marginal impacts of greenhouse gas emissions increase with the concentration of greenhouse gases in the atmosphere (Kuik et al. 2007). Therefore, it is commonly assumed that because of the expected increases in concentration over time, the present value of emissions also increases over time. That is, all else being equal, the present value of emissions in 2010 will be higher than the present value of emissions in the year 2000. The literature reports annual increases in the marginal costs of CO_2 emissions range between 1 and 2 percent (Rouggarden & Schneider 1999). Annual increases of marginal costs for other greenhouse gases may differ in relation to their expected lifetime in the atmosphere. The qualifier "all else being equal" used above primarily relates to the vulnerability of natural and economic systems and their ability to adapt to a changing climate.

Recently, there has been a flurry of research projects on the 'social cost of carbon' (SCC) in the United Kingdom. The social cost of carbon is the social cost of the emission of one tonne of CO_2 at a particular date; hence it is another word for the marginal (social) cost of CO_2 emissions. It is measured as the present value of the impacts of one tonne of CO_2 over its lifetime in the atmosphere. Large-scale projects included the SCC-project of Watkiss, Downing and others and, most recently, the Stern Review. The specific objectives of the SCC project were to review the previous use of the SCC values in policy assessment, and the possible approaches for future assessment, taking into account the factors that influence the values; to undertake expert stakeholder consultation, to obtain their views on how such analysis should be undertaken, and on the uses of SCC estimates in policy assessment in the face of uncertainty; to develop a series of case studies to demonstrate the various approaches for including SCC estimates in policy decision-making; and to make recommendations (Kuik et al. 2007).

The review argues that "using typical assumptions about discounting and aggregation", some central estimates of the marginal damage cost of carbon dioxide emissions may be lower than the "illustrative value" of £70/tC (€28/tCO_2) that is currently used by the UK government (Watkiss 2006). This reflects a trend in the literature towards lower SCC values in recent years. However, the literature studies do not cover all the impact categories of climate change, and most researchers consider the possibility of negative surprises to be more likely than positive ones. The authors therefore assessed the coverage of the valuation studies to investigate the extent to which they may under-estimate the total SCC. The studies were compared against a risk matrix, in relation to the uncertainty of climate change impacts and the uncertainty in valuation. Mapping the literature studies onto this matrix, the research found that very few studies cover any non-market damages, or the risk of potential extreme weather (floods, storms, etc). None covers socially contingent effects, or the potential for longer-term effects and catastrophic events (Watkis et al. 2005).

The Stern Review (Stern 2006), was another major review of the social costs of carbon. The review assessed the economics of moving to a low carbon economy, focusing on a medium to long term, plus the potential of different approaches to adaptation and lessons for the UK, in the context of climate change goals. From a review of the evidence, the report concludes that the benefits of strong and early action far outweigh the costs of not acting. Climate change will affect the basic elements of life for people globally – including access to water, food production, health and environment, potentially affecting hundreds of millions of people. Using the results from an integrated assessment model (the PAGE model), the review estimated that the total damage costs of climate change could be at least 5% of global GDP each year, now and forever. If a wider range of risks and impacts is taken into account, the estimates of damage could rise to 20% of GDP or more. The review suggested a SCC of € 85 per ton of CO_2, which is considerably higher than the UK government's "illustrative value" of € 28 per ton, and also far out in the right tail of Tol's probability density function (see above). In contrast to these high costs of inaction, the costs of action – reducing greenhouse gas emissions to avoid the worst impacts of climate change – can, according to Stern, be limited to around 1% of global GDP each year. The numerical results of studies into the external costs of greenhouse gas emissions remain speculative, but they can provide insights on signs, orders of magnitude, and patterns of vulnerability. Results are difficult to compare because different studies assume different climate scenarios, make different assumptions about adaptation, use different regional disaggregation and include different impacts. The Nordhaus and Boyer (1999) estimates, for example, are more negative than others, partly because they factor in the possibility of catastrophic impact. The Mendelsohn et al. (1996) and Tol (2002) estimates, on the other hand, are driven by optimistic assumptions about adaptive capacity and baseline development trends, which results in mostly beneficial impacts. According to Tol, the current generation of aggregate estimates may understate the true cost of climate change because they tend to ignore extreme weather events; to underestimate the compounding effect of multiple stresses; and to ignore the costs of transition and learning. However, these studies may also have overlooked positive impacts of climate change and not adequately accounted for how development could reduce impacts of climate change.

The cost of greenhouse gas emissions is an important component of the total external cost of electricity production. In the framework of the CASES project two approaches were followed to assess global warming. With the first methodology the quantifiable marginal damage costs for a unit of emissions were evaluated for CO_2, CH_4 and N_2O, while with the second one the marginal avoidance costs of GHG emissions were evaluated based on meta analysis. Damage costs mean the costs that are caused to environment including human health because of GHG emission releases. Avoidance costs are the costs related with implementation of GHG emission reduction measures. Since cost data is not site specific, the same values for all EU-27 member states were applied. Data is time specific, hence different values have to be used for the different periods. In Table 3 the damage costs and avoidance costs up to 2010 are presented (Porchia & Bigano 2007).

Tab. 3: Marginal costs of GHG emissions in Euro2005 per ton

	Damage costs	**Avoidance costs**	**Damage costs**	**Avoidance costs**
Period	2000-2009	2010	2020-2029	2030
CO_2	20,66	19	18,12	30
CH_4	747,26	399	555,57	636
N_2O	27202,90	5890	21596,68	9387
SF_6	1260,64		1130,36	

As one can see from Table 2 the damage and avoidance costs of CO_2 emissions are in quite good agreement. Ideally these costs should be equal as Pareto optimum in GHG emission level is being achieved then marginal damage costs are equal to the marginal GHG emission reduction costs. External costs of other

GHG emissions are in quite different ranges. The highest damage costs are associated with N_2O emissions; however damage costs of N_2O emissions are almost 5 times higher than avoidance costs. This is related to very high uncertainties in GHG external costs assessments. The uncertainty is characterized in terms of geometric standard deviations σg and multiplicative confidence intervals: if cost has been estimated to be μg (geometric mean or median) with geometric standard deviation, the probability is approximately 68% that the true value is in interval [$\mu g/\sigma g^2$; $\mu g * \sigma g^2$]. For the external cost of climate change the uncertainty σg based on literature review is about 5.

In CASES project for total external costs of electricity supply, damage costs were selected. In Table 4 these costs are presented for the main energy generating technologies including fossil, nuclear and renewables.

Tab. 4: European marginal damage costs of GHG emissions in 2005-2010 per technology, Ec/kWh

	Marginal damage cost
Nuclear	0.0599
Heavy oil condensing power plant	0.9311
Light oil gas turbine	1.8409
Hard coal condensing power plant	3.5581
Hard coal IGCC without CO_2 capture	3.6321
Lignite condensing power plant	3.8477
Lignite IGCC	3.8899
Natural gas combined cycle	1.8385
Natural gas turbine	2.7465
Hydropower run of river 10 MW	0.0541
Hydropower run of river <100 MW	0.0386
Hydropower run of river >100 MW	0.0348
Hydropower dam	0.0684
Hydropower pump storage	0.0617
Wind on-shore	0.0412
Wind of-shore	0.0322
Solar PV roof	0.2428
Solar PV open space	0.4647
Solar thermal parabolic trough	0.419
Natural gas CHP with extraction condensing turbine	1.6786
Hard coal CHP with extraction condensing turbine	3.1328
Natural gas combined cycle CHP with backpressure turbine	1.8550
Hard coal CHP with backpressure turbine	3.3858
Straw CHP with extraction condensing turbine	4.9348
Woodchips CHP with an extraction condensing turbine	4.8010
Natural gas fuel cell- MCFC	0.9031
Natural gas fuel cell – SOFC	0.7290
Biogas fuel cell – MCFC	1.4104

As one can see from the data presented in Table 4 the highest marginal costs of GHG emissions were obtained for lignite and hard coal condensing power plants. Very high marginal damage costs were obtained for straw and woodchips with extraction condensing turbine. The lowest European marginal damage costs are associated with electricity generation at off shore wind mills and other renewables. The nuclear technologies also have quite low marginal damage costs of GHG emissions.

Conclusions

1. Energy efficiency improvements are the most efficient GHG emission reduction measures as the increase of electricity price by 3 cents/kWh provides for cost saving of about 1 cent/kWh. In 2016 even higher savings equal to GHG mitigation costs can be achieved by implementing energy efficiency measures. The second cheapest options are nuclear energy and carbon storage however because of political reasons extension of nuclear is still limited in many countries and carbon capture and storage technologies needs special geological conditions for installations and therefore costs of such technologies depends on the uniqueness of each site geological conditions.
2. Analysis of low carbon energy costs conducted by Vattenfall in Sweden indicated that there are about 7 Gt of negative and zero cost opportunities for GHG emission reduction up to 2030. About 27 Gt of GHG emissions can be abated at costs bellow 40 EUR/ton. The most efficient GHG emission reduction measures defined are: insulation improvements in buildings, fuel efficient commercial vehicles, efficient lighting and air conditioning, reduction of stand by losses. Small hydro and nuclear are also cheap GHG emission reduction options. Carbon capture, solar energy, wind, new coal are GHG emission reduction options implying lower than 20 EUR/ton costs for GHG emission abatement.
3. By comparing marginal damage costs of GHG emissions with marginal GHG abatement costs one can notice that no more than 20 EUR/t CO_2 eq GHG emission reduction options should be applied up to 2010 as GHG emission damage cost are calculated to be about the same range. This guarantees the optimal GHG pollution level. In 2030 marginal GHG emission reduction costs would increase and reach 30 EUR/tCO_2 though GHG emission marginal damage costs would decrease.
4. The highest marginal costs of GHG emissions were obtained for lignite and hard coal condensing power plants. Very high marginal damage costs were obtained for straw and woodchips with extraction condensing turbine. The lowest European marginal damage costs are associated with electricity gen-

eration at off shore wind mills and other renewables. The nuclear technologies also have quite low marginal damage costs of GHG emissions.

5. The most efficient way to reduce GHG emissions in Lithuania is to implement measures in the fuel combustion sector. The GHG emission reduction potential in 2008-2012 in this sector makes about 18,3 Mt. The average GHG emission abatement costs are in range from 0,6-34 EUR/tCO_2 eq in this sector. In other GHG emission sectors GHG emission reduction potential in 2008-2012 is only 2,6 Mt/year and GHG emission reduction costs are significantly higher. In Lithuania the most efficient GHG emission reduction option is implementation of energy saving measures in all sectors of the economy. The cost of GHG emission reduction is 0,6-6 EUR/t CO_2 eq.
6. Comparing GHG emission reduction costs in Lithuania with damage costs obtained in CASES project one can notice that just in fuel combustion sector in Lithuania it is worth to implement GHG emission abatement measures as in other sectors GHG emission reduction efforts cause significantly higher costs then GHG emission damage costs.

References

Behrens, A. and Egenhofer, C. (2007), "An initial assessment of policy options to reduce – and secure against – the costs of energy insecurity." In: CASES WP5 Report (2) on Policy Assessment of Energy Security Measures Incorporating Externality Measures, Centre for European Policy Studies, Brussels.

European Commission (2007), An Energy Policy in Europe. COM (2007) 1 final, Brussels.

Hogg, D. (2000), "The limitations of economic instruments as stimuli for technical change, technological change and innovation." In: Andersen, M.S. and Sprenger, R.U. (ed.), Market-based instruments for environmental management. Politics and Institutions, Edward Elgar, Cheltenham, pp. 87-92.

Kammen, D. (2007), „Opportunities for GHG reduction." House of Representatives Committee on Oversight and Government Reform no. 8, pp. 1-13.

Kammen, D.M. and Pacca, S. (2004), "Assessing the costs of electricity." Annual review of Energy and Environment no. 29, pp. 301-344.

Kuik, O., Brander, L., Nikitina, N., Navrud, S., Magnussen, K. and Fall, E.H. (2007), "A Database of Studies on Energy-related External Costs due to Land Use Changes, Acidification and Eutrophication, Visual Intrusion and Climate Change." In: CASES WP3 External Costs Estimate Database, Institute for Environmental Studies, Brussels.

Kuriyan R. and Kammen, D. (2008), "How to use technology to spur development." Issues in science and technology, pp. 70-74.

Malaman, R. and Pavan, M. (2002), "Market-based policy approaches for end-use energy efficiency promotion." In: Proceedings of the IEECB'02 Conference, ADEME, Nice.

Mendelsohn, R.O. and Schlesinger, M.E. (1999), "Climate-Response Functions." Ambio no. 28, pp. 362-366.

Ministry of Environment (2008), Institutional capacities strengthening for the implementation of Kyoto protocol requirements in Lithuania, Vilnius, Final report.

Nemet, G.F. and Kammen, D.M. (2007), "U.S. energy research and development: declining investment, increasing need, and the feasibility of expansion." Energy policy 35 no. 1, pp. 746-755.

Nordhaus, W.D. and Boyer, J.G. (2000), Warming the World: Economic Models of Global Warming, MIT Press, Cambridge.

Porchia, R. and Bigano, A. (2008), "Development of a set of full cost estimates of the use of different energy sources and its comparative assessment in EU countries." In: CASES WP6 report. Total Energy Costs in EU Countries, FEEM, Milano.

Roughgarden, T. and Schneider, S.H. (1999), "Climate change policy: quantifying uncertainties for damages and optimal carbon taxes." Energy Policy no. 27, pp. 415-429.

Stern, N. (2006), The Economics of Climate Change, HM Treasury, Cambridge University Press, Cambridge.

Streimikiene, D. and Bubniene, R. (2004), "The challenges of Kyoto commitments for Lithuanian energy sector." In: ECEMEI. 3rd European congress Economics and management of energy in industry, CENERTEC, Lisbon, pp. 49-59.

Tol, R.S.J. (1999), "The Marginal Damage Costs of Greenhouse Gas Emissions." The Energy Journal no. 20, pp. 61-81.

– (2005), "The Marginal Damage Costs Of Carbon Dioxide Emissions: An Assessment Of The Uncertainties." Energy Policy no. 33, pp. 2064-2074.

Tol, R.S.J. and Dowlatabadi, H. (2001), "Vector-borne Diseases, Climate Change, and Economic Growth." Integrated Assessment no. 2, pp. 173-181.

Watkiss, P. (2006), The Social Costs of Carbon (SCC) Review – Methodological Approaches for Using SCC Estimates in Policy Assessment. Final Report to Defra, London.

Watkiss, P., Downing, T., Handley, C. and Butterfield, C. (2005), The Impacts and Costs of Climate Change. AEA and Stockholm Environment Institute, Report prepared for the DG Environment, Brussels.

Acknowledgement

This article presents the results of European Commission FM 6 project under thematic priority: Sustainable energy Systems CASES (Cost Assessment of Sustainable Energy Systems) www.feem-project.nt/cases) but views expressed herein are those of authors and can therefore in no way be taken to reflect the official opinion of the European Commission. The project was concluded in September 2008.

Chapter 8

Campaign Strategies by Environmental NGOs in the Negotiations on Climate Change

Evangelos I. Manolas

Abstract

Climate change is a global issue with important environmental, social, economic and political dimensions. This paper outlines the campaign strategies of environmental NGOs regarding international negotiations before, during and after Kyoto, covering the period from the negotiation of the Framework Convention on Climate Change in 1992 to the Conference of the Parties 11 (COP11) / Meeting of the Parties 1 (MOP1) in 2005 when the Kyoto Protocol came into force.

The paper concentrates on the activities of the World Wide Fund for Nature (WWF), Greenpeace, and Friends of the Earth (FoE) as these are widely recognised as the most important environmental NGOs in the climate change negotiations but mention is made, whenever necessary, on other NGOs involved in the process, and in particular, the Centre for International Environmental Law (CIEL), Environmental Defence (ED), the Foundation for International Law and Development (FIELD) and Ozone Action. Ten campaign strategies are discussed using information from websites, press services, reports, or journal articles.

This paper attempts to fill an information gap on NGO campaign strategies using the issue-area of climate change negotiations. As the impact of NGOs is a contestable matter and as the factor NGO strategies is always regarded as an important factor in assessing NGO effectiveness, the effort made in this paper may in the future contribute to more systematic analyses of NGO effectiveness on international negotiation outcomes in particular and NGO effectiveness in general.

Introduction

The Framework Convention on Climate Change (FCCC), the core legal instrument at the heart of the global climate regime, was negotiated as part of the 1992 United Nations Conference on Environment and Development (UNCED). It was negotiated in numerous rounds termed Intergovernmental Negotiating Committees or INCs. Since the FCCC's coming into force in 1994, there have been twelve annual Conferences of the Parties (COPs), and two Meetings of the Parties to the Kyoto Protocol (MOP), as well as a host of smaller meetings organized around specific issues and occasionally involving the adjustment and elaboration of the treaty commitments. The Kyoto Protocol, a treaty negotiated under the FCCC, and opened for signature in 1997, came into force at COP11/ MOP1 in 2005. Environmental NGOs have played a role in all of these conclaves (Raustiala & Bridgeman 2007).

The NGOs involved in the climate change negotiations as outlined above are not a homogenous group. On the one hand we have the traditional 'activist' groups; on the other hand, there are more 'pure' research-based groups with legal and/or technical expertise to promote environmental goals. NGOs that are activist are the World Wide Fund for Nature (WWF), Greenpeace, and Friends of the Earth (FoE). Important advisory organisations or 'think tanks' include the Centre for International Environmental Law (CIEL), Environmental Defence (ED), the Foundation for International Law and Development (FIELD) and several others.

The campaign activities of environmental NGOs in the climate change negotiations may fall under the dichotomy insider-outsider or reformist-revolutionary although mention is often made in the literature on NGOs pursuing dual strategies depending on the issue. According to Andresen and Gulbrandsen (2003) an NGO can pursue an *insider strategy,* seeking to attain influence by working closely with negotiators and governments by providing solutions and expert advice. They also engage in knowledge construction, producing research-based reports and papers on particular topics or offering innovative concepts and ideas (Gough & Shackley 2001). NGOs can also pursue an *outsider strategy* which means putting pressure on negotiators, governments and target groups through campaigning, letters of protest, rallying, direct actions, boycotts and even civil disobedience. With regard to climate change negotiations the tactic here is to influence public opinion in order to induce states to be more flexible in certain positions.

According to Beder (1991) the dichotomy revolutionary-reformist is defined as a tension arising for environmental groups from "the differing ideologies, but also from the practical effects of differing strategies". She observed that the more radical (or revolutionary) strategies "foster[ed] a sense of urgency and crisis so that people will cry out for change" while negotiation (or reform) „diffus[ed] that sense of crisis ..." (Beder 1991). More recently, Maddison and Scalmer (2005) interviewed an environmentalist who described revolutionaries as those who "campaign from the outside" and reformists as those who "work through engagement and partnership with government" (Hall & Taplin 2007).

This paper concentrates on the activities of WWF, Greenpeace, and FoE as these are widely recognised as the most important environmental NGOs in the climate change negotiations but mention is made whenever necessary on CIEL, ED, FIELD and Ozone Action. For each, archival information was reviewed on their strategies in the negotiations on climate change from websites, press services, reports, or journal articles. The paper aspires to fill an information gap with regard to NGO campaign strategies.

Campaign strategies

Knowledge construction

The term knowledge construction refers to the initiation of creative concepts, ideas and directions. Ideas are crucial in any human endeavor.

One way in which NGOs engage in knowledge production is through production of research papers aiming to illuminate or introduce new evidence on a particular topic. The respectability of these publications depends on the credibility and apparent neutrality of the authors. It is not uncommon for NGOs to commission established academics to produce research papers or for particular individuals within an organization to gain a respect within the scientific community regardless of that organization's media targeted campaigning activities (Gough & Shackley 2001): WWF has commissioned relatively weighty reports on the effects of climate change and biodiversity (Malcolm & Markham 2000) and extreme weather events (Vellinga & Verseveld 2000) while Greenpeace's *Pacific in peril* reviews the biological, economic and social impacts on Pacific coral reefs (Hoegh-Guldberg et al. 2000). FoE was one of the first NGOs to publish a report in 1988 on the scientific awareness of climate change (Biliouri 2000).

Another way in which NGOs engage in knowledge construction is by participation in academic or policy-inspired exercises. An example on climate change policy is the involvement of Greenpeace International in the development of the 'safe landing concept' by the Dutch research organization RIVM (Rijksinstitut voor volksgezondheit en milieu: National Institute of Public Health and Environment) (Gough & Shackley 2001). At the Vienna meeting CIEL and WWF presented a joint paper, proposing the main elements of a compliance system for the climate regime – based on lessons from other relevant international regimes. In particular, and this is another example of NGOs engaging in knowledge construction, CIEL and WWF jointly introduced the idea of a dual approach to compliance, which refers to a combination of a facilitative body available to assist parties to comply with their commitments and a judicial like enforcement body (Morgan & Porter 1999).

Environmental NGOs are also widely credited with coining the term 'hot air' in reference to proposals that would enable a country whose greenhouse gas emissions were below its legally binding limits to trade the difference. Environmentalists argued:

This problem is particularly acute in the case of Russia. If it received the same target as all of the other Annex I countries and this target was less than a five to ten percent reduction in emissions from 1990 levels, the volume of 'paper' carbon credits would be enough to eliminate the effect of the emission reduction commitment (Potential Loopholes in the Kyoto Protocol, 1997).

The term 'hot air' has since become a part of the jargon of international climate change negotiations and continues to be a central part of discussions on emissions trading. Many participants and observers to the negotiations note that without environmental NGOs, the issue of hot air would not have been raised (Report of the Third Conference of the Parties to the United Nations Framework Convention on Climate Change: 1-11 December 1997). Establishing jargon is a way for NGOs to influence how negotiators and observers perceive various issues and proposals in a negotiation (Bettelli 1997).

Forming alliances with other environmental NGOs

Environmental NGOs coordinate their participation in the climate change negotiations under the umbrella of the Climate Action Network (CAN). CAN which was formed in 1989 is a worldwide network of over 365 NGOs working to promote government and individual action to limit human-induced climate change to ecologically sustainable levels CAN members work to achieve this goal through the coordination of information exchange and NGO strategy on international, regional and national climate issues. CAN has seven regional offices which co-ordinate these efforts in Africa, Central and Eastern Europe, Latin America, North America, South Asia and Southeast Asia ("About CAN", 2008). CAN was the voice of the environmental community during the Kyoto Protocol negotiations. Members met daily during each negotiating session. These meetings were an important forum for sharing information, debating issues, and focusing lobbying efforts (Betsill 2002).

Being the recognised umbrella NGO in the international negotiations CAN unites activist and advisory environmental NGOs in one network. CAN is split into a number of working groups according to issue area, and there is a separate group on compliance issues. Over time CAN is said to have developed into a well functioning body, characterised by good procedures, open discussions and loyalty by member organisations. Although CAN is more important for the less resourceful groups than for the major ones, the CAN network usually is an effective way of communicating NGO positions with one voice during the climate negotiations (Andresen & Gulbrandsen 2003).

CAN works as an umbrella organization because environmental NGOs actively trying to affect the course of the negotiations assume similar positions. After all, despite their differences and although at times they compete for press coverage, environmental NGOs are united in their demand for environmental effectiveness. They are also aware that when they disagree on the best position to take, they send an inconsistent message that can be easily disregarded by statesmen (Corell & Betsill 2001).

Greenpeace, WWF, and Friends of the Earth International take leading roles within CAN. At the negotiations, CAN's co-ordinators and the heads of Greenpeace, WWF, and FoE climate campaigns plot their strategy in morning meetings.

Known informally as the 'politburo', this group excludes 'rank and file' environmental NGO representatives, and makes key decisions about how to manage environmental NGOs' collective campaign. During afternoon general meetings, which are not open to the public and are only open to academic observers if they are vouched for by a member of CAN, environmental NGO representatives update one another regarding specific aspects of the negotiations and discuss positions and strategy. Though members of the politburo generally direct these meetings, strategy is agreed by consensus. An important aspect of these meetings is determining what should be written in the next day's ECO Newsletter, and which country should be awarded the Fossil of the Day for impeding the negotiations (Lisowski 2005).

It must be added that within CAN a strategic division of labour is not absent. For example, CAN has left most of the responsibility for the work with the compliance system to a small group of experts. Most of the environmental NGOs have neither been very interested nor knowledgeable on the issue. The Compliance Working Group is the smallest one in CAN, usually composed of some 20 members and mostly chaired by CIEL, illustrating the rather low-key technical and complex nature of the issue. Among the large traditional activist groups, WWF has been most active on the compliance issue while organisations like Greenpeace and FoE have been less involved. To some extent this reflects a *strategic division of labour* between the NGOs. Some NGOs work mainly on the mechanisms, some on compliance and some mainly on external issues (Andresen & Gulbrandsen 2003). This type of work is important for reasons of speed and effectiveness in decision-making.

Publishing newsletters

Environmental NGOs use the quotidian ECO Newsletter to communicate their positions on issues under negotiation, their opinions on the negotiations as they unfold, and most importantly, their support or opposition for potential compromises or specific texts under negotiation. ECO is published in hardcopy at the negotiations, and is available to the public through CAN's website (Lisowski 2005).

CAN is recognised, even by official delegates, to contain valuable information and to enhance the transparency of the process (Newell 2000). ECO is part of the tactics by NGOs in their 'corridor lobbying' (Newell 2000) as it is very difficult to get in officially. In pushing for progressive decision-making during world climate change negotiations the members of CAN vote for the country that made the 'worst' input to the negotiations. The three 'winners' then get the 'Fossil-of the Day' prize. At the end of each negotiation session the overall laggard is determined by summing up the daily awards. This event has become an expected and recognised event. Some countries even try to lobby CAN to either give a 'fossil' to another country, or plead that they do not get one themselves (Anon. 2001).

Other indications that ECO is both widely read at the COPs and that its contents are taken seriously come from Betsill (2000) who documents several occasions during which national delegations asked to respond in writing to criticisms of their negotiating positions published in a previous *ECO* (Betsill 2000).

Participation in government delegations

Members of NGOs have appeared on several government delegations and have acted as consultants for governments. Some examples:

Several countries, including Canada, Denmark, Portugal, India, Indonesia, and the Philippines have invited NGO representatives to participate in the climate change negotiations as members of their official delegations (Lisowski 2005).

The Swiss Government regularly reserves three seats on its negotiating team at climate negotiations for NGO delegates: one for environmental groups, one for academics, and one for business NGOs (Raustiala & Bridgeman 2007).

FIELD, Greenpeace and WWF have all helped the Alliance of Small Island States (AOSIS) with policy advice and scientific backup in the climate negotiations (Newell 2000) and FIELD lawyers have frequently been accredited as members of small islands delegations (Oberthür et al. 2002). The tiny member governments of AOSIS, often lacking much indigenous expertise about climate change and the policy possibilities, are widely acknowledged to have played an important role in the compliance negotiations in conjunction with FIELD (Raustiala & Bridgeman 2007).

Rectifying informational asymmetries

Environmental NGOs rectify informational asymmetries between negotiators by analysing proposals and positions, and providing advice. Environmental NGOs bring experts of all kinds to the negotiations, including international lawyers, economists, atmospheric scientists, and carbon sinks experts. These individuals analyse negotiating texts and proposals on an ongoing basis. Their views are then personally conveyed to negotiators, and occasionally distributed in hard-copy or used in ECO articles. Over time, certain individuals and organizations gain reputations for providing accurate, useful information.

By rectifying informational asymmetries, environmental NGOs clearly empower negotiators. Meanwhile, by raising public awareness, environmental NGOs introduce a measure of public scrutiny to the negotiations. Both activities increase the transparency of the negotiations, and both are likely to affect the activities of negotiators. Negotiators acting on information provided by environmental NGOs, and under public scrutiny, are likely to make different decisions than they would otherwise. In particular, the additional transparency can make it difficult for some negotiators to compromise and bargain. EU negotiators at COP-6 Part I, for instance, prevaricated and then essentially refused to compromise on their demand for "supplementarity", a quantitative cap on emissions trading which would ensure a certain level of domestic emissions reductions. Environmental NGOs had publicised supplementarity as a key means to preserve the Kyoto Protocol's environmental integrity, and they followed negotiations on supplementarity assiduously. Though EU negotiators reportedly knew their demand for supplementarity was "completely unacceptable and probably unworkable", EU negotiators could not drop it without a significant concession from other negotiating parties (Lisowski 2005).

Unofficial interaction with governments

When negotiations took place behind closed doors, this meant that environmental NGOs had to use more indirect strategies to keep abreast of the progress of the talks and to promote their agenda. However, these limitations on NGO access to governments did not prove insurmountable. During the Kyoto Protocol negotiations, virtually all environmental NGO representatives as well as state delegates carried cellular telephones. It was frequently reported that delegates would step out of closed negotiations to call a particular environmental NGO and ask its opinion on an item under debate. Though not physically in the room, this enabled environmental NGOs to participate in the negotiations (Betsill 2002). Also, before and during the negotiations many informal meetings between NGO representatives and government negotiators were organized in order to discuss specific negotiating texts. In addition, NGO representatives were often observed in conversation at the negotiations with key negotiators noting the strategic, scientific, economic, political or even moral rationales for their views on particular issues (Lisowski 2005).

Cooperation with business NGOs

Business entities are not necessarily anti-environment. Business-NGOs have diverse views on environmental issues and some of them support environmental protection policies since they are compatible with their profit-seeking strategies. In this framework, environmental NGOs have certainly tried to involve business NGOs in the COP negotiations (Giorgetti 1999; Alkoby 2003).

Greenpeace addressed the European insurance and re-insurance industries, which has been receptive to their advice. Greenpeace organized numerous conferences to address the issue with the risk sector, published numerous papers, and researched the issue extensively. WWF approached the building, transportation and heavy industry (e.g., chemical and energy intensive industries) sectors to create a critical mass of support for measures to curb climate change. Ozone Action contacted the tourist industry (Giorgetti 1999).

Providing images

Exploiting strong media images to communicate a clear message to the wider public and politicians is another tactic used by NGOs. Credibility is gained not through esoteric debate but by using marketing techniques and powerful visual imagery to 'sell' climate change – both for Greenpeace International and WWF websites use photographs of polar bears to represent their respective climate change campaigns. Other popular images include icebergs calving, retreating glaciers, parched soils, floods and montages creating analogues between for example, London or Dublin and Siberia or Canada to indicate the possible effects of disruption to the thermohaline circulation. FoE has used pictures and 'quotes' from ordinary people around the world describing their experiences during extreme weather events („Counting the human cost of climate change", 2001; Gough & Shackley 2001).

Media engagement

CAN, Greenpeace, WWF and FoE court the press throughout the negotiations. Their public affairs managers provide information to the press on a daily basis, and often host press conferences. Their members frequently send editorials analysing the progress of negotiations to national newspapers, and, in addition to publishing ECO and the Fossil of the Day Awards online, their websites feature updates on the negotiations, sometimes on a daily basis. Finally, these organisations draw attention to climate change and the negotiations with installations. At COP-6 Part I in The Hague, for instance, environmental NGOs sandbagged a portion of the conference centre to highlight the danger posed by rising sea levels. At COP-6 Part II, environmental NGOs unveiled a large globe carved out of ice, which proceeded to melt in the afternoon sun. These installations provided a dramatic backdrop for television reports (Lisowski 2005).

Direct action

Direct action means the range of protests from marches and demonstrations to the actions of individual-members of NGOs (Jeffrey 2001). Examples include: Greenpeace's occupation of the island of Rockall in a campaign which, in addition to climate change, cited conservation of whales, birdlife, and other marine species to prevent further oil exploration in the Atlantic frontier (Rose 1997). Greenpeace's blockade of the Hemweg power station in Amsterdam, preventing coal from being loaded into the power station (Gough & Shackley 2001). The climbing by three activists from Greenpeace of a 190-meter tall chimney of a German coal power plant in Frimmersdorft (near Cologne) the day before COP1 began and spent twelve days on the chimney. The goal of the activists was to focus attention on the source of a chief greenhouse gas, CO_2 (Greenpeace Scales Chimney of Coalfired CO_2 Factory, 1995; Raustiala & Bridgeman 2007). Another example is when fifty activists chained themselves to the coaches and limousines used by delegates to reach the convention center in Berlin to protest the GCC and the slow pace of negotiations at the climate change conference (Jaura 1995). The FoE giand dyke, built from sandbags decorated with letters from supporters, surrounded the conference center hosting the COP6 negotiations in the Hague (Gough & Shackley 2001). At COP 3, Japanese environmental NGOs covered bushes outside the conference hall with messages, such as "Are U going to save us, Al?," and Greenpeace erected a dinosaur made out of used oil cans and spare car parts (Boulton 1997).

At COP 4, FoE, Greenpeace, and Oilwatch (an international network of NGOs monitoring the activities of oil companies) arranged the following media event in order to show that, at least in the context of the United Nations climate negotiations, environmental groups can challenge oil companies in their efforts to direct state decision-making:

One person, dressed as the Exxon oil corporation tiger, led on puppet-strings a suited figure representing the United States. The Exxon tiger was stuffing money into US pockets, and the US was chanting, "Oil good, Exxon good". The goal was to portray the United States government delegation at COP 4 as a puppet of oil interests. The procession toured around the conference centre, gathering a following of delegates and observers to the climate negotiations, until it stopped in front of the meeting room pre-assigned to conference attendees from business and industry groups. Initially, the industry representatives clustered around the open door of their meeting room and peered out at the spectacle. They then retreated into their room, closed the door, and continued to trade insults and comments with the crowd outside, through the closed door. A few minutes later, the procession disbanded (Pulver 2005).

A regular protest activity at the COPs was the presentation of the 'fossil of the day' by the CAN. The 'fossil of the day' is awarded to government delegations that are seen as blocking progress at the climate change negotiations. At every COP, an exhibit is set up to publicly track the number of fossil awards per country (Orr 2006). According to a European activist,

Giving the Fossil is a pretty effective thing. Yesterday, Brazil was upset that it has the same number of fossils as Australia, and the head of the delegation tried to talk to some of us about taking away some of their fossils from the main foyer. We said no. At another COP, we caught a Saudi delegate trying to steal some of his fossils, needless to say Saudi Arabia got another fossil award that day and we wrote a big article about it in ECO. Even little things can make a big difference (Orr 2006).

Such protest activities provide opportunities for the public to participate in the policy process, often have an educational component and draw media attention to government decisions or lack thereof (Orr 2006).

Conclusions

This paper concentrated on the activities of the three most important environmental NGOs involved in the negotiations on climate change namely Greenpeace, WWF and FoE, but mention was made whenever necessary on other NGOs involved in the process and, in particular, CIEL, ED, FIELD and Ozone

Action. Ten campaign strategies were discussed using information from websites, press services, reports, or journal articles. These strategies were knowledge construction, forming alliances with other environmental NGOs, publishing newsletters, participation in government delegations, rectifying informational asymmetries, unofficial interaction with governments, cooperation with business NGOs, providing images, media engagement and direct action. The findings show that these strategies may be insider-outsider or reformist-revolutionary but all of the three most important NGOs examined – which are traditionally thought of as activist – undertook at least one activity within the insider or reformist category either at an organizational level or under the auspices of CAN.

The findings in this paper also suggest that NGOs have no hesitation in using dual strategies when they perceive that these may be the best way towards achieving campaign goals. This is supported by the research of Maddison and Scalmer (2005), whose research found that in some situations the division between 'reform' and 'revolution' becomes irrelevant because "if the revolution cannot be made, then the stark difference with reform drops away". Instead, as shown in this paper, these two strategic approaches can be effectively used in combination. As Maddison and Scalmer say

> "revolutionary and moderate groups ... can operate as two wings of a common campaign. Here, radicals demand attention and foment controversy. ... In contrast, reformists offer a sane, rational alternative. ... Governments fear the extremists and meet with the reformists. Strategic divisions within the movement can become, briefly, a kind of political resource" (Maddison & Scalmer 2005).

Take (1999) also points out that cooperation activities are not necessarily less successful than protest-based activities:

An increasingly co-operative stance on the part of NGOs need not lead to the kind of co-option by state or international actors which is feared in various quarters, and thus to a decrease in the importance of NGOs. It may just as easily smooth the path to new strategies for exerting influence, and thus open up additional opportunities for influence (Take 1999).

The fact is that NGOs are here to stay. Perhaps, the challenge is figuring out how to incorporate NGOs into the international system in a way that takes account of their diversity and scope as well as their various strengths and weaknesses.

Many scholars agree that NGOs do make a difference in global environmental politics (e.g., Chatterjee & Finger 1994; Princen & Finger 1994; Wapner 1996; Raustiala 2001), but it is contestable to what extent they are actually successful in achieving the goals they set (Betsill & Correll 2001; Betsill 2002; Andresen & Gulbrandsen 2003; Schreck 2006; Hall & Taplin 2007; Raustiala & Bridgeman 2007). As the strategies NGOs use is always one of the factors singled out in assessing effectiveness, this paper made an effort to fill the information gap with regard to NGO campaign strategies on environmental issues by concentrating on the issue-area of the negotiations on climate change. All of the ten NGO strategies which were selected in the paper were accompanied by actual examples in order to rivet the attention of the reader to a larger issue, provide focus, illustrate application, clarify and crystallize (Goldsmid & Wilson 1980). Such an effort may in the future contribute to more systematic analyses of NGO effectiveness on international negotiation outcomes in particular and NGO effectiveness in general.

References

„About CAN" (2008), Available www.climatenetwork.org/about-can.

Andresen, S. and Gulbrandsen, L.H. (2003), The Role of Green NGOs in Promoting Climate Compliance, FNI Report, No 4.

Anon. (2001), "What is 'Fossil-of-the-day' about", available www.fossil-of-the-day.org/about.html.

Beder, S. (1991), "Activism Versus Negotiation: Strategies for the Environmental Movement." Social Alternatives vol. 10 no. 4, pp. 53-56.

Betsill, M. (2000), Greens in the Greenhouse: Environmental NGOs, Norms, and the Politics of Global Climate Change, Ph.D. Dissertation, University of Colorado.

– (2002), "Environmental NGOs Meet the Sovereign State: The Kyoto Protocol Negotiations on Global Climate Change." Colorado Journal of International Environmental Law and Policy vol. 13, no.1, pp. 49-64.

Bettelli, P., Carpenter, C., Davenport, D. and Doran, P. (1997), Report of the Third Conference of the Parties to the United Nations Framework Convention on Climate Change: 1-11 December 1997. Earth Negotiations Bulletin 12 (76).

Betsill, M. and Corell, E. (2001), "NGO Influence in International Environmental Negotiations: A Framework for Analysis." Global Environmental Politics vol. 1 no. 4, pp. 65-85.

Biliouri, D. (2000), The Role of NGOs in Promoting Climate Change: An EU Perspective. Climate Change Communication Conference, available http://dsp-psd.pwgsc.gc.ca/Collection/En56-157-2000E.pdf.

Boulton, L. (1997), "Kyoto Draws Together a Diverse Cast." Financial Times, Dec. 6, at 3.

Chatterjee, P. and Finger, M. (1994), The Earth Brokers: Power, Politics and World Development, Routledge, London.

Corell, E. and Betsill, M. (2001), "A Comparative Look at NGO Influence in International Environmental Negotiations: Desertification and Climate Change." Global Environmental Politics vol. 1 no. 4, pp. 86-107.

„Counting the human cost of climate change" (2001), Earth Matters no. 48, Friends of the Earth International, Amsterdam.

Giorgetti, C. (1999). "From Rio to Kyoto: A Study of the Involvement of Non-Governmental Organizations in the Negotiations on Climate Change." New York University Environmental Law Journal vol. 7, pp. 201-245.

Goldsmid, C.A. and Wilson, E.K. (1980), Passing on Sociology: The Teaching of a Discipline, American Sociological Association / Teaching Resources Center, Washington, D.C.

Gough, C. and Shackley, S. (2001), "The Respectable Politics of Climate Change: The Epistemic Communities and NGOs." International Affairs vol. 77 no. 2, pp. 329-45.

Greenpeace Scales Chimney of Coalfired CO_2 Factory (Mar. 27, 1995), available www.greenpeace.org/~climate/berlin1995/index.html#report1.

Hall, N.L. and Taplin, R. (2007), "Revolution or inch-by-inch? Campaign approaches on climate change by environmental groups." Environmentalist no. 27, pp. 95-107.

Hoegh-Guldberg, O., Hoegh-Guldberg, H., Stout, D.K., Cesar, H. and Timmerman, A. (2000), Pacific in peril: biological, economic and social impacts of climate change on Pacific coral reefs, Greenpeace Australia Pacific, Sydney.

Jaura, R. (1995), Environment: Protests Shake Up the U.N. Climate Meet in Berlin. International Press Service, March 30, available in 1995 WL 2260055.

Jeffrey, D.W. (2001), "The Roles of Environmental Non-Governmental Organisations in the Twenty-First Century." Biology and Environment: Proceedings of the Royal Irish Academy vol. 101B no. 1-2, pp. 151-156.

Lisowski, M. (2005), "How NGOs Use their Facilitative Negotiating Power and Bargaining Assets to Affect International Environmental Negotiations." Diplomacy and Statecraft no. 16, pp. 361-383.

Maddison, S. and Scalmer, S. (2005), Activist Wisdom: Practical Knowledge and Creative Tension in the Life of Social Movements, UNSW Press, Sydney.

Malcolm, J. and Markham, A. (2000), Global warming and terrestrial biodiversity decline, Report prepared for WWF Switzerland.

Morgan, J.L. and Porter, S.J. (1999), Compliance Institutions for the Kyoto Protocol: A Joint CIEL/WWF Proposal, Discussion draft, prepared for the fifth conference of the parties.

Newell, P. (2000), Climate for Change. Non-state Actors and the Global Politics of the Greenhouse, Cambridge University Press, Cambridge.

Oberthür, S., Buck, M., Müller, S., Pfahl, S. and Tarasofsky, R.G. (2002), Participation of Non-Governmental Organisations in International Environmental Governance: Legal Basis and Practical Experience, Ecologic, Berlin.

Orr, S.K. (2006), "Policy Subsystems and Regimes: Organized Interests and Climate Change Policy." The Policy Studies Journal vol. 34 no. 2, pp. 147-169.

Potential Loopholes in the Kyoto Protocol (1997), ECO, Climate Action Network. Bonn G.B.R.), Oct. 27, at 3, available www.climatenetwork.org/eco/eco4_agbm8.html.

Princen, T. and Finger, M. (1994), Environmental NGOs in World Politics. Linking the Local and the Global, Routledge, London.

Pulver, S. (2005), "A Public Sphere in International Environmental Politics: The Case of the Kyoto Protocol Negotiations." 2005 Berlin Conference on the Human Dimensions of Global Environmental Change: International Organizations and Global Environmental Governance, Berlin.

Raustiala, K. (2001), "Non-state Actors in the Global Climate Regime." In: Luterbacher, U. and Sprinz, D.F. (eds.), International Relations and Global Climate Change, MIT Press, Cambridge.

Raustiala, K. and Bridgeman, N.L. (2007), "Non-state Actors in the Global Climate Regime." UCLA School of Law, Public Law & Legal Theory Research Paper Series, Research Paper no. 07-29.

Report of the Third Conference of the Parties to the United Nations Framework Convention on Climate Change: 1-11 December 1997, Earth Negotiations Bulletin, International Institute for Sustainable Planning, New York, December 13 at 15.

Rose, C. (1997), Putting the lid on fossil fuels: why the Atlantic should be a frontier against oil exploration, Greenpeace UK, London.

Schreck, S.C. (2006), "The Role of Non-governmental Organizations in International Environmental Law." Gonzaga Journal of International Law vol. 10 no. 2, pp. 252- 268.

Take, I. (1999), "NGOs as Strategic Actors in International Politics", Paper prepared for the Workshop on 'Environmental Protest in Comparative Perspective' at the 27th Joint Sessions of ECPR Workshops in Mannheim, 26-31 March, available www.essex.ac.uk/ecpr/jointsessions/Manpapers/w21/take.pdf.

Vellinga, P. and Verseveld, W.J. (2000), Climate Change and extreme weather events, Report prepared for WWF Switzerland.

Wapner, P. (1996), Environmental Activism and World Civic Politics, State University of New York Press, Albany.

Chapter 9

The Role of Sustainable Buildings in Climate Change Mitigation

A.C. Warnock, C.G. Warnock

Abstract

Transforming the built environment will be critical in mitigating global emissions of green house gases. Within OECD states, buildings (in terms of their construction, operation and de-construction) are major emitters of green house gases. Thus far, market forces alone have failed to ensure the widespread adoption of sustainable buildings and thus a catalyst in the form of government intervention is necessary. This paper considers examples of law and policy from around the world that aim to promote a sustainable built environment. To set the contextual framework, the paper details the high-level political and scientific acknowledgment that sustainability in buildings and construction is a critical element in addressing climate change and thereafter provides an example and brief description of an exemplary sustainable building. The body of the paper describes both central and local government initiatives achieved through the conduit of fiscal policy, planning controls, certification schemes and regulation. Particular emphasis is placed upon the necessity to bring existing buildings into the sustainability equation. It should be noted that the focus of the paper is primarily upon building practices prevalent in developed nations, although this of course has a resonance for developing nations that may be adopting similar, equally unsustainable approaches to construction.

Introduction

The inherent limits of the atmosphere and its susceptibility to human influence can be neatly illustrated. If it were possible to liquefy the atmosphere, its volume would equate to $1/500^{th}$ that of the world's oceans.[1] The actions of the six billion human inhabitants and our carbon-based economies are adding approximately two parts per million (ppm) of carbon dioxide equivalent to the atmosphere each year and this is increasing annually. In 2005, there were approximately 379 ppm of carbon dioxide equivalent in the atmosphere; once the concentration exceeds 450 ppm scientists have calculated that there is a 20% chance of reaching a tipping

---

1 Speech given by Professor Tim Flannery (Macquarie University, Sydney, author of *The Weather Makers: The History and Future Impact of Climate Change* Text Publishing, Melbourne, 2005) to the University of Otago on 7th August 2007.

point from which there is no return.[2] Given this fraught backdrop, calls to reduce greenhouse gas emissions in the region of 50-80% from 1990 levels by 2050 seem eminently appropriate and in order to achieve such dramatic reductions, there must be a sea change in the way that we live. Clearly, all possible methods and mechanisms to reduce emissions must be seized upon and it is becoming increasingly apparent that the built environment – the way in which we build and how we use buildings – has a critical role to play. Focusing upon the structural element of buildings, and primarily upon law and policy to promote sustainable buildings, this chapter considers the ways in which buildings might become part of the solution as opposed to exacerbating the problem of climate change.

The role of buildings in climate change mitigation

Buildings and the building industry are responsible for significant environmental damage. By way of example, the building industry alone consumes some 40% of the world's energy, 25% of forest timber and 16% of the world's fresh water (Dimson 1996). Given the enormity of this damage, without radical reform of the construction industry, it may become impossible to 'meet the needs of the present without compromising the ability of future generations to meet their own needs' (WCED 1987). As a result, global interest in promoting *sustainable* building and construction is gaining momentum. However, it is the issue of climate change that is proving to be the most convincing catalyst for change as the development and consequent use of buildings accounts for approximately 30% of global greenhouse gas emissions (Laurenzi 2007). The Intergovernmental Panel on Climate Change (IPCC) notes that globally the highest rate of greenhouse gas emissions from all the sectors studied are linked to commercial and residential buildings and sees the 'greening of buildings' as a critical means to addressing both the mitigation of and adaptation to climate change. Within the Fourth Assessment Report of the IPCC, pre-eminent emphasis is placed upon the necessity to increase the performance of both new and existing buildings (IPCC 2007). In terms of introducing measures that will result in significant emissions savings, the building sector is a relatively easy sector to address compared to, for example, agriculture or transport. There are three primary reasons for this. Firstly, developing high performance buildings is one of the most cost-effective measures available to reduce emissions; it is 'low hanging fruit' in terms of mitigation (McKinsey 2007). Secondly, the greening of buildings

---

2 Flannery, supra note 1. For further information see IPCC, 2007: Summary for Policymakers. In: *Climate Change 2007: The Physical Science Basis. Contribution of Working Group I to the Fourth Assessment Report of the Intergovernmental Panel on Climate Change* [S. Solomon, D. Qin, M. Manning, Z. Chen, M. Marquis, K.B. Averyt, M. Tignor and H.L. Miller (eds.)] Cambridge University Press, Cambridge, United Kingdom and New York, NY, USA.

promises a win-win scenario whereby the cost of the initial capital expenditure is recouped over time (Eurima 2005). Thirdly, the technology and knowledge already exist to construct buildings that are considered 'carbon neutral' or 'zero carbon' as the following case study demonstrates.

How to build zero carbon homes: a case study from the UK

Within the UK, and in its most simplistic terms, a zero carbon building is defined as a building that meets its operational energy demands in a way that does not produce carbon dioxide. To achieve a zero carbon building, the problem of emissions must be attacked from two perspectives. The need for energy must be reduced to the smallest reasonable level without compromising modern acceptable lifestyle standards and the remaining energy required must be produced in a way that doesn't compromise the future health of the planet. The UK's commitment to reducing carbon has produced a new system for assessing the impact that houses have on the environment that takes into account these considerations. The Code for Sustainable Homes looks at building developments holistically to assess their impact in terms of energy and carbon dioxide emissions; water management; management of buildings during construction and use; the health and well being of occupants; materials used; the reduction of pollution during construction and subsequent use, and the enhancement of ecology (CSR 2007). The highest level of measurable sustainability – level 6 – will be awarded to sustainable, zero carbon developments.[3] Industry scepticism abounds as to whether this level of performance is achievable however zero carbon developments, and proof as to their possibilities for the future, already exist.

The Building Research Establishment (BRE) Innovation Park in Watford, Hertfordshire, boasts the country's first two Code for Sustainable Homes level 6 buildings (BRE 2008). The site was devised by BRE as a showcase for modern living with sustainability at the core. At the time of writing, the park contained six houses, a school and a visitors centre all set within a bio-diverse landscape, designed by MacFarlane Wilder along the principles of Sustainable Urban Drainage.[4] The development is a kind of eco-hamlet with houses achieving anywhere from code level 4 to 6. The Innovation Park has developed through a collaboration of building professionals, manufacturers and construction companies. With 200 technologies on show, the development is effective in demon-

3 Code Levels range from 1 to 6. A Code level 1 house must have energy efficiency 10% better than the 2006 Building Regulations requirement, level 3, level 4 and level 5, 25%, 44% and 100% better respectively. Level 6 will produce zero carbon.

4 Sustainable Urban Drainage promotes rainwater being absorbed into the local landscape or holds it in retention ponds, sumps and streams to mitigate the effects of flash flooding on the public drainage system.

strating that the innovations already exist to build practical and affordably sustainable homes and the Kingspan Lighthouse, completed in June 2007, is triumphant in being the first building in Britain to achieve Code for Sustainable Homes level 6 (Kingspan 2007).[5]

An exploration into modern living, the Kingspan Lighthouse is an elegant example of how re-thinking the housing typology will bring zero carbon homes within reach. Designed by architects Sheppard Robson, and with engineering by Arup, it is shaped like a billowing sail. The house is 'up-side-down' in organisation, with bedrooms at entrance level and double height, open plan, living accommodation opening on to a terrace at first floor. A mezzanine deck over the living space provides a home office facility, with the option for extending to provide further bedroom accommodation in lieu. The decision to up-end the typical housing model was driven by a need to reduce the percentage of window-to-wall space and to limit heat loss. Therefore, bringing light though the roof into the lofty upper floors is a rational design response to the difficult relationship between light levels and heat loss. Taking a section through the building demonstrates that it is organised around a full height stair void, capped by a wind catcher with south facing glass and aluminium louvers. This provides a chimney at the heart of the dwelling, promoting airflow to cool in summer and ventilation throughout the year. The wind catcher forces daylight deep into the heart of the building, reducing the need for artificial light. The dramatic roof, which arches from apex to terrace, yields a large surface area on which an extensive solar array can happily sit without dominating the look of the building. The roof's geometry is governed by a 40degree angle towards the South to maximise outputs from the combination of solar thermal and photovoltaic panels. The roof, which is over-clad with sweet chestnut slats, incorporates retractable shutters that prevent over-heating in the summer by offering dappled shade to the living room windows or alternatively, unshaded the glazing promotes passive solar gain in the winter. The form and organisation of the Kingspan Lighthouse therefore support passive design which reduces energy demand: the wind catcher passively cools and ventilates, the orientation and glazing accords with passive solar heating, whilst the roof cladding provides passive shading to prevent overheating in summer.

Generating sufficient energy to meet current demands via micro generation, whilst continuing to use typical building construction, would be impracticable. Energy efficiency is therefore key to the Kingspan Lighthouse's success in becoming net carbon zero. A Structural Insulated Panel system (SIP), which was fabricated off-site, forms the shell of the house. The system comprises of 284mm

---

5 As well as being net carbon zero, the Kingspan Lighthouse achieved Code for Sustainable Homes Level 6 based on other qualities such as water saving features, quality of internal environment and use of materials which are rated in BRE's *Green Guide to Housing Specification* (BRE 2008).

thick Oriented Strand Board (a sheet product made from shredded timber) bonded with zero-ozone depleting potential, rigid Urethane insulation. The SIPs are lightweight and can be assembled on site in a matter of days. As a factory made product, with specially designed jointing details, the system enables the building to be incredibly airtight. This is crucial to the energy efficiency of the dwelling: much heat loss is through gaps in building envelopes and low-level draughts create thermal discomfort, adding to the perceived need for heating. Through using this system of fabrication, and by installing triple glazed, gas filled windows, the heat lost from the building's envelope is around 60% less than a typical home built under the current building regulations. Through the benefits of enhanced natural lighting, thermal efficiency and passive cooling, the basic architectural design of the Lighthouse goes a significant way to reduce the energy demands of the home and coupled with energy saving devices such as A++ rated appliances, 100% compact fluorescent lighting and smart metering (that allows occupants to monitor energy consumption), the level of renewable technology required to power and heat the home becomes commercially and practically achievable.

Domestically, by far the most voracious consumers of energy constitute heating and hot water systems, taking nearly 75% of the total share (UK DCLG 2006). The technology to address space and water heating showcased on the Kingspan Lighthouse is largely familiar. Heat loss parameters are maintained through inclusion of a whole house ventilation system with 88% heat recovery. This system maintains air quality without sacrificing the efficiency of the building envelope. Low-level space heating, and a proportion of the hot water demands, are met by an automatic wood pellet (biomass) boiler. Biomass is considered carbon neutral, as burning the wood is offset by the process of growing the timber. Hot water and power is provided by the 4m2 solar hot water array and 46m2 of photovoltaic panels on the roof. Perhaps the most untested innovation in the Kingspan Lighthouse is the Phase Changing Material (PCM) used to line the underside of the roof. PCM is a product developed by BASF and is a technology with selective thermal mass. It is made from gypsum-based board, incorporating tiny polymer spheres containing paraffin wax. The curved ceiling is clad using PCM and on a hot day, the PCM absorbs heat. As it does so, the paraffin within the spheres turns to liquid. At night, as the building cools, assisted by the windcatcher, the liquid solidifies, emitting heat into the space beyond. Traditionally, thermal mass in buildings has been provided using material with high-embodied energy, such as concrete, therefore developing lightweight building systems, which respond slowly to thermal input, could be a significant advance forward.

The Kingspan Lighthouse proves that environmental systems, construction methods and an innovative approach to design can produce a high quality dwelling that permits a modern lifestyle whilst producing zero carbon emissions. However, as a one-off house the building costs approximately 40% more than a typical house, with the solar array alone costing GBP£40,000. This stumbling block, and the role that economies of scale might play in reducing the cost, has been noted by the design team and as a consequence the Kingspan Lighthouse has been developed to provide a template either for a single dwelling or a multiple unit development where district heating and an on-site wood fuelled Combined Heat and Power plant (CHP) would provide a more economic solution for generating electricity and heating. In addition, the design team have calculated that on an estate consisting of up to 250 units the installation of wind turbines to supplement the power provided by CHP could be commercially justified.

In principle, zero carbon housing has therefore been proved to be achievable and importantly, in April 2008, the Barratt Green House, designed by Gaunt Francis Architects and constructed by Barratt Development plc, became the second house to achieve Code level 6. This is the first home in the Innovation Park to be undertaken by a mainstream house builder. Whilst the economies of scale required for developing ever more cost effective ways of generating renewable electricity remain an issue, the involvement in the debate of major players in the house building industry hails the beginning of a new expectation for sustainable housing in the UK.

Will the market deliver sustainability in buildings?

There can be little doubt that buildings developed to a high standard of sustainability cost less to run and as a consequence should, theoretically, prove to be more valuable than a standard building. One might expect that market demand for such a product would be high. Despite this, the vast majority of new buildings are not built to high performance standards and researchers have noted that this flows from a combination of factors. Primarily, developers have little interest in creating high performance buildings. As developers (in most instances) are not the end users of the product, asides from complying with mandatory building regulations, they have little interest in the running costs of the building and in the absence of client demand are unlikely to incorporate 'green' features into the construction process. The majority of developers operate to maximise profits and thus there may seem little point to them in incurring additional time and effort up-skilling or increasing build costs by including renewable energy technology. To complicate matters, the vast amount of information, often contradictory, as to what constitutes a sustainable building can obfuscate client demand (Warnock 2007). Members of the UK Sustainable Building Task Group note that there is a market failure in respect of sustainable buildings but its members do not believe that consumers are acting in an economically perverse way, rather this situation is explained by the fact that the "market lacks the features of choice, reliable labelling, product information and comparability which characterises consumer-led markets in most other goods" (Harman 2005: 14). The Task Group also noted the economies of scale conundrum: without significant client demand (which reduces the proportion of research costs in the final price of a product and fosters competitive market conditions) 'green' technologies and building practices can prove expensive, as the Kingspan Lighthouse case study demonstrates. It was anticipated by the Task Group that the cost of adopting higher standards would decrease sharply as volume increases. Clearly a catalyst is required to promote the metamorphosis towards sustainable building and it seems inevitable that this must come in the form of state intervention.

The necessity for state intervention

Given the critical role that the built environment will play in climate change mitigation, the Intergovernmental Panel on Climate Change has urged all governments to introduce policies and remove market barriers to promote sustainable construction and buildings (IPCC 2007). In countries around the world, plans are a pace to introduce policy and increasingly stringent legislation to promote sustainability in buildings with a particular focus being placed upon minimising greenhouse gas emissions from new buildings. For example, within the Asia-Pacific region, the Chinese Vice-Minister of Construction has recently acknowledged that there must be a wholesale national review of the law and policy to promote energy efficient 'green buildings' (Marrakech 2007). In the UK, the Government has announced the intention that all new-build homes will be zero carbon emitters by 2016 and proposes to use three main policy levers to achieve this target: the incremental ratcheting upwards of the energy efficiency standards in the Building Regulations, the planning system and the Code for Sustainable Homes (UK DCLG 2007a). From May 2008, all new houses in the UK must be assessed and given a Code for Sustainable Homes level rating with the obligation rising steadily towards an anticipated level 6 by 2016. Thus from that date, zero carbon development may be the norm. Clearly, building legislation and regulations will drive the sustainability agenda for new buildings but economic instruments are also important; financial incentives encourage innovation and result in higher levels of sustainability than mere regulation alone (Warnock 2007). Within North America and Europe a myriad of fiscal tools are used to promote sustainable buildings including tax incentives and disincentives, grants, reduced and subsidised loan rates, state guarantees to banks, the release of state land for sustainable developments and so on (Warnock 2006).

However, another issue is apparent. A cursory look at statistics reveals that the approach taken to *existing* buildings will be critical in progressing sustainability within the built environment. For example, in New Zealand, where this issue is presently taxing policy makers, there are approximately 1.6 million dwellings; approximately 1.04 million homes were built before 1977 and of these, a third have wholly inadequate insulation (Amitrano 2006). In contrast, only 25,000 new homes, or thereabouts, are built each year (EECA 2007).

Bringing existing buildings into the sustainability equation

Addressing the energy efficiency of the existing building stock will be critical in order to maximise the mitigation opportunities from the built environment. As the authors of the United Nations Environment Programme Report on Buildings and Climate Change note,

> "[i]n Finland a study calculated the overall impact of energy-efficient buildings on energy consumption … Assuming that all new houses built from 2006 onwards were to be passive houses with an energy consumption of 30kWh/m^2, the total energy saving compared to Business as Usual would be 15% in year 2003. The relatively low percentage is due to the building stock's long life cycle. From these calculations we can draw the conclusion that in order to achieve significant results in the following decades much emphasis needs to be put on the existing building stock and its renovation, at least in the more developed countries" (UNEP 2007b).

Within European states in particular, various mechanisms are used to ensure that existing buildings become part of the sustainability equation and meet updated standards of energy efficiency. Such measures encompass regulation, financial incentives and disincentives, voluntary mechanisms and the provision of information to influence the market. Given, the critical importance of this issue, some of the most pertinent measures are considered below.

Command-control measures

As the catalyst for promoting green buildings is climate change, a general approach taken by states within the European Union (EU) is to enshrine in legislation national reduction targets for greenhouse gas emissions and then to develop sectoral policies to achieve these targets. By way of example, France has launched the 'Factor 4' programme that aims to reduce national greenhouse gas emissions by 75% between 1990 and 2050. To achieve the necessary radical cuts in the building sector, the 'PREBAT Programme (2005-2009)' has been launched and brings together industry and government to develop mechanisms to achieve the target. The modernisation of existing buildings is a critical component (UNEP 2007b).

More focused action within the EU has been co-ordinated by the European Parliament and on 1st January 2003, the Directive on the Energy Performance of Buildings 2002/91/EC (the Directive) was introduced (EPB 2002). The Directive harmonizes minimum mandatory requirements within the EU member states in relation to sustainable construction and buildings. In relation to existing buildings, the Directive provides for: minimum energy performance requirements for major renovations of existing buildings larger than 1000 m^2; energy performance certificates for all buildings; regular inspections of boilers/ heating and air conditioning systems; for energy savings targets to be ratcheted up periodically and rules concerning tendering, education and the dissemination of information. It is possible for individual member states to introduce more stringent national requirements and many have done so. The authors of the 2007 UNEP report have affirmed that regulations should as far as possible cover the energy use over the entire life span of buildings and be applicable to new buildings as well as existing ones (UNEP 2007b). Whilst it is practically difficult to achieve the same level of energy efficiency in old buildings as in new-build, most EU states require specific alterations to buildings below the EU Directive threshold of 1000 m^2 to meet *revised* performance standards.

There are many examples of additional command-control measures that have been implemented to encompass existing buildings. Over 60 local authorities in Spain have solar ordinances, mandating solar water heating in both new build and renovation projects that encompass water heating (BEA 2007). In the Netherlands, the Government aims to bring all houses up to 1975 standards of insulation by making this a mandatory requirement when a house is sold and providing a rates subsidy as a fiscal incentive (Halliday 2004). Particular states have begun to consider how planning regulations might be used to actively encourage the greening of buildings and in the UK a more permissive regime for micro-generation equipment attached to domestic properties is to be implemented (UK DCLG 2007b).

Economic measures

In addition to including renovated buildings within revised regulatory standards, the leading 'green building' nations have introduced a raft of economic measures to promote energy efficiency in existing buildings. These take the form of both constraining measures (such as taxes, fees, price levies) and enabling measures (tax rebates, subsidies, grants). The priority given by such nations to promoting this goal can be demonstrated by the budgets provided. By way of example, Finland (with a population of 5.2 million) has a current annual budget of 17 million euros set aside to promote energy-efficient renovations of residential buildings (UNEP 2007b).[6] The German Government has announced a 20-year programme to bring all pre-1978 housing stock up to contemporary energy standards and has provided a significant budget to achieve this end (House of Commons 2006). Specific economic policies of note include the German 'green audits' consisting of a grant of up to 250 euros to fund an 'on-site' consultation with an architect or other professional to detail what energy efficiency improvements could be made to a building. In addition, the German Federal Government, in conjunction with the KfW Forderbank, has undertaken a large-scale research and educative project that has resulted, to date, in the renovation of approximately 2,230 housing

---

6 Grants are offered for changing heating systems from oil to solar, installing wood pellet heating system, connecting to district heating, building ground heat pump system and so on.

units. The renovated units are on average 50% more energy efficient than the German Energy Conservation Regulations require of new buildings. The project serves to highlight best practice case studies and demonstrates how different buildings can meet present day energy efficiency standards (German FMTBUA 2007). Other examples of notable economic measures include the preferential lending rates offered by the Husbanken (State Housing Bank) in Norway to disadvantaged members of the community undertaking renovations that meet environmental criteria. The criteria are established in environmental design guides established by the Bank (Husbanken 2007).[7] A public-private partnership between the Piedmonte Region, Italy and local banks offers preferential loans in relation to new or renovated buildings that attain a high performance rating on the Protocollo ITACA (a rating tool based on the Green Building Challenge assessment tool) (iiSBE Italia 2007). Outside Europe, the United States, Canadian and Australian federal and state governments offer panoplies of financial incentives (Industry Canada 2006). By way of example, the Photovoltaic Rebate Program funded by a partnership between the Victoria State and the Australian Federal Governments offers rebates of up to AUD $1,500 towards the installation of solar water heating systems and AUD $4,000 to householders, schools and community buildings who install photovoltaic systems. This programme is also intended to provide a boost to the Australian photovoltaic industry.[8]

Voluntary measures

In terms of voluntary models and partnerships between industry and government, the Dutch 'National Packages for Sustainable Buildings' scheme is of particular note. The scheme has lead to the production of manuals detailing sustainable building measures for both new and exiting buildings. The aim is to ensure that all actors agree as to the measures and standards that should be adopted to create a sustainable building and to increase the capacity of builders to provide this service (UNEP 2007b).

Labelling

Arguably the most significant policy introduced in the EU to date has been the requirement for the labelling of buildings. It has long been acknowledged that improvements cannot be made to the existing building stock until the present energy efficient status of property is measured. As the UK Sustainable Building Task Group noted the lack of labelling and product information in the housing market in

7 Further, see SBIS *Policies and Programs* for Norway, available at www.sbis.info/database/dbsearch/details/polprogdetails.jsp?index+4 (note: all ULRs are as at 31 March 2008).

8 See Photovoltaic Rebate Programme, Government of Australia available at www.greenhouse.gov.au/renewable/pv/index.html.

particular has contributed to the apparent market failure in green building (Harman 2005). The EU Directive mandates Energy Performance Certificates (EPC) for all buildings and member states have until 2009 to fully implement the terms of the Directive. An EPC must report on the present performance of the building (grading the building on a scale of A-G based on present building regulations) and also contain recommendations for improvements in the energy performance of that building. An EPC is required for all new buildings and where an existing building is to be sold or let. The vendor, landlord or builder is responsible for providing and bearing the costs of the EPC. The Directive provides that an EPC may be valid for ten years but individual states can set more stringent regulations in this regard. The UK, for example, is proposing that an EPC must be less than twelve months old at the date of sale or lease. Within the UK, the Energy Performance of Buildings (Certificates and Inspections) Regulations 2007 (SI 2007/991) provide that local authorities will be responsible for enforcing the EPC regulations and that local authority officials will be able to issue penalty charges for non-compliance.

It is anticipated that the introduction of mandatory certification will have far reaching effects. In Japan, where mandatory reporting of energy conservation measures for non-residential buildings has been in place since 2003, 'compliance' with voluntary energy-efficiency regulations has risen from 34% in 1999 to 74% in 2004 (Laurenzi 2007). In addition, given purchasers' fears with regards to fuel insecurity and the expectation of increasingly stringent climate change policies, energy efficient homes will arguably be at an advantage in the market; people will actively seek them out and they will be more valuable. Energy inefficient homes will be devalued. Although empirical evidence to support such a contention is not yet available in terms of the residential market, research in relation to the letting of commercial buildings supports this assumption (Jones Lang LaSalle 2007) (Myers 2007). Labelling will certainly affect the type of buildings that housing associations and public bodies purchase or lease and will influence the investment decisions of the trustees of investment and pension funds bound by certain fiduciary duties (Freshfields 2005). Energy Performance Certificates may be used as a point of negotiation in property sales and leasing agreements (as surveys are at present). Purchasers may negotiate for a reduced price or insist that the seller incorporate some of the recommended energy efficient features prior to sale.

A logical next step is for labelling simply to become the first stage in a legislative process that requires the recommended improvements to be carried out. Alternatively, the EPC may provide a means of differentiating between buildings for the purposes of rates or taxes. Given the commitment within the EU to reduce carbon dioxide emissions by at least 60% by 2050 the necessary political will is present and in a market of buoyant property prices, requiring energy efficient property is not a difficult political argument to make.

Local authorities role in promoting sustainable buildings

The approaches mentioned thus far have focused upon the role of central government or upon public-private initiatives. Given the clear need for state intervention in the promotion of 'green buildings' an additional question emerges: does local government have a role to play in promoting sustainability in new and existing buildings? In many nations the answer is a resounding 'yes'; local authorities may have a pivotal role to play in this regard. However, whilst a collation of present local authority practices from around the world will provide many useful examples of the role of local government in promoting sustainable building, it is of course important to note that all jurisdictions operate different regulatory frameworks. Local authorities that achieve the greatest environmental results often have powers to impose direct amendments to building regulations (for example, Malamo in Sweden, Vancouver in Canada and municipal governments in Germany) or have considerable latitude in directly influencing the performance of sustainable buildings via the planning approval process (for example in the UK where the national building regulations contain a minimum standard of performance and there is no bar on authorities requiring higher standards). Asides from the power to directly amend building regulations there are however a myriad of alternate ways in which local authorities can promote sustainable buildings including their role as educators, funding bodies and 'exemplary citizens' adopting high standards for public buildings. Some of the most innovative methods do however involve the authorities roles and responsibilities in the planning arena and to illustrate this contention a selection of techniques taken from the Canadian, British, German and American jurisdictions are included below.

Sustainability checklist

For local government purposes, a useful tool in promoting sustainable building is a 'sustainability checklist'. Essentially, a sustainability checklist is a framework for discussing the standards of sustainability that a particular development might achieve. A checklist should contain a condensed version of the community's goals and ideally should follow a relatively simple format that developers can complete with ease. Developers use the checklist to identify community goals for development and to report which goals the proposal will meet, to what degree and how. Checklists can require developers to report upon the sustainability of the proposed development in terms of the design and structure of individual buildings, in addition to addressing wider site issues. Many local authorities in different jurisdictions use sustainability checklists. Within British Columbia, Canada, sustainability checklists have been described as follows:

> "A sustainability checklist is a non-regulatory measure that can help encourage energy efficient or green buildings. Completion of the checklist is required as part of a development permit application and/or an application for an amend-

ment to the zoning bylaw (rezoning). Completion of the checklist is mandatory, but pursuing specific actions is not mandatory. For this reason, the checklist is not regulatory in nature and is very developer friendly. The checklist is part of the municipal development review process and helps council to ensure development proceeds in accordance with the community's values" (and this is achieved by linking the checklist to objectives specified in planning documents)" (Community Action on Energy and Emissions 2007: 35).

In the UK, the *London Plan* requires the completion of a sustainability checklist for regionally significant projects that will be considered by the Mayor of London, and many Boroughs within Greater London also have versions of sustainability checklists that are required in relation to smaller developments. The Ealing Borough Council advises that their checklist will be used to help determine planning applications and it has the legal status of a 'material consideration' which the local authority is entitled to take into account in making decisions (Ealing 2007). The drafters of the *London Plan* describe the 'Regional Sustainability Checklist for Developments' as follows,

> "The checklist is a tool which looks at the sustainability of developments and is used by decision makers within the planning approval process, whether they are in the developer team, the local authority, the client team or are potential purchasers. It looks at a range of sustainability issues by expressing the standards that are outlined in the London Plan's Supplementary Planning Guidance (SPG) on Sustainable Design and Construction by a series of questions. The checklist also indicates the level of performance that the development achieves for each issue. A high level summary highlights how the development performs in each of the ten sections, and also indicates issues which are deemed not applicable or where the minimum standard has not been achieved. The checklist provides a simple way to help deliver sustainable design and construction policy (and standards) through the planning system in a regionally specific but nationally consistent way" (Mayor of London 2008).

It is important that checklists are available to development teams at the very beginning of the development process and may, accordingly, be used to guide the design and planning for the development. Checklists can provide a useful template for structuring discussions between developers and local authorities, for guiding agreements as to where good or best practice standards would be appropriate rather than minimums, and for agreeing in advance which issues may not apply. The checklist may also play an important role as a foundation document for fast-tracking the planning process or for obtaining various financial incentives for green building. Examples of specific sustainability checklists may be found on local authority websites around the world.[9]

---

9 See for example: City of Port Coquitlam, BC, Canada, *Sustainability Checklist* available at www.city.port-oquitlam.bc.ca/__shared/assets/Sustainability_Checklist2040.pdf. City of New Westminster, BC, Canada, *Smart Growth Development Checklist* available at http://66.51.172.116/AboutUs/Issues/AlternativeDevelopmentStandards/NewWestminster

Fast Tracking Planning Approvals

Many local authorities have a process linked to a sustainability checklist or rating tool assessment that enables the fast tracking of development permits and planning approval.[10] Research for the City of Toronto suggested that,

> "developers prefer a process that speeds up planning approvals, as compared to cash incentives, which typically represent a very small percentage of the total project costs [as] a faster approvals process can reduce the financial risks and costs associated with project delays. A fast track process for green buildings also has the added benefit of developing greater in-house expertise within a municipality to negotiate with developers for green features" (Toronto 2006: 56-57).

Zoning for Green Developments

Around the world, examples of local authorities setting aside land specifically for green developments abound. The regulatory framework dictates to what degree zoning can provide for green development. By way of example, in Canada local authorities may, given particular circumstances, establish comprehensive development (CD) zones within which authorities have a great deal of latitude towards land uses and approaches that do not accord with existing zoning schedules.[11] Alternatively, development permit areas (DPA) may be created (i.e. areas within which a developer must obtain a permit before the land is subdivided or building commences).

In CD zones, authorities can mandate that all buildings achieve a high standard of sustainability and may, for example, utilise a rating tool as a measure. The United States Green Building Council has spearheaded the *LEED for Neighborhood Development* programme[12] in which over 238 local authorities in Canada, USA, Mexico, Korea, the Bahamas and China have established 'green develop-

---

SmartGrowthChecklist/tabid/135/Default.aspx. Note: as it may be appropriate for there to be different checklists proportional to the scale of the development, the City of New Westminster is currently working on another checklist designed specifically for single-family developments.

10 See for example: County of San Diego, California, US, *Green Building Incentive Programme*, www.sdcounty.ca.gov/dplu/docs/DPLU%20273.pdf fast tracking saves 7-10 days on average; City of Issaquah, Washington, US, *Sustainable Building Incentives* www.ci.issaquah. wa.us/page.asp?navid=327. Note: It is considered preferable to use a well-known national standard or rating tool as opposed to one created by the locality. It is also important that any tool is not unnecessarily prescriptive and should not stifle innovation.

11 For a simple explanation of CD zones in the Canadian context see West Coat Environmental Law *Urban Growth and Development* "Smart Bylaws Guide – Mixing Uses – Comprehensive Development Zoning" available at www.wcel.org/issues/urban/sbg/Part3/mixeduse/CD-Zoning.htm.

12 US Green Building Council, *LEED for Neighborhood Development* available at www.usgbc.org/DisplayPage.aspx?CMSPageID=148. A list of participating pilot authorities is available at www.usgbc.org/ShowFile.aspx?DocumentID=3546.

ment zones'.[13] Building in such zones must reach a particular sustainability rating on the LEED scale (although the *LEED for Neighborhood Development* tool encompasses more than just the sustainability of individual buildings).

Within DPA, development permit guidelines (akin to 'permitted status activities' within plans) may also be used to promote sustainable building, although there are greater limitations in terms of what might be achieved in DPAs as opposed to CD zones. The development permit guidelines can include requirements about landscaping, drainage systems, the siting, form, exterior design and finish of buildings but cannot promulgate energy efficiency requirements unrelated to form and character (although guidelines may indicate that these are desirable) (Rutherford 2006). To maximise sustainability, however, it is preferable for a building to use solar passive architecture if possible (a planning issue) as opposed to relying upon compensatory measures, for example, insulation (an issue for building regulations). The City of Richmond in British Columbia has introduced mandatory permit guidelines within DPAs to provide for solar access, minimum north-south spacing, orientation of windows, operable windows and other facets of natural ventilation; the District of Saanich guidelines also provide for energy efficiency via shading, natural ventilation and building orientation (Community Action on Energy and Emissions, 2007). These are prime examples of the planning process working in tandem with building regulations to maximise the sustainability of buildings.

Applications for re-zoning and planning permission

In many nations, local authorities have considerable scope to negotiate with developers before an application for planning permission or rezoning is submitted. Negotiations must seek to further the community's main objectives as expressed in the local authority plan.[14] It is important to note that pre-application negotiations will not guarantee the approval of a consent authority, but will help to secure staff members' favourable recommendations to the Council. In the UK, local authorities are able to exercise considerable flexibility in negotiating 'plan-

---

13 Outside Canada, another well-known example of a green development area concerns Vauban in Frieburg, Germany. Only highly energy efficient buildings are permitted in the zone; there are also a number of "plus energy houses" that create more energy than they need. Vauban is "car free" and parking free". Cars must be parked in community car parks on outskirts of development and distances are carefully calculated so that for most short journeys it proves quicker to walk, cycle or take public transport. There is a biogas plant for sewage, co-generation plant and effective community grey water recycling systems. See Vauban District, Freiburg *English Introduction*, available at www.vauban.de/info/abstract.html.

14 See for example Municipality of Bowen Island Policy on "Green Building Standards for Residential Re-zoning March 2007", available at www.bimbc.ca/files/policies/Green%20 Building%20Standards2.pdf.

ning obligations'. Planning obligations constitute pre-application negotiations that become legally enforceable once agreed to, incorporated into a deed and planning permission has been granted. Although recent government guidance in the UK has redefined the ambit of planning obligations (UK DCLG 2007c), bilaterally agreed obligations are still significantly more flexible than 'conditions' imposed upon grants of planning permission for development.[15] Central government in the UK has said as follows:

> *Local requirements for sustainable buildings*
>
> 30. Planning authorities, developers and other partners in the provision of new development should engage constructively and imaginatively to encourage the delivery of sustainable buildings. Accordingly, planning policies should support innovation and investment in sustainable buildings and should not, unless there are exceptional reasons, deter novel or cutting-edge developments. Planning authorities should help achieve the national timetable for reducing carbon emissions from domestic and non-domestic buildings.
>
> 31. There will be situations where it could be appropriate for planning authorities to anticipate levels of building sustainability in advance of those set out nationally. When proposing any local requirements for sustainable buildings planning authorities must be able to demonstrate clearly the local circumstances that warrant and allow this. These could include, for example, where:
>
> - there are clear opportunities for significant use of decentralised and renewable or low-carbon energy; or
> - without the requirement, for example on water efficiency, the envisaged development would be unacceptable for its proposed location.
>
> 32. When proposing any local requirement for sustainable buildings planning authorities should:
>
> - focus on the development area or site-specific opportunities
> - specify the requirements in terms of achievement of nationally described sustainable building standards, for example in the case of housing by expected identified housing proposals to be delivered at a specific level of the Code for Sustainable Homes;
> - ensure the requirement is consistent with their policies on decentralised energy; and
> - not require local approaches for a building's environmental performance on matters relating to construction techniques, building fabrics, products, fittings or finishes, or for measuring a building's performance unless for reasons of landscape or townscape" (UK DCLG 2007c, paras 30-31).

Thus there is considerable scope for pre-application negotiations in the UK to focus upon achieving sustainable buildings.

---

15 Conditions on planning permissions in the UK and New Zealand are legally constrained and must accord with the tests set down in *Newbury DC v Secretary of State for the Environment* [1981] A.C. 578.

In Canada, "local government administrative staff can, on a re-zoning application, suggest that a section 219 Land Title Act covenant be registered, requiring the incorporation of certain green building or development features, as a condition of staff giving their recommendation to the re-zoning and development proposal. Staff cannot promise the re-zoning will be granted, as staff cannot bind council to a decision and council cannot close its mind before a re-zoning hearing, but staff can promise their support of the proposal – a significant hurdle for any developer" (Rutherford 2006: 27).[16] Once again, to all intents and purposes the 'consideration' for the agreement is the staff members support for the application.

Planning Gains

To provide an added incentive for green development, local authorities may offer specific planning gains to the developer, for example unit 'density bonuses' or a relaxation of the maximum site coverage rules. Researchers in Canada have noted that a good planning gains regime

> "(1) establishes a maximum 'uplift' in the OCP (Official Community Plan), following a community dialogue on the subject; (2) uses the OCP to identify areas where density bonuses would be desirable and to prioritise desired community amenities; and (3) establishes a transparent and public accounting mechanism to calculate the value of the bonus and what the bonus will buy in terms of community amenities [which include green buildings]" (Rutherford 2006: 27).

In terms of unit 'density bonuses', developers are granted permission to build more units than the zoning would permit on the condition that they build sustainably. Particular regard is paid to the provision of sustainable solutions to minimise an increased load on infrastructure.[17] The City of Burnaby in British Columbia has a 'green building bonus' that affords percentage increases in density proportionate to the increase in building performance (Community Action on Energy and Emissions 2007). A rating tool is used for this purpose. Density bonuses are also commonly used in the UK to address the wider ramifications of sustainable development and local authorities may ensure that a set percentage of the development constitutes 'affordable housing'.

---

16 This approach has been endorsed by the Court in *Burnaby (City) v Marando*, 2003 BCCA 400 (CanLii) as cited in Rutherford (2006).

17 In particular, sustainable drainage systems to deal with water run-off; extremely efficient water fittings; micro-generation or CHP plants to reduce the pressure for increased off-site power generation; green roofs or the provision of covenanted areas of indigenous vegetation (if possible) to facilitate wildlife corridors (a critical issue in climate change mitigation, see Warnock, A C and Wheen, N, "Climate Change, Wildlife Movement and the Law: A Case Study from New Zealand" *Commonwealth Law Bulletin* (2008) vol 34 III (forthcoming)) etc.

In other examples of green building bonuses, developers are permitted to increase the site coverage of a building over and above the maximum site density rules on the basis that the building meets certain sustainability indicators (Community Action on Energy and Emissions 2007). This bonus is used in Germany, particularly with regards to green roofs. Green roofs help replace the vegetative footprint destroyed by the building, manage water run off, minimise the necessity for the creation of and maintenance of stormwater infrastructure, absorb carbon, help preserve energy, reduce the urban heat island effect, increase the longevity of roofing systems, provide sound proofing, habitat for birds and business opportunities for local businesses. The majority of German cities have promulgated regulations or provide incentives for green roofs (Industry Canada 2006) and one incentive is to permit developers to use green roofs as mitigation for the loss of open space on the site. Maximum site coverage can be renegotiated where provision is made for green roofs and green roofs can be secured by a covenant.[18]

Covenants

If local authorities sell or lease land for development, the authority may seek covenants on the land (that run with the land) mandating green building. Given covenants are achieved by agreement between the buyer / developer and seller they constitute an extremely flexible tool and offer considerable scope for promoting sustainable building. Depending upon the respective regulatory framework, it may also be possible to secure covenants against private land in return for tax or rate reductions.[19]

Review Bylaws

Local authorities should also undertake an audit of bylaws or planning controls that might impose unintended hurdles to sustainable building. A common difficulty concerns the use of grey water systems but research in Canada has noted other specific technical requirements of zoning bylaws that create disincentives to green buildings such as,

> *Floor space ratio*: most governments limit density by having a maximum floor space ratio that calculates space to include space taken up by heaters, thus en-

18 On the issue of green roofs, the City of Port Coquitlam has a green roof bylaw amendment that rather than focusing on roofing materials "prescribes the roofs performance vis-à-vis run-off and landscaping criteria". Green roofs are required for larger buildings and the bylaw has also secures the maintenance of the roofs (Rutherford 2007: 29).

19 In Canada, other means of achieving security is to take a bond from the developer (that may be returned on an incremental basis to ease the financial burden on the developer) or to impose a temporary "no build" covenant that is released upon fulfilment or performance of all required green infrastructure installations (Rutherford 2007: 63).

couraging least efficient space saving technologies like baseboard heaters. Floor space ratio is also calculated from the outer wall, discouraging thick energy efficient wall assemblies. Local government can adopt policies / bylaws that: exclude spaces occupied by highly energy efficient heating and cooling apparatus from the floor space ratio.

Building setbacks: Building setback requirements may act as a disincentive to the construction of thicker energy efficient walls or for energy efficient retrofits, such as new exterior cladding that pushes the exterior wall beyond the setback area, requiring exceptional approvals.

Unnecessary restrictions: Zoning bylaws sometimes unduly restrict passive solar heating or natural ventilation features, creating an impediment to the implementation of "easy" energy efficiency features. Bylaws might be amended to allow access to the sun and daylight (thereby enabling passive solar), and to encourage natural ventilation (e.g. building height restrictions might be relaxed) (Rutherford 2006: 38-39).[20]

Option to lease roof space to utility companies

Many German local authorities enable property owners to lease their roof space to utility companies for the installation of micro-generation (windmills or photovoltaic systems) which can often prove expensive for private householders to install themselves. Thus, micro-generation is encouraged by public policy and by ensuring that planning controls do not hinder this development. The micro-generation units are connected to the main electricity grid and limit the necessity for the development of large off site power plants (Industry Canada 2006).

Service Area bylaws

In British Columbia, the *Community Charter* permits municipalities to "provide any service that the council considers necessary or desirable, and [they] may do this directly or through another public authority or another person or organization ... a bylaw can be used to establish service areas for particular types of energy services (e.g. a hydronic district heating system) and *to require buildings within the service area to connect to the energy service*" (italics added) (Community Action on Energy and Emissions 2007: 34). To similar effect, in the UK, local authorities are permitted to "set a target percentage of the energy to be used in new development to come from decentralised and renewable or low-carbon energy sources where it is viable" and to specify such in the plan (UK DCLG 2007c: 16).

20 See also Trenouth, C. 'District Plans, Barriers to Sustainable Residential Development?' (2007) *Planning Quarterly*, December 2007: 18-21, New Zealand.

Relief of taxes and rates

A perverse disincentive to building sustainably is that the resulting structure might be of a higher value than a traditional building and (despite the fact that the building may impose less of a burden on infrastructure) may accordingly attract higher taxes or rates. In order to "acknowledge the additional private expense incurred for the public good", British Columbia permits local authorities to introduce tax exemption bylaws for buildings that reach specified performance standards or incorporate certain features (such as photovoltaic systems).[21] Exemptions are for set periods only (a maximum exemption of 10 years is permissible), which reduces the need for continual local authority monitoring of the situation, and authorities are free to establish the amount and the extent of the tax relief available. This practice is also apparent in the USA and in many German cities where, for example, a 100% local utility charge is levied for impervious roofs and a reduced rate for green roofs (Industry Canada 2006).

Relief of Development charges

More discreetly targeted financial incentives can be achieved by reductions in development charges. Development charges are generally levied to help fund the additional burdens on infrastructure created by development. Research undertaken in Canada suggests that significant infrastructure cost savings are associated with green development and that accordingly green developments are providing an important public service (Rutherford 2007). The City of Toronto is in the process of undertaking a quantification of the public benefits flowing from green developments in order to achieve a fiscally neutral policy for development charge reductions. Researchers in Toronto discovered that a reduction in development charges was the second most preferred incentive for developers (after fast tracking applications) (Toronto 2006).

Relief of building permit fees

Given the high degree of autonomy that many authorities have to set administrative fees for processing permits, reductions or exemptions to such fees are often used as a tool to incentivise green building. Again, this can either be based upon receipt of a satisfactory sustainability checklist (and potentially secured by a covenant), refund post-inspection, rating tool certification or installations of specific features. There are many examples: the County of San Diego in California does not charge a fee for a building permit and plan check for the installation of

---

21 British Columbia Ministry of Community Services (2007), "Revitalisation Tax Exemptions: A Primer on the Provisions in the Community Charter", p. 2, available at www.cserv.gov.bc.ca/lgd/gov_structure/library/community_charter_revital_ tax_exemptions.pdf.

photovoltaic systems.[22] The District of Saanich, British Columbia, "gives priority processing and offers a grant based on building permit fees of up to thirty percent to residential applications achieving energy efficiency and green building standards … Saanich also pays for a contractor to serve as an energy adviser, directing builders to available technologies and incentives and assisting them with applications" (Community Action on Energy and Emissions 2007: 40).[23]

Financial grants

Direct financial grants to promote green building are particularly prevalent in the USA. These grants may be awarded on a competitive basis to promote innovation[24]or simply to projects that achieve certain ratings on performance tools. For larger projects over US$5,000,000, Seattle City Council offers technical assistance and direct monetary grants if the building achieves a minimum of a LEED certified rating via the *LEED Incentive Program*. The grant may be increased in proportion to the LEED rating achieved. Seattle has brought together a number of interested organizations, including utility companies, in order to offer this incentive package.[25] The City of Toronto in Canada has specific regulations for green roofs requiring them on certain developments[26] but also provides direct monetary subsidies that compensate for the increased installation costs of a green roof. This subsidy is provided by the municipal water department and based on an assessment of the infrastructure costs that have been avoided.

In addition to the specific mechanisms mentioned above there are a myriad of other ways that local authorities can promote sustainable, energy efficient building developments. Certainly, building regulations and the planning regime should work in tandem to maximise the performance of buildings and it appears clear that local authorities have a critical role to play in promoting sustainable building.

Conclusion

Although the scientific knowledge foreseeing climate change has been in the public domain for over a century, it is only relatively recently that mankind has

---

22 County of San Diego, California, US, Green Building Programme, available at www.sdcounty.ca.gov/dplu/greenbuildings.html.

23 See also District of Saanich, *Green Building Rebate Programme* available at www.gov.saanich.bc.ca/business/development/greenbuilding/GreenBuilding.html.

24 See for example, Portland, Oregon, "Green Investment Fund" available at www.portlandonline.com/osd/index.cfm?c=ecbde.

25 Seattle City Council, US, "LEED Incentive Program" available at www.cityofseattle.net/light/conserve/sustainability/leed/cv5_lds1.htm.

26 *City of Toronto Act 2006*, s 108. See also City of Toronto, Living in Toronto, Environment, "Green Roofs" available at www.toronto.ca/greenroofs/index.htm.

accepted the need to act. The role that the built environment will play in mitigating greenhouse gas emissions is critical and given the longevity of the building stock in most nations, action to significantly change the way that we build must be taken now. Government intervention is necessary to catalyse this sea-change. Assuming that the necessary political will exists, the mechanisms to ensure that new-builds achieve zero carbon status appear to be relatively simple; regulation will be the driver whilst additional financial incentives will maximise innovation. Further, local government, in terms of its planning function will have an important role to play. However, achieving an energy efficient built environment that minimises buildings' contribution to climate change will fail without addressing the role of existing buildings. Whilst this chapter highlights a number of mechanisms to bring existing buildings into the sustainability equation, given the recent introduction of most of these policies, it is difficult to gauge their likely success. Clearly, ongoing monitoring of these methods must be undertaken and the most successful policies replicated to ensure that buildings achieve their full potential in the mitigation of climate change.

Finally, in his book entitled *The Architecture of Happiness* the philosopher Alain de Botton wrote, "[a]rchitecture even as its most accomplished will only ever constitute a small and imperfect … protest against the state of things". Let us hope that we can prove him wrong.

References

Amitrano, L., Page, I. and Kirk, N. (2006), Market Segmentation of New Zealand's Housing Stock. Beacon Pathways, Report PR106, New Zealand.

BEA (2007), Barcelona Energy Agency, Barcelona Thermal Ordinance (OST), available at www.barcelonaenergia.com/eng/operations/ost.htlm.

BRE (2008), Building Research Establishment website: www.bre.co.uk and see Peter White at BRE, 'Blog offsite 2007' available at http://offsite2007.com/progress-blog.jsp.

Carols Kucharek, J. (2007), "The shape of homes to come." Building Design Magazine, 15th June 2007.

Community Action on Energy and Emissions (2007), Energy Efficiency and Buildings: A Resource for BC's Local Governments, British Columbia, Canada, available at www.bcclimateexchange.ca/doc/FBC_Manual_ONLINE.pdf.

Culnane, M. (2007), "Put on the Best Performance" *Solicitors Journal Property*, June 2007 pp. 6-10, London, UK

De Botton, A. (2006), The Architecture of Happiness, Penguin Books, UK.

Dimson, B. (1996), "Principles and Challenges of Sustainable Design and Construction." UNEP Industry and Environment vol. 19 no. 2, April-June 1996.

Ealing Borough Council, London, UK (2007), Supplementary Planning Guidance 1: Sustainability Checklist, available at www.ealing.gov.uk/ealing3/export/sites/ealingweb/services/environment/planning/planning_policy/sustainable_planning/_docs/01sustainability.pdf.

EECA (2007), New Zealand Energy Efficiency and Conservation Authority available at www.eeca.govt.nz.

EPB (2002), Directive 2002/91/EC of The European Parliament and of the Council of 16 December 2002 on the energy performance of buildings, Official Journal of the European Communities 4.1.2003, L1/16.

Eurima (2005), Cost Effective Climate Protection in the EU Building Stock available at www.eurima.org/downloads/leaflet_with_keypoints.pdf.

Freshfields Bruckhaus Deringer (2005), A legal framework for the integration of environmental, social and governance issues into institutional investment, 2005, UNEP FI, available from UNEP FI, Publications, at www.unepfi.org/fileadmin/documents/freshfields_legal_resp_20051123.pdf.

German Federal Ministry of Transport, Building and Urban Affairs (2007), Existing Low Energy Houses, available at www.bmvbs.de/en/Building/Climate-change-and-energy-effi-,2832/Existing-Low-Energy-Houses.htm.

Halliday, S. (2004), Sustainable Construction and the Regulatory Framework – A Thinkpiece, Gaia Research, Edinburgh, UK.

Harman, J. and Benjamin, V. (2005), Better Buildings – Better Lives, Sustainable Building Task Group Report prepared for the review of the UK Building Code, London, UK.

House of Commons Environmental Audit Committee (2006), Sustainable Housing: A Follow-up Report, Fifth Report of Session 2005-2006 HC 799, HMSO, London, UK.

Husbanken (2007), "Husbanken Basic Loans", available at www.husbanken.no/Toppmeny/English/Basic%20loans.aspx.

IISBE Italia (2007), available at www.iisbe.org.

Industry Canada (2006) Sangster, W. Benchmark Study on Green Buildings: Current Policies and Practices in Leading Green Building Nations. Industry Canada Energy and Environmental Industries Branch, August 2006.

IPCC (2007), Climate Change 2007: Mitigation of Climate Change. Contribution of Working Group III to the Fourth Assessment Report of the Intergovernmental Panel on Climate Change. In: Metz, B., Davidson, O.R., Bosch, P.R., Dave, R. and Meyer, L.A. (eds.). Cambridge University Press, Cambridge, UK and NY, USA.

Jones Lang LaSalle (2007), Future Proofing New Zealand Commercial Property, available at www.joneslanglasalle.co.nz/en-gb/research/ researchabstract? artid =2469.

Kennett, S. (2007), "Carbon Copy." Building Services Journal, July 2007.

Kingspan Group plc (2007), Level 6 Net-Zero Carbon House Fact File, available at http://kingspanoffsite.com and http://kingspanlighthouse.com (note: The Kingspan Lighthouse team includes Sheppard Robson, Arup, MacFarlane Wilder, Davis Langdon, CCB Evolution, Kingspan Off-site, Chorus Group).

Laurenzi, M. (ed.) (2007), Building Energy Efficiency: Why Green Buildings Are Key to Asia's Future Asia Business Council, Hong Kong.

Marrakech Taskforce on Sustainable Buildings and Construction (2007), Buildings for a Better Future. Ministry of the Environment, Finland.

Mayor of London (2008), London Sustainability Checklist, available at www.londonchecklist.co.uk/page/faq/what#what.

McKinsey Global Institute (2007), Curbing Global Energy Demand Growth: The Energy Productivity Opportunity, May 2007 report, available at www.mckinsey.com/mgi/publications/Curbing_Global_Energy/index.asp.

Myers, G., Reed, R. and Robinson, J. (2007), "The Relationship between Sustainability and the Value of Office Buildings", paper given at the 13th Annual Pacific Rim Real Estate Conference, 21-24 January 2007, Curtin University of technology, Perth, Australia. Paper on file with the author.

Olcatyo, R. (2007), "Scaling New Heights in Zero-Carbon Housing." Building Design Magazine, 31st August 2007.

Rutherford, S. (2006), The Green Buildings Guide: Tools for Local Governments to Promote Site Sustainability. West Coat Environmental Law, British Columbia, Canada available at www.wcel.org/resources/publications/ default.cfm.

– (2007), The Green Infrastructure Guide: Issues, Implementation Strategies and Success Stories. West Coat Environmental Law, British Columbia, Canada, available at www.wcel.org/resources/publications/default.cfm.

Toronto (2006) City of Toronto. Making a Sustainable City Happen: The Toronto Green Development Standard Discussion Paper, June 2006, City of Toronto, Canada available at www.toronto.ca/environment/greendevelopment.htm# why.

UK Department of Communities and Local Government (2006), Building a Greener Future: Towards Zero Carbon Development; Consultation Paper, December 2006, available at http://communities.gov.uk/archived/ publications/ planningandbuilding/buildinggreener.

– (2007a), Building a Greener Future; Policy Statement, July 2007, available at www.communities. communities.gov.uk/ pub/ 970/ BuildingaGreenerFuture-policystatement_id 1511970.pdf>

– (2007b), Changes to Permitted Development – Permitted Development Rights of Householder Microgeneration Consultation paper on proposed regulations under Town and Country Planning (General Permitted Development) Order 1995 (SI 1995/418), 2007, available at http://communities.gov.uk/ archived/ publications/ planningandbuilding/changespermitted.

– (2007c), Planning Policy Statement: Planning and Climate Change Supplement to Planning Policy Statement 1, December 2007, available at http://communities.gov.uk/publications/planningandbuilding/ppsclimatechange.

UNEP (2007a), 'Urban Issues' webpage available at www.unep.org/urban_environment/key_programmes/index.asp.

– (2007b), Buildings and Climate Change: Status, Challenges and Opportunities, available on the website of the Marrakech Task Force on Sustainable Buildings and Construction at www.environment.fi/default.asp? contentid=233068 & lan=EN.

Warnock, A.C. (2005), "Sustainable Construction in New Zealand." New Zealand Journal of Environmental Law 9, p. 337.

– (2006), "Sustainable Construction in New Zealand? Regulatory and Policy Lessons from Europe." Global Built Environment Review 109, University of Lancashire, UK.

– (2007), "An Overview of Integrating Instruments to Achieve Sustainable Construction and Buildings." Management of Environmental Quality: An International Journal vol. 18 no. 4, p. 427.

World Commission on Environment and Development (WCED) (1987), Our Common Future, OUP.

Chapter 10

A Spatially Explicit, Indicator-based Methodology for Quantifying the Vulnerability and Adaptability of Natural Ecosystems

Bálint Czúcz, Gergely Torda, Zsolt Molnár, Ferenc Horváth, Zoltán Botta-Dukát, György Kröel-Dulay

Abstract

Ecosystems contribute inconspicuously, yet fundamentally, to human well-being by supplying vital goods and services, including genetic resources, habitat maintenance and climate and runoff regulation. The combined effects of climate change and other global change drivers may impose dramatic impacts on species and ecosystems worldwide, with potentially detrimental consequences on human society. In this chapter we present a vulnerability assessment for the natural and semi-natural ecosystems of Hungary, calculating the local exposure, sensitivity and adaptive capacity of different habitat types. Exposure was calculated using six different global climate model (GCM) outputs comprising of four different models and three emission scenarios, providing a cross-section of the climatic and socio-economic uncertainties within the projections. To estimate the sensitivity of habitats, four types of climate sensitivity were identified and estimated either quantitatively or semi-quantitatively. Adaptive capacity of habitat occurrences was assessed using landscape ecological evaluation of the quality and distribution of habitat patches. Three potential indicators of adaptive capacity were identified, describing (1) the potential resilience of the individual habitat patches, (2) the local refuge-providing ability of the landscape, and (3) the connectivity and permeability of the landscape. By combining results of exposure, sensitivity and adaptive capacity, climatic vulnerability maps of natural ecosystems were produced. This case study, prepared for the Hungarian National Climate Change Strategy, provides the first example of a methodology to give quantitative estimation of the potential climatic vulnerability and adaptive capacity of ecosystems based on a detailed habitat database.

Introduction

The combined effects of climate change, associated disturbances (e.g., flooding, drought, wildfire, insect outbreaks, ocean acidification) and other global change drivers (especially land-use change, pollution and over-exploitation of resources) will seriously challenge most species and ecosystems worldwide during the course of this century (Fischlin et al. 2007). Major losses in biological diversity seem to be inevitable, resulting in a general decrease in the level of goods and services supplied by ecosystems, especially the regulating and supporting services, including genetic resources, habitat provision and climate and runoff regulation (Fischlin et al. 2007; Duraiappah et al. 2005). The erosion of local and global biodiversity prognosticates an increased likelihood of "ecological surprises" (Fischlin et al. 2007; Burkett et al. 2005), which might impose detrimental consequences on human welfare.

To address the pervasive risks and uncertainties pertinent to climate change, impact adaptation and vulnerability (CCIAV) studies have surged in this decade. Most studies for estimating climate impacts on ecosystems and biological diversity can be characterized by using either a correlative, or a mechanistic approach (Fischlin et al. 2007). Mechanistic models are based on the current understanding of the dynamics of energy, water, nutrient and carbon dynamics among major species groups and their physical environment. Such models are unrivalled in representing vegetation transformations on a global scale; however, for biodiversity on lower spatial levels the application of these data-hungry models is rather limited. Correlative models, on the other hand, are generally based on more readily available species distribution data, producing projections on future distribution by assuming a statistical relationship between actual distribution and climatic needs of the species (Guisan & Zimmermann 2000). Nonetheless, due to the limitations of the model structure and the data sources applied, there is generally no straightforward way to take processes of natural adaptation into account. Such models usually apply two extreme assumptions of non-adaptive systems (usually referred to as "no dispersal" of species) or perfectly adaptive systems ("universal dispersal"), as a generalized "interval projection". Thus, studies employing correlative models can be regarded as simple "impact assessments" (the first stage of CCIAV assessments sensu Füssel & Klein 2006), and according to Rothmann and Robinson (1997) the evolution towards realistically adaptive agents is inevitable as assessments become increasingly integrated. A further issue is raised by uneven data coverage between taxonomical groups, showing significant imbalance towards easily observable, popular and charismatic species groups, however, modelling studies that include large numbers of species are generally accepted to reveal general tendencies for biodiversity.

The ecosystem vulnerability model, presented here, is fundamentally based upon established techniques of correlative species distribution models, but extending the framework into a "first generation vulnerability assessment" (the second stage of CCIAV assessments sensu Füssel & Klein 2006). The principal goal of such assessments is to understand the *vulnerability* of the investigated systems as a function of *exposure* to climatic forcing, the inherent *sensitivity* of the exposure units, and their capacity of autonomous adaptation (Metzger el al. 2005):

$$\begin{aligned} \text{vulnerability} &= \text{f(exposure, sensitivity, adaptive capacity)} = \\ &= \text{f(potential impact, adaptive capacity)}, \end{aligned}$$

Potential impact means the impacts that may occur without adaptation (considered here as a function of *exposure* and *sensitivity*), whereas adaptive capacity accounts for the ability of a system to adjust to climate change to moderate potential damages (IPCC 2007, Glossary).

The main initiative objective of this modelling study was to describe imminent regional impacts of climate change on biodiversity for the Hungarian National Climate Change Strategy[1]. The spatial scope of the study was set to the country of Hungary (~93 000 km^2), situated in Central Europe. On a temporal scale, the study focuses on the relatively near future (time horizon set to 2025, with an outlook for 2050) according to the basic interest of stakeholders reflected in the design principles of the National Climate Change Strategy. These circumstances determined three important underlying concerns of our analysis design. Accordingly, we intended to implement a methodology, that is (1) policy relevant, (2) realistic and (3) appropriate for the addressed spatial and temporal scales.

Most models that account for ecological impacts of climate change work either with species or with major biomes as exposure units. However, neither of these approaches seems to be appropriate for regional or national level ecosystem vulnerability assessments; there are too few biomes and too many species which lack reliable data on life traits and distribution. Accordingly, we selected an ecosystem classification with a high thematic resolution and local relevance with respect to the study area: the major habitat types of Hungary, building upon the Hungarian National Habitat Classification System (Bölöni et al. 2007). In this sense, habitats (habitat types) are more than just the spatial locations of species occurrences – habitats can rather be defined as communities of species living together, determined by trophic linkages and/or similar environmental (e.g. climatic) needs. Consequently, while species react individually to the external changes, climate impacts on species sharing similar climatic needs will presumably be similar.

Owing to the completeness and the local ecological relevance of the classification system used, we considered habitats superior to species for estimating overall ecological vulnerabilities on a regional level. The main source of habitat distribution data in Hungary is the MÉTA database, a country-level, vegetation database, relying on recent field survey results (Molnár et al. 2007; Horváth et al. 2008). Habitat data were collected on a network of 35 ha hexagonal grid cells. Data for each grid cell comprise a list of the occurring habitat types, along with their estimated area and habitat quality (naturalness), as well as several other attributes. Altogether 86 different types of natural and semi-natural habitats were distinguished, with a detailed Habitat Guide (Bölöni et al. 2007) to assist the more than 200 surveyors, and standardize the mapping procedure, performed between 2003 and 2006.

1 The National Climate Change Strategy, accepted unanimously by the Hungarian Parliament on 17 March 2008, can be downloaded from http://klima.kvvm.hu/documents/14/nes_080219.pdf (in Hungarian).

With respect to the policy goals, the available data sources and the characteristics of first generation vulnerability assessments, two main objectives were set:

- to give a comprehensive evaluation of the vulnerability of all habitat types on a national level, and
- to provide a spatially explicit evaluation of the potential impacts and adaptation capacity for the most vulnerable habitats.

Consulting with the national nature conservation authorities, a further, third goal was also added:

- to compile a synthesis of practical policy recommendations tailored for Hungary.

To meet these objectives, we developed a new, spatially explicit, indicator-based methodology for quantifying the vulnerability of natural and semi-natural ecosystems, based on the MÉTA database. The remainder of this chapter is organized according to the construction of this methodology, following the outlined structure of first generation vulnerability assessments. First we summarize the changes in the climatic drivers to which ecological systems are exposed (climatic *exposure*). Next we analyze the changes that climatic exposure might cause to ecosystems (*potential impacts*), identifying and quantifying all the relevant *sensitivities* of the different habitat types. Further, we introduce a conceptual framework to calculate landscape ecological indicators of the *adaptive capacity*, and give general policy recommendations based on the lessons learned and literature survey. Finally we conclude this chapter by placing this methodology in the context of current developments in CCIAV research.

Exposure to climate change

In the CCIAV jargon, exposure generally refers to "the nature and degree to which a system is exposed to significant climatic variations" (IPCC 2001, Glossary). Understanding the character and magnitude of climatic changes that the investigated system will probably be exposed to is the starting point of most climate impact studies. To get a reliable picture on the direction and uncertainties of the climatic changes that Hungary is currently facing, we resorted to climatic scenarios from the third phase of the Coupled Model Intercomparison Project, run by the World Climate Research Programme, available from the IPCC Data Distribution Centre (Meehl et al. 2007)[2]. In order to allow for both climatic and socio-economical uncertainties of the plausible futures, a range of six climate change scenarios were selected, representing a combination of four different global climate models (GCM) and three emission scenarios. The main characteristics of the projected monthly temperature and precipitation values for the region of Hungary are summarized in Table 1.

---

[2] www.ipcc-data.org.

Tab. 1: Main characteristics of the climate change scenarios used (annual, summer and winter temperature [°C] and precipitation changes [%]) for the region of Hungary, grouped by GCMs (a) and emission scenarios (b).

	2025			2050			2025			2050		
a)	T_a	T_s	T_w	T_a	T_s	T_w	P_a	P_s	P_w	P_a	P_s	P_w
HADCM3 A2	1.4	1.6	1.2	2.6	2.9	2.3	-2.4	-8.2	3.1	-3.6	-13.4	7.0
CNCM3 A2	1.3	1.6	1.1	2.3	2.4	2.1	-4.0	-8.3	0.6	-5.6	-9.6	-0.8
CSMK3 A2	1.2	1.2	1.2	1.7	1.8	1.5	4.6	4.7	4.1	-1.5	0.4	-3.3
GFCM21 A2	0.9	1.1	0.7	1.8	2.1	1.6	0.7	-1.9	4.5	-5.2	-14.2	6.2
b)	T_a	T_s	T_w	T_a	T_s	T_w	P_a	P_s	P_w	P_a	P_s	P_w
HADCM3 A2	1.4	1.6	1.2	2.6	2.9	2.3	-2.4	-8.2	3.1	-3.6	-13.4	7.0
HADCM3 A1B	1.6	1.7	1.6	2.9	3.3	2.6	1.4	-3.3	5.6	-1.1	-10.9	9.4
HADCM3 B1	1.4	1.6	1.1	2.4	2.6	2.3	-0.2	-1.5	0.4	-4.7	-12.4	3.5

The low resolution GCM outputs were downscaled spatially using the simple pattern scaling approach proposed by Parry and Carter (1998), using high resolution present climate surfaces (temperature from the WORLDCLIM database and precipitation from the Hungarian Meteorological Service – see Hijmans et al. 2005 and HMS 2001) adjusted by the difference (or ratio) between future projections and control runs with the respective GCMs. We calculated with 30-year climatic means for the periods 1960-1990 (the baseline), 2010-2040 (for 2025) and 2035-2065 (for 2050). To get more biologically meaningful variables, from the downscaled means we calculated twelve simple bioclimatic indices, which are frequently used in ecological applications (bio1, 4 and 10-19 of the WORLDCLIM database). The habitat distribution models introduced in the following section relies on this set of bioclimatic variables.

Habitat sensitivity and potential impacts

According to the conventions of the climate impact research community, the sensitivity of a system designates the degree to which it is affected either adversely or beneficially by climate change (IPCC 2007, Glossary). In the case of habitat types, climatic sensitivity can be interpreted as the aggregated sensitivity of the specialist species (species with specific habitat requirements) to changing macroclimate conditions. Based on the assumption that the present day distribution of habitats reflects the fundamental climatic requirements of the adherent organisms, the direct sensitivity of the ecosystems can be modelled with correlative bioclimatic distribution models. However, this approach has several limitations (see e.g. Pearson & Dawson 2003 for a detailed discussion), from which we highlight the following four:

- specific dispersal strategies and limitations are hardly possible to take into account in correlative models;

- there are many indirect feedback mechanisms involved in shaping species and community responses (e.g. biotic interactions between species; but also complex physical or socio-economical feedback mechanisms);
- communities never reach a real climatic equilibrium due to the complexity of interactions, large-scale environmental changes (i.e. biotic succession and climatic alterations) and evolution;
- extrapolation to future climatic conditions not occurring in the calibration region of the model can result in biologically meaningless results.

If the objects of analysis are not individual species but communities, one further issue of concern arises. Since different species respond idiosyncratically, present communities might increasingly disassemble as climate change proceeds (Thuiller 2004). Disintegration starts when the changes are too fast for the species to follow, or the available habitat patches are too small or degraded. Accordingly, at least for the starting phase of the changes, it is reasonable to assume, that climatic conditions that have for long supported the existence of a habitat, will still remain available for the majority of its species, providing ground for the community to survive. This problem is closely related to the first limitation mentioned above (the uncertainty pertinent to species dispersal). One of the main goals of the adaptive capacity component of our vulnerability model is to address this issue, and provide indicators of habitat availability and connectivity.

Traditional species distribution models excel in grasping direct relationships between the species and their environment (direct sensitivity). Indirect linkages, especially when spatially or temporally remote impacts are involved, are hardly possible to reckon with in correlative models. Unfortunately, this is typical for climate change driven physical mechanisms (e.g. changes in fire or flood regimes, termed as "associated disturbances" by Fischlin et al. 2007), leading to significant changes in the physical environment with important impacts on ecosystems (indirect sensitivities). Indirect sensitivities are very difficult to model, but it is possible to evaluate them qualitatively based on expert knowledge on a national level, which is generally a desirable practice in advanced CCIAV assessments (Rothmann & Robinson 1997).

Keeping in mind these considerations and our main policy goals, we proceeded in the following way:

- we first evaluated the (direct) climatic sensitivity of all habitat types using correlative distribution models;
- we identified 12 common habitat types with high direct sensitivity, and gave spatially detailed evaluation of the potential impacts (changes in occurrence probabilities) for these habitat types; and
- we gave a comprehensive qualitative evaluation (with the indirect sensitivities included) on a national level for all habitat types, aggregated from model results and qualitative expert analysis.

Fig. 1: Actual distribution, and modelled occurrence probabilities of oak-hornbeam woodlands for the Southern-Transdanubia (Dél-Dunántúl) region of Hungary

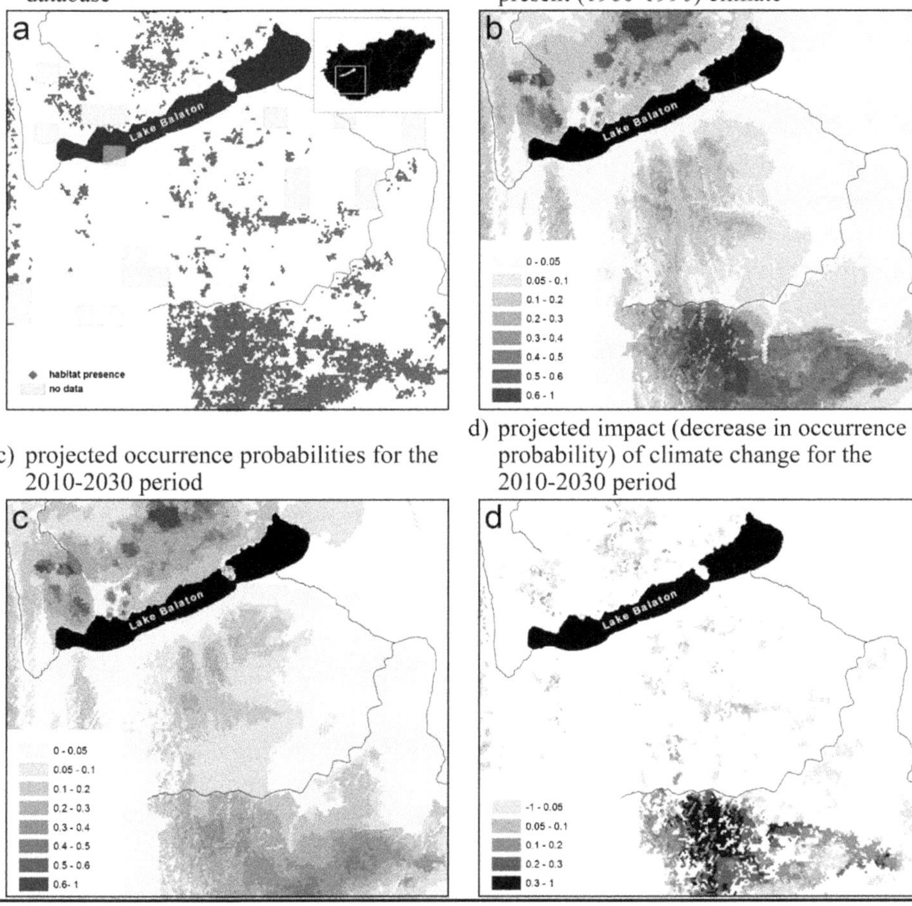

a) actual distribution according to the MÉTA database

b) projected occurrence probabilities for the present (1960-1990) climate

c) projected occurrence probabilities for the 2010-2030 period

d) projected impact (decrease in occurrence probability) of climate change for the 2010-2030 period

To estimate the direct sensitivity of the habitats and the potential impacts of climate change thereon, we used conditional inference-based decision tree models (Hothorn et. al. 2006). Apart from the bioclimatic indices, we also considered non-climatic variables (soil and distance from water bodies) as model predictors. Climatic sensitivity of the habitat types was interpreted on the basis of the predictive power of different models including either only climatic, non-climatic or both types of predictors.[3] Spatially explicit potential impact maps for the most climate sensitive habitat types were calculated on the basis of the models including all

3 The list of the predictors and the evaluation metrics for climate sensitivity can be found in Czúcz et al. 2007.

predictors. All the calculations were performed in an ensemble-like framework, based on twelve random bootstrap model runs for each of the aforementioned six climate change scenarios, using the means as consensus predictions, and the range of predictions indicating uncertainties. Figure 1 shows the mean projection for mixed oak-hornbeam woodlands, the most widespread forest type of Hungary.

To accomplish the comprehensive qualitative evaluation, we first enumerated the most important mechanisms originating sensitivities, which were hidden from the statistical model:

- Warming and drying can lead to increasing frequency of wildfires, which may affect many habitat types, especially forests already at the edge of their climatic tolerance.
- Changing river flow regimes are projected to decreasing flood frequency with increasing flood levels – both of these can adversely affect riverine habitats.
- Habitats attached to springs and upwelling waters from aquifers (fens and partly other types of wetlands) can be seriously affected by decreasing precipitation sums. This relationship works primarily on large spatial and temporal scales (Ruprecht & Botta-Dukát 2000).
- Upwelling waters play a moderately known, but important role in the formation and sustenance of alkaline soils and the related habitat types. Changing hydro-geological structures might on the long term harm salt steppes harbouring high portion of the remaining ecological capital in Hungary.

The latter two processes are not likely to exert considerable impact on a limited time scale as ours. Still, we included these, since due to the slow feedback cycles and the large inertia of the systems, it is important to be conscious about these feedbacks well before they start to evoke problems.

Besides the physical mechanisms of climatic sensitivity, there are several habitat types, which will definitely be affected by the direct impacts of climate change, however, due to the limitations of the study area and habitat database, these relationships could not be recognized by the model:

- Weak relationships can only be detected for prevalent habitats by the statistical model. For extremely rare habitat types even strong relationships may remain unnoticed.
- Habitats occurring in the warmest/most arid regions of the country require extrapolation from the model. (This problem could theoretically be handled by including data from a wider geographical range.)

All natural and semi-natural habitat types were evaluated qualitatively with respect to each hidden sensitivity mechanism by a group of five independent experts on a five grade scale. In order to bring expert-based and quantitative results to a common platform, we derived ordinal-scale potential impact estimates from the modelled direct impacts, aggregated on the national level. In Table 2 we present the complete list of projected climate impacts on the natural and semi-natural habitat types of Hungary.

Tab. 2: Estimated level of climate change potential impacts (PI) on the natural and semi-natural habitat types in Hungary

PI	Habitat types (MÉTA code, Natura2000 code*)
5	Euhydrophyte communities of fens (A4, 3160)[f], Transition mires and raised bogs (C23, 7140)[f], Rich fens (D1, 7230)[f], Calluna heaths (E5, 4030)[a], Alder and ash swamp woodlands (J2, 91E0)[f,a], Acid beech woodlands (K7a, 9110)[a], Closed lowland steppe oak woodlands (L5, 91F0)[d,b,c], Open sand steppe oak woodlands with openings (M4, 91I0)[d,b,c], Acid coniferous woodlands (N13)[a]
4	Festuca rubra hay meadows and related communities (E2, 6510)[a], Salt meadows (F2, 1530)[a,f], Closed rock grasslands, species rich Bromus pannonicus grasslands (H1, 6190)[b], Birch mire woodlands (J1b)[f], Lowland oak-hornbeam woodlands (K1a, 91F0)[a], Limestone beech woodlands (LY3, 9150)[b], Open loess steppe oak woodlands with openings (M2, 91I0)[b,c,d], Open salt steppe oak woodlands with openings (M3, 91I0)[b,c,d]
3	Oligotrophic reed and Typha beds of fens, floating fens (B1b)[f], Tussock sedge communities (B4, 7230)[f], Soft and hard water flushes (C1)[f], Molinia meadows (D2, 6410)[f], Mesotrophic meadows (D34, 6440)[f], Water-fringing and fen tall herb communities (D5, 6430)[f,a], Cynosurion grasslands and Nardus swards (E34, 6520)[c,a], Bromus erectus-Brachypodium pinnatum xero-mesophilous grasslands, dry tall herb communities and forest steppe meadows (H4, 6210)[a], Open vegetation of shadowed rocks (I4)[a], Willow mire woodlands (J1a, 91E0)[f], Riverine oak-elm-ash woodlands (J6, 91F0)[e], Oak-hornbeam woodlands (K2, 91G0)[a], Beech woodlands (K5, 9130)[a], Closed and mixed steppe oak woodlands on foothills (L2x, 91I0)[d], Ravine woodlands (mesic rock woodlands rich in Acer pseudoplatanus) (LY1, 9180)[b], Poplar-juniper steppe woodlands (M5, 91N0)[d,b,c]
2	Salt marshes (B6, 1530)[f], Arrhenatherum hay meadows (E1, 6510)[a], Artemisia salt steppes (F1a, 1530)[f], Tall herb salt meadows and salt meadow steppes (F3, 1530)[f], Dense and tall Puccinellia swards (F4, 1530)[f], Annual salt pioneer swards of steppes and lakes (F5, 1530)[f], Open sand steppes (G1, 6260)[c], Closed sand steppes (H5b, 6260)[c], Riverine ash-alder woodlands (J5, 91E0)[a], Turkey oak -sessile oak woodlands (L2a, 91M0)[a], Closed acid oak woodlands (L4a, 91M0)[a], Continental deciduous steppe thickets (M6, 40A0)[a]
1	Standing water communities with Trapa, Lemna, Salvinia and Ceratophyllum (A1, 3150), Euhydrophyte communities with Nymphaea, Nuphar, Utricularia and Stratiotes (A23, 3160), Slowly running water communities with Potamogeton and Nymphoides (A3a, 3150), Athalassal saline euhydrophyte communities (A5, 1530), Eu- and mesotrophic reed and Typha beds (B1a), Glyceria, Sparganium and Schoenoplectus beds (B2), Water-fringing helophyte beds with Butomus, Eleocharis and Alisma (B3), Non-tussock beds of large sedges (B5), Tall herb communities of floodplans and marshes (D6, 6430), Achillea salt steppes on meadow solonetz (F1b, 1530), Calcareous open rock grasslands (G2, 6190), Acid open rock grasslands (G3, 6190), Calcareous rock steppes (H2, 6240), Slope steppes on stony ground (H3a, 6240), Closed steppes on loess, clay, tufa (H5a, 6240), Amphibious communities on river gravel and sand banks (I1), Semi-desert vegetation on loess cliffs (I2), Riverine willow scrub (J3, 91E0), Riverine willow-poplar woodlands (J4, 91E0), Acid oak-hornbeam woodlands (K7b, 91G0), Closed termophilous oak woodlands (L1, 91H0), Turkey oak–pedunculate oak woodlands (L2b, 91M0), Open acid oak woodlands (L4b, 91M0), Slope woodlands (LY2, 9180), Mixed relic oak woodlands on rocks (LY4, 9150), White oak scrub woodlands (M1, 91H0), Continental deciduous rock thickets (M7, 40A0), Thermophilous woodland fringes (M8, 6210), Calcareous Scots pine woodlands (N2)

a: direct sensitivity modelled – b: direct sensitivity unmodelled (extremely rare habitat) – c: direct sensitivity unmodelled (too close to southern/xeric distributional limit) – d: indirect sensitivity (wildfires) – e: indirect sensitivity (river flow regimes) – f: indirect sensitivity (upwelling water). *Natura2000 codes are only provided for unambiguous matches

(according to the Hungarian National Habitat Classification System, Bölöni et al. 2007; 5: most vulnerable → 1: least concern)

Indicators of adaptive capacity

Estimating the adaptive capacity of the exposure units is generally the most important, but also the most problematic part of traditional climate change vulnerability assessments (Grothmann & Patt 2005). For most CCIAV studies adaptation and adaptive capacity mostly refers to *planned adaptation* based on deliberate policy decisions, whereas the spontaneous adjustment (= *autonomous adaptation*) of the agents in reaction to the persistent changes is generally considered under the scope of sensitivity (Carter et al. 2007). Notwithstanding the common application, in this chapter adaptive capacity refers to autonomous adaptation caused by organism, species and ecosystem level responses to the changing external conditions. However, autonomous adaptation is mostly influenced by the distribution and quality of the habitats and the level of additional anthropogenic pressures (e.g. pollution), which are largely determined by the forms and intensity of human land use. Thus, autonomous ecological adaptation acts, in fact, as a feedback mechanism from land management and natural resource allocation towards keystone ecosystem services, such as climate and runoff regulation, pest control, and especially the maintenance of genetic resources. In this sense understanding autonomous ecological adaptation is an important step in developing comprehensive socio-economic adaptation strategies.

As it is common practice in the evaluation of complex systems (e.g. Haddad 2005; Yohe & Tol 2002), we proposed a framework of adaptive capacity indicators that account for the most important coping options or mechanisms of the system. The proposed indicators aim at estimating the chances that the species inhabiting the studied habitat types can avoid degradation (and the resulting erosion of genetic diversity) caused by climate alteration. To this end, we distinguished three basic components (mechanisms) of ecological resilience to climate change:

- *ecosystem resilience*: habitat occurrences in more favourable (more natural) condition are more likely to be resilient to climatic stresses;
- *refuge-based adaptation*: withdrawing of species into local microclimatic refuges in a generally unfavourable environment;
- *migration-based adaptation*: large-scale movement of species following macroclimatic changes (the shift of climatic zones) by migration.

There are no sharp boundaries between the three conceptual mechanisms of adaptation (e.g. adapting by migration is equal to finding a distant zonal refuge). In effect, all three forms of adaptation are the manifestations of the structure, ordination and coordination of living systems, on different levels and spatial scales. Although occurring on different spatial scales, all the mentioned adjustment processes rely upon the resilience resulting from the spatially structured diversity of the systems, making it possible to dampen external changes by internal restructuring. Naturally, different spatial scales imply different temporal scales. Accordingly, there will be a presumable shift in the availability and importance of adaptation mechanisms from local resilience towards large-scale migration, which will remain the only option for more and more species to avoid extinction as climate change becomes increasingly severe.

Although migration can be accompanied with local extinctions, on larger scales it can moderate biodiversity loss, if it helps to avoid global extinctions. Migration can also help to sustain local ecosystem services by letting transformed habitats be colonized by new arrivals. Migration can, naturally, have negative aspects as well (e.g. the migration of invasive species introduced from distant localities), but climate induced distribution changes of native species should still be accepted as an important natural adjustment mechanism of ecosystems. Unfortunately, the dispersal capacity of most species is poorly known; several studies suggest it to be inadequate to track climatic changes in a landscape that is heavily fragmented by humans (e.g. Opdam & Wascher 2004; Skov & Svenning 2004). However, there is some paleoclimatic evidence suggesting that in extreme situations, species are able to migrate substantially faster, governed by rare, long distance dispersal events and subsequent local colonization (Bacles et al. 2006). Early Pleistocene-Holocene vegetation history suggests that both refuge- and migration-based adaptation played an important role in that species were capable of surviving significant changes in climate.

The proposed indicators for the outlined three mechanisms are based on the basic limiting factors of the processes. Local ecosystem resilience (determined by the compositional, structural and functional diversity of the habitat – see. e.g. Noss 1990) is limited primarily by patch size and local land use intensity. According to this, we estimated local resilience for each habitat patch in each grid cell with the "natural capital" of the habitat (the product of patch area and habitat quality from the MÉTA database – for a justification of this approach see e.g. ten Brink 2000 or Parkes et al. 2003). Concerning the other two mechanisms, while migration-based adaptation requires large connected habitat networks, refuge-based adaptation is limited mainly by the local/regional availability of appropriate recesses. Based on this, to estimate local refuge-based adaptive capacity we used a habitat diversity index (the naturalness-weighted Shannon diversity of the habitat patches in the grid cell)[4], whereas local migration potential was assessed by a landscape connectivity index (ESLI patch connectivity, see Vos et al. 2001 and Swihart & Verboom 2003). Landscape connectivity was calculated for each cell considering all similar habitat types in the neighbourhood of the focal grid cell, assuming an effective dispersal distance of ~1-8 km.

Results of adaptive capacity calculations are shown in Figure 2 for two important and typical habitat types. To ease interpretation, we classified the calculated indicator values into five ordinal categories based on the habitat type specific distribution of the values (for habitat resilience), or the analysis of model landscapes (for local diversity and landscape connectivity).

---

4 The details can be found in Czúcz et al. 2007: 224-225.

Fig. 2: Adaptive capacity indicators for two major habitat types: oak-hornbeam woodlands (a, c, e) and mesotrophic meadows (b, d, f) in the Southern-Transdanubia (Dél-Dunántúl) region of Hungary

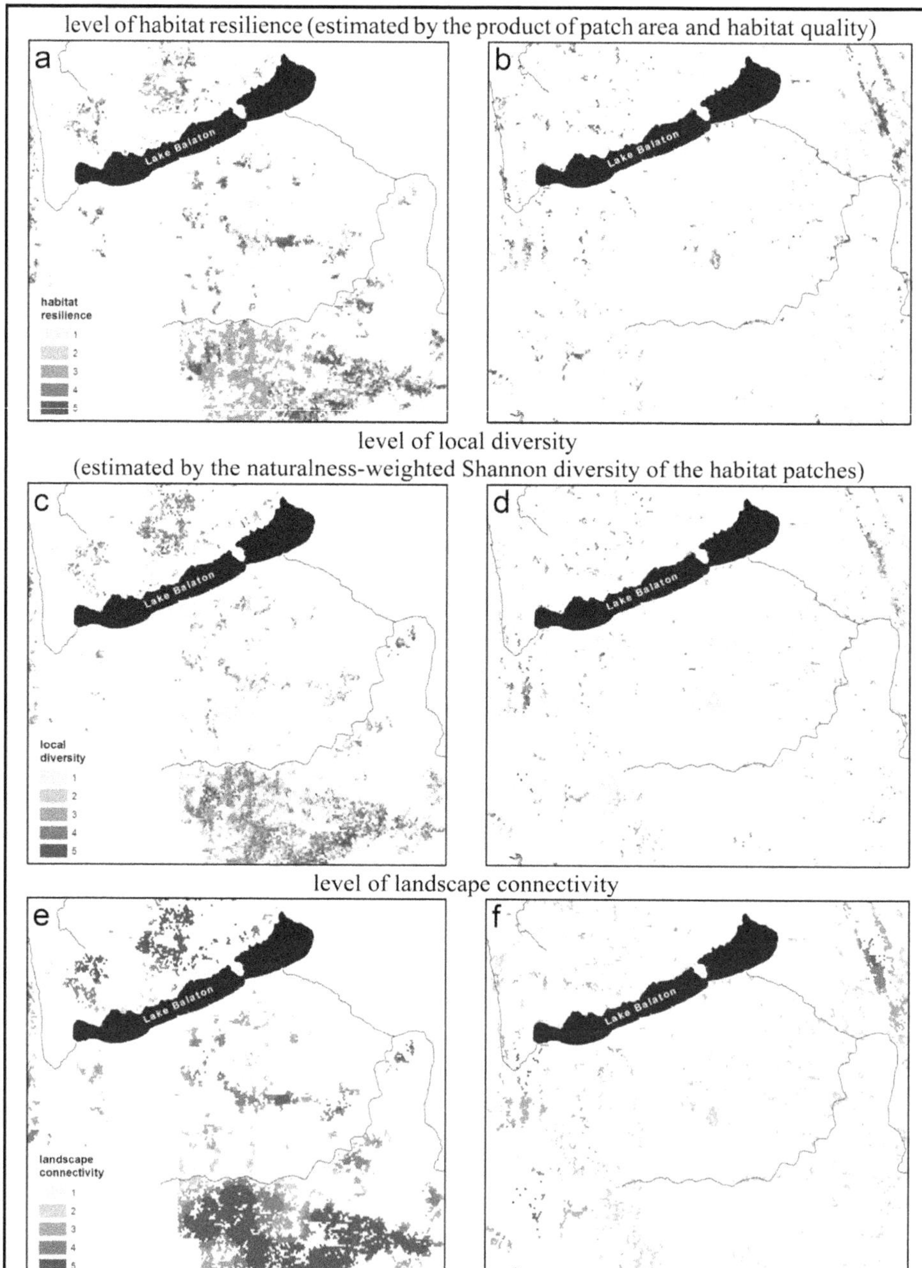

(estimated by the ESLI patch connectivity index – see Vos et al. 2001)

Vulnerability and policy recommendations

Although the individual indicators are often more meaningful for practical purposes than aggregated ones (Patt et al. 2005), composite indicators can also be informative for policy decisions, for example by distinguishing especially "hopeful" and "hopeless" situations. Areas highly exposed (i.e. high potential impact) to climate change with low adaptive capacity can be identified as particularly vulnerable sites, deserving urgent policy attention. Areas of high exposure and high adaptive capacity can be seen as potential targets for monitoring. Such an analysis is given for oak-hornbeam forests in Figure 3, highlighting "high vulnerability areas" (high exposure and generally low adaptive capacity), and "high adaptivity areas" (high exposure, and either generally high or varying adaptive capacity), which are potential targets for monitoring of climate impacts and autonomous adaptation processes.

The identification of vulnerable situations can be of particular interest to authorities and organizations engaged in nature conservation. Such analyses may form the basis for:

- objective evaluation of current reserve network, defining spatial priorities,
- identification of potential migration networks, and
- determination of unsustainable conservation targets in the face of climate change.

Current conservation practices are generally considered to be poorly adapted to the challenges caused by climate change (Fischlin et al. 2007). To provide some information on the relationship between the vulnerability of oak-hornbeam woodlands, and the current protection status of the different areas, we placed the outlines of the Natura 2000 ecological network on Figure 3. Interestingly, examining the pattern of high vulnerability and high adaptivity areas, we can see, that the former ones seem to avoid, while the latter ones seem to fit within the Natura 2000 areas. This has definitely a lot to do with the designation priorities of Natura 2000 areas, providing justification for the delineation of the network, at least for oak-hornbeam woodlands. Although the illustration is informative, just one vulnerability map of a single habitat type cannot tell much about the general conservation potential of the landscape. Nonetheless, mass evaluation of similar maps has the potential to provide significant assistance to the development of conservation and land-use strategies.

Besides providing factual risk indicators, the whole process of vulnerability analysis can also enhance general understanding of the linkages and mechanisms of the complex systems. This understanding can help us to find appropriate points for monitoring and intervention, to enhance adaptation and mitigate adverse side effects. Since undisturbed ecosystems are dynamically stable, self-regulating systems, the best available adaptation policy approach for nature conservation is enhancing the autonomous adaptive capacity of ecosystems (Fischlin et al. 2007). In accordance with the conceptual model that was used in the "adaptive capacity" part of the vulnerability analysis discussed above, this can be done on three levels:

- enhancing the local natural status of the habitats (compositional, structural and functional diversity) by reducing additional stresses and degrading land use practices (drainage, over-harvesting, over-grazing, etc.);
- preserving and increasing local/regional landscape diversity;
- preserving and increasing the connectivity and permeability of the landscape for the species of natural habitats on a regional, national or preferably international level.

Fig. 3: Highlighted areas of beech-hornbeam woodlands, and the Natura 2000 ecological network in Hungary.

a) "high vulnerability areas" with high exposure (the projected decrease in occurrence is more than 10%) and a generally low adaptive capacity (all three adaptive capacity indicators have low scores)

b) "high adaptivity areas" with high exposure, and a generally high adaptive capacity (recommended for monitoring)

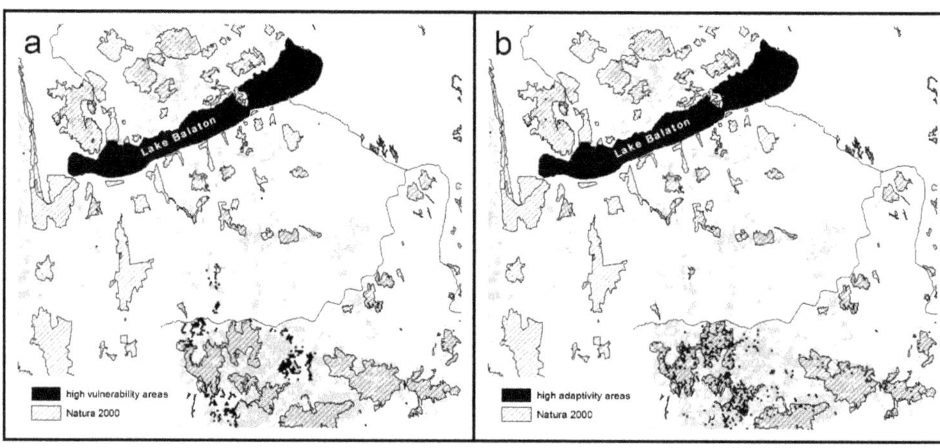

Under stable environmental conditions, many species and habitats can be preserved by simply creating reserves of appropriate dimensions. In a changing climate where species need to track the changes, the state of the agricultural-urban matrix that surrounds natural ecosystems will have crucial impact on the success of adaptation. Thus, under advancing climatic changes, biodiversity in densely populated areas, such as Central Europe, will not be retainable without integrating the aspects of conservation into the policies of several sectors (e.g. agriculture or water management). There is already growing evidence that environmental impacts may result from perverse or unintended effects of policies from other sectors (Fischlin et al. 2007, based on Chopra et al. 2005). Accordingly, conservation policies should increasingly focus on managing areas outside the protected reserves (von Maltitz et al. 2006). Fortunately, the existence of policy side-effects is getting increasingly recognized and addressed in several sectoral policies (e.g. EU Water Framework Directive, Agro-environmental schemes, Pro Silva forest management, etc.), but further advancements are still desirable (table 3).

Tab. 3: A list of cross-sectoral policy recommendations prepared for the Hungarian National Climate Change Strategy to minimize climate change impacts on biodiversity

Sectors	Policy recommendations
Nature conservation	• prepare priority lists of climate sensitive habitats and species • design and perform necessary restoration activities • improve water retention of reserves wherever possible • preserve / enhance mosaicity of habitats and successional states wherever possible • introduce / increase network concept in reserve selection • elaborate a concept / measures for protecting networks (of hedges, tree rows, roadsides) in agricultural / industrial / urban landscapes • improve monitoring activities
Water management	• develop / focus on retention oriented water management policies (instead of current drainage oriented policies) • introduce ecological aspects in the management regime of reservoirs and floodplain areas • follow the prescriptions and recommendations of the EU Water Framework Directive
Forestry	• emphasize natural forest management techniques providing continuous forest cover • introduce different regulations for forests with semi-natural structure and composition and timber / biofuel plantations • acknowledge forests of low canopy closure (forest–steppe mosaics) as valid management targets
Agriculture	• maintain / reintroduce elements of traditional landscape management (mowing, grazing) • provide buffer zones around areas of high conservational value • promote low intensity agricultural techniques • increase the heterogeneity of agricultural landscapes with networks of tree lines and hedges
Transportation	• increase the number of green bridges, ecoducts and other types of wildlife crossings on national motorways • introduce / maintain strips of semi-natural vegetation (hedges, forests) on the margin of major roads

(Czúcz et al. 2007)

Conclusions

At first sight the survival of the species might seem to be of marginal relevance for mankind, compared to economic and social challenges and conflicts shaping our everyday life. Nevertheless, ecosystems contribute inconspicuously yet fundamentally to human well-being and significant impairment of ecosystems may lead to ecological surprises with detrimental side-effects for human society. Ecological changes induced by anthropogenic climate change have the potential to impose major negative feedback on society, which makes it important to reckon with the ecological impacts of climate change in adaptation (or mitigation) strategies. Accordingly, ecological systems should be regarded as an important subsystem of socio-economic systems (coupled by ecosystem services) and general CCIAV models should incorporate appropriate sub-models for estimating climate change impacts on ecosystems.

Recognizing the complexity of interactions between biophysical and socio-economical systems, climate change impact, adaptation and vulnerability studies are increasingly shifting from linear to more complex chains of analysis, capable of addressing policy relevant questions (Rothman & Robinson 1997; Füssel & Klein 2006). Both Chapter 2 and Chapter 4 of the Fourth Assessment Report of the IPCC underline the importance of cross-sectoral, integrative studies and set these as research priorities (Carter et al. 2007: 162; Fischlin et al. 2007: 249). However, even the most complex models consist of relatively simple constituent sub-models, and the development of new methods to address specific climate change problems is also considered a research priority by IPCC (Carter et al. 2007: 161). We believe that the simple methodology to perform regional or national level vulnerability assessment for natural ecosystems facing climate change presented here constitutes such a relevant sectoral improvement, and we hope that it might also contribute to national or regional integrative adaptation strategy assessments, by improved representation of local level biodiversity feedback on human society.

References

Bacles, C.F.E., Lowe, A.J. and Ennos, R.A. (2006), "Effective seed dispersal across a fragmented landscape." Science vol. 311, p. 628.

Bölöni, J., Molnár, Zs., Illyés, E. and Kun, A. (2007), "A new habitat classification and manual for standardized habitat mapping." Annali di Botanica vol. 7, pp. 105-126.

ten Brink, B.J.E. (2000) "Biodiversity indicators for the OECD Environmental Outlook and strategy – A feasibility study", RIVM, available: www.rivm.nl/bibliotheek/rapporten/402001014.pdf.

Carter, T.R., Jones, R.N., Lu, X., Bhadwal, S., Conde, C., Mearns, L.O., O'Neill, B.C., Rounsevell, M.D.A. and Zurek, M.B. (2007), "New assessment methods and the characterisation of future conditions." In: Parry, M.L., Canziani, O.F.,

Palutikof, J.P., van der Linden, P.J. and Hanson, C.E. (eds.), Climate Change 2007: Impacts, Adaptation and Vulnerability. Contribution of Working Group II to the Fourth Assessment Report of the Intergovernmental Panel on Climate Change, Cambridge University Press, Cambridge, pp. 133-171.

Chopra, K., Leemans, R., Kumar, P. and Simons, H. (2005), Ecosystems and Human Well-being, vol. 3: Policy Responses, Island Press, Washington, DC.

Czúcz, B., Kröel-Dulay, Gy., Rédei, T., Botta-Dukát, Z. and Molnár, Zs. (2007), "Climate change and biological diversity – explorative analysis for a more effective adaptation strategy in Hungary" (in Hungarian with English summary), Institute of Ecology and Botany of the Hungarian Academy of Sciences, Vácrátót, Hungary, available: www.botanika.hu/download-01/NES.

Duraiappah, A., Naeem, S., Agardi, T., Ash, N., Cooper, D., Díaz, S., Faith, D.P., Mace, G., McNeilly, J.A., Mooney, H.A., Oteng-Yeboah, A.A., Pereira, H.M., Polasky, S., Prip, C., Reid, W.V., Samper, C., Schei, P.J., Scholes, R., Schutyser, F. and van Jaarsveld, A. (2005), Ecosystems and Human Well-being: Biodiversity Synthesis. Island Press, Washington, DC.

Fischlin, A., Midgley, G.F., Price, J.T., Leemans, R., Gopal, B., Turley, C., Rounsevell, M.D.A., Dube, O.P., Tarazona, J. and Velichko, A.A. (2007), "Ecosystems, their properties, goods, and services." In: Parry, M.L., Canziani, O.F., Palutikof, J.P., van der Linden, P.J. and Hanson, C.E. (eds.), Climate Change 2007: Impacts, Adaptation and Vulnerability. Contribution of Working Group II to the Fourth Assessment Report of the Intergovernmental Panel on Climate Change, Cambridge University Press, Cambridge, pp. 211-272.

Füssel, H.M. and Klein, R.J.T. (2006), "Climate change vulnerability assessments: An evolution of conceptual thinking." Climatic Change vol. 75 no. 3, pp. 301-329.

Grothmann, T. and Patt, A. (2005), "Adaptive capacity and human cognition: The process of individual adaptation to climate change." Global Environmental Change vol. 15 no. 3, pp. 199-213.

Guisan, A. and Zimmermann, N.E. (2000), "Predictive habitat distribution models in ecology." Ecological Modelling vol. 135 no. 2-3, pp. 147-186.

Haddad, B.M. (2005), "Ranking the adaptive capacity of nations to climate change when socio-political goals are explicit." Global Environmental Change vol. 15 no. 2, pp. 165-176.

Hijmans, R.J., Cameron, S.E., Parra, J.L., Jones, P.G. and Jarvis, A. (2005), "Very high resolution interpolated climate surfaces for global land areas." International Journal of Climatology vol. 25 no. 15, pp. 1965-1978.

HMS (2001), Magyarország Éghajlati Atlasza [Climate Atlas of Hungary] (in Hungarian), Hungarian Meteorological Service, Budapest.

Horváth, F., Molnár, Zs., Bölöni, J., Pataki, Zs., Polgár, L., Révész, A., Krasser, D. and Illyés, E. (2008), "Fact sheet of the MÉTA Database 1.2." Acta Botanica Hungarica (in print).

Hothorn, T., Hornik, K. and Zeileis, A. (2006), "Unbiased recursive partitioning: A conditional inference framework." Journal of Computational and Graphical Statistics vol. 15 no. 3, pp. 651–674.

IPCC (2001), Climate Change 2001: Impacts, Adaptation, and Vulnerability. Contribution of Working Group II to the Third Assessment Report of the Intergovernmental Panel on Climate Change. McCarthy, J.J., Canziani, O.F., Leary, N.A., Dokken, D.J., White, K.S. (eds.), Cambridge University Press, Cambridge.

– (2007), Climate Change 2007: Impacts, Adaptation and Vulnerability. Contribution of Working Group II to the Third Assessment Report of the Intergovernmental Panel on Climate Change. Parry, M.L., Canziani, O.F., Palutikof, J.P., van der Linden, P.J. and Hanson, C.E. (eds.), Cambridge University Press, Cambridge.

von Maltitz, G.P., Scholes, R.J., Erasmus, B. and Letsoalo, A. (2006), "Adapting conservation strategies to accommodate impacts of climate change in southern Africa." Working paper no. 35, Assessments of Impacts and Adaptations to Climate Change (AIACC), Washington DC.

Meehl, G.A., Stocker, T.F., Collins, W.D., Friedlingstein, P., Gaye, A.T., Gregory, J.M., Kitoh, A., Knutti, R., Murphy, J.M., Noda, A., Raper, S.C.B., Watterson, I.G., Weaver, A.J. and Zhao, Z.-C. (2007), "Global climate projections." In: Solomon, S., Qin, D., Manning, M., Chen, Z., Marquis, M., Averyt, K.B., Tignor, M. and Miller, H.L. (eds.), Climate Change 2007: The Physical Science Basis.Contribution of Working Group I to the Fourth Assessment Report of the Intergovernmental Panel on Climate Change, Cambridge University Press, Cambridge.

Metzger, M.J., Leemans, R. and Schröter, D. (2005), "A multidisciplinary multi-scale framework for assessing vulnerabilities to global change." International Journal of Applied Earth Observations and Geoinformation vol. 7 no. 4, pp. 253-267.

Molnár, Zs., Bartha, S., Seregélyes, T., Illyés, E., Botta-Dukát, Z., Tímár, G., Horváth, F., Révész, A., Kun, A. and Bölöni, J. (2007), "A grid-based, satellite-image supported, multi-attributed vegetation mapping method (MÉTA)." Folia Geobotanica vol. 42, pp. 225-247.

Noss, R.F. (1990), "Indicators for monitoring biodiversity: a hierarchical approach." Conservation Biology vol. 4 no. 4, pp. 355-364.

Opdam, P. and Wascher, D. (2004), "Climate change meets habitat fragmentation: linking landscape and biogeographical scale levels in research and conservation." Biological Conservation vol. 117 no. 3, pp. 285-297.

Parkes, D., Newell, G. and Cheal, D. (2003), "Assessing the quality of native vegetation: the 'habitat hectares' approach." Ecological Management and Restoration vol. 4, supp. 1, pp. S29-S38.

Parry, M.L. and Carter, T. (1998), Climate Impact and Adaptation Assessment: A Guide to the IPCC Approach, Earthscan, London.

Patt, A., Klein, R.J. and de la Vega-Leinert, A. (2005), "Taking the uncertainty in climate-change vulnerability assessment seriously." Comptes Rendus Geosciences vol. 337 no. 4, pp. 411-424.

Pearson, R.G. and Dawson, T.P. (2003), "Predicting the impacts of climate change on the distribution of species: are bioclimate envelope models useful?" Global Ecology and Biogeography vol. 12 no. 5, pp. 361-371.

Rothman, D.S. and Robinson, J.B. (1997), "Growing pains: A conceptual framework for considering integrated assessments." Environmental Monitoring and Assessment vol. 46 no. 1, pp. 23-43.

Ruprecht, E. and Botta-Dukát, Z. (2000), "Long-term vegetation textural changes of three fen communities near Cluj-Napoca (Romania)." Acta Botanica Hungarica vol. 42, pp. 265-284.

Skov, F. and Svenning, J.C. (2004), "Potential impact of climatic change on the distribution of forest herbs in Europe." Ecography vol. 27 no. 3, pp. 366-380.

Swihart, R.K. and Verboom, J. (2003), "Using ecologically scaled landscape indices to assess biodiversity consequences of landuse decisions." In: Swihart, R.K. and Moore, J.E. (eds.), Conserving Biodiversity in Agricultural Landscapes: Model-based Planning Tools, Purdue University Press, West Lafayette, IN, USA.

Thuiller, W. (2004), "Patterns and uncertainties of species' range shifts under climate change." Global Change Biology vol. 10 no. 12, pp. 2020-2027.

Vos, C.C., Verboom, J., Opdam, P.F.M. and Braak, C. (2001), "Toward ecologically scaled landscape indices." The American Naturalist vol. 157 no. 1, pp. 24-41.

Yohe, G. and Tol, R.S.J. (2002), "Indicators for social and economic coping capacity: Moving toward a working definition of adaptive capacity." Global Environmental Change vol. 12 no. 1, pp. 25-40.

Chapter 11

Understanding Farming Community Concerns about Adapting to a Changed Climate

Frank Vanclay, Peat Leith, Aysha Fleming

Abstract

In this chapter we seek to enhance understanding of the social basis of 'resistance' to climate change amongst the farming community. We consider the importance of this in terms of extension (practice change) in relation to climate change. Extension is a well developed field of critical enquiry and practice targeted at creating change in agricultural thinking and practice. As such, extension can play an important role in enabling adaptation of agriculture management to climate change. Nevertheless, many argue that such adaptation is not occurring as fast as it should despite acknowledgements that required change is rarely well defined. We detail a variety of social issues associated with 'resistance' to adaptation in agricultural industries generally. In so doing we provide theoretical and practical guidance for the design of extension programs oriented to addressing issues surrounding climate change.

Introduction

Climate change implies that changes in behaviour are required, both to reduce the severity of climate impacts in the future and to prepare for the changes that are now inevitable. Yet, as many of the local and regional impacts of climate change are uncertain, so are the practices and technologies that will be required to adapt to them and to reduce vulnerability to their impacts. In part this uncertainty leads to resistance to implementing change. Resistance, however, does not only result from a lack of certainty about future climatic conditions. Adoption of new knowledge, practices and technologies among farmers relies heavily on the relationships between proponents and potential users, both personally and institutionally. Put simply, adoption relies on trust. Relationships are thus key elements in the social context of resistance. To detail the social context of adoption and resistance of new ideas, technologies and practices among farmers, we draw on theoretical and empirical social research, much of which was conducted in agricultural regions of Australia. Many of these regions have already had to adapt to highly variable and sometimes changing climate regimes. We conclude our chapter by suggesting three general principles that will assist policymakers to develop programs to promote climate change adaptation.

Understanding the adoption of new practices

There is a very large body of scholarship on innovations and their diffusion. This scholarship crosses disciplinary boundaries, especially those of rural sociology, anthropology, agricultural extension, development studies, psychology, human geography, management, marketing, communication studies and the emerging fields of innovation studies and social marketing. It also spans many decades, starting well before the classic text of Everett Rogers (1962) and incrementing exponentially since. Rogers produced new editions of his text each decade, the last (the fifth edition) appearing in 2003, one year before his death. Later versions have been somewhat critical of earlier versions as his thinking developed, both reflexively and in response to the many commentators. Despite the various criticisms, however, the essence of the model remained the same over time and has become quite entrenched in thinking around the study of adoption of innovations.

Diffusion is defined as "the process by which an innovation is communicated through certain channels over time among the members of a social system" (Rogers 1983: 15). An innovation is defined as "an idea, practice, or object that is perceived as new by an individual or other unit of adoption. It matters little, so far as human behaviour is concerned, whether or not an idea is 'objectively' new as measured by the lapse of time since its first use or discovery" (Ibid: 11). The distribution of adoption over time approximates a normal curve, or if expressed cumulatively, a sigmoid or S-shaped curve. Adopters can be classified according to their positioning in the adoption curve (see figure 1).

Fig. 1: The classification and labelling of adopters in the Rogers model

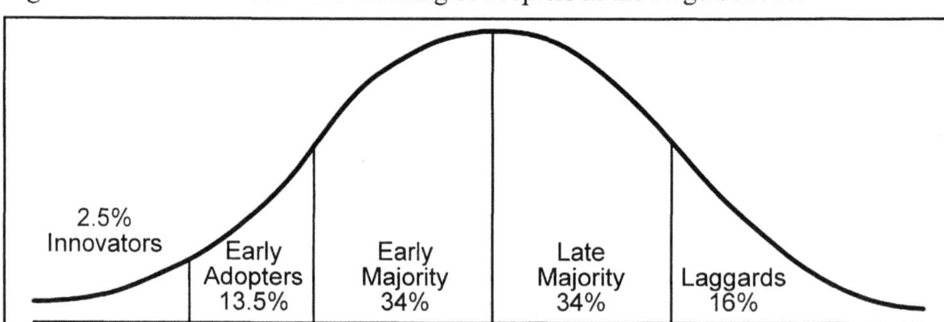

Source: Rogers 1983: 247

According to Rogers and his many followers, there are social characteristics that can be ascribed to the adopter categories in Figure 1. Innovators – that is strictly the first 2.5 percent of adopters – tend to be venturesome, risk-taking, and have a burning desire to experiment. Rogers describes early adopters as being 'respectable', the early majority group as 'deliberate', the late majority as 'skeptical', and the laggards as 'traditional' (Rogers 1983: 248-250). While research in the field of innovation studies tends to concentrate on innovators and early adopters, much of the research in rural sociology and anthropology has looked at the late majority and so-called laggards. Invariably, it is discovered that these groups have very good reasons for not adopting, prompting Vanclay (2004) to suggest that there is no such thing as barriers to adoption, only legitimate reasons for non-adoption.

Research into the social dimensions of farmers in Australia reveals that some form of lay understanding of the diffusion model is widespread, perhaps because of the hegemonic influence of agricultural extension agencies with staff trained in that tradition (Vanclay et al. 1998; 2006). However, there is much confusion and the words are not used in the precise ways intended by Rogers. For example, many farmers in Australia desire to be what they call 'innovators' and may claim to be so, but they actually mean the 'early adopters' in Rogers' terms (Mesiti and Vanclay 2006). This social research has tended to reject much of the deterministic and simplistic prescriptions of the naïve diffusion of innovations approach arguing that the adoption process is a much more complex social process (Vanclay 1992; 2004; Vanclay & Lawrence 1994; 1995). The diffusion model has been criticised by many other writers for a range of reasons (see for example Buttel et al. 1990). Not all of the criticisms raised are necessarily valid of Rogers himself, as over time he has developed a well-elaborated thinking which addressed many of the issues raised. Therefore, despite all the criticism, the model remains an important conceptual tool (Pannell et al. 2006).

An important understanding is that adoption is not a single act, but rather part of a communication process. Typically, it involves many actors and is not the result of the uni-directional transfer of knowledge from scientist to farmer. Conferring with others is part of the validation of a decision for many people. The legitimacy of the source of information, and the social circles in which farmers move and through which they come to hear of new ideas or practices are significant issues affecting the likelihood of adoption. Another important aspect is that adoption depends on the socio-psychological characteristics of individuals and their emotional state and mental wellbeing as well as the socio-cultural and socio-economic context in which they are in. Individuals who are alienated or demotivated are unlikely to have much interest in change. The process of promoting change needs to be as much concerned about these broader issues and not just about the adequacy of the information or the effectiveness of the communication process.

One of the less controversial aspects of the model is the understanding that adoption is a process of learning and decision making, involving a number of steps or phases, which potentially can occur concurrently or be revisited. Thus adoption is not a single act. Even with innovations that require only a one-time change in behaviour, there is a long process of decision making and information evaluation during which time a person's views (propensity to adopt) can waiver. Behaviours that must continue to be repeated are subject to ongoing validation and individuals may revert to former practices if there is not ongoing benefit.

This process of adoption is represented as a sequence of steps or stages or phases in different ways by different writers. Rogers (1983), for example, lists them as: (1) knowledge, (2) persuasion, (3) decision, (4) implementation, and (5) confirmation. In the agricultural extension field, they are often presented as being: awareness raising (gaining an awareness of the issue); information seeking (to find information to address the issue); decision making (to decide to take action in relation to the issue); and action taking (the implementation of decisions made, and any adjustment that need to made to make it work).

Too often, organisations promoting practice change have viewed communication as a simple top-down process, with farmers perceived as being eager recipients passively waiting for the next gift of knowledge from science. In this construction of their worldview, non-adoption is inconceivable and could only be explained by a failure in the communication process (leading to the need to improve extension practice), or by a belief that there were a range of barriers to adoption which must be overcome. These organisations and their staff frequently assumed non-adopters were ignorant or incompetent. The reality – as established by research into the non-adoption of recommended practices by farmers – is that these farmers frequently understood what was being promoted and knew the reasons why they were being promoted; but they actively contested either the problem, the scientific account of the problem, or that the solutions being proposed would effectively resolve to the problem (Lockie et al. 1995; Vanclay 2004). Agencies promoting change need to presume that farmers are intelligent individuals, who have their own legitimate understanding of issues. Such understandings tend to relate to the social and political aspects of issues as well as to the biophysical ones.

A further criticism of adoption research and the field of agricultural extension is their individualistic focus and latent assumption of a male farmer. The reality is that the vast majority of farms in Australia, and in most places in the world, are family farm businesses. Decision making, including the adoption of new practices, takes place in a shared negotiation space where the views of all parties have standing.

Resistance to change

If adapting to change was easy, the world would not have the many problems it has. Adapting to change is actually difficult because it takes people outside what is known, familiar and comfortable – and because people are socio-cultural, emotional beings not rational, calculating automatons. Furthermore, and perhaps conversely, often there are very good reasons not to adopt the various new practices or technologies that are promoted by governments and other organisations, at least from the perspectives of those not adopting. Based on a range of surveys with farmers about their perceptions of new environmental management practices, Vanclay (1992; 2004) considered there were about 12 generic reasons for not adopting whatever practice was being promoted (see Box 1).

Box 1: Twelve generic reasons for farmer non-adoption

1. Too complex. In general terms, the more complex the innovation, the greater the resistance to adoption. Complexity makes the innovation more difficult to understand, and generally requires greater management skills. This increases the risk associated with the innovation.
2. Not easily divisible into manageable parts. Divisibility allows for partial adoption and/or trialing. Farmers can adopt that part of an innovation that they like or that is consistent with other farming objectives. Obviously, therefore, the more divisible into component parts an innovation is, the more likely it is to be adopted.
3. Not compatible with farm and personal objectives. Farmers are more likely to adopt innovations that are compatible with other farm and personal objectives.
4. Lack of flexibility. Farmers like flexibility because it means that they can change commodities in response to market and climatic conditions. Farmers are quite likely to resist the adoption of new technologies that restrict their flexibility.
5. Not profitable, at least in farmers' perception.
6. Capital outlay is too high. In addition to the economics of the innovation – in terms of whether or not the innovation will increase profit – it is necessary also to consider the availability of capital required to adopt the new technology. Many farmers are risk averse and dislike borrowing money.
7. Too much additional learning is required – there is an intellectual cost to adoption.
8. Risk and uncertainty is too great.
9. Conflicting information.
10. Don't see that there is a problem (lack of appreciation of the problem).
11. Lacking the physical infrastructure.
12. Lacking the social infrastructure. In the same way that a physical infrastructure exists as a mechanism to encourage production of particular crops, and inhibit others, a social infrastructure also exists. The social infrastructure refers to the social networks of farmers, which provide a knowledge bank for farmers to utilise. The accumulated knowledge of other farmers is usually regarded as a more important source of information than extension services. Except for a few maverick farmers, no individual farmer wants to be the only one doing a particular activity because they would have no social support to discuss their problems.

Source: modified from Vanclay 1992 and 2004

Understanding adoption as a process involving a number of steps is important. There can be resistance to adoption at each stage. That is why simply providing information does not necessarily change behaviour. The often presumed 'information deficit model', where people not adopting were presumed to lack appropriate information, has been thoroughly criticised by many (Irwin & Wynne 1996; Vanclay & Lawrence 1995; Wynne 1991; 1992).

Effective extension – "the process of enabling change in individuals, communities and industries involved in the primary industry sector and with natural resource management" (SELN 2006: 4) – is a capacity building process not just information provision, and needs to consider what is needed to promote change at each stage in the process. There are various expressions of this. The UK Department of Environment, Food and Rural Affairs (DEFRA), for example, expresses this as the

Four E's: enable, engage, exemplify and encourage (see figure 2). The Four E's has become the basis of a major discussion paper of the UK Cabinet Office into achieving culture change, with a clear realisation that the achievement of behaviour change in relation to socially-desirable behaviours requires a multi-dimensional approach and an appropriate blend of policy instruments (Knott 2008).

Fig. 2: DEFRA's behaviour change model

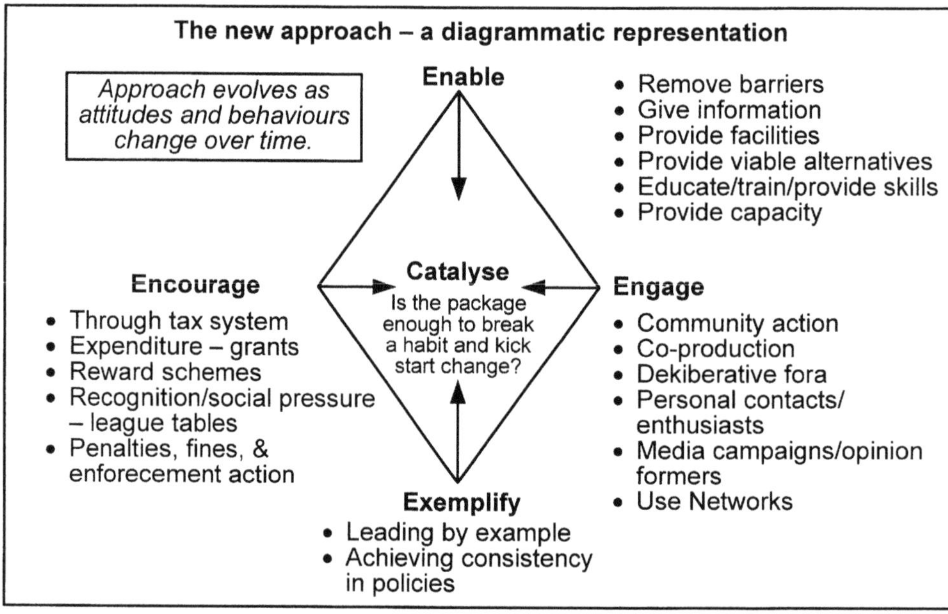

Source: DEFRA no date, p. 2

Embracing a social understanding of resistance

Enabling change requires consideration of the social and cultural basis of resistance. Resistance to changing behaviour should be viewed through a more reflexive lens, rather than by assuming resistance is a pathological problem to be overcome. Resistance should be taken as active engagement with knowledge and technologies. From such a perspective, resistance might be usefully considered as a resource which can inform extension and, in turn, the transformation of action. Viewing resistance as a process of evaluation and validation among active and interactive individuals within particular cultural, economic and geographical settings brings us to focus on the importance of building appropriate mechanisms and institutions through which governments and NGOs can interact with such active subjects, so that the information provided to farmers is valuable and valued, and can become useful and useable. This task is by no means straight forward. However, we argue that an obligatory first step is to ensure that extension, particularly relating to climate change vulnerability and adaptation, needs to be underpinned by a social view of resistance.

We are all too aware that knowledge does not travel in straight lines in these post-modern times. Most people suffer from information overload. Everywhere there is a crisis of legitimacy. When contrary perspectives abound in all forms of media, individuals and communities must constantly ask who can and should be trusted to provide information that is not marred by inaccuracy, or made illegitimate by perceived vested interests. The trustworthiness of information, actors and organisations is also underpinned by their ability to address the concerns of the people they are informing. In the context of climate change, these concerns are magnified by the style and content of the information which often includes carefully caged statements about an uncertain future at different scales. Scenarios or predictions at decadal or longer timescales evoke legitimate skepticism, and questions about how such predictions can be used to inform decisions at a local or regional scale are often difficult to answer. However, even early studies on public perceptions of climate change have noted that despite a lack of technical scientific understanding about the drivers of anthropogenic climate change, there was a widespread sense of impropriety among respondents of our profligate use of fossil fuels and of our use of the atmosphere as a commons for waste disposal (Kempton 1997). Such research opens up a raft of questions about how public knowledge is created and whether it is either desirable – or even possible – to separate out the technical or natural from the social, moral and political.

Efforts to make sense of how knowledge travels reveals that actors do not just evaluate how useful a piece of information appears, its salience or relevance, they continually negotiate the legitimacy and credibility of information and of the actors proffering it (Cash et al. 2003). Legitimacy in this context refers to the institutional arrangements and processes which lead to information or knowledge being generated. Credibility refers to the acceptability of the methods by which the information was generated. Clearly salience, credibility and legitimacy are highly interactive – increasing one can either increase or reduce the others. For example, were one to claim that over the next decade the autumn rainfall for a particular agricultural region would drop by 25 percent as a result of climate change, this would constitute a very salient piece of information for local farmers; yet it is unlikely to be viewed as credible among researchers used to dealing with scenarios rather than predictions. If such a claim were made in the public domain by some renegade climate researcher, others would come forward to counter the claim, the ensuing debacle might undermine the legitimacy of climate scientists more generally, especially if a media or public narrative developed along the lines of 'scientists can't agree amongst themselves'.

This brief, yet unfortunately familiar scenario points to the imperative of not only producing information that is salient, credible and legitimate for particular people, but the necessity of managing the way information is generated and publicised more broadly within and among public, political and scientific domains. Public discourse never really was controllable. In the context of climate change, with the diverse interests at stake in responding and adapting to climate change, it is little wonder that the basic acceptance of anthropogenic climate change as a real and current concern remains in question for many people. How to build a legitimate form of expert advice is a central challenge in the post-modern world. Given the enormity of this challenge, we propose three basic principles we believe are useful in the context of communicating about adaptation to climate change: Firstly, problems will need to be defined through meaningful participation across scientific and lay cultures. Secondly, the places and contexts of peoples' lives are central to their ability to adapt, reduce vulnerability and become more resilient – therefore these places and contexts need to become forefront in the development of regional and local strategies for climate change adaptation. Thirdly, adaptation is not just about having the right information – that information is contextualised by the stories we tell ourselves and those we are told to believe about how we should live – that is, narratives are important. These principles are developed further below.

Principle 1: Problem definition must be a participatory process

There is a need for climate adaptation programs to develop ways of defining the issue which span scientific and lay cultures. Through participative work, researchers, policy people and wider publics will be connected in meaningful ways via appropriately devised programs. Extension theorists have proposed a variety of different models of participatory research. These tend to suggest that any vestiges of a top-down model of risk assessment need to be removed. In the context of climate change, it is likely that many of the impacts will not come from the sky or the sea, but from the likely massive changes in the global economic system, from conflict over food and water security and from unprecedented numbers of environmental refugees. Technical solutions are only part of the answer to most problems; this is all the more so when problems are complex and involve overlapping scientific, political, social and moral dimensions. Also, people rarely consider knowledge as being removed from the interests of its sources. Thus, as is the case for political organisations, applied scientific and extension organisations need to be closely engaged with those who might be able to apply their knowledge. Their interests need to be aligned and this requires a variety of measures which will ensure that what is said by scientists and extensionists is close to what is heard.

Developing such clarity requires the building of appropriate institutions and processes such that individuals from scientific and lay communities come to understand one another's concerns, interests and understandings. For example, climate applications research around the world has increasingly tended towards participatory research such that, for instance, farmers help to define the research aims of projects after discussing with climate scientists and agronomists what might be achievable. It is through forming associations with communities that scientists begin to understand the problems farmers face and vice versa (see Box 2). Knowledge and uncertainty as well as identities and interests are better understood through such dialogue.

Box 2: Different ways of thinking

The ideal that farmers' decisions should be underpinned by scientific data has been achieved in Australia on occasion through the adoption of a risk management ethos. In particular, Decision Support Systems (DSSs) based on computer models of agricultural systems have been developed and extended to farmers. These provide decision makers with probabilities, for instance, of the likelihood of achieving below or above average yield as a result of tactical actions and seasonal rainfall probabilities (Hammer et al. 2000). The thing about probabilities is that they are difficult to verify over short timeframes unless those probabilities are very high. Most agricultural technologies, such as herbicides or machinery, have a highly visible impact – that is whether they work or not is obvious. These technologies can easily be tested by using them in a prescribed way and observing outcomes. DSSs, on the other hand, are not like this. Judgments must be made on the part of farmers as to whether such technologies are sufficiently useful and useable. From research involving farmers, scientists and extension staff, it is apparent that when people adopt these tools, they tend to first require that the models are aligned more closely with farmers' understanding of the agricultural system, or vice versa. Only once models become trusted to adequately represent the system will decisions be made on the basis of the probabilities they produce. This trust would appear to be conditional and dependent on the ability of scientists to demonstrate how well they understand the agricultural system. That is, trusting the model implies trusting the knowledge and skill of those who produced it. This sort of trust is developed through dialogue, which has led some systems modellers to suggest that DSSs are better considered as Discussion Support Systems (Nelson et al. 2002), in that the models themselves provide a focal point where extension, staff, researchers and farmers can discuss the pros and cons of different management options.

Often through such dialogue it becomes clear that the problems facing farmers are much messier than those that scientists are trying to resolve. Rather than trying to maximise farm profit, farmers are often looking to sustain a way of life that can continue across generations with immediate income being just one component, albeit an integral one to achieving this goal (Vanclay 2004). Because climate change information is often highly complex and contingent, it is likely to need to be described repeatedly in different ways such that different people can come to appreciate what is being suggested and the implications of this information, as well as to consider possible strategic actions that might reduce vulnerabilities or improve resilience. Everyone is aware, for example through their interpersonal relationships, that the way a message is framed has important implications for the way it is received and ultimately understood. Just as a good manager will talk to different staff in different ways depending on the historically generated dynamics of their interaction, a similar process should be emphasised in extension. Respected extensionists have earned trust through talking with people at an appropriate level of detail, through listening carefully and absorbing the knowledge of farmers as well as the manner in which it is relayed.

Thus they develop relationships which turn these actors into gatekeepers between communities, and between knowledge produced in fields and kitchens and that which is produced in laboratories or through modelling. From time to time, climate scientists themselves can and do take on this sort of role, learning how to negotiate different knowledge cultures through doing interdisciplinary and participatory research. Such work combined with their histories and backgrounds enables them to represent the climate system in the contexts of people's lifeworlds, livelihoods and identities. This requires a sense not only of the relationships between climate and the people with whom a scientist is trying to communicate, but also that the climate communicator is aware of the legitimacy of the narratives of identity and management that they are proposing to farmers.

At a broader institutional scale, the necessity of framing problems in ways that end up producing useful and useable knowledge and technologies has an important structural corollary: organisations responsible for developing research programs need to be accountable to the diverse cultures they serve (public, scientific and political) – that is, they should be 'boundary organisations' (Guston 2001).

Box 3: The Signs of the Seasons

In the semi-arid rangelands of Australia, decision-making is complex and works at a variety of levels. At the simplest level, one might describe the calculus of managing land as simply as balancing the total stocking rate of any tract of land against the amount of available feed. Yet, a family managing a property or a number of properties will have different goals, histories and ways of knowing and thinking about the work of managing land and stock, about what a certain paddock or a particular bloodline of stock can tolerate, and in what conditions they will thrive. One thing that is common among graziers, however, is that climate is not discussed just in terms of meteorological parameters such as rainfall, wind, temperature, and humidity. Rather, the way these elements play out in the landscape and in the behaviour of stock and people gives as much indication of how people think about climate as does their assiduous maintenance of property rainfall records. For example, in interviews some wool growers mentioned in passing that they always used to sheer their sheep in the middle of summer, but that this had changed in the last decade because it was now simply too hot at that time of year. A couple mentioned that when they were growing up, the end of the rainy season was always demarcated by the annual local Show Day, but that it was the case that the rainy season has not lasted that long since the 1970s. Because the seasons play out in the landscape, the social memory of climate is built around local and regional ecologies, the movement and behaviour of native animals and stock. These are the signs of the seasons. They may not be accurate as ways of predicting or describing the climate in scientific terms, but they connect people with places, and as such, they serve an important role for rangelands graziers in putting current conditions into the context of what has gone before. From interviews with graziers in the Australian rangelands in 2004, even those who were sceptical about climate change often related that old patterns did not hold as well as they had previously. Here was a hint of the work of verifying climate change at this very local level through observation. In such a variable environment, however, it is justifiably difficult to be persuaded that the climate is changing when dry periods have previously continued for decades. In such situations, the local stories that include the changing ecologies of places might be a meeting point between scientists and farmers, especially where such ecological shifts are congruent with those predicted and monitored by scientists themselves. Here, partnerships can emerge whereby local ecological knowledge feeds scientific knowledge, and vice versa.

Principle 2: An understanding of place, context and culture is central to adaptation

The imperative of balancing salience, credibility and legitimacy can be addressed by grounding scientific knowledge, practice and people in the issues facing particular communities of practice and place. Science for adaptation must become a form of extension, in that scientists must be able to understand and internalise the place and context-specific problems which are likely to be faced by particular individuals and communities. In part, this can be achieved through discussions about the places in which people live, which provide the means by which their climate and its meaning comes to be understood. Climate is not only mediated by scientists and data; it is also curated through experience, the transmission of place-specific knowledge and through social memory (McIntosh, Tainter & McIntosh 2000). For example, changes in seasons are seen in the responses of plants and animals to different patterns of temperature and rainfall, and these come to be interpreted as signals of the local climate (see Box 3). The role of science and scientists in the constitution of places is rarely made explicit, yet it is important that scientists are aware that 'place' is quite different from the spatial field they tend to think about when conducting experiments or modelling studies. Place is constituted through talk and experience (Vanclay, Higgins & Blackshaw 2008), as is the climate of a place (Leith 2006). Yet, especially for farmers (and perhaps to a lesser degree, other people who work outdoors), place and climate are bound together in stories. Local grounding of the broader societal discourse about climate change may happen through stories about place which gradually come to be attached to conceptions of climate change. While climate scientists are wary of suggesting that any extreme conditions in any given season are the *effect* of climate change per se, they are more likely to suggest that *the tendencies of the seasons are in line with what would be expected if the models were performing well, thus representing a partial validation of the models.* There is clearly a gulf between the language of climate modellers and the language by which the climate of a place is gradually constituted. This chasm can and should be bridged by the dynamic process of building relations between people and landscape, by joining scientific and local stories together, and by gradually making scientific and lay knowledge more salient, credible and legitimate in relation to one another.

Because the meaning of climate is socially negotiated within particular cultural settings, the importance of climate impacts (including the secondary and tertiary impacts of climate change and variability) will necessarily vary from place to place, and from one context and culture to the next. Thus, places, cultures and contexts become guides in the development and implementation of projects aimed at building capacity for adaptation to global change. Extension relevant to climate change thus becomes a means of engaging with problem formulations and their resolution in particular places and cultures. Such extension grounds social and biophysical research and researchers within a particular region, local, industry or community.

Principle 3: Storytelling is effective

Box 4: Learning Together in the Mallee Region of Victoria, Australia

The researchers had established something they thought would interest the farmers of the Mallee region, with whom they were working. The hourglass for the project was still half-full, and here was something of a revelation. It was always going to be partial knowledge – they were, after all, talking about the climate! Through their own regionally funded Research, Development and Extension organisation, the farmers had contracted the research. The scientists would work to make seasonal rainfall conditions somewhat less uncertain. Among the first things the scientists noticed after they had set up their software to dredge the historical record, was this: the best rain did not come from cold fronts off the Southern Ocean as everyone had previously thought. Instead, a large-scale synoptic pattern known as a 'cut-off low' seemed to be associated with the majority of the region's useful rainfall during the winter cropping season. To the farmers, this was a revelation which would change the way they would look at synoptic charts in the years to come. Some cut-off lows, however, seemed to be more 'productive' of rain than others, as the farmers might have put it. Here was a new riddle. The scientists – an oceanographer, a synoptic climatologist, and an extension agent – drove for many hours to get to the Mallee. They reported to the assembled farmers that while the current growing season had at least its fair share of cut-off lows, for some reason these cut-offs did not appear to be producing rain. Later, in conversation, a farmer commented to one of the scientists that it had been an exceptionally cold winter. The scientists wondered if this was relevant to the unproductive cut-offs. They traced it back through a re-analysis of the records and found another element in the contingency of weather-making: cut-off lows, historically, had produced more rain when they occurred in the presence of a strong north-south temperature gradient. This year was not only cold in the Mallee, but to the north of it as well. Another fragment of the picture of the regional climate had emerged. However, in addition to this increased technical understanding, something else – perhaps something more important – had also happened. The scientists and farmers had realised that their shared interest in the climate was mutually beneficial. They realized that if they would listen carefully to one another, they would each learn more about how climate works, and how to live with it, than they could on their own.

Source: Adapted from a story told by Peter McIntosh (CSIRO) and used with permission. See also Pook et al. 2006.

We have always relied on stories. They help us decide how to live. Adapting to climate change, reducing vulnerability and building resilience within socio-ecological systems will be no different. Yet the narratives of techno-science in previous eras relied heavily on ideas of certainty and risk. Climate change science relies instead on narratives of uncertainty and consensus, which infers that we accept the stories that are settled-on within a group. While contemporary climate change narratives do provide ample room for debate, this debate appears counter to public views of science as being something to do with finding out facts, or truth. In the context of the crisis of legitimacy, the most effective forms of extension appear to be those which are interdisciplinary and collaborative, which embrace all the forms of uncertainty that play on the decisions faced by farmers and community members. In such participatory knowledge-making, applied researchers articulate the farmers' concerns in a way that sheds new light on them. These forms of research can reconstitute the climate via changing the ways by which it is discussed, and thus the ways by which people consider and manage variability and change. They also build a sense within research communities of what it means to make knowledge that is salient and legitimate rather than just doing work that is methodologically credible (see Box 4).

Thinking about the process of communication and extension of climate change in terms of narratives, as well as in the usual terms of communicating science, opens up possibilities for building more reflexive forms of scientific practice. Listening to the stories people tell makes it abundantly clear that people address problems in different ways and think in different ways. What people know through local experience and through building a shared social memory, often through stories about place, serves a role that science simply cannot fill. Thus scientific knowledge must work in conjunction rather than competition with other forms of knowledge.

Conclusion

Climate change is real and we all need to change our behaviour to reduce the severity of climate impacts in the future and to prepare for the changes that are now inevitable. But this will not be achieved simply by increasing the amount of information available. Resistance to change is also real, and rather than patronising or vilifying those who will not adopt what is deemed to be appropriate, a more considered approach that accepts resistance as legitimate is needed. Understanding the perceptions and concerns of people and engaging with them is important.

While the field of community-based social marketing (see McKenzie-Mohr & Smith 1999) provides much guidance to assist in influencing the general public, farmers are a specific group with specific interests, especially in terms of adapting to climate change. Vanclay (2004) identified some 27 principles to assist in the promotion of environmental management to farmers which would also be of assistance in the promotion of practice change with respect to climate adaptation. Perhaps the most important of these principles is that farmers are not homogenous. Even within the same commodity and agricultural region, there are differing enterprise objectives, differing management strategies, differing social and economic contexts, differing decision-making processes, differing personal objectives, differing information-seeking behaviours, differing preparedness to accept risk, differing learning styles and so on. This diversity is much more complex than the adopter categories of the diffusion of innovations theory.

Instead of focussing solely on the one-way provision of information that calls for a rational response to technical data, a more comprehensive approach to enabling change is required, one that considers both the rational and emotional elements of decision-making and behaviour change at the family farm level. Programs should comprise of identifying and enacting the enabling conditions that:

- Create a suitable environment/milieu conducive to encouraging change through participative processes that assist in identifying problems and solutions

- Assist in developing the motivation and enthusiasm to embrace the possibility of change, as well as to appreciate its risks and benefits
- Provide the resources and means to allow change to happen, such as providing incentives or compensation; and
- Facilitate access to the information and knowledge needed to know how to make the change.

Each one of these enabling conditions is important. Addressing any one without considering the others will only lead to frustration.

We also suggest that a genuine strategy of community engagement is needed. In this chapter, we articulated three principles to assist in the development of such a genuine strategy of engagement in relation to climate change. Firstly, identification of adaptation options needs to be developed through meaningful participation across scientific and lay cultures. Secondly, we suggest that the places and contexts of peoples' lives are central to their ability to comprehend and to adapt, and therefore they need to be prominent in the regional and local strategies for climate change adaptation and in their development. Thirdly, information is negotiated and mediated through stories. The telling of stories in various forms can assist greatly in people coming to understand, and in developing interest in responding. Stories can be a powerful force for change.

References

Buttel, F., Larson, O. and Gillespie, G. (1990), The Sociology of Agriculture, Greenwood Press, New York.

Cash, D.W., Clark, W.C., Alcock, F., Dickson, N.M., Eckley, N., Guston, D.H., Jaeger, J. and Mitchell, R.B. (2003), "Knowledge systems for sustainable development." Proceedings of the National Academy of Sciences of the United States of America vol. 100 no. 14, pp. 8086-8091.

DEFRA (no date), Changing policy through policy making. United Kingdom Department of Environment, Food and Rural Affairs, London www.sustainable-development.gov.uk/what/priority/pdf/change-behaviour-model.pdf.

Guston, D.H. (2001), "Boundary organisations in environmental policy and science." Science, Technology and Human Values vol. 26 no. 1, pp. 399-408.

Hammer, G.L., Nicholls, N. and Mitchell, C. (eds.) (2000), Applications of Seasonal Climate Forecasting in Agricultural and Natural Ecosystems: The Australian Experience. Kluwer, Dordrecht.

Irwin, A. and Wynne, B. (eds.) (1996), Misunderstanding Science, Cambridge University Press, Cambridge.

Kempton, W. (1997), "How the public views climate change." Environment vol. 39 no. 9, pp. 12-21.

Knott, D. (2008). Achieving Culture Change: A Policy Framework, A discussion paper by the Strategy Unit, UK Cabinet Office, London.

Leith, P. (2006), Conversations about climate: Seasonal variability and graziers' decisions in the eastern rangelands, University of Tasmania, Hobart.

Lockie, S., Mead, A., Vanclay, F. and Butler, B. (1995), "Factors encouraging the adoption of more sustainable crop rotations in South East Australia: profit, sustainability, risk, and stability." Journal of Sustainable Agriculture vol. 6 no. 1, pp. 61-79.

Mesiti, L. and Vanclay, F. (2006), "Specifying the farming styles in viticulture." Australian Journal of Experimental Agriculture vol. 46 no. 4, pp. 585-593.

McKenzie-Mohr, D. and Smith, W. (1999), Fostering Sustainable Behaviour: An Introduction to Community-Based Social Marketing, New Society Publishers, Gabriola Island.

McIntosh, R.J., Tainter, J.A. and McIntosh, S.K. (2000), "Climate History and Human Action." In: McIntosh, R.J., Tainter, J.A. and McIntosh, S.K. (eds.), The Way the Wind Blows: Climate History and Human Action. Columbia University Press, New York, pp. 1-41.

Nelson, R.A., Holzworth, D.P., Hammer, G. and Hayman, P. (2002), "Infusing the use of climate forecasting into crop management practice in North East Australia using discussion support software." Agricultural Systems vol. 74, 393-414.

Pannell, D., Marshall, G., Barr, N., Curtis, A., Vanclay, F. and Wilkinson, R. (2006), "Understanding and promoting adoption of conservation practices by rural landholders." Australian Journal of Experimental Agriculture vol. 46 no. 11, pp. 1407-1424.

Pook, M.J., McIntosh, P.C. and Meyers, G.A. (2006), "The synoptic decomposition of cool season rainfall in the south-eastern Australian cropping region." Journal of Applied Meteorology and Climatology vol. 45 no. 8, pp. 1156-1170.

Rogers, E.M. (1962), Diffusion of Innovations. Free Press, New York.

– (1983), Diffusion of Innovations. 3rd edn. Free Press, New York.

– (2003), Diffusion of innovations. 5th edn. Free Press, New York.

SELN (2006), Enabling change in rural and regional Australia: The role of extension in achieving sustainable and productive futures, State Extension Leaders Network www.seln.org.au.

Vanclay, F. (1992), "Barriers to adoption: a general overview of the issues." Rural Society vol. 2 no. 2, pp. 10-12.

– (2004), "Social principles for agricultural extension to assist in the promotion of natural resource management." Australian Journal of Experimental Agriculture vol. 44 no. 3, pp. 213-223.

Vanclay, F., Higgins, M. and Blackshaw, A. (eds.) (2008), Making Sense of Place, National Museum of Australia Press, Canberra.

Vanclay, F., Howden, P., Mesiti, L. and Glyde, S. (2006), "The social and intellectual construction of farming styles: Testing Dutch ideas in Australian agriculture." Sociologia Ruralis vol. 46 no. 1, pp. 61-82.

Vanclay, F. and Lawrence, G. (1994), "Farmer Rationality and the adoption of environmentally sound practices: a critique of the assumptions of traditional agricultural extension." European Journal of Agricultural Education and Extension vol. 1 no. 1, pp. 59-90.

– (1995), The Environmental Imperative. Central Queensland University Press, Rockhampton.

Vanclay F., Mesiti, L. and Howden, P. (1998), "Styles of farming and farming subcultures: Appropriate concepts for Australian rural sociology?" Rural Society vol. 8 no. 2, pp. 85-107.

Wynne, B. (1991), "Knowledges in context." Science, Technology and Human Values vol. 16 no. 1, pp. 111-121.

– (1992), "Public understanding of science research: new horizons or hall of mirrors?" Public Understanding of Science vol. 1 no. 1, pp. 37-43.

Chapter 12

Risk Assessment and Local Government, Tasmania: Applying an Interdisciplinary Approach to Climate Change Adaptation

Melissa Nursey-Bray, Tony Ferrier

Abstract

Climate change is a global problem occurring at multiple scales and its resolution demands interdisciplinary solutions. In Tasmania, Australia, the State of the Environment Report for (RPDC 2004) identified evidence that sea scales are rising and will continue as a result of climate change. This constitutes risk for local government in regard to protecting public infrastructure and providing appropriate development control. Planning decisions need to be well informed and good natural resource hazard information is essential. Kingborough Council, Tasmania, has developed an inter-disciplinary risk assessment tool for dealing with climate change at local government scale. This risk assessment tool was specifically designed to be interdisciplinary in both nature and implementation. The evolution of this tool brought many different practitioners together, in an attempt to forge working links across different knowledge domains, both social and scientific in order to develop effective decision making about climate change on the ground. The project involved all sections of the Kingborough Council and is being run in synchronicity with the State and Federal government Climate Change adaptation programs. This chapter outlines the progress of this initiative and the lessons learned for building future interdisciplinary projects when dealing with climate change at local government scales.

1. Introduction

> "Globally, 1.2 billion people (23% of the world's population) live within 100 km of the coast, and 50% are likely to do so by 2030. These populations are exposed to specific hazards such as coastal flooding, tsunamis, hurricanes, and transmission of marine-related infectious diseases. For example, today an estimated 10 million people experience coastal flooding each year due to storm surges and landfall typhoons, and 50 million could be at risk by 2080 because of climate change and increasing population densities. More and more, adaptive responses will be required in coastal zones to cope with a plethora of similar hazards arising as a result of global environmental change" (Adger et al. 2005: 78).

Climate change is a global problem that affects everyone. The science has established that the impacts of climate change will be widespread. Coastal communities will be no exception. However, the costs of addressing climate change will not be evenly distributed, and in coastal contexts is likely to fall disproportionately on local government, industries, communities, and workers. Local governments in particular are going to be disproportionately affected by climate change, not only physically, but in terms of the responsibilities they will shoulder to manage this issue. As noted by the Local Government Association of Tasmania:

> "Local government provides for the health, safety and welfare of its community and if a council cannot show that it has taken preventative action against any threat to the health, safety and welfare of its community, it faces the possibility of liability costs – costs which can be reduced if a council identifies the threats to its community and implements appropriate strategies to prevent these threats" (Local Government Association of Tasmania 2004).

Developing an understanding of the risks posed by climate change is part of the agenda for adaptation reform for local government. In Australia, local governments have been active in initiating adaptation policy responses to climate change. For example, as far back as 1989, a coalition of 15 councils was formed in NSW called the Sydney Coastal Councils (SCC) group. This group was designed to 'promote co-ordination between member councils on environmental and natural resource management issues relating to the sustainable management of the urban coastal environment' (SCC 2005). SCC has since agreed on a collective approach to climate change. Gold Coast Council in Queensland has similarly incorporated climate change within its disaster management program, providing a series of disaster management tools for application by the public and flood management planning undertaken (Betts 2001; Boak et al. 2001). In the Australian state of Victoria, a coalition of Victorian local governments formed in June 2004 established a regional framework for local stakeholders to work together on greenhouse gas abatement and climate change projects (WPGA 2006). Darwin City Council is developing climate change risk and adaptation mechanisms within its environmental management plan. In South Australia, the Local Government Association, in conjunction with the State Government has established a Mutual Liabilities Scheme (MLS), monies from which will be used to produce 25 local government adaptation plans by 2009-2010. The City of Onkaparinga (also in South Australia), has determined to impose a 10% tax on rate payers. This reform will generate just under a million dollars a year which will be specifically directed into climate change adaptation and mitigation strategies in the region.

This chapter reports on the development of an internal decision making tool for the local government of Kingborough Council, Tasmania, Australia. A decision to implement and develop the decision making tool according to the principles of interdisciplinarity, facilitated not only production of the tool, but the involvement of the wide range of sectors and individuals needed to effectively implement it. This chapter will outline the progress of this initiative and lessons learned for building future projects that deal with climate change at local government scales.

2. Method

Case study approach

In 2005-2006, research was conducted to assess and analyse the flexibility within local government regions in Tasmania, to react to climate change impacts (Nursey-Bray 2006; 2008). The following objectives were considered: (a) to determine what institutional arrangements and legislative frameworks exist within Australian local governments to provide for management of climate change? (b) to determine local government perceptions about the impact of climate change on local communities and; (c) to determine whether the relationship between existing institutional arrangements and legislative frameworks in relation to climate change and other natural resource management arrangements can provide the opportunity for local government to build capacity to provide for management of climate change. This research in turn led to a collaboration with Kingborough Council to develop an internal mechanism that would enable staff to identify the risks of climate change in their area. Funds were subsequently sought and obtained to do this work. The project had two aims (i) to develop an inter-disciplinary risk assessment methodology for dealing with climate change for the local government of Kingborough, and (ii) to develop educational tools in relation to sea level rise and coastal vulnerability.

In this context, this research took a qualitative case study approach. The case study approach is one commonly used by social scientists as it provides an empirical basis for investigating current issues in real life contexts. It facilitates understandings between the boundaries of the object of inquiry and its context, and enables the use of multiple sources of evidence, thus building research credibility and viability (Hamel 1983; Yin 1984). Information for the project was collected through a process of semi-structured qualitative interviews, focus groups and meetings. Interviews were then conducted using both snowball and purposive sampling techniques (Denzin & Lincoln 1998). The project involves all sections of the Kingborough Council and is being run in synchronicity with the State and Federal government Climate Change adaptation programs.

3. Assessing climate change risk: Kingborough Council

The Kingborough municipality is located immediately south of Hobart in southern Tasmania. It comprises an area of about 720 square kilometres and is home to over 32,000 people. Approximately half of the municipality is on the Tasmanian "mainland" and half is made up of Bruny Island. The population density is generally quite low with most people living in the north within close commuting distance of Hobart.

The municipality includes some of the most beautiful and diverse scenery in Tasmania with stunning coastal landscapes and large areas of rural and naturally vegetated bush lands. The total measurable extent of the coastline is about 350km, all of which is appreciated and enjoyed by the local residents and visitors to Kingborough. Almost all of the main settlements are located on the coast and these include (from north to south) Taroona, Kingston Beach, Blackmans Bay, Margate, Snug, Kettering and Woodbridge. On Bruny Island there are a number of smaller coastal settlements, with the two largest being Adventure Bay and Alonnah.

Kingborough has been experiencing one of the fastest growth rates in Tasmania for the last 20 years. This has accelerated in recent years with increased commercial activity responding to the recent extensive residential development. This relatively rapid rate of growth is imposing major pressures on public infrastructure, natural resources and the existing character of locally treasured places. The State of the Environment Report for Tasmania (RPDC 2004) identified evidence that sea scales are rising and will continue as a result of climate change. This constitutes risk for local government in regard to protecting public infrastructure and providing appropriate development control. Planning decisions need to be well informed and good natural resource hazard information is essential.

There is an urgent need to better understand the impact this development is having on important environmental values and to have in place the necessary and associated management actions and regulatory regimes.

In this regard, the Kingborough Council has prepared a State of Environment report and is always reviewing its existing planning scheme. The planning scheme is relatively new (gazetted in mid-2004) and has reasonably effective environmental protection provisions. The Kingborough Planning Scheme is the primary regulatory mechanism by which Council can take account of climate change. The standards that are included within the planning scheme should be designed in a precautionary manner to guard against the future impacts of such phenomena as sea level rise.

In the absence of any mapping that specifically identifies areas at risk, the scheme defines high risk areas as being any land 30 metres from the high water mark or less that 3 metres above AHD. Moderate risk areas are between 30 and 100 metres from HWM or less than 3 metres above AHD. All such development within these identified risk areas is discretionary and must meet some fairly stringent requirements. Permitted development must be 30 metres back from the rear of any sand dune.

As well as local planning scheme provisions, there is at a State government level, a State Coastal Planning Policy. This may soon be replaced by a new Policy that has various additional provisions to take account of future sea level rise. The recommended standard within the draft Policy is similar in effect to the existing standard within the Kingborough Planning Scheme. It however requires the prior mapping of a "storm surge hazard area" and risk is defined by the "type of coastline". Both types of standards are trying to achieve the same outcome – that is, not allowing future development to occur in high risk locations without sufficient safeguards in place.

While Kingborough Council, in its role as a planning authority, requires private developments to comply with such standards, it has for some time been concerned about the need to incorporate climate change factors within its day-to-day operations. That is, in programming the future maintenance and upgrade of public infrastructure, it is necessary to better understand the future additional risks that result from climate change. However the Council had no real idea how to do this or, in other words, did not know how an appropriate risk assessment methodology should be applied.

Within Kingborough Council there is an understanding that recurrence intervals for flooding and inundation will change over time as a result of climate change and that this will impact upon design standards. It has however been difficult to know where to go to seek relevant information on how to determine the relative increased scales of risk and to know then how this can be utilized in adjusting the future capital works and maintenance programs. Such information would of course also be very relevant to the above mentioned planning and regulatory controls administered by Council, as well as any public awareness programs.

In recent years, Kingborough Council officers have been working on a number of strategic high priority infrastructure projects – such as a long term wastewater treatment strategy, assessing infrastructure capacity (i.e water, sewerage, stormwater, roads etc), headworks charges on development, settlement planning (mainly in regard to identifying and designing suitable future residential and commercial development areas), water conservation and asset management. These various projects are at different stages of development and completion. The future risks associated with climate change impact upon all such projects and the council has been seeking a suitable methodology to help in developing the right responses.

For Kingborough an assessment of risk will inevitably involve having mapped those actual areas that are at risk across the municipality from flooding, storm surge and bushfire – all of which are influenced by climate change. In the case of sea level rise for example, it would identify "storm surge hazard areas" (based on Sharples 2005) and could refer to more recent Coastal Values mapping that has provided more details on foreshore geomorphology and vegetation conditions. There are however significant limitations due to a lack of detailed topographic information that would enable a more accurate estimation of actual areas likely to be impacted upon by inundation.

Nonetheless with what information is available, Kingborough Council has conducted some preliminary investigations into the economic loss to Council owned assets that might be caused solely by predicted sea level rises. This is based upon the 800mm rise over 100 years.

Tab. 1: Estimated Economic Loss to Council (Kingborough) Owned Assets from predicted sea level rise (Kingborough Council 2007)

Replacement costs for assets for Kingston Beach
• Road and bridges – $4.2 mill. • Sewerage – $2.5 mill. • Buildings – $14.5 mill. • Stormwater – $0.6 mill. • Water – $22 mill.
Replacement costs for assets for Blackmans Bay
• Sewerage – $1.4 mill. • Buildings – $0.6 mill. • Stormwater – $0.3 mill. • Water – 0.1 mill. • Roads – $1.3 mill
Replacement costs for assets for Snug
• Sewerage – $0.4 mill • Stormwater – $0.23 mill. • Water – $0.1 mill. • Roads and bridges – $0.55 mill • Buildings – $0.55 mill.
Total Cost is in the order of $50 million

It is stressed that these costs only relate to public infrastructure and a great deal of private land (mainly residential and commercially developed) would obviously be impacted upon. The information also only relates to these three locations and there will be other (less developed) areas which will be affected.

The above figures provide some idea of why Kingborough Council needs to take the issue of sea level rise very seriously. They do not however provide any starting point as to how the council should actually respond. Council appreciates that more detailed mapping is necessary and that improved planning scheme provisions will need to be developed. It is also appreciated that internal asset management procedures must incorporate the changing natural conditions that make past assumptions redundant. This improved information and an increased level of knowledge is critical in moving forward from here, in that available resources are extremely limited and the council must spend public money wisely.

The adoption of the best available risk assessment methodology is thus very important. Such a methodology should be able to provide a Council, such as Kingborough, with the necessary practical guidelines to follow and enable priorities to be set and optimal work programs implemented. Local Councils will want a very pragmatic and relevant process that is easily understood and applied. This may be a difficult ask in regard to such a complex and varied phenomena as climate change, but is necessary if it is to be actually adopted.

4. The risk assessment tool

We provide the detail to set the context for understanding the rationale underpinning the development of a risk assessment tool for Kingborough. Kingborough Council's situation is not unique. As noted earlier, local governments across Australia face similar challenges, albeit in different geographical regions. The process of developing a risk assessment tool is also overlain with extra issues. Risk assessment involves not only the identification of the actual risk, but also an identification of how the risk is perceived by the community and what is an 'acceptable' or 'dangerous' level of risk. Key challenges for policy makers in the context of climate change include the short-term planning framework that they operate within, time lags between cause (emissions) and effect (climate change), and how to motivate stakeholders to adapt. Establishing how to assess the risks associated with the impacts from climate change are vital. As a first step, and to determine the most appropriate risk assessment tool or metric, a review of existing models was undertaken. Carter et al.'s (1999) impact assessment framework emerged as pivotal, identifying a useful starting point for consideration of a risk assessment process:

Fig. 1: Seven steps of impact assessment (Carter et al. 1994).

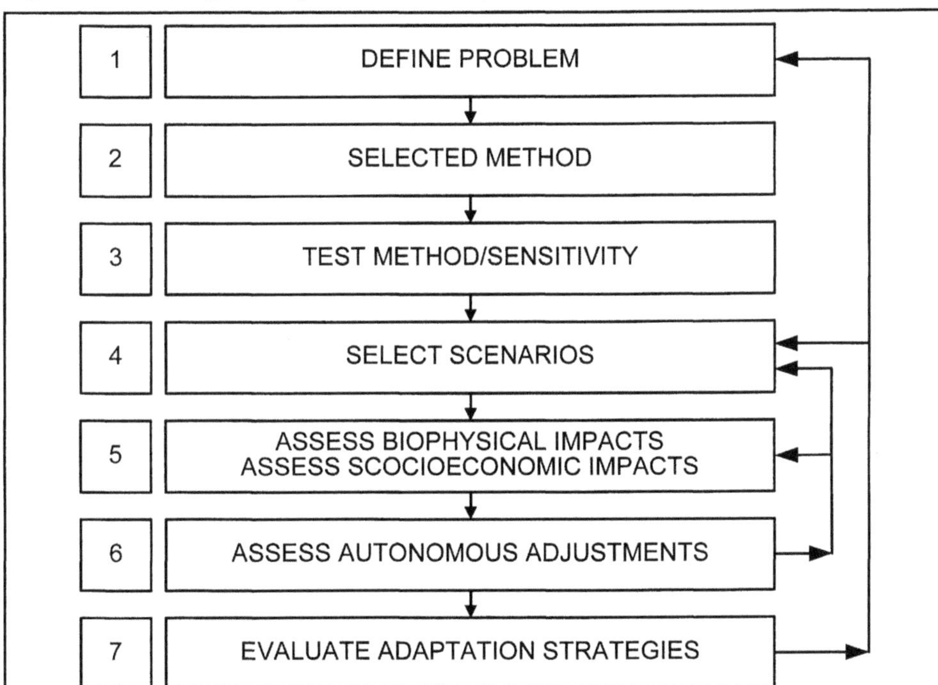

Building on this model, Jones (2001) considers impact assessment in the context of environmental risk assessment and risk management. His framework assesses the impacts of climate change on individual *exposure units*. The exposure units address the issue of uncertainty, identify those that are potentially vulnerable to climate change and establish how to incorporate the findings within a planning response. Windows of adaptation are defined as the gap between the identification of an acceptable level of risk and its exceedance. Accordingly, Jones (2001) suggests the following risk assessment framework:

Fig. 2: Risk Assessment Framework[1]

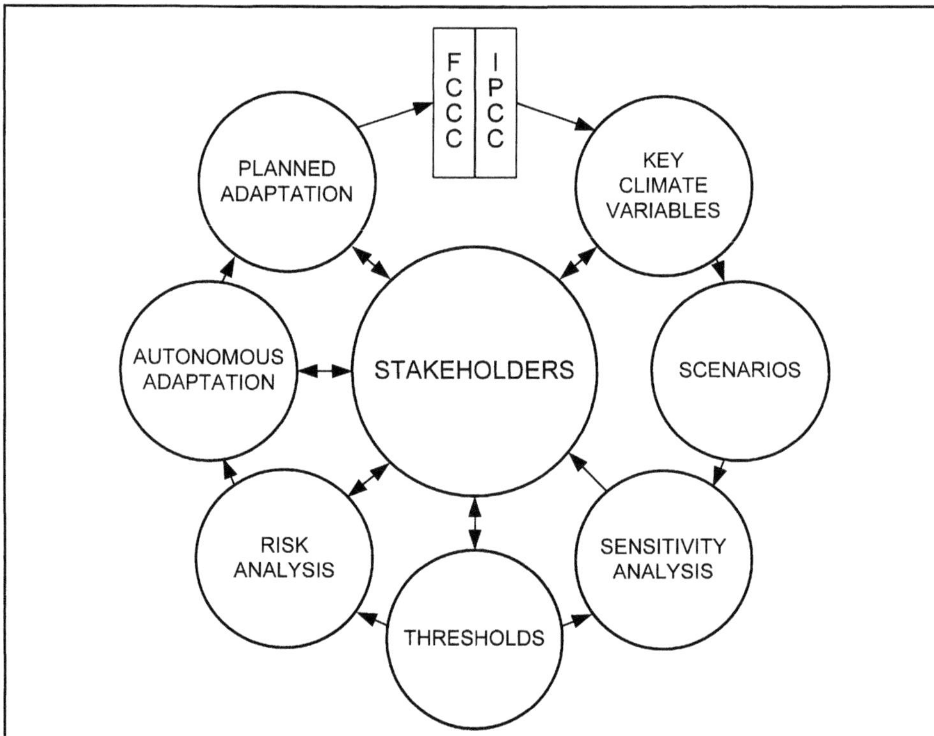

Source: Jones 2001

Conceptually, this type of representation of risk is useful for local government when considering the need to keep the tool (i) relevant, (ii) pragmatic, (iii) simple and (iv) interdisciplinary; all criteria Kingborough staff advanced as important in any risk assessment process.

Therefore, Jones (2001) suggests the following steps that need to be taken when assessing risk:

---

1 In the diagram the FCCC is the Framework Convention for Climate Change and the IPCC the Intergovernmental Panel for Climate Change.

1. Identify the *key climatic variables* affecting the exposure units being assessed;
2. Create *scenarios* and/or projected ranges for key climatic variables;
3. Carry out a *sensitivity analysis* to assess the relationship between climate change and impacts;
4. Identify the impact *thresholds* to be analysed for risk with stakeholders;
5. Carry out *risk analysis;*
6. Evaluate risk and identify feedbacks likely to result in *autonomous adaptations*; and
7. Consult with stakeholders, analyse proposed adaptations and recommend *planned adaptation* options.

Thus the tool simply takes the staff through a process of iteratively determining risk in various scenarios. Scenarios were established based on meetings with staff across different areas, including environment health, asset services and planning. It is designed so it is consistent with the recommended framework for risk management provided by the Australian and New Zealand Standard AS/NZS 4369 Risk Management conceptually shown in Figure 3 below:

Fig. 3: Australian and New Zealand Standard AS/NZS 4369 Risk Management

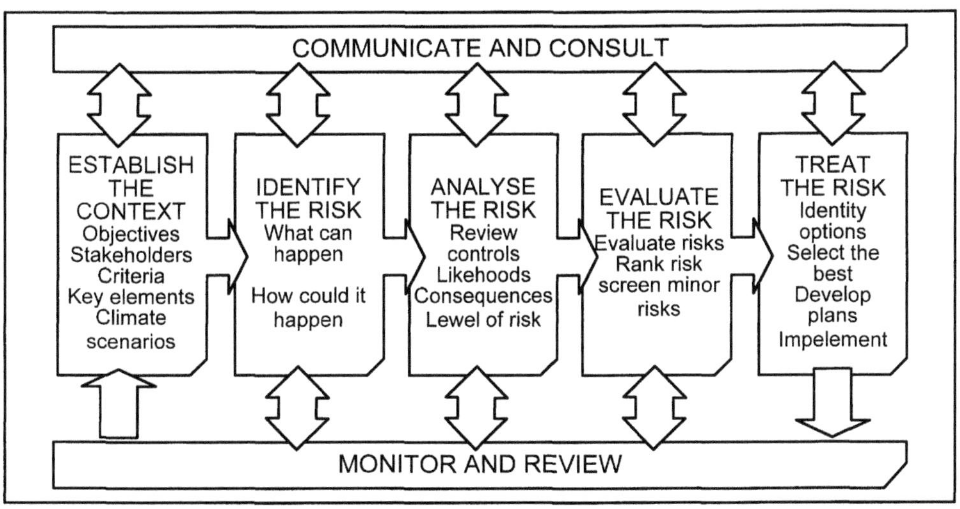

Source: Marcus Jacobs Associates 2007

In terms of context, the risk tool is also underpinned by an assumption that the worst case predictions scenarios of the IPCC and local science predictions are the basis for decision making. It then asks staff to determine what information they need to make the decision. If the relevant staff member feels there is not enough information to make a decision, the risk tool advises that the decision should be either a negative or suspended pending further information. However if it is considered there is enough information, staff are guided through a decision making process which allows them to consider the (i) risks attached to the project, (ii) the likelihood of that risk manifesting during the life of the project (i.e. its sensitivity), (iii) the adaptive capacity to deal with that risk, and (iv) the vulnerability therefore of the council to that risk. An assessment of the level of risk – extreme, high, medium or low is then made. As such, council staff will be able to make broad decisions based on an appropriate assessment of risk.

The tool is now in the process of being trialled and will be, if endorsed by the elected Council members, adopted through the planning scheme and hence become a statutory planning instrument. The characteristics of this tool are forged by the very nature of its requirement to be interdisciplinarity – that is applicable in cross disciplinary contexts. Of necessity this makes it a simpler tool, but workable in a real life situation. Nonetheless, achieving a truly interdisciplinary tool was fraught with particular issues that needed resolution prior to completion.

5. The process: Lessons learned in development of the risk assessment tool

One lesson from the process was the identification of the need to deal with uncertainty. While climate change per se is accepted within local government as a reality, there is still uncertainty as to how much the impact will be manifest at local scales. Uncertainty in fact fashioned local government discourse about action on climate change, such as the issue of how to manage for what one respondent called the 'time lag effect'. Many staff members pointed out the disjuncture between the need to make immediate decisions in the context of uncertain but potential future climate effects/impacts (and their scale). In the words of one planner: "The time lag thing is very difficult – it is really hard to make decisions in terms of effects that might take 10s of 100s of years perhaps to mature or manifest. So a move in the right direction is going to be very slow." Uncertainty is one of the key difficulties for policy makers in assessing how to deal with the uncertainty of exactly how climate change will manifest and therefore the extent, diversity, regularity, distribution and magnitude of its impacts (Ha-Duong et al. 2007; Yohe & Neumann 1997). Uncertainty is a key theme in the global discourse about climate change, and debate rages in discussions on how fast change will occur, at what scale, how catastrophic it will be, and on the accuracy of climate predictions (Carter et al. 1999). Uncertainty arises from insufficient, inaccurate or unavailable data, external developments and cross-boundary issues, and the unpredictability of human behaviour (Westmacott 2001). As Jones (2001) notes on the importance of uncertainty within policy planning:

> "Uncertainty management is the raison d'être of risk assessment, extreme care must be exercised through out an assessment, so that uncertainties are identified, the nature of their propagation throughout the assessment is understood and that they are communicated as part of the results." (Jones 2001)

How then does one deal with uncertainty? Dessai and Hulma (2007) charge policy makers with the following questions: How much climate change uncertainty do we want to adapt to? How can we trade-off robustness with increasing cost? Are robust adaptation options socially, environmentally and economically acceptable? How do climate change uncertainties compare with other uncertainties (e.g., changes in demand)? How much public money should be invested to research the largest scientific (tractable) uncertainty to try to reduce it?

There are ways to reduce this uncertainty (Yohe et al. 2004), and Kingborough did so by adopting the IPCC estimates of confidence in projected changes in extreme events and other climate variables. While these estimates are obviously a rough guide for local scales, nonetheless, they provide a good basis for precautionary planning. In terms of scope, and in the absence of discrete data sets, they give a wider berth for integrated planning.

Another challenge was the need to integrate the science with the social dimensions in relation to dealing with climate change. For example, Clarence Council, another local government within Tasmania, has spearheaded a coastal vulnerability assessment, by undertaking a combined socio-economic and scientific program. In this case, they had the advantage of having secured funding to do this. Their experience tells us that understanding the scientific parameters of how climate change will manifest at local scales is crucial yet the reality is, that most councils, like Kingborough will not be able to afford to do this kind of fine scale work. How then does one ensure the necessary integration of all the disciplines in the knowledge sets so as to develop an effective risk tool? In the process of answering this question, we identified that in fact there is a huge bank of information available to local government across all disciplines to support decision making. Crucially however, most staff were either unaware of its existence, or only aware of the information in their own area of expertise. One of the outcomes then of the risk assessment project was the collation of different data sets across different disciplines in relation to climate change. The process of developing the risk tool then provided an opportunity to 'join the dots' between each discipline and knowledge set within the process developed. As such, the final decision making tool allows the user to access information across many paradigms and areas.

Linked to this challenge is the reality that how ratepayers or stakeholders perceive risk and the actual nature of the risk can be two different things. In order to ensure that all parties are 'singing from the same hymn sheet', how climate change risk is understood by different sectors and interests within the community is vital (Dake & Wildavsky 1990; Dake 1992; Slovic & Peters 1998; Lowenstein et al. 2001; Leiserowitz 2006). For example, a case study of American risk perceptions highlights the critical influence of risk perception. A risk perception index using nine variables[2] showed that while Americans overall viewed global climate change as a moderate risk, they were more concerned about standards of living, water shortage and disease. More importantly, the use of the index highlighted a trend for Americans to discount local impacts, seeing them as 'unlikely', therefore disconnecting them from the possibility of planning for them. This finding has policy implications for municipalities attempting to develop momentum for local climate policy, and highlights the value of undertaking such studies prior to decision making and program development.

Moreover, understanding the relationship between risk perceptions and the willingness to address climate change is important (O'Connor et al. 1999; Wildavsky & Dake 1990). Studies of risk assessment show that behavioural intentions can be predicted when risk perception is defined as the 'perceived likelihood of negative consequences to oneself and society from one specific environmental phenomenon: global warming' (Wildavsky & Dake 1990). This is an issue often overlooked in studies that focus on the relationship between environmental beliefs and how they predict behavioral intentions as they rarely include specific risk perceptions as independent variables or criteria within vulnerability assessments.

As such, in the context of developing decision making tools at local scales it is clear that there needs to be more work done on integrating community perception of risk with council assessment of risk and then this needs to be communicated. In the case of Kingborough Council, the communication of risk was also an internal issue. We found that staff across divisions did not necessarily construct the risk to the council equally. Moreover, how the risk of climate change was constructed by staff and by elected members differed slightly. Moreover we identified some resistance to the idea that *climate change* needed management, largely due to the resource implications this would have (rather than climate skepticism). In response, we decided that the best approach would be to 'mainstream' the issue and develop the risk tool as a decision making process for *existing* issues, but ones that would also manifest as climate change impacts (Adger et al. 2005; Arvai et al. 2006; Gupta & Asselt 2006; Smit & Wandel 2006). As Lynch et al. (2007) note:

> "when climate change is considered in concert with existing decision processes and structures, that is, 'mainstreaming' climate change adaptation … practical implementation is more likely to ensue."

Examples include fire management, water issues, flooding management, erosion and bank stabilization. The tool was refined to extend and build on existing risk

---

2 These include: holistic concern; assessments of the severity of current and future impacts of global climate change on human health (mortality and morbidity); measures of local and global impacts of climate change on standards of living, water shortages and rates of serious disease; the seriousness of the threat to non-human nature; the seriousness of the current impacts of climate change around the world, and the scale of concern. (Leiserowitz 2006).

management processes to additionally meet the challenges of climate change. For example, there may be a temporal adjustment to planning within existing flood management programs to account for the nature of climate change, i.e planning for a 1 in 10 year flood rather than a 1 in 50 year flood. Another mechanism used to 'mainstream' the issue of climate change was to modify the terminology used when talking about it. We discovered that the use of the term 'risk assessment tool' alienated some staff who simply 'switched off' because in their view risk management was the function of emergency services or occupational heath and safety. However, when we started to sue the term 'internal decision making tool' to assess the impacts of climate change, a much more encompassing and interdisciplinary term, staff across all divisions felt it was relevant to them. The State government also signaled greater interest in this project when constructed as an internal process. Thus the change in terminology created spaces for interdisciplinary interaction and innovation to occur. This brings us to the final issue that was of importance in the evolution of this project, that is, the question of how to communicate it to the community, to staff and departments, and to the elected council. These are issues of governance as well as communication but bear a real weight on the issue overall.

Effective communication between stakeholders and between stakeholders and policy makers is important to the success of any program. Demerritt and Langdon (2004) argue that effective communication is essential in forging links between local government and the community on climate change. In a case study of local authorities within the UK in 2003, they surveyed environmental officers in every Local Authority in England and Wales to assess the reception and response of local government to the information being provided through the UK Climate Change Program. Results found that over three quarters of respondents felt they did not have access to the best information about the impacts of climate change on their areas. The survey also recorded high scales of stress, cynicism, and futility among these officials. Demerrit and Langdon (2004) conclude that communication about climate change is more than making locally specific information about climate change impacts available, and will not motivate appropriate action unless confidence is built up at local level. Moser and Dilling (2007) show the multiple aspects of communicating climate change in the US. These studies highlight the importance of effective communication about climate change. Of further interest, Hwacha (2005: 512) discusses deliberative dialogue as a means to facilitate not only effective communication but to harness the advantages of an interdisciplinary approach:

> "[a] structured facilitation process ... engages stakeholders [and] through a shared exploration of different perspectives, participants thoughtfully discuss a complex issue in potentially new ways that tend to break away from habitual positions or 'stuck' and pre-determined solutions. Deliberative dialogue builds on participants' knowledge and experiences to find common ground from which alternative strategies or policies can be pursued."

In relation to Kingborough, staff will need to be explicit about how they decide to communicate the outcomes of this risk management/ internal decision making exercise. This will be necessary in order to address liability and compensation issues both now and in the future. The development of this communication tool is ongoing but as noted earlier the process will need to facilitate dialogue about council's intent between the council itself and ratepayers, within the staff and various departments and between staff and its' elected officers.

6. Discussion: Climate Change and Interdisciplinarity

Adger (2006) specifically notes that "Climate change represents a classic multi-scale global change problem in that it is characterized by infinitely diverse actors, multiple stressors and multiple time scales". In this context building diversity into management approaches is important, as climate change is a problem particularly suited to the practice of interdisciplinarity. Nonetheless, the concept of interdisciplinarity is a loaded one and means many things to different people (Klein 1990). For clarification, this chapter is based on Klein and Newell's (1996) definition of inter-disciplinarity which is:

> "A process of answering a question, solving a problem, or addressing a topic that is too broad or complex to be dealt with adequately by a single discipline or profession ... Interdisciplinary studies draws on disciplinary perspectives and integrates their insights through construction of a more comprehensive perspective" (Klein & Newell 1996: 395).

One characteristic of inter-disciplinarity is integration, whether that is the integration of data sets or the merging of theoretical and practical approaches. In developing a decision making tool to assess risk within local government, the use of interdisciplinary knowledge sets enabled a synthesis of the patterns in power and knowledge located across departments within local government. While an integration of content or paradigms was not always possible, the decision making tool can now assist in achieving an interdisciplinary management *process* (Stostak 2007). Moreover, one of the weaknesses of disciplinary approaches is that as disciplines vie with each other for precedence, the determination of *who*, *what, why* and *how* to integrate becomes very problematic. This decision making tool, by providing or catalyzing reflection on participants own issues of concern provides clues as to how to prioritise issues and across which management area. In the context of climate change where the intersection between science and social science is vital to the implementation of effective adaptation strategies, decision making tools such as this have the potential to play an important role.

Another characteristic of the inter-disciplinary environment is its capacity to motivate individuals to make connections, hitherto possibly not seen, between different knowledge sets and disciplinary orientation. Being able to make these connections implicitly encourages innovation and the development of new ways of seeing and doing. Again, the development of this tool promoted face to face interaction between different staff from different areas, and different researchers from different disciplines. As such, an interdisciplinary decision making tool has the facility to assist institutions in forging hitherto unsuspected connections, and to *become* a resource in itself.

Using interdisciplinarity as a rubric for understanding how to adapt to climate change at local scales is also crucial to overcoming the dilemmas faced by local government in relation to dealing with problem of climate change at local scales. The key value of an interdisciplinary approach in relation to risk assessment has been that the integration has been achieved across disciplines and sets of individuals working within a particular area.

This is not an end point but a good start to the process of collaborative endeavour that the issue of climate change will necessarily force on professionals and societies as they attempt to adapt to change and build social ecological resilience[3] and adaptive capacity (Holling 1996; Holling et al. 1998; Ludwig et al. 1997; Peterson et al. 1998)[4].

7. Conclusion

The link between research and policy in relation to climate change is important (Kerkhoff 2005) As Vogel et al. (2007) note the multi-disciplinary nature of climate change merits an inter-disciplinary response, so that early engagement with researchers will ensure that responses are credible and knowledge based, and hence planning outcomes are achieved. Future research could work with local governments to monitor the effectiveness of climate change strategies that are being initiated, and foster interdisciplinary action research projects, such as this one, that will help local governments build climate reform agendas. Thus the link between research and policy becomes an active rather than passive relationship. We have used interdisciplinarity both in principle and practice to progress a risk assessment goal, but there are many other contexts in which interdisciplinary practice can be applied and relevant in the context of climate change.

This chapter has shown how the integration of disciplines and knowledge bases can create the conditions where inter-disciplinarity can be used as a tool for forging action on climate change. While much research work still needs to be

---

3 After Folke (2006) we understand social-ecological resilience to be: '(i) the amount of disturbance a system can absorb and still remain within the same state or domain of attraction, (ii) the degree to which the system is capable of self organisation (versus lack of organisation, or organisation forced by external factors), and (iii) the degree to which the system can build and increase the capacity for learning and adaptation.'

4 For further information on resilience per se and literature, the web site for the Resilience Alliance and the online journal Conservation Ecology are a very useful starting point.

done in this area, this research has highlighted that it is only through building on multiple dimensions and drawing on interdisciplinarity that the issue of climate change can be dealt with effectively and have multiple benefits once applied. While there is a clear need for good scientific work to be undertaken so as to determine the parameters of the problem at local scales, suggested solutions will also need to incorporate the best that interdisciplinary practice has to offer in order to build the institutional strategies and political solutions that are needed. Only then can the social, cultural and economic factors profoundly influencing a problem of this magnitude, be resolved at local scales.

References

Adger, W.N., Arnell, N.W. and Tompkins, E.L. (2005), "Successful adaptation to climate change across scales." Global Environmental Change 15, pp. 77-86.

Adger, W.N., Hughes, T., Folke, C., Carpenter, S. and Rockstrom, J. (2005), "Social-ecological Resilience to Coastal Disasters." Science, 12 August 2005, 309 no. 5737, pp. 1036-1039.

Arvai, J., Bridge, G., Dolsak, N., Franzese, R., Koontz, T., Luginbuhl, A., Robbins, P., Richards, K., Smith Korfmacher, K., Sohngen, B., Tansey, J. and Thompson, A. (2006), "Adaptive Management of Global Climate Problem: Bridging the Gap between climate research and climate policy." Climatic Change, 78, pp. 217-225.

Betts, H. (2001). The incorporation of climate change in flood plain planning at the Gold Coast City Council, City of the Gold Coast, Australia: Gold Coast City Council, 14pp.

Boak, E.H., Jackson, L.A., McGrath, J.E. and Brosnan, M.P. (2001), "An overview of Gold Coast coastal management, 1960-2001." Paper presented at Coasts and Ports 2001, 25-28 September, 2001, Gold Coast, Queensland.

Carter, T., Hulme, M. and Viner, D. (eds.) (1999), "Representing uncertainty in climate change scenarios and impact studies." Proceedings of the ECLAT-2 Helsinki Workshop, 14-16 April 1999, Climatic Research Unit, Norwich, UK, 128pp.

Dake, K. (1992), "Myths of nature: Culture and the social construction of risk." Journal of Social Issues vol. 48, pp. 21-27.

Dake, K. and Wildavsky, A. (1990), "Theories of risk perception: Who fears what and why?" Daedalus. 119, pp. 41-60.

Demeritt, D. and Langdon, D. (2004), "The UK Climate Change Programme and Communication with Local Authorities." Global Environmental Change, 14, pp. 325-336.

Denzin, N.K. and Lincoln, Y.S. (eds.) (1998), Handbook of Qualitative Research. Sage, London.

Dessai, S. and Hulme, M. (2007), "Assessing the robustness of adaptation decisions to climate change uncertainties: a case-study on water resources management in the East of England." Global Environmental Change 17 (1).

Folke, C. (2006), "Resilience: The Emergence of a perspective for social-ecological systems analyses." Global Environmental Change 16, pp. 253-267.

Gupta, J. and Asselt, H. (2006), "Helping Operatonalise Article 2: A transdisciplinary methodological tool for evaluating when climate change is dangerous." Global Environmental Change 16, pp. 83-94.

Ha-Duong, M., Swart, R., Bernstein, L. and Petersen, A. (2007), "Uncertainty management in the IPCC: agreeing to disagree." Global Environmental Change 17 (1).

Hamel, J. (with *Dufour, S. and Fortin, D.*) (1993), Case study methods. Newbury Park, CA: Sage.

Holling, C.S. (1996), Engineering resilience versus ecological resilience, pp. 41-44. In: Schulze, P. (ed.), Engineering within ecological constraints. Washington (DC). National Academy.

Holling, C.S., Berkes, F. and Folke, C. (1998), Science, sustainability and resource management, pp. 342-362. In: Berkes, F. and Folke, C. (eds.), Linking Social and Ecological Systems. Management Practices and Social mechanisms for Building Resilience. Cambridg eUniversit Press.

Hwacha, V. (2005), "Canada's Experience in Developing a National Disaster Mitigation Strategy: A Deliberative Dialogue Approach." Mitigation and Adaptation Strategies for Global Change 2005, 10, pp. 507-523.

Jones, R. (2001), "An Environmental Risk Assessment/Management Framework for Climate Change Impacts Assessments." Natural Hazards 23, pp. 197-230.

Kerkhoff, L. (2005), "Integrated research: Concepts of Connection in Environmental Science and Policy." Environmental Science and Policy 8, pp. 452-463.

Klein, J.T. (1990), Interdisciplinarity. History, Theory and Practice, Wayne State University press.

Klein, J.T. and Newell, W.H. (1996), Advancing interdisciplinary studies. In: Gaff, J.G. & Ratcliff, J. (and associates) (eds.), Handbook of the undergraduate curriculum, pp. 393-395. San Francisco: Jossey-Bass.

Leiserowitz, A. (2006), "Climate Change Risk Perception and Policy Preferences: The role of Affect, Imagery and Values." Climatic Change (2006) 77, pp. 45-72.

Local Government Association of Tasmania (LGAT) (2004), Forging Links. Emergency Management and Local Government, An Information Kit, Tasmanian State Emergency Service & Local Government Association of Tasmania.

Loewenstein, G., Weber, E., Hsee, C. and Welch, E. (2001), "Risk as feelings." Psychological. Bulletin 127, pp. 267-286.

Lynch, A.H., Tryhorn, L. and Abramson, R. (2007), Working at the boundary: facilitating interdisciplinarity in climate change adaptation research, Bulletin of the American Meteorological Society 89, pp. 169-179.

Ludwig, B., Walker, B. and Holling, C.S. (1997), Sustainability, stability and resilience. Conservation Ecology (online) 1 (1): 7, available from the internet. www.consecol.org/vol1/iss1/art7.

Marcus Jacobs Associates (2007), Climate Change Impacts and Risk Management, A Guide for Business and Government, Australian Greenhouse Office, Department of Environment and Water Resources.

Moser, S. and Dilling, L. (2007), Creating a Climate for Change. Communicating Climate Change and Facilitating Social Change, Cambridge Press.

Nursey-Bray, M. (2006), Institutional Capacity and Climate Change: Local Government in Tasmania, Research Report no. 1, 2006, Australian Maritime College, Tasmania.

– (2008), Inter-disciplinarity as a tool for action on climate change: Applying social science methodology to climate change initiatives within local governments, Australia, The International Journal of Interdisciplinary Social Sciences vol. 2.

O'Connor, R., Bord, R. and Fisher, A. (1999), "Risk Perceptions, General Environmental beliefs and Willingness to Address Climate Change." Risk Analysis, 19 (3), pp. 461-471.

Peterson, G., Allen, C.R. and Holling, C.S. (1998), "Ecological Resilience, biodiversity and scale." Ecosystems vol. 1, pp. 6-18.

Resource Assessment and Planning Commission (2004), State of the Environment Report, Tasmania, Tasmanian State Government.

Sharples, C. (2006), Indicative Mapping of Tasmanian Coastal Vulnerability to Climate Change and Sea-Level Rise: Explanatory Report (Second Edition); Tasmania.

Slovic, P. and Peters, E. (1998), "The importance of worldviews in risk perception." Risk Decision and Policy 3, pp. 165-170.

Smit, B. and Wandel, J. (2006), "Adaptation, Adaptive Capacity and Vulnerability." Global Environmental Change 16, pp. 282-292.

Sydney Coastal Council Group (SCC) (2005), Sydney Coastal Council Strategic Plan 2005-8. Information found on web site, www.sydneycoastalcouncils.com.au/, accessed Nov 1 2007.

Szostak, R. (2007), "How and why to teach interdisciplinary research practice." Journal of Research Practice vol. 3 (2), Article M17. Retrieved [date of access], from http://jrp.icaap.org/index.php/jrp/article/view/92/89.

Vogel, C., Moser, S., Kasperson, R. and Dabellk, G. (2007), "Linking vulnerability, adaptation and resilience science to practicea: Pathways, players and partnerships." Global Environmental Change vol. 17, pp. 349-364.

Westmacott, S. (2001), "Developing Decision Support Systems for Integrated Coastal Management in the Tropics: Is the ICM Decision-Making Environment Too Complex for the Development of a Useable and Useful DSS?" Journal of Environmental Management 62(1), pp. 55-74.

WPGA (2005), Climate Change Impacts and Adaptation in Western Port Region Needs Analysis Report, The Regional Development Company Pty Ltd Assisted By Marsden Jacob Associates.

Yin, R.K. (1984), Case study research: Design and methods. Newbury Park, CA: Sage.

Yohe, G. and Neumann, J. (1997), "Planning for sea-level rise and shore protection under climate uncertainty." Climatic Change 37(1), pp. 243-270.

Yohe, G., Andronova, N. and Schlesinger, M. (2004), "To hedge or not against an uncertain climate." Science 306, pp. 416-417.

Chapter 13

Positioning Oneself and Others to Resolve Climate Change

Lionel Boxer

Abstract

For humanity to change course and undo the damage done by industrialization it will be necessary to have a unified global vision. Achieving such an amalgamation may seem unlikely and, in the shadow of the current aggregate paradigm of our species, perhaps even beyond unlikely. From the perspective of our history, to unite six billion human beings would be a daunting task that would require mega personality and super powered charisma. However, the way we have seen grass root change appear in recent years implies that it is no longer a single personal force that unites people. Rather it is a noble ideal effectively communicated, replicated and translated into different contexts and cultures.

Facebook and Myspace are two internet social networking facilities that have brought together millions of minds into one interconnected repository. However, they remain disconnected from the rest apart from the networks they created. Most remain within a relatively closed network pod. Yet, there are some who have achieved links between these pods of familiarity. Might it be possible to create an underlying mood to undo the damage done by industrialization and disseminate it through these networks to reach the multimillion participants lurking in the space?

It has been shown that strong leaders can create a unified underlying mood that they align with their organizational goals (Boxer 3002a). This is done by establishing a clear vision and explaining to people various rules that will enable them to align with that vision. By attracting and maintaining a core of likeminded people, it is possible to enact the underlying mood. Those who are within the organization and not yet aligned with this underlying mood can be coached to conform and others who are unable to accept such alignment can go their own way or be sent away. This concept will be explored in more detail later in the chapter.

Whereas a single entity may have a common underlying mood, it will remain a challenge to replicate that underlying mood in adjoining communities and an especially daunting challenge doing so across dispersed communities. Yet this does occur. Subcultures existed around the world even before the internet. Consider how the Beatles aroused an entire generation in countless nations from England to Australia and America to Hong Kong. I recall traveling from Toronto through the United States to Australia to South East Asia, to Europe and on to London in 1964 and observing a consistency of urgency to adopt the Beatlemania that was underlying the subculture inspired by four simple punks from an irrelevant English town. Perhaps an idea of one became four, which influenced or impacted a generation around the globe. Of course, the Beatles had the support of commercial radio and the promise of profits from the sale of recorded music. However, this phenomena also occurs without such an infrastructure.

Consider how terrorist networks achieve successful promulgation of ideals and objectives around the world without any infrastructure (Guidère & Morgan 2007). While this is an extreme case, the same has been observed in recent social movements dealing with feminism and sustainability (Boxer 2003a). These two movements have been shown to be inspired in part by Foucauldian thought which have led to techniques of disruption and influence that are useful to change management.

This chapter develops a framework for aligning an underlying mood with a specific set of objectives and disseminating that underlying mood to other groups. It draws on research with chief executive officers (CEOs) who were known for their success in dealing with sustainability issues and subsequent work that has led to the development of an approach to lead the sort of change that is necessary to undo the damage done by industrialization (Boxer 2003a).

In doing so, Positioning Theory – in part inspired by Foucault – will be shown to be the foundation of the framework. The framework will then be derived with further benefit from Foucauldian ideas. Then dealing with the challenge of dissemination will be explored. That is, the use of networks to disrupt obsolete ways and instigate necessary new ways. This is the new style of inspiration that contemporary leaders must employ.

Leadership at All Levels

To deal with issues like reversing climate change, the practice of leadership is required throughout all levels of society. Business issues such as sustainability, safety and quality have been shown to all be obligatory and externally imposed (OEI), with common implementation challenges (Boxer 2003a). It has been shown that these sort of OEI issues tend to not fit naturally with people's daily routines and – on first consideration – may even obstruct individual and collective goals that are brought about by a key performance indicator (KPI). Hence, determined conscious effort is necessary to inspire change.

Quality management provides a useful and established model on which to base initiatives such as resolving climate change challenges. Porter (2006a: 6) points out that "3M originated 'Total Quality Environmental Management', the environmental correlate of TQM" and, citing Berry & Rondinelli (1998) (p. 49) who also notice similarity between environmental and quality issues, provides further substantiation to support the correlation between environmental and quality issues. Rusinko (2005) argues that, because of the common beliefs underlying quality and sustainability paradigms, quality initiatives can help organisations achieve a stronger form of sustainability. She explores how most of Deming's 14 points (Deming 1986 and Walton 1989) for shifting the quality paradigm are equally applicable to shifting the sustainability paradigm. Because the quality movement has already established the path of paradigm shifts and skill development, the sustainability paradigm can follow this path and adopt the same skills. Eisen and Mulraney (1992), Boxer (1993), and Kekäle (1999) suggest that leadership and culture can obstruct the implementation of quality initiatives. Based on this, findings regarding the implementation of quality could be applied to sustainability. This insight, experience implementing a range of OEI initiatives, and research with CEOs from a range of industries led Boxer (2003a) to conclude that dealing with sustainability issues requires personal leadership. Furthermore, strong and focused leadership is necessary at every level throughout an organisation by not only the CEO, but also subordinate leaders (Slim 1957). I have observed several organisations merging their quality, sustainability and – sometimes – safety approaches into a single paradigm or an integrated management system. They have ensured success by taking steps to align their organisational culture with the quality, sustainability and safety requirements peculiar to their organisation. These steps will be explored after positioning theory is introduced.

Influence Understood With a Positioning Framework

Positioning is the discursive production of selves or a study of how ordinary conversation leads individuals to creation themselves and those they speak with. It also concerns the development and evolution of rules of how people relate to one another. It will be seen that through a positioning framework is will be possible to understand influence that is required to deal with climate change. Every conversation contributes to the positioning of oneself and the others they are in contact with, as well as the underlying mood of the society they live in. This idea will become clearer in the following sections.

The Importance of Mood

One can easily *commit* to behaving in a more ecologically sustainable way that embraces all aspects of sustainability and contributes to the resolution of climate change challenges. Yet, achieving more ecologically sustainable *behaviour* requires commitment as well as – a more challenging – shift of behaviour. To perceive and articulate this difficulty it is helpful to frame the change towards more sustainable behaviour in terms of realignment of a residual underlying mood. Here mood is viewed as being a measure of that aspect of culture that defines the way individuals understand the way they should behave as well as how they can expect others to behave.

Having observed a range of success achieved by a variety of clients to implement planned change, research was undertaken to explore why some organisations were more successful at introducing more ecologically sustainable behaviour (Boxer 2003a) than others. CEOs known for having implemented sustainability initiatives were interviewed and the transcripts subjected to a positioning theory (Harré & Davies 1990) framework (Boxer 2003a). Foucauldian concepts – some of which had influenced positioning theory – were harnessed to develop a concept of social flux (now referred to as underlying mood). It was shown that CEOs, who are able to implement sustainability initiatives, do so by realigning the underlying mood with the sustainability issues at hand. In doing so, they remove the residual underlying mood that obstructs the sustainability initiatives and replace it with a mood supportive of sustainability. This realignment was shown to be achieved through discursive action – or ordinary conversation – that introduced a new line of thought about sustainability between the CEO and individuals or groups of people.

That same framework describing how some CEOs deal with sustainability issues can be used to help other leaders understand what they need to do to deal with sustainability issues such as climate change.

Subsequent sections of this Chapter discuss challenges implementing sustainability issues, positioning theory and underlying mood theory, applying this theory to change oneself, to team, society and the globe in regards to climate change.

Positioning Theory Constructionist Framework

Figure 1 encompasses the various components composing a framework that will be used throughout this Chapter. The theoretical grounding of this framework will be discussed in a subsequent section. Appreciating the interconnectedness of the several parts of the framework, it should be clear that a change to one component will affect the other components.

The top portion of Figure 1 represents the principle concern of positioning theory; a representation of "what happens when people engage in conversation?" It depicts the leader and team members, each being a moral agent with a particular knowledge and directing a personal force. How two or more people interact in conversation is described by the tri-polar conversation that is composed of the storyline followed during the conversation, the position each person brings to the conversation, and the speech acts that each person employs during the conversation. Each conversation affects both the position of each person involved as well as the aggregate underlying mood of the organisation. This mood becomes a field of power that influences future conversations, positions that people assume.

A vice president of a global resource business, commenting on Figure 1, indicated that there may be more universal application of this theory – and its implications –

and that the framework would prove equally useful in the case of a "sustainability inclination" as an "un-sustainability inclination". The notion that a leader may or may not be moral or that there are varying degrees of morality is real, but consideration of such a parameter is not included in the scope of this Chapter.

Fig. 1: Social Constructionist Model

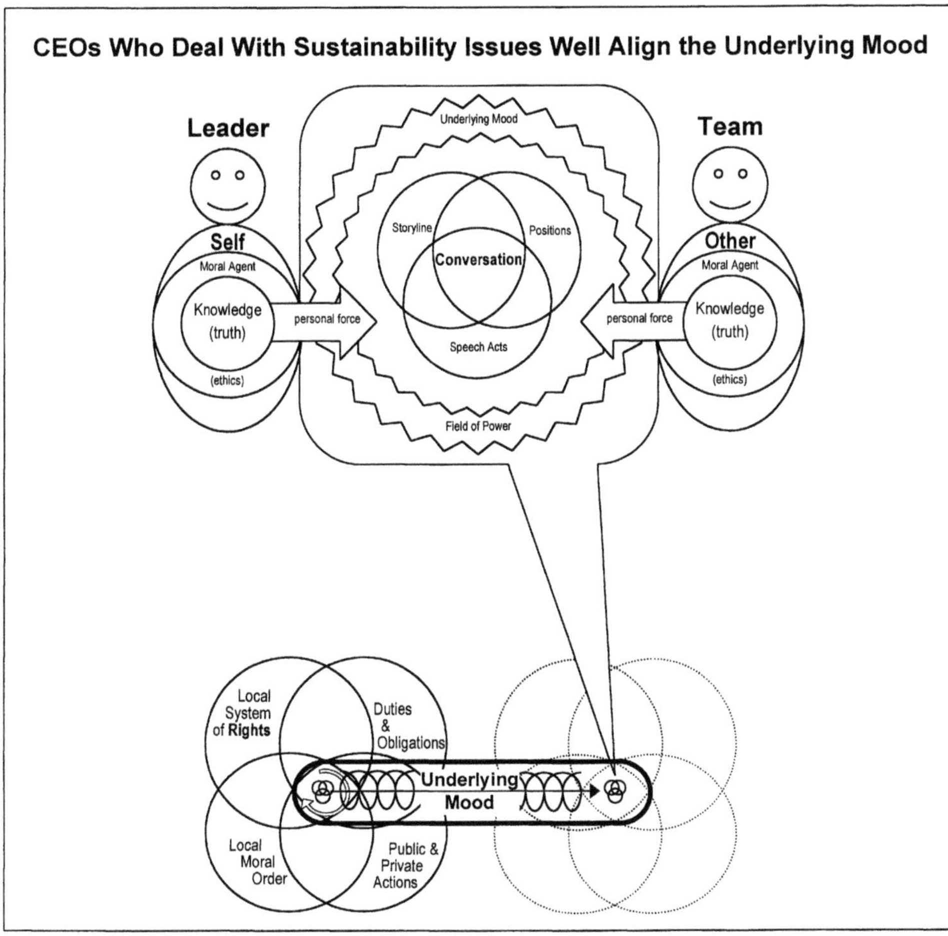

The bottom portion of Figure 1 is composed of several parts. At the left hand side are four interlaced circles that depict four components of the underlying mood (rights, duties, moral order and actions). The tri-polar conversation model is shown at the centre of these interlaced circles, implying that conversations occur in this underlying mood. The spiralling shaft extending to the right represents the ongoing conversations that iteratively contribute to the evolution of the underlying mood. At the right hand side are the four interlaced circles at a future moment in time, as influenced by the aggregate of conversations that have occurred in the interim.

This framework suggests that, while on the one hand, the underlying mood can have an effect on conversations, the reverse is also true. That is, the framework suggests that each conversation can affect the underlying mood. While one conversation may not make a noticeable impact on the underlying mood of a large organisation, an aggregate of many identical conversations will likely noticeably change the underlying mood. It is this sort of aggregate impact that leaders need to create if they are to deal with sustainability.

A Theory of Underlying Mood

This section provides detail of the theoretical grounding of the social constructionist framework introduced in the previous section. Heidegger (1956) describes the whole of being as "stimmung" or "mood" and, it is in this sense that mood is here understood to be an indicator of culture. This is especially interesting if culture is defined in Schneider's (1988: 353) terms:

> "Culture refers to: (a) the values that lie beneath what the organisation rewards, supports and expects; (b) the norms that surround and/or underpin the policies, practices and procedures of the organisations; (c) the meaning incumbents share about what the norms and values of the organisation are."

Culture can be described in terms of this Heideggerian "mood" of values, norms and meanings if there are some measurable parameters to define mood. It is here that discursive action (and the resulting positioning taking place within the underlying mood) becomes important. Informed by the Harré and van Langenhove (1999: 3) treatment of "moral order", "rights", "obligations" and "acts", Boxer (2005a; 2003a) shows that the moral order, rights, obligations and acts are four mutually interdependent components that define a larger social order or culture. Such a concept of culture could provide a discursive way of defining, measuring and thinking about the culture of an organisation that – in turn – could enable management education to better prepare people for dealing with sustainability. It was with this interest in understanding how senior leaders used their influence to effectively implement sustainability initiatives that mood was identified as an enabler in dealing with issues that affect climate change.

A number of CEOs were invited to reflect on their experience dealing with sustainability issues (Boxer 2003). These leaders were selected from multinational petrochemicals, local municipalities, world-class hospitality, tertiary education, manufacturing, and service. Interviews were conducted and recorded, transcripts produced, and these texts subjected to the positioning theory framework depicted in Figure 1. Following a series of codification and reflection with participants based on this conceptual framework, a theory of underlying mood was more fully developed. This framework can be used as a basis for management education to map out the various skills required to deal effectively with sustainability issues and how those skills interdependently contribute towards necessary leadership behaviour. To understand the framework more clearly, positioning theory will be explored in relation to climate change.

Confronting the climate change issue

Confronting oneself

If achieving sustainability is considered in terms of dealing with selfishness the challenge becomes one of inspiring people to move beyond themselves and into what is possible between people. How may we elevate people to times when their pursuits engage with their values, strengths and capacity without thinking about themselves? This may be when they reflect on their experience and focus on the in between space that occurs in a community. In that space, where the underlying mood is defined and experienced they experience a transpersonal space, which offers unique insight. This only occurs in a dynamic of human interactivity and never occurs in the static of singular human isolation. The singular only offers a formula for incompetence, because everything is continually changing outside the self and without community the self is unaware of change. Selves do not experience the change in the same way as a community does. So, a healthy and useful self is one that is always moving, growing, learning, changing and open. Might adopting such a perspective be a first step to dealing with selfishness and, hence, achieving a sustainable approach to climate change?

An open person may be more sensitive to collective issues such as climate change if the self is immersed in a community that cares about climate change. With such awareness, an individual may be more likely to explore the challenges of dealing with climate change and consider possible solutions. An accommodating underlying mood will facilitate the development of solutions as well as their implementation and execution. At this stage, a deeper understanding of positioning theory is helpful.

Positioning theory

While Harré is known as a realist and retains that perspective, in later works he (Harré & Secord 1972; Harré 1979; Harré 1983; Davies & Harré 1990; Harré & Gillett 1994; Harré & van Langenhove 1999) draws on discursive practice to understand the way people construct realities. Positioning theory, however, is about personal construction of self and others. Personal construction of self implies agency and agency implies realism. Harré developed positioning theory in part from being a student of the founder of speech act philosophy, J.L. Austin, himself a student of Wittgenstein. It is from this paradigm that Davies and Harré (1990) deliver the seminal work on positioning theory. From their work it can be suggested that people can learn to alter their positioning and that of others, as well as influence mood.

Porter (2006a: 499) suggests that "attending to members' value systems and identifications may be one way of improving the likelihood that an organization will productively coevolve with the natural environment". Through demonstrating how people create themselves and others through discourse, positioning theory enables the exploration of value systems and identifications. In doing so, positioning theory may help to develop the best approach for achieving the ecologically sustainable behaviour of individuals and collective organisations, because it considers local context and how people's values influence the way sustainability issues are dealt with.

So, the way in which a social construction of selves occurs provides the dynamic to explain what happens when people communicate. In constructing themselves and others (those other people with whom they are interacting), people position themselves and others in a democratic way that is limited by the social conventions already in place. In more rigid social situations, established conventions limit the positions that individuals can define for themselves and others. For example, in an army, soldiers wear some badges on their clothing that clearly establish their level in the hierarchy and other badges to distinguish their particular expertise. In this context, while an untrained recruit wears a uniform with no embellishments, superior officers are conferred with precise ranks and clear insignia that denote those ranks, qualifications and awards that they have achieved. In this way, a soldier entering into a social encounter will immediately know their place in the hierarchy and whether or not the others involved are competent to inform or instruct them. The army falls at one end of a continuum of social rigidity, where there is typically little opportunity to reposition oneself from the forced positioning imposed by the discursive practices of the institution.

In a formal organisation such as a business, the "boss" is positioned as such. Similarly, subordinate leaders and staff are positioned in their place. While business people may not be positioned as immovable and finite as soldiers in an army, there are conventions that have been established and those who remain in the business will learn to respect those conventions. Regardless, in a business there is room for positioning to occur during every social encounter.

Each person brings with them certain knowledge – or perception of what the truth is – and this establishes them as moral agents with a framework of ethics that they have established over time. Their agency imposes a personal force on any social encounter with another person. As this personal force is exuded it confronts the underlying mood, which is a field of power. Where the personal force of one individual may not have an impact on the underlying mood, collectively, many people can initiate a re-alignment of the underlying mood.

Components of Positioning

As depicted in Figure 1, positioning occurs under the influence of three components: position, storyline and speech acts. Each person brings a preconceived position and, while that position may not be accepted by all involved, this initial position influences the conduct of the encounter. The strength of their personal agency will have some influence over how their position may evolve over the course of each social encounter. A storyline is established within which the people involved follow. Throughout the encounter people engage various speech acts in the utterances they introduce. It is through the initial positions, storyline and speech acts that conversation unfolds and positioning occurs. At the end of an encounter, those people involved in the conversation may find themselves with unchanged positions or with new positions. This dynamic will be repeated during subsequent encounters.

As opposed to the static concept of an appointed role (such as CEO or manager or receptionist), the negotiated position is dynamic. One can establish a powerful position in one encounter only to be reduced in the next encounter. On another dimension, a person can find themselves in very different positions with every other person they encounter. This is reflected in Figure 1 by the spiralling underlying mood. The culture impacts conversation and perhaps positions. Conversely, a changed position can alter the sort of conversation that takes place and perhaps influence the culture. If a conversation respects the established rights, duties, moral order and actions of the organisation then the underlying mood will not be affected. However, if the conversation is subversive in relation to the established underlying mood then any one or all of these components of the underlying mood may change and disrupt the underlying mood. Positioning occurs in this mood.

Place at which positioning occurs

The *place* at which positioning occurs – the underlying mood – introduces four components that become important elements in explaining what leaders need to do when implementing ecologically sustainable behaviour. These are the four interlaced circles in Figure 1 including rights, duties, moral order and actions. It is suggested (Boxer 2003a) that this *place* is a representation of culture or organizational climate.

Rights are those things that enable people to behave in a certain way. If someone behaves contrary to established rights then it is apparent that they are breaching the accepted conventions of the organisation. However, in those situations where conventions of the organisation are not in line with evolving broader societal expectations the accepted rights may be inappropriate. For example, a person working in an organisation in which pouring sump oil in the gutter is routine will assume that they have the right to continue doing so, but the contemporary sensibilities of broader society would denounce such behaviour. Perhaps Weick

(1996) has this in mind when he suggests that sometimes it is sensible and necessary to "drop old tools" and abandon them in favour of more appropriate ones.

Duties in some ways mirror rights. In every society people have a duty to perform or behave in certain ways. People have duties to the organisation, duties to others, and duties to broader society. Any "new tools" (Weick 1996) would need to incorporate any duties expected by the organisation or broader society.

The **moral order** encapsulates all the rules and conventions that influence how a member of a society or organisation determines and defines appropriate rights and necessary duties. Every group – large and small – has a unique moral order which may or may not be aligned with the moral order of broader society and those of related stakeholders. Such variance may be for good reasons. For example, ordinary citizens do not have a right to kill others, but under certain circumstances soldiers do.

Actions are a representation of one's interpretation of the moral order, and could be referred to as behaviours in motion. These are influenced by rights, duties and the moral order. Actions that are aligned with rights, duties and the moral order reinforce the organisation and create harmony. Conversely, actions that are misaligned with rights, duties and the moral order undermine the organisation and create disharmony.

Challenges in the community

As with any action, there is a reaction to the sustainability movement. For example, a Victorian State Labor government once banned an accredited journalist from attending press releases, because the journalist was perceived to be an activist, who asked disruptive questions (Age 2001). The following Sub-Sections consider three possible ways of rejecting the movement.

Challenging Sustainability Zealots

Landsburg (1993) complains that environmentalists harness mental programming techniques, such as rote repetition and other intrusive myths, superstitions and rituals. He views the resulting environmentalism as naïve. While his argument lampoons social trends and his own defiance, Landsburg demonstrates questionable processes that occur in the sustainability discourse. He develops examples in which conveniently applied unscientific bias and selective logic are shown to form the basis for several environmental and social messages of the sustainability movement.

Scientific Critique

Broecker (2001) and Harris and Rieber (1996: 8) argue that measures used to gauge environmental issues may be 'based on incomplete information or incorrect premises'. They go on to say (p. 11), 'the assumptions necessary to demonstrate [the validity of current measures] are extremely strong and bear no resemblance to man's actual experience'. With this sort of uncertainty, it can be expected that there will be diverse opinions about environmental and social issues among senior managers, who make collaborative decisions. At the same time, Harris and Rieber (1996: 20) stress that they do not view environmental considerations as unimportant or that they have an indifferent reaction to depletion of critical resources. Hence, how managers resolve conflicts arising from TBL is a subject worthy of study. Rowell (1996: 135) refers to these scientific arguments as 'counter-science'. The clash between '(c)onservationist (radical) Greenspeak … and sophisticated exploitationist (conservative) Greenspeak' is raised by Harré, Brockmeier and Mühlhäuser (1999: 2) and the wider discourse and debate surrounding this contrast is explored in detail.

While Lomborg (2001) does not reject outright the sustainability message, he does challenge the focus of effort. Through a thorough examination of costs and benefits, he demonstrates that many high profile environmental efforts are mis-spent. His criticism follows a similar line to Landsburg (1993), but it does come from a scientific basis. Although Andersen (1999) and Sørensen (1998) disagree with its validity, Lomborg (2001) has provided a scientifically-sound sceptical view of the assumptions underlying much of the sustainability movement. Lomborg concludes that, on a global scale, the situation is far better than being suggested by many activists, but he acknowledges there remain many local problems to be resolved. His message is not to ignore the sustainability movement, but to focus effort on achievable objectives.

Just Marketing

Despite representations by 'ethical' entrepreneurs such as Roddick (1991), critics (Petean 1996; Entine 1995, Greenpeace) suggest that sustainability is simply a form of marketing. They insist that products made available by such companies have been created – as have the markets for the products – by the hype of the sustainability issue. Roddick's (1991: 256) concluding statement could be interpreted as an indication of her self-interest.

> 'Make no mistake about it – I'm doing this for me'.

Roddick does indeed profit from the products her business creates and sells. In that way, what she is doing could be seen to be done to increase her personal financial wealth. However, if viewed in the context of sustainability – her book implies that this is the perspective Roddick comes from – it is conceivable that

Roddick means that she is improving the environment and society so she creates a better world for herself – and others.

Sustainability Discourse

Harré provides two profound works that address both the topic of this chapter and the methodology of inquiry. Of interest here is Harré, Brockmeier and Mühlhäuser (1999), that introduce and explore discourse on environmentalism. Three categories of discourse are identified, written, spoken and pictorial. Harré et al. (1999: 43-50) suggest that there is a modern surrogational discourse, that is based on three-dimensional space and straight-line, one-way-only time in the traditional Western sense, and a non-surrogational discourse, that is based on moral ideas.

The same authors warn (p. ix): '(w)e perceive, in the increasing greening of English and other Western language, a kind of linguistic *Ersatzhandlung* (redirection), with the very real danger of talk replacing action'. Perhaps this is an example of the sort of 'modification in the rules of formation of statements which are accepted as scientifically true' that Foucault (1980: 112) raises. Harré, Brockmeier and Mühlhäuser (1999: 4) are concerned that discursive phenomena have made environmental issues the 'crisis of our times'.

> "It comes about through a shift in our ways of seeing and assessing what we see, made possible by the taking up into our discursive resources new vocabularies, new judgemental categories, new metaphors and analogies that have promoted awareness of much that was previously overlooked."

Should there then be different discourses within organizations that deal more effectively with sustainability issues? If there is a dominant discourse, perhaps that explains why these organizations deal better with sustainability (Rost 1991).

Change initiatives do not happen as a due process following the crafting of strategy and policy (Vallabhaneni 2005; Manners 2002). "Without effective implementation, even the most creative strategy will fail" (Vallabhaneni 2005: 7). Wright (2002) identifies lack of leadership as a major challenge and barrier to implementation of sustainability plans. It is suggested that, in considering climate change, there should be a component dedicated to understanding how leaders deal with implementation of plans for sustainability initiatives. More specifically, leaders need to be able to inspire others to deal with sustainability issues in a way that these plans are implemented as intended. A social constructionist approach is helpful to understand this.

Steven Mayne (www.crikey.com.au/), a shareholder activist has explained in private conversation and on various public media that individuals realize they are able to best exercise influence at shareholder meetings, whereas collectives work in and through larger and more diverse public forums. To borrow from marketing (Kotler, Clark & Adam 2001), individuals tend to use a 'rifle' approach and Non-Governmental Organizations (NGOs) use a 'shotgun' approach; NGOs appear to deploy individuals on specific tasks. Regardless, they are reacting against government of one sort or another.

Where specific activism focuses on specific issues (such as the practices of a business), there is a broader threat to unity and utility. Business can no longer carry on as in the past, not necessarily because it is unsustainable to do so, but because activists are influenced by an ideal of *natural capitalism* (Hawken, Lovins & Lovins 1999) and empowered with a self-perception that they can make a difference. That is, they position themselves as such and – in doing so – they are aware that power flows through them (Foucault 1980). The following Sub-Sections examine possible sources of activism and explore how that phenomenon is focused to influence business leaders to deal with a variety of issues such as sustainability.

A Foucauldian Perspective on Activism

Cheney (1995) observed that Foucault was a key figure at the time social activism was gaining such momentum that it was able to disengage the United States of America from the Vietnam War. Although not directly involved in activism, Foucault's much quoted analysis of the exercise of power in society can be seen as a way to explain how social order or domination is maintained.

Governmentality (Foucault 1979) – the ease of which society is governed and utilized in productive endeavours – has declined in recent years. Foucault's (1978) interest in docility and utility led to his concept of governmentality; a less governmental society is less docile and of less benefit to the production of wealth. Foucault (1978: 86) shows that exertion of power is tolerable only if its mechanism is concealed. With increasing transparency and a decline in the impact of disciplinary processes (Foucault 1977b) these mechanisms are increasingly apparent. In this light, Ferentzy (2000: 222) draws on Foucault to suggest that some beliefs are not as controlling as they once were. O'Farrell, (1997) and McInlay and Starkey (1998) similarly draw on Foucault to suggest that people no longer put up with being told they must comply with things they do not believe to be true.

Foucauldian (1978: 89) philosophy can explain that contemporary Western society's control 'is not ensured by right but by technique, not by law but by normalization, not by punishment but by control, methods that are employed on all levels and in forms that go beyond the state and its apparatus'. Collectively, the many diverse groups that are springing up throughout the world appear to struggle against a common hegemony of Foucauldian bio/disciplinary power, but they are committed to anonymity and unanimity amongst themselves. This too is explained by Foucault (1991: 76-78), in which he employs an 'event' concept as a way to break up or disrupt well-established and taken-for-granted continuities in traditional history, and as a tool to ensure that historical events are seen and grasped in their unique singularity and not forced onto pre-established patterns of historical understanding. In recent years significant activism has focused on an event and taken the first letter of the month and date. For example S11 2000, when the WTO was under siege in Melbourne. Not all dissent is radical nor is it violent, but if it is collective even passive dissent is activism.

The recipe for group activism involves assembling a critical mass that embraces the same new mode of thinking (Schiffman 1991: 67). While not all activists understand Foucault's ideas, they are caught up in a momentum of dissent. Where there has been a tradition of radicalism devoted to resolving inequalities such as feminism (Weedon 1987), sexual orientation (Warner 1993) and racism (Minwalla 1997), the technologies of radicalism are available to the sustainability movement. For example, Warner (1993: xxvi) introduces his book of collected articles on queer politics which he argues:

> "rejects a minorizing logic of toleration or simple political interest-representation in favour of a more thorough resistance to regimes of the normal".

Radical thought realizes the ability of 'governmentality' to be achieved from a distance. Combining Latour's (1987: 219-232) idea of 'action at a distance' and Foucault's (1979) idea of 'governmentality', Rose (1989; 1996) and Kendall (1997) write about 'government at a distance'. Kendall (1997: 90) specifically refers to the 'efficacy' with which Australia was governed as a colony without 'physical presence'. Banerjee (1999: 12) raises concerns that non-Western societies and their forms of knowledge have been marginalized. For example, 'based on British common law, Australia became a colony of England because it was settled as *terra nullis* or land belonging to no one'. He goes on to call for change to this 'government at a distance' and other inequalities. Perhaps a decrease in governmentality may increase a society's tendency to be influenced by activism.

Shareholder Activists

Consider the impact one woman had on a major Australian mining business regarding their Jabaluca mine (ABC 2001). Erica Ford decided that she would raise opposition and successfully changed the company's strategy. As a shareholder she confronted the board, assembled a 'critical mass' of supporters (North Ethical Shareholders), manoeuvred herself to a position of authority, took the company to court, and stopped something that she believed was wrong. At the time of this writing she is involved in other campaigns. Perhaps this may appear to be "crazy", but in a free world people have a right to voice their opinions and behave in ways that others may perceive to be "crazy".

While members and officers of NGOs may take part in shareholder activism, they rarely identify their affiliations or 'badge themselves'. This appears to be a

Foucauldian disruptive tactic borrowed from radical feminists and others (Calás & Smircich 1992; Parker 2002), in which individuals see value in remaining separate, or in small groups, and uniting for a purpose when appropriate. This appears to be an effective strategy, as it makes shareholder activism a more challenging opposition for business leaders. Businesses are under siege from unpredictable assaults of a variety of groups with different agendas, but common cynicism directed towards corporations. They appear to unite in this opposition under the banner of shareholder activism.

The 2001 merger of a major Australian business (BHP) with a major South African business (Billiton), despite major reservations of shareholders, provides anecdotal evidence that corporations can act at variance with shareholders' desires. In reaction to this sort of behaviour, Steven Mayne operates an Australian internet web site (http://crikey.com.au) that enables him to act in an individual capacity to alter the behaviour of corporations and demand a balanced holistic approach and responsibility and accountability, as is reflected on by CEO Morgan (1998.) Mayne takes steps to get elected to the boards of the companies that he targets to dilute control (Age 2002). He claims to have been successful in 2000, having apparently forced a former board member to step down from his board position with Telstra, due to alleged conflicts of interest (Workers Online 2000; Grealy 2000). This was apparently accomplished by threatening to raise these alleged conflicts of interest in his campaign. His tactics include targeting companies that do not meet his standards of performance, and buying shares to enable him to attend shareholder meetings so that he may influence the discourse.

Ferentzy (2000: 234) draws on Foucault (1978: 26) to show how new developments only become legitimate issues if sufficient conflict can make the issue important. Even then it is not legitimate until the dominant power is comfortable that it can be administered and controlled. Foucault implies that 'conflicts were necessary' and 'economic emergencies had to arise' for instigators of change to have their issues legitimized. More important, he shows how 'a whole technology of control' is needed to be in place and concealed. This permits change to take place while observing the new process and exploiting the new way to a similar degree as the old way was exploited. Pickles (1996: 24) agrees that this is a 'complex of discourse, practice and institutional ensemble … affecting changes in the modalities of power'.

Activism perhaps has a role in bringing urgent issues to the surface, but perhaps it is necessary for activists to work with business to implement changes effectively. This may be to the benefit of the activists, as they will be able to influence the sort and extent of controls that are imposed on issues.

Avoiding the Wrath of Crazy People

Shareholder activism appears to be having results. Packard and Reinhardt (2000: 130) suggest that those companies that do not consider and appropriately deal with risks relating to climate change may come under scrutiny from their shareholders. Beets (1999) explains, many companies are becoming more responsive to investor's concerns about the environment. Konar and Cohen (1997) demonstrate how businesses that reduced emissions also achieved significant increase in market value. Whether or not this demonstrates a causal relationship between environmental responsibility and market share or that these companies are just managed well – and that environmental responsibility is simply a symptom of good management – is not important. It could be interpreted as that contemporary good management embraces environmental (and social) responsibility. Elkington (1998) – formerly an activist and now a contributor to corporate board deliberation – explains that the awareness of this phenomenon has led to sustainability becoming an issue that business needs to incorporate into strategic plans and operations. Then CEO of Ford, Jack Nassur (1999) explains to staff how he was joined by his then chairman and the then CEO of BP Amoco to discuss sustainability with John Elkington.

Confronting the community

There are several levels of the community to be confronted if climate change is to be dealt with.

Confronting others in your local community

When recounting how they dealt with sustainability issues, senior managers have referred to instances in which they have challenged obstacles and reinforced achievements. In commenting on these two sorts of activities, a common theme has been their interactions with people. People have been shown to present obstacles to dealing with sustainability issues as well as needing to have the sustainability message reinforced. In dealing with those who present obstacles and require reinforcement of the sustainability message – which reasonably could include most people – senior managers engage in a variety of discursive actions. It appears that those senior managers who are able to deal with sustainability well do so because they have determined what obstacles are presented and who is presenting them.

Senior leaders, by virtue of their appointment, would be expected to have a general perspective and, in doing so, they each might be expected to traverse the range of technical, consultative, dialectic and transpersonal characteristics that Ling (1998, 156-158) used to categorize his participants. Instead, CEOs of Boxer (2003a) demonstrated behaviour on a different plane – one of diversity that is enabled by the ability to traverse a power/parity dimension. Power can be seen to give way to parity when moving from a technical to transpersonal orientation. This relationship uncovers a sensitivity of Ling's approach in that it appears to distinguish between lower and senior level managers.

So, inspiring those in your local community may require local sensitivity to contextual transpersonal factors.

Having learned from the global feminist movement of the incredible influence of small discussions in localities, sustainability activists compete with global forces by engaging in telling small stories at the local level. They create varying degrees of resistance in order to express hopes for new power relations. They realise they can be autonomous subjects rather than objects of development. They are well guided to operate their local resistance in certain global ways through an aggregate of local effort. They engage in deconstruction and practices of *re-subjectivation*. They create breaches for unequal power to be introduced with the objective of destabilizing the hegemony of unsustainable practices. As deconstructionists, activists see a dominant discourse that lacks concern for sustainability. Because they realise its colonising power, activists stand outside the main stream for a more transparent perspective. Activists liberate society from the non-sustainable practices by creating new discourse that enables re-positioning of individuals into a more sustainable approach and assume power in new forms. It becomes a project of *resubjectivation*; that is new institutions and practices are created for people to align with.

A senior manager in a hospitality complex reflects on how activists converging on his site had influenced his planning. These people assumed a *right* to interfere with proceedings to make their sustainability message heard. He realised the danger presented by this activism:

> We are so big and our core business attracts a lot of media attention. It means that we have a responsibility that goes far and beyond what other organisations might.

However, he acknowledged the opposing *right* of shareholders to maximise their return: Everyone wants to maximise their own position including making money, status, and all the benefits, but there are those people with a sense of community who say, 'well there is a bigger thing here' and they have major life goals that are above personal gain.

Accepting the *right* of his tenant shop owners to engage in the core business, he introduced measures to ensure compliance with his rules. By doing so he avoided the wrath of activists:

> Human nature is to try to find a short cut. So why can I not push the boundaries of my service to the point where if it is something that I can solve and that I can take responsibility for and deliver them a service that frees them of that obligation, then I am in complete control. Compliance is my issue – not theirs – they can focus on the business of making money.

His *actions* reinforced the *duty* he assumed to facilitate the sustainable way for others to follow:

> I want to know who it is – not to go about accusing people – so I can find out who it is, learn what is going wrong, and then improve it.

A university CEO explained how influence leaders (in some cases activists) were helpful in understanding issues. He demonstrated the *duty* he assumed to understand issues that may lead towards the sustainable way:

> I am very assiduous at picking influence leaders, who are conducting debates in their own right or are involved in sub debates or subsets going on in the university. So, you are working with multiple levels, multiple conversations, conversation forms, both formally and informally right across the community.

The petrochemical CEO explained his reliance on younger people, who may be idealistic about sustainability issues, to drive the sustainability movement within his organisation. Perhaps their youthful enthusiasm naturally inclined them towards activism – albeit respectfully and with deference.

He noticed that they assumed a *duty* to live the sustainable way, whereas older employees might assume a *right* to avoid the sustainable way:

> Often you will find that this is led particularly by young people. We have a number of processes that enable young people to be very active – we have a thing called Project Better World for example which involves allowing people to go off and spend time with Earth Watch; and we have a fairly significant environmental sponsorship that we will be announcing in the not too distant future which will also enable people to get involved.

The CEO of a municipality has similarly harnessed the enthusiasm of activists. Staff and councillors, who assume a *right* to question inappropriate situations and a *duty* to take *action* can create the *moral order* necessary for the sustainable way:

> I would not consider my understanding anywhere near as detailed as my staff or some of the councillors – I mean they are very progressive. I am involved in a few (sustainability forums), but (I learn) mainly through my staff and the councillors. Two people who have been influential are (a manager) who has been a very strong advocate for sustainability for a long time and one of our councillors, who I learn a lot from just by listening to him and the passion.

Those CEOs who deal with sustainability well realise that they do not have a *right* to ignore activists. In fact they realise their *duty* to embrace and involve activists already employed in their own organisation. By establishing a *moral order* that respects the views of those passionate about sustainability, they encourage *action* that contributes to organisation-wide sustainable behaviour.

Boxer (2003a) has shown how senior leaders influencing long-term changes to the way their organizations have approached sustainability or other OEI issues such as climate change. It has been determined here that their success can be explained by their ability to alter what is described here as underlying mood.

It can be said that reversing climate change will require leaders to realize the need to alter the underlying mood and an ability to do so through discursive action as shown in Figure 1. It is put forth that their actions – to vary story-lines, positions and speech acts – have altered the underlying mood, which is a reverse to the implications of the conjoined situation implied in Figures 1.

Assuming a position that enabled them to change the way people view sustainability issues, and positioning others as being aware of the need to deal with sustainability issues and willing to do so, senior managers appear to have entered into discursive action. They appear to have adopted a story-line regarding the need to deal with sustainability issues and made use of speech acts. This process affects them and others such that the new way of approaching sustainability issues results in repositioning. However, this behaviour does not deal with the long-term change. For this to occur, the components of the social constructionist model need to be altered.

There is a need for action to alter the moral order, local system of rights, duties and obligations, and public and private actions. These changes lead to an altered underlying mood that is conducive to dealing with sustainability issues. As such, the underlying mood can be altered and consequently the discursive action can be altered. With these fundamental changes to the way people conduct themselves, new positions can be assumed without constant reinforcement.

Confronting the world

Global level dynamics are far more complex that local, wider and national community. Governments need to engage in diplomacy. Political leaders need to become statesmen and engage in international power flows. In dealing with climate change they still need to align the underlying mood by engaging in personal conversations. Yet, the opportunity may be lost due to the inability to deal with the truth in complex diplomacy.

Foucault's concept of parrhesia (Foucault 1985) may provide insight into how leaders deal with issues by speaking the truth. It could be said that leaders who effectively deal with sustainability issues identify the truth that needs to be dealt with and then get on with dealing with that truth. Foucault's parrhesia provides a framework for understanding the frankness and the difficulties faced by those who deal with issues by speaking the truth. Coupled with Foulkes' (1989) gracious talk, parrhesia could provide useful insight into what stops people from confronting important issues.

Aligning the Social Order through Freedom of Speech

That global leaders need to take steps to confront the underlying mood when dealing with sustainability issues could suggest that they are sensitive to the components of the underlying mood. They assume a duty to engage in a freedom of speech about what needs to be dealt with. Namely, the perceived rights of people, the duties people should undertake to perform, the underlying moral order, and their own actions as senior managers are all dealt with by senior managers who do deal with sustainability issues. This suggests that the answer of dealing with issues lies in aligning the underlying mood to best support resolution of issues. Such a freedom of speech could be explained by Foucault's ideas, such as those expressed in parrhesia (Foucault 1985).

The four components of the social order could be perceived to alter the social order in a similar way as levelling control knobs on a physical platform. In other words, four adjustments need to take place. First, the rights that people perceive they have must to be confirmed and corrected. Second, with an appropriate system of rights in place, people need to understand the duties that they are required to perform. Third, there needs to be an appropriate moral order in place that supports rights and duties. Fourth, senior managers need to ensure that their own actions and the actions of others are congruent with what is expected from their subordinates. These four components interact to place the social order in an appropriate plane of reference for sustainability issues to be dealt within.

Implications for Practicing Global Leaders

Organizations at all levels have difficulty with dealing with sustainability and other OEI issues. It could be that some senior managers may overlook their obligation to lead. While Deming (Deming 1986: 248; Walton 1989: 69-70; Gabor 1990: 22-23) explains that leaders should not regulate *people*, he could be interpreted as suggesting that leaders should regulate the *underlying mood*, to direct people towards effectively dealing with issues. In not dealing with the underlying mood in this way, global leaders may be undermining their intentions to deal with sustainability and other OEI issues such as climate change.

For successful resolution of sustainability such as climate change, leaders at all levels could emulate the way shown here to deal with an inappropriate underlying mood. It may well be that those organizations and nations that do not deal with sustainability simply overlook the underlying mood and fail to align it appropriately. However, it may not be politically prudent to tell constituents that they do not have the right to behave in ways that contribute to climate change challenges. It may be far more expedient to let people carry on and it may be more fun to join in with their fun. As Deming explained during a presentation, "It is not necessary to change. Survival is not mandatory"; if humanity insists it has a right to do stupid things (Cipolla 1987) then extinction may be an unavoidable outcome.

Conclusions

Evidence of positioning is shown to occur in sustainability discourse. That positioning contributes to the way sustainability is dealt with and how further iterations of positioning occur in subsequent discourses. An aggregate of all discourses contribute to the underlying mood in a society. Unless steps are taken to alter discourse, situations are shaped by the past so that the future continues to be a reflection of the past. That is, the underlying mood does not change.

In this Chapter, it has been shown that people need to engage in contextually relevant and unique discourses about climate change for climate change to be dealt with effectively. The residual underlying mood reflects the position of each stakeholder group and the perceptions of individuals in those stakeholder groups. Observing this discourse and the consequential positioning leads to understanding that can be used to better understand issues and how each group treats the ideas and other stakeholders. With detailed contextual data, such an analysis can provide leaders with increased intelligence to enable more appropriate decision-making that could deal with climate change.

It is one thing to network a community together, but another to align their underlying mood. Such alignment is achieved by focused discursive action – as opposed to aimless chatter – and sufficient influence. For example, through his film, *An Inconvenient Truth*, Al Gore directs his discourse to networks and teams of local communities to change the underlying mood of diverse societies so that they recognize the reassuring lie they have been living. He ensures this is replicated by recruiting and training ambassadors of his cause.

Sustainability is not necessarily new, although there is a contemporary enthusiasm for more appropriate stewardship of environmental, social and economic resources. Furthermore, politics that occur at all levels can affect how the rules and resources are managed. Not only is there increased awareness of sustainability, but also some elements of society demand resolution of sustainability issues as well addressing the climate change challenge.

External influence of activism, with the potential to shut down an organization, has made the sustainability issue obligatory. Hence, leaders need to address sustainability issues and demonstrate both their commitment to do so as well as produce tangible results. With the urgency created by the obligatory nature of the issue and the external influence, internal confrontations will likely occur regarding implementation of related policy. The problem for leaders at all levels involves engaging in discourse that resolves internal confrontations regarding sustainability issues so that they can be appropriately dealt with in society. It has been suggested here that leaders engage in discourse on sustainability in their everyday work and that a framework for analysing their discursive action can be found in positioning theory.

In the light of conflicting objectives, leaders' commitment to sustainability has been considered. It has been shown that, while there is appreciation of the principles of sustainability, there is also a reluctance to let sustainability obstruct the opportunity to make a profit. The likelihood of conflict introduces the prospect of discursive turmoil surrounding these issues. This has led to consideration of how people talk about sustainability and climate change. What has been shown? There is a need to engage in a determined discourse aimed at resolving climate change challenges.

If there is a collective effort to disrupt inappropriate behaviours then climate change will be dealt with. Otherwise, our society may not survive.

References

ABC (2001), "Taking care of business." An Australian Story. Australian Broadcasting Corporation, broadcast 9 Aug 2001.

Age (2001), "Govt Bans Crikey.com.au Editor", 30 Apr 2001, www.theage.com.au/news/2001/04/27/FFXKZ97EZLC.html, accessed 7 Nov 2002.

– (2002), "Crikey: Mayne nearly made News", 9 Oct 2002, www.theage.com.au/articles/2002/10/09/1034061249271.html, accessed 7 Nov 2002.

Andersen, M.S. (1999), "Bjørn Lomborg: The true state of the world." Politica no. 1, www.au.dk/~cesamat/debate.Politica.pdf, accessed 27 Mar 2002.

Banerjee, S.B. (1999), "Whose Mine is it Anyway? National Interest, Indigenous Stakholders and Colonial Discourse: The Case of the Jabiluka Uranium Mine", Paper presented at the Critical Management Studies Conference (Post Colonial Stream), 14-16 Jul 1999, Manchester.

Barnse, R.C. (2003), "Leading = influencing: a simple equation: Influence as the essence and foundation of leadership." Air & Space Power Journal, May. www.airpower.maxwell.af.mil/airchronicles/cc/barnes.html, accessed: 19 Jan 2007.

Beets, S.D. (1999), "Corporate environmental reports: The need for standards and an environmental assurance service." Accounting Horizons, Jun 1999, vol. 13 iss. 2, pp. 129-146.

Berry, M.A. and Rondinelli, D.A. (1998), Proactive corporate environmental management: a new industrial revolution, Academy of Management Executive 12(2), pp. 38-50.

Boxer, L.J. (1993). Overcoming common barriers to TQM. In: TQM'93, AIC Continuous Quality Improvement in a Service Environment, Fourth Annual Event, The Sydney Marriott Hotel, 17-19 Feb 1993.

– (2003a), Using positioning theory to understand how senior managers deal with sustainability issues, unpublished PhD Thesis, Business Faculty, Royal Melbourne Institute of Technology.

– (2003b), Assessment of quality systems with positioning theory. In: Harre, R. and Moghaddam, F.M. (eds.), The Self and Others, Praeger, CT, Westport.

– (2003c), A positioning theory method for culture analysis and development. In: Boxer, L.J. (ed.), Excellence in the face of crisis: proceedings of the 4th international conference for the advancement of organisational excellence, 20-22 Oct, Melbourne, Intergon.

– (2005a), The Sustainable Way, Melbourne, Brolga.

– (2005b), "Discourses of change ownership in higher education." Quality Assurance in Education 13(4), pp. 344-352.

– (2007), "Sustainability Perspectives." Philosophy of Management 6(2), pp. 87-97.

Broecker, W.S. (2001), "Was the medieval warm period global?" Science vol. 291, pp. 1497-1499.

Calás, M.B. and Smircich, L. (1992), "Re-writing gender into organizational theorizing: Directions from feminist perspectives." In: Reed, M. and Hughes, M. (eds.), Rethinking Organization: New directions in organization theory and analysis, Sage, London, pp. 227-253.

Cheney, L.V. (1995), Telling the Truth: Why our culture and our country have stopped making sense, and what we can do about it, Simon and Schuster, New York.

Cipolla, C.M. (1987), "The Basic Laws Of Human Stupidity." Whole Earth Review, Spring 1987, pp. 2-7. www.mentalsoup.com/mentalsoup/basic.htm, accessed 25 May 2006.

Davies, B. and Harré, R. (1990), "Positioning: The discursive production of selves." Journal for the Theory of Social Behaviour 20(1), pp. 44-63. www.massey.ac.nz/~ALock/position/position.htm, accessed 11 Sept 1999.

Deming, W.E. (1986), Out of the Crisis, Boston, MA., MIT Press.

Dewey, J. (1938), Experience and Education, New York, Macmillan.

Elkington, J. (1998), Cannibals With Forks: The Triple Bottom Line of 21st Century Business (Conscientious Commerce), New Society Publishing, 1998.

Entine, J. (1995), "The messy reality of socially responsible business." BWZine, www.betterworld.com/BWZ/9512/cover2.htm, accessed 26 Mar 2002.

Ferentzy, P. (2000), "The Addiction Concept: How the language of sin was replaced by that of disease." Unpublished doctoral thesis, York University, Toronto.

Foucault, M. (1972), The Archaeology of Knowledge, (trans.) A.M. Sheridan Smith, London, Routledge.

– (1975), I, Pierre Rivière, having slaughtered my mother, my sister, and my brother ...: A case of parricide in the 19th century, trans, by F. Jellinek, Pantheon, New York.

– (1977), Discipline and Punish: The birth of the prison, (trans.) A. Sheridan, Pantheon Books, New York.

– (1978), History of Sexuality vol. 1, Random House, New York.

– (1979), "Governmentality." Ideology and Consciousness vol. 6 no. 3, pp. 5-21.

– (1980), Power / Knowledge: Selective interviews and other writings 1972 – 1977, (ed.) Gordon, C. (trans.) Gordon, C., Marshall, L., Mephan, J. and Soper, K., Pantheon, New York.

– (1985). In: Pearson, J. (ed.), "Discourse and Truth: The problematization of parrhesia." Six lectures given by Michel Foucault at the University of California at Berkley, Oct-Nov 1983, http://foucault.info/documents/parrhesia, accessed 28 Jan 2003:

– (1991), "Question of method." In: Burchell, G., Gordon, C. and Miller, P. (ed.), The Foucault Effect: Studies in governmentality: with two lectures by and an interview, (trans.) by Colin Gordon, Harvester Wheatsheaf, London, pp. 73-85.

Foulkes, F. (1989), The Letter of Paul to the Ephesians: An introduction and commentary, Inter-Varsity Press, Leicester, England.

Eisen, H. and Mulraney, B.J. (1992), Impediments to the adoption of modern quality management practices, working paper, School of Management, David Syme Faculty of Business, Monash University, March.

Gabor, A. (1990), The Man Who Discovered Quality: How W. Edwards Deming brought the quality revolution to America, Times Books, New York.

Gellerman, S.W. (1986), Why "good" managers make bad ethical choices, Harvard Business Review, Reprint 86402.

Glaser, J. (2006), The new leader: from telling to asking, Consulting Today: Developing Leaders, pp. 15-17.

Grealy, M. (2000), "Mansfield Announces Telstra Board Changes." Letter to Company Announcement Office, Australian Stock Exchange, 28 Sep 2000, www.telstra.com.au/investor/docs/tls24.pdf, accessed 11 Nov 2002.

Greenpeace, What's Wrong With the Body Shop? A criticism of 'green' consumerism, www.mcspotlight.org/beyond/companies/bs_ref.html.

Guidère, M. and Morgan, N. (2007), Le Manuel de recrutement d'Al-Qaïda (Broché), Editions Le Seuil, Paris.

– (1979), Social Being, Oxford, Blackwell.

– (1983), Personal Being, Oxford, Blackwell.

– (2002b), email to L. Xxxxx, 15 Aug.

Harré, R., Brockmeier, J. and Mühlhäuser, P. (1999), Greenspeak: A study of environmental discourse, Sage, London.

Harre, R. and Gillett, G. (1994), The Discursive Mind, London, Sage.

Harré, R. and Secord, P.F. (1972), The Explanation of Social Behaviour, Oxford, Blackwell.

Harré, R. and van Langenhove, L. (eds.) (1999), Positioning Theory: Moral contexts of intentional action, written by Berman, L., et al. Oxford, Blackwell Publishers.

Harris, D.V. and Rieber, M. (1996), "Commentary and critique of 'Accounting for mineral resources: Issues and BEA's initial estimates'." Renewable Resources vol. 5 no. 1, International Association for Mathematical Geology, New York.

Hatcher, J.A. and Bringle, R.G. (1997), "Reflection: bridging the gap between service and learning." College Teaching 45(4), pp. 153-156.

Hawken, P., Lovins, A.B. and Lovins, L.H. (1999), Natural Capitalism: The next industrial revolution, Earthscan Publications, London.

Heidegger, M. (1956), Essence and Being, London, Vision Press.

Kekäle, T. (1999), "The effects of organizational culture on success and failures in implementation of some total quality management approaches." Bristol Business School Teaching and Research Review 1, www.uwe.ac.uk/bbs/trr/Issue1/IS1-1_1.htm, accessed 17 Jan 2006.

Kendall, G. (1997), "Governing at a distance: The colonisation of Australia." In: O'Farrell, C. (ed.), Foucault the Legacy, Queensland University of Technology, Kelvin Grove, Queensland.

Konar, S. and Cohen, M. (1997), "Does the Market Value Environmental Performance?" Vanderbilt University, Nashville, www.vanderbilt.edu/VCEMS/papers/mkt_value.html, accessed 5 Aug 1999.

Kotler, P., Clark, E. and Adam, S. (2001), Marketing, Pearson Education Australia, Frenchs Forest, New South Wales.

Landsburg, S.E. (1993), The Armchair Economist: Economics and everyday life, The Free Press, New York.

Latour, B. (1987), Science in Action. Open University Press, Milton Grove.

Learner, M. (1975), "The shame of professions." Saturday Review vol. 3 no. 3, pp. 10-12.

Ling, I. (1998), "The Role of the Curriculum Coordinator: An exploration through discursive practice", unpublished thesis, submitted in partial fulfilment of the requirements for the degree of Doctor of Education in the Faculty of Education at The University of Melbourne.

Lomborg, B. (2001), The Skeptical Environmentalist: Measuring the real state of the world, Cambridge University Press, Cambridge.

Manners, E. (2002), The role of participatory democracy in achieving environmental sustainability: A review of community engagement in regional transport planning in South East Queensland, dissertation for the degree of Master of Environmental Science, Griffiths University.

McInlay, A. and Starkey, K. (1998), Foucault, Management and Organization Theory: From panopticon to technologies of self, Sage, London.

Morgan, H. (1998), "Sustainable Development: an emerging part of business strategy." World Council of Sustainable Development, 1 July 1998, www.wbcsd.ch/Speech/s41.html, accessed 7 Sep 1999.

Nasser, J. (1999), Let's Chat About the Business, A weekly email to all Ford employees throughout the world, 19 Nov 1999, www.dearborn2.com/jnchat, accessed 21 Nov 1999.

O'Farrell, C. (ed.) (1997), Foucault the Legacy, Queensland University of Technology, Kelvin Grove.

Packard, K.O. and Reinhardt, F. (2000), "What every executive needs to know about global warming." Harvard Business Review July-Aug 2000 vol. 78, iss. 4, pp. 129-135.

Parker, M. (2002), "Queering management and organisation." Gender, Work and Organisation vol. 9 no. 2, pp. 146-166.

Petean, S. (1996), "Broken Promises", www.brazzil.com/p16dec96.htm, accessed 10 May 2002.

Pickles, J. (1996), "Representations in an electronic age." In: Pickles, J. (ed.), Ground Truth: The social implications of geographic information systems, Guilford Press, London, pp. 1-20.

Porter, T.B. (2006a), "Coevolution as a research framework for organizations and the natural environment." Organization and Environment 19(4), pp. 479-504. www.coba.usf.edu/jermier/TOC19four.html, accessed 20 Jan 2007.

– (2006b), Inside greening: the role of middle managers' strategic processes, attitudes, and behaviours in corporate environmentalism, unpublished PhD dissertation, Isenberg School of Management, Graduate School, University of Massachusetts.

Porter, M.E. and van der Linde, C. (1995), "Green and Competitive: Ending the Stalemate." Harvard Business Review 73 (5), pp. 120-134.

Prince, H.T. (2001), Leadership Programs in Higher Education II. Models of Ethical Leadership for a Changing World, Centre for Ethical Leadership, Lyndon B Johnson School of Public Affairs, The University of Texas at Austin. www.utexas.edu/ lbj/ research/ leadership/ publications/ conference1/ HigherEducationII.html, accessed 24 Jan 2007.

Roddick, A. (1991), Body and Soul: Profit with principles – the amazing success story of Anita Roddick and The Body Shop, Crown Publishers, New York.

– (2000), Business As Unusual: The triumph of Anita Roddick, Thorsons, London.

Rose, N. (1989), Governing the Soul: The shaping of the private self, Routledge, London.

– (1996), Inventing Ourselves: Psychology, power, and personhood, Cambridge University Press, Cambridge.

Rost, J.C. (1991), Leadership for the Twenty-First Century, Praeger, New York.

Rowell, A. (1996), Green Backlash: Global subversion of the environmental movement, Routledge, New York.

Rusinko, C.A. (2005), "Using quality management as a bridge in educating for sustainability in a business school." International Journal of Sustainability in Higher Education 6(4), pp. 340-350.

Scarnati, J.T. (1999), "Beyond technical competence: the art of leadership." Career Development International 4(6), pp. 32-33.

Schiffman, J. (1991), "Fight the power: Two groups mobilize for peace." In: Burawoy, M. et al. (eds.), Ethnography Unbound: Power and resistance in the modern metropolis, Berkeley, University of California Press, pp. 58-79.

Schneider, B. (1988), Notes on climate and culture. In: Lovelock, C. (eds.), Managing Services, Prentice-Hall, Englewood Cliffs, NJ., pp. 352-366.

Slim, W. (1957), "Leadership in Management: fourth William Queale Lecture to the Australian Institute of Management." Australian Army Journal, September 1957, pp. 5-13.

Sørensen, H. (1998), Lomborg's Claims are Untrue and Dangerous, Jyllands-posten, 15 Mar 1998, www.au.dk/~cesamat/debate.Sorensen.pdf, accessed 27 Mar 2002.

Vallabhaneni, S.R. (2005), Wiley CIA Exam Review vol. 4, Business Management Skills, 3rd Edition, New York, Wiley.

Wakin, M.W. (1981), Ethics of leadership. In: Taylor, R.L. and Rosencach, W.E. (eds.), Military Leadership: In Pursuit of Excellence, New York, Westview Press, originally in Buck, J.H. and Korb, L.J. (eds.), Military Leadership, London, Sage, pp. 95-111.

Walton, M. (1989), The Deming Management Method, London, W.H. Allen.

Warner, M. (1993), Fear of a Queen Planet: Queer politics and social theory, University of Minnesota Press, Minneapolis.

Weedon, C. (1987), Feminist Practice and Poststructuralist Theory, Basil Blackwell Ltd, Oxford.

Weick, K.E. (1996), "Drop your tools: an allegory for organizational studies." Administrative Science Quarterly 41, pp. 301-313.

Workers Online 2000, "The Riddles of Democracy at Telstra." Workers Online, iss. no. 80, 1 Dec 2000, http://workers.labor.net.au/80/c_historicalfeature_telstra.html, accessed 7 Nov 2000.

Wright, T.S.A. (2002), "Definitions and frameworks for environmental sustainability in higher education." International Journal of Sustainability in Higher Education 3(3), pp. 203-220.

Chapter 14

Climate Change and the Impacts on Flood Events at the Fränkische Saale – An Interdisciplinary Case-Study

Michael Belau, Jane Korck, Hans Weber

1. Abstract

Today's climate researchers predict increasing severe flood events as a consequence of climate change. How can societies adapt to these challenges induced by global warming? This question is gaining importance both at EU level and at global level given the alarming trends that are being observed. It is also indicated by the pilot project "Fränkische Saale" in the frame of the EU-project ESPACE European Spatial Planning Adapting to Climate Events (www.espace-project.org). Through the analysis of possible climate change impacts on water resources in the socio-economically vulnerable Main River basin within this pilot-project it was possible to show the way towards dealing with this alarming development.

Within the project, a regional climate model, a water balance model, statistical and hydraulic calculations and the analysis of the socio-economic effects of past flood events in the case study area were combined to enable the description of the anticipated influence of climate change on land use, infrastructure and property, focusing on the quantification of climate change effects on the water basin level. By considering the physical, economic and social impacts of severe flooding in the case study area an interdisciplinary and holistic assessment of the possible impacts of climate change was achieved.

Regional flood maps with high resolution show the increasing flood risk and the resulting increase in damage potentials as a consequence of climate change. They also build the basis for the analysis of concrete protection measures, backed up by cost-benefit calculations for different options and a Decision Support System. Taking lessons learnt within ESPACE this information was disseminated in the frame of a communication strategy focussing on both people affected directly and decision makers.

This article gives a brief summary of the outcomes of the ESPACE project and focuses on the aims, methods and results of the Case Study Fränkische Saale.

2. Introduction

Europe is already experiencing the effects of climate change like the devastating heatwave of 2003 or the repeated floods in different regions in recent years. The predicted increase in the risk of a changing climate together with a greater frequency of extreme weather events, has put climate change firmly on the agenda across Europe. Scientists, policy makers and politicians are all in agreement that a fast and efficient adaptation to these changing conditions is one of the biggest challenges facing Europe to ensure that our social, economic and environmental systems are best-prepared for the unavoidable risks. The Stern Review on the Economics of climate change highlights that we now know enough about the issue to assess and plan for the risks (Stern 2006). Spatial planning 1 was identified in this review as one of four fundamental areas that will play a crucial part in helping communities 'adapt' to climate change.

Against this background, the European project ESPACE European Spatial Planning Adapting to Climate Events (www.espace-project.org North West Europe INTERREG IIIB Programme) set out to influence the philosophy and practice of spatial planning by recommending how adaptation to climate change can be incorporated into spatial planning systems. Concentrating on water management issues, it was the first project of its kind to focus on increasing the awareness of the need to adapt to the impacts of climate change and to begin to provide some of the necessary policy guidance, tools and mechanisms to incorporate adaptation into planning systems and processes.

Recognising that climate change is a global issue transcending political boundaries, ESPACE was founded by a transnational group of 10 partners spanning four North West European countries and bringing together representatives from all levels of civic society. A range of actions were undertaken by partners to identify how to best adapt to climate change. These include actions looking at the most effective ways of raising awareness, the role of behaviour change in adapting to climate change, policy review through the use of innovative techniques and the development of tools and models.

As the only German Partner in the ESPACE project, the BayLfU concentrated on the question of how to adapt to the consequences of climate change in the field of flood protection. The central element of the BayLfU contribution to the ESPACE project is the case study "Fränkische Saale" which is described in chapter 4 of this article.

---

1 Spatial planning is "a process that assimilates and interprets evidence-based knowledge to inform those activities that aim to ensure spatial development takes place in an appropriate way, from a functional, social, economic and environmental point of view." Working definition of the ESPACE Project 2007.

3. Outcomes of ESPACE – the transnational strategy

The outcomes – especially the key transnational findings and lessons learned throughout the duration of the project – focused on:

- How to manage climate adaptation through spatial planning
- Developing and applying adaptation policies and strategies in spatial planning
- Methods so that the planning process can foster and promote adaptation
- Creating capacity and understanding with key stakeholders of how spatial planning mechanisms can be influenced to address climate adaptation
- Awareness of climate adaptation issues and creating 'agency for change'.

These lessons have been used to develop a common transnational strategy that outlines the ESPACE project recommendations for incorporating climate adaptation into spatial planning (ESPACE 2007b). The strategy includes a number of key recommendations for individuals and organisations involved in spatial planning processes, for example:

- Adopt two main approaches to deliver a climate-adaptive spatial planning system:
  - approaches based on change management – to create the governance and management required to address adaptation in spatial planning
  - Risk management processes – to integrate climate risks into the process of policy – making and identify the most appropriate adaptation policies and measures.
- Commit to taking account of changing climate risks over the whole of the decision 'lifetime' (e.g. the lifetime of a built development) rather than over the spatial plan period.
- Create a conscious balance of awareness, agency and association, to encourage, inform and reinforce long-lasting actions on climate change adaptation. Foster 'climate adaptation champions' who have a key role in creating and sustaining momentum.

4. Case study Fränkische Saale – Climate Change and Flood Risk

Background

The focus of the ESPACE (European Spatial Planning Adapting to Climate Events) project was on how adaptation to climate change can be incorporated into spatial planning systems with a special focus on water management issues.

In Germany, spatial planning is the responsibility of the German Länder (states). At the highest level, there is the State Development Programme (Landesentwicklungsprogramm LEP) followed by the regional plans for the 19 planning areas in Bavaria. The guiding principle of these plans is the development and preservation of equal and healthy living and working conditions in all parts of the state. Within this context, the principle of sustainability at the regional level is implemented by means of the regional plans (RP), which describes the fundamentals of the LEP in concrete terms and focuses on the respective region, taking into account the regional requirements, interests and particularities for the regions in Bavaria.

The two plans mentioned above are the basis for all sectoral planning (masterplans). In water resources management there are additional strategic programmes and river catchment plans. One important part of these plans is flood protection which is developed and executed in flood management plans (Hochwasseraktionspläne). This fulfils in advance the requirements of the new EU Floods Directive.

In general, the fundamental strategy for these plans in Bavaria is the Integrated Flood Risk Management approach (3 pillar strategy). The underlying principle is to take into account all feasible options concerning:

- natural retention within the river catchment and the valleys
- infrastructural flood protection (structural and non-structural measures)
- additional precautionary measures by the affected groups and people.

Fig. 1: The area of the case study Fränkische Saale

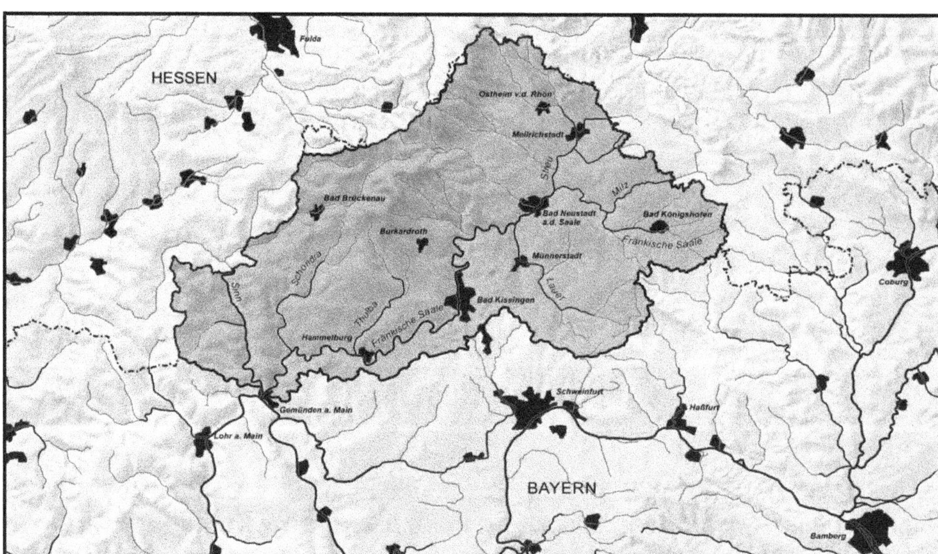

Within the policies mentioned above the aspects of climate change impacts have not been considered adequately until now. Against this background the Case Study Fränkische Saale was carried out within the European project ESPACE and concentrated on adaptation to the consequences of climate change in the field of flood protection. Guidance on how to take climate change into account in the flood protection planning process was based on an interdisciplinary and holistic assessment of the impacts of climate change on vulnerability to flooding.

The case study area is the catchment of the river Fränkische Saale (2700 km^2) within the northern part of the river Main catchment, as shown in Figure 1.

Aims and Steps

The main aims of the Case Study Fränkische Saale were to:

a) Quantify the possible effect of climate change on the water regime, especially on flood events, in the case study area
b) Assess the impacts of climate change on flood events beyond sectoral boundaries by looking at the physical, economic and social impact of (more) severe flooding
c) To develop general guidance on adaptation to climate change in the flood protection planning process based on the findings in a) and b).

Methods and Results

a) Quantifying the possible effect of climate change on the water regime of the case study area

A flowchart showing all necessary steps to estimate the quantitative impacts of climate change on the water regime and especially flood events is shown in Figure 2. The flowchart demonstrates the application of the "with-and without principle" which is the comparison of scenarios "with climate change" (marked in yellow) and "without climate change" (marked in blue). It is only through this comparison that a systematic assessment and quantification of the impact of climate change is possible.

As shown in the flow chart (figure 2), the assessment of possible impacts of climate change on the water regime requires an assessment of how crucial parameters like temperature and precipitation will change. This data is supplied by so called climate models. Global climate models simulate the climate at a low resolution so that the data they provide has to be transformed in to high resolution data using regional models (downscaling). As there is no optimum method for downscaling global climate model data and as different global and regional climate models produce different results, different downscaling methods applied to different GCMs (Global Circulation Models) based on various different emission scenarios produce a wide range of climatological data sets for a possible future climate. In climate research it is now becoming more widespread that a so called "ensembles" approach is used to produce a kind of "average scenario" using data from several different model runs.

Fig. 2: Methodological Procedure

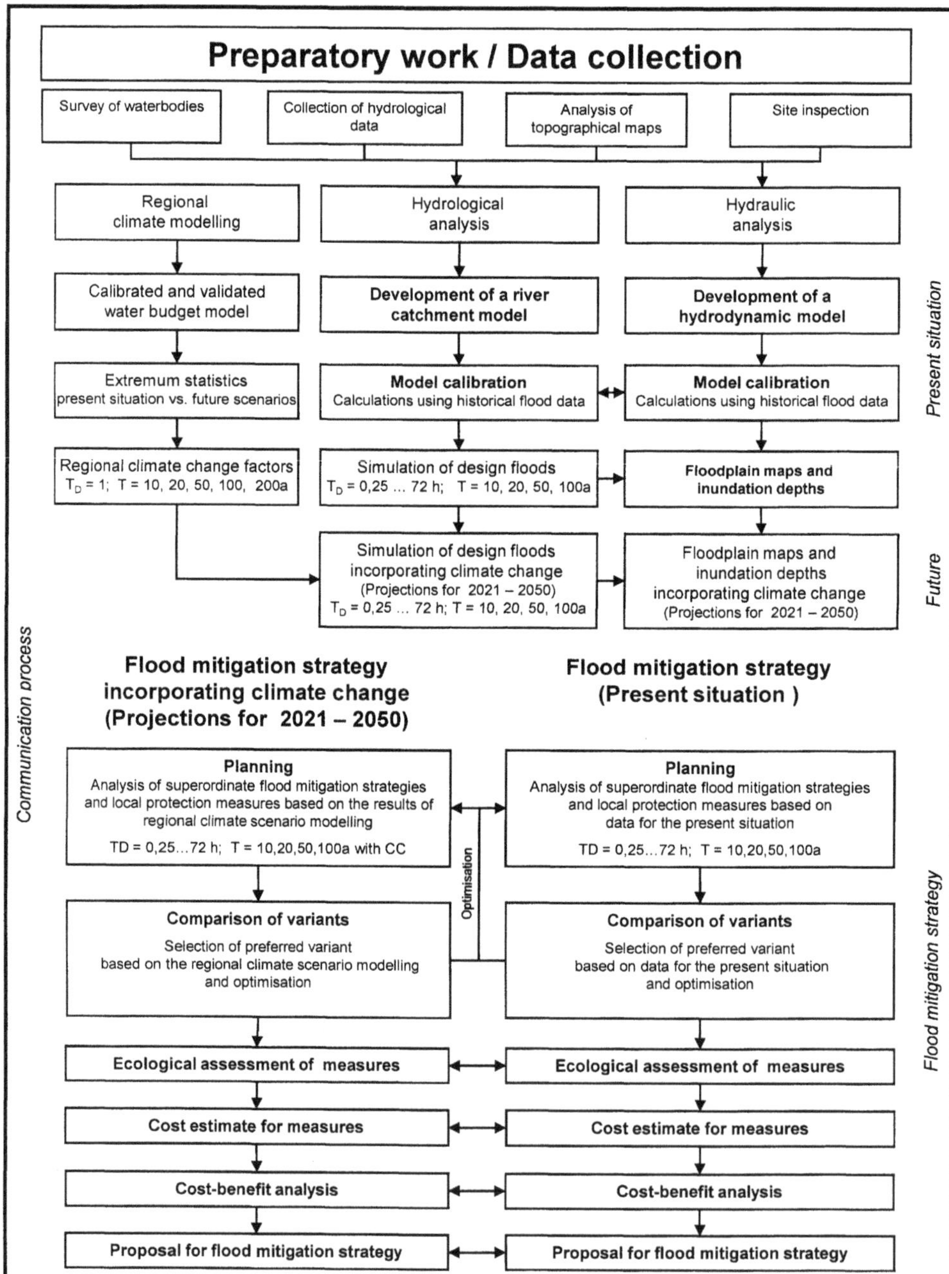

Alternatively, one model run based on one greenhouse gas emissions scenario can be chosen and its results defined as the projection on which adaptation measures can be based. This was the method applied in the case study Fränkische Saale the chosen models (global: ECHAM4/regional: Enke emissions scenario: IPCC B2) supplied the input data for the water balance modelling. The climate scenario was validated for the period 1971-2000 (control run) and supplied a projection for a 30 year period (2021-2050). The results with comparatively high spatial and temporal resolution include changes in temperature, precipitation, humidity, sunshine duration and wind speed. Figure 3 shows the results for the River Main catchment of which the case study area is a sub catchment (the river Fränkische Saale is a tributary to the river Main). The dark blue areas in Figure 3 (bottom map) show that the projection shows a significant increase in precipitation in winter months. These data sets serve as input for the water balance modelling taking climate change into account.

Fig. 3: The differences in precipitation for the river Main catchment, decade 1971-2000 vs. 2021-2050

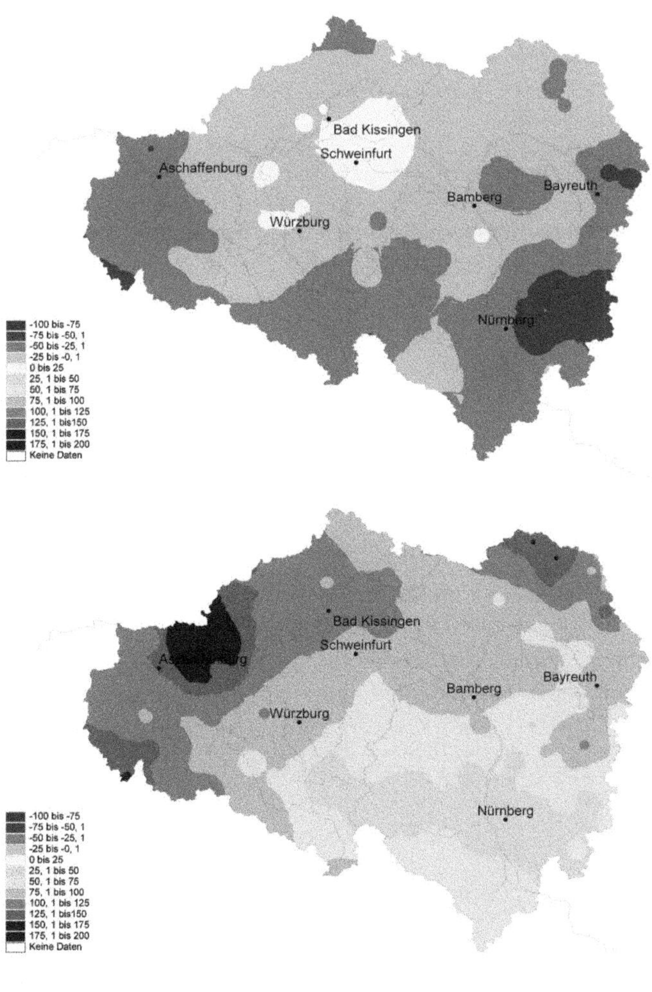

A calibrated water balance model for the case study area for the existing climate conditions with measured climate data is another key prerequisite for the assessment of climate change impacts on water balance components. Model requirements are high: forecast models need a physically based process description. Extreme value statistics are used to derive flood discharges based on the outputs of water balance modelling. Characteristic flood events with different probabilities of occurrence are the key information for spatial planning in catchment flood management. In the case study the water budget model ASGi was used.

Based on the regional climate modelling, the model system ASGi with its water balance model WASIM-ETH delivers a sufficient process description of the water balance in the case study area taking account of the changing climate conditions until 2050. In the first step, the water balance models for the catchment were established, calibrated and validated at a daily temporal scale (period 1961-2000) and a 1000m spatial scale.

b) Assessing the impacts of climate change on flood events beyond sectoral boundaries by looking at the physical, economic and social impact of (more) severe flooding

Based on the results of hydrological and hydraulic modelling "with climate change" and "without climate change" it is possible to derive the increase in vulnerability due to climate change. The aggravated flood conditions lead to a higher hydraulic load which in many cases leads to a rise in damage potential. An interdisciplinary and holistic approach to the assessment of the impacts of climate change on vulnerability to flooding was achieved by looking at the physical impact, the economic impact and the social impact of (more) severe flooding.

Physical Impact

Figure 4 for example shows an increase and a more pronounced winter peak in the mean monthly flood discharges until 2050 at the gauge "Bad Kissingen" which is close to the exit gauge of the case study area catchment.

Using hydrological statistics, characteristic flood events with and without climate change impacts have been calculated for the Main catchment with its subcatchments. Climate change factors for floods with different probabilities of occurrence (2 to 100 years) were derived and are shown in Figure 5.

Fig. 4: Impacts of climate change on the middle monthly flood discharge, decade 1971-2000 vs. 2021-2050

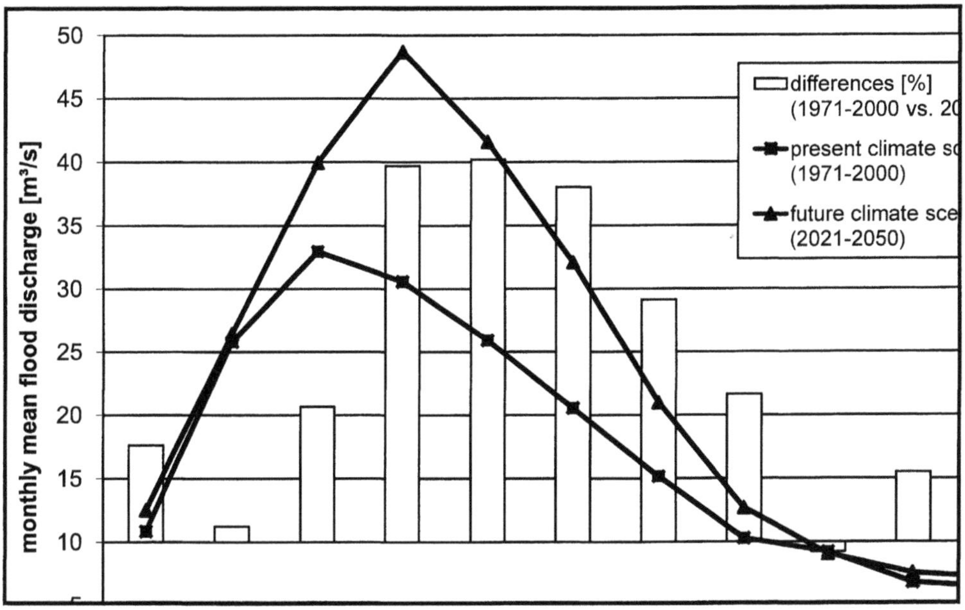

Fig. 5: Climate change factors for different probabilities of occurrence for the Fränkische Saale river

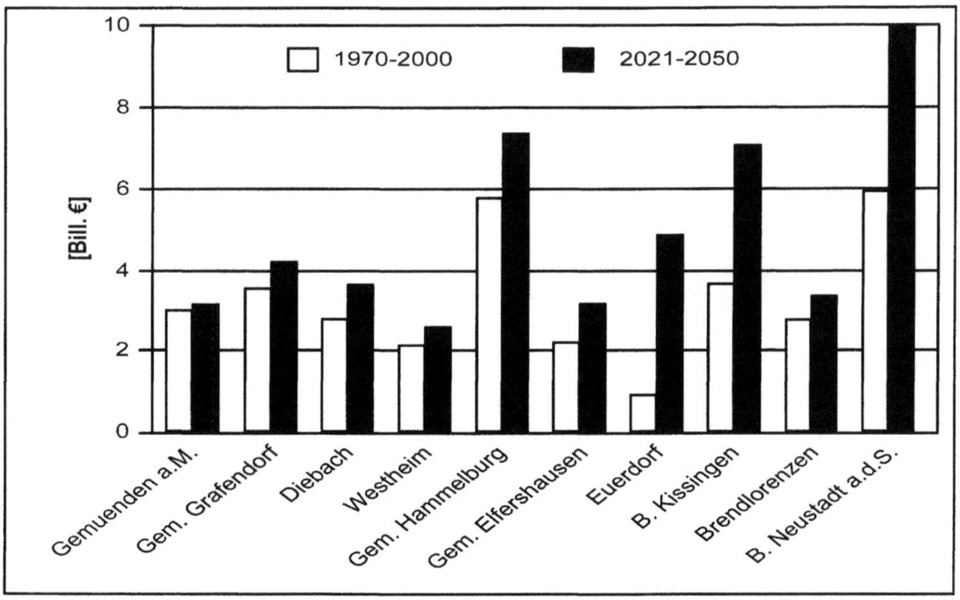

The results from the hydrological statistics are the input for hydrodynamic modelling of water levels, discharge and flow velocity. These results are then combined with digital topographic models of the catchment to simulate the physical impact of the calculated floods with different probabilities of occurrence (with and without climate change impact). These results are illustrated in flood maps, figure 6 shows one of a total of 274 flooding maps, which are also offered on our website www.klimaprojekt-espace.bayern.de.

Fig. 6: The darker the blue, the greater the probability of flooding

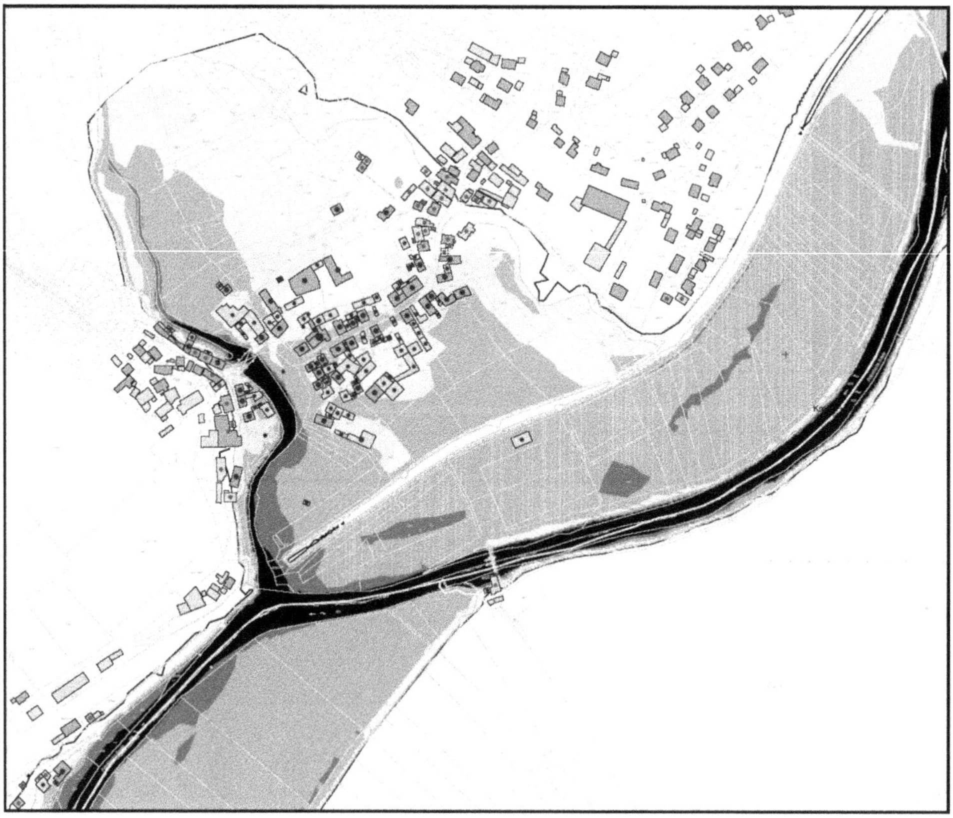

Economic Impact

Within the vulnerability analyses, damage potentials with and without climate change were calculated. A comparison of the physical impacts calculated with the hydrodynamic model with land use, infrastructure and property values gives insight into the level of vulnerability for a particular area.

This allows the determination of damage potentials with and without taking climate change into account. Figure 7 and Figure 8 show the increase of damage potentials for a 100-year flood due to climate change.

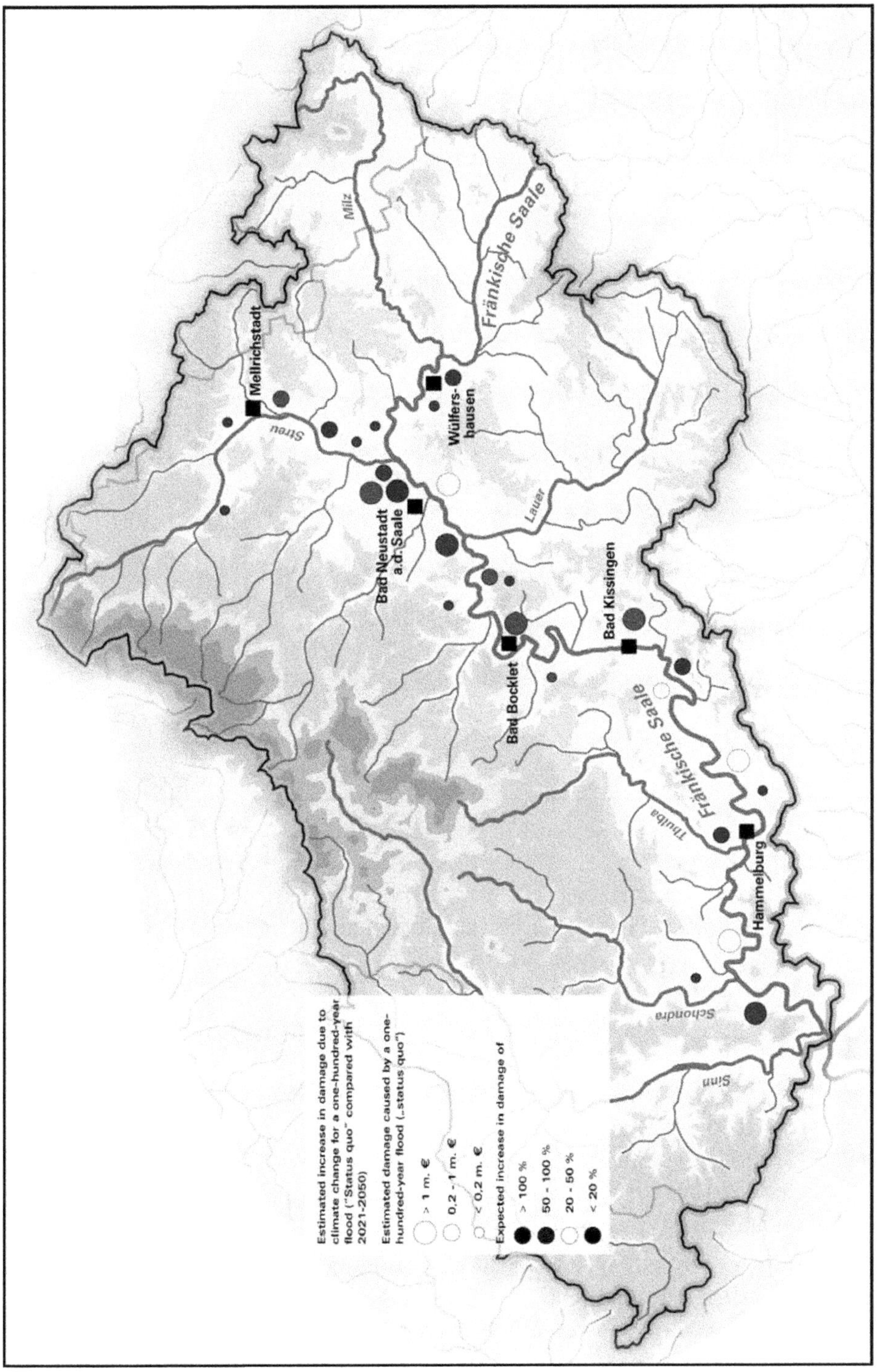

Fig. 7: The increase of damage potentials for a 100-year flood due to climate change – spatial distribution

Fig. 8: The increase of damage potentials for a 100-year flood due to climate change – statistic distribution

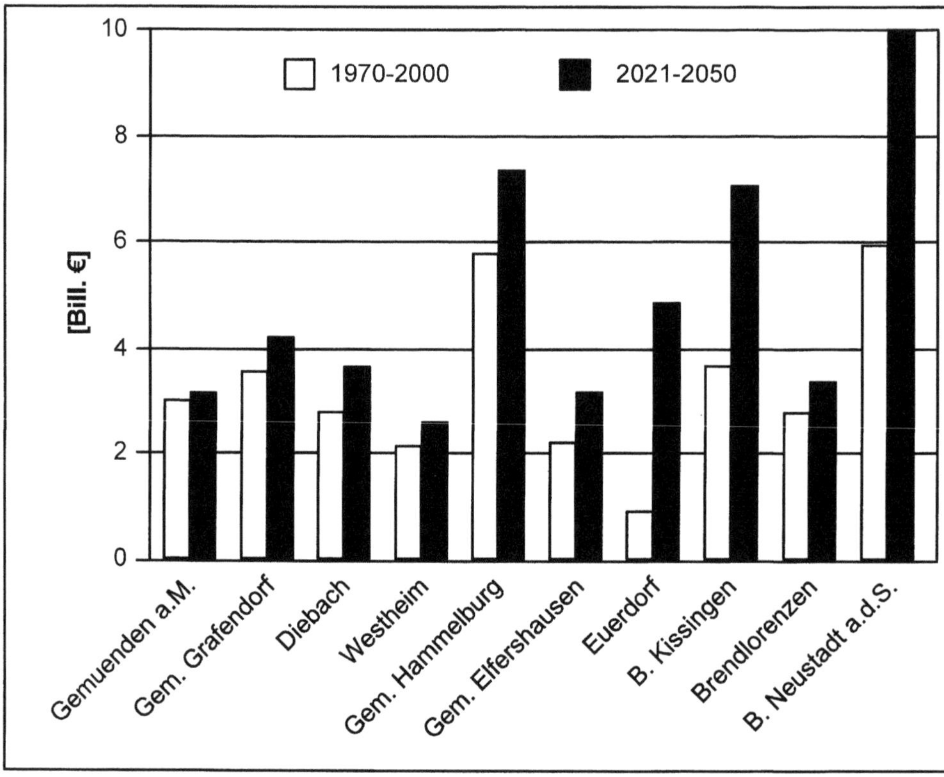

An assessment of the social effects of severe flooding was achieved through the detailed analysis of the social effect of past flood events. To collect the necessary data affected businesses, property owners and private households were questioned systematically and the results analysed statistically. One interesting result was that the economic vulnerability of private households was particularly high. Simple precautionary measures have been proven to reduce vulnerability significantly, and indeed those persons questioned who had experienced the lowest level economic damage during recent flood events were those who had taken simple precautionary measures based on experience.

This shows that knowledge about risk potential is in many cases the key to reducing actual loss or damage. The best way to achieve the necessary level of awareness among those at risk and among the general public decision makers is through a comprehensive communication strategy.

Against this background, the focus of the communication strategy within the Case Study was the communication process with stakeholders. Key elements of the communication tool are websites offering (detailed spatial) flood protection information and possible precautionary measures, meetings with the stakeholders and the distribution of information material (Leaflets etc.).

The key aims are: Raising Awareness, Decision Support System and self-precaution measures – following the convention that those who are well-informed can support measurements and protect themselves better. Future floods will occur more frequently and be more severe. Therefore, it makes sense to adapt to the new situation today and take precautions. This does not only apply to the government and local authorities, but to each and every individual.

Individuals can only make specific, purposeful provisions if they know what the potential flood risk is. To this end, the website "Climate Change and flood protection" provides the necessary information for the area of the Fränkische Saale: www.klimaprojekt-espace.bayern.de. It explains important facts about climate change and the flood situation and describes the planned protective measures. Maps showing flood areas and flood depths based on new climate change related calculations help to make better flood risk assessments possible.

Fig. 9: Rasing awareness in the next generation: Opening of a climate change exhibition for students

Fig. 10: Stakeholder dialog in Bad Kissingen

c) *To developing general guidance on adaptation to climate change in the flood protection planning process based on the findings in a) and b)*

Besides the quantification of possible climate change impacts and raising awareness to encourage adaptation another main aim was to look at adaptation to climate change at the operational level by developing an approach for planning and decision making for flood related issues that takes climate change into account. Communication with all the parties involved has to accompany all phases of the flood protection planning process to ensure decision-making and implementation of an optimal mix of measures concerning mitigation and adaptation to climate change. These general principles based on the results of the first two steps of the Case Study Fränkische Saale are shown in the following diagram, the "climate-proof" planning circle which contains the important steps towards optimised adaptation to climate change in the frame of the above mentioned interdisciplinary approach.

The steps towards optimised adaptation showed in figure 11 have to be embedded in a comprehensive all-encompassing communication strategy using adequate tools which were developed and tested in the frame of the case study "Fränkische Saale".

Looking at the past, public authorities put most emphasis on technical solutions, neglecting communication with the affected people. This led to underestimation of residual risks and a lack of knowledge about simple precautionary measures. Lessons learnt from the past, communication has to take place in every phase of the planning process and should be implemented in the flood protection plans.

Fig. 11: The "climate-proof planning circle" for flood protection measures taking climate change into account

The main overall objectives for the communication strategy are: raising awareness, change of behaviour, coordinated planning, implementation of optimal measures and cooperation. Concerning recommendations for such a comprehensive all-encompassing strategy we established two important points for effective communication: First, the dialogue is only successful when it is possible to induce strong signal effects which highly exceed the normal information "noise". Secondly, the dialogue must induce a "stereo-effect" by communicating the facts in different ways and modified forms. The communication process with all parties involved should accompany every phase of the process and lead to an optimal mix of measures concerning mitigation and adaptation to climate change.

Concerning the application of the policies it is also important to communicate the limits of protection by technical infrastructure also in the context of climate change and to encourage a paradigm shift to an increased self-dependent precaution. This involves raising awareness which in many cases automatically leads to a change in behaviour.

5. Summary and outlook

The efficiency and reliability of the management of water resources systems taking climate change into account depend, to a large extent, on the quality of the hydrological information on which the planning and management of such systems is based. In this context, one of the main tasks in hydrology is to investigate the impacts of climate change on the spatial and temporal distribution of the surface water at all scales and the resulting economic impacts as an important basis of decision making in the field of flood protection. This is in order to provide data for a sustainable management of water resources, as well as for the planning, construction and operation of hydraulic structures and systems, taking into account climate change.

Against this background, the main aims of the Case Study Fränkische Saale in the frame of the EU-project ESPACE were first to quantify the possible effect of climate change on the water regime, especially on flood events. Second, the impacts of climate change on flood events beyond sectoral boundaries by looking at the physical, economic and social impact of (more) severe flooding were assessed. The corresponding investigations are based on regional climate modelling, water balance modelling, extreme value statistics, hydraulic calculation, estimation of climatic factors, cost-benefit calculation and analysis of alternative measurements. In addition, communication instruments were developed and used in stakeholder dialogues to raise awareness and change behaviour.

Based on the analysis of these results, recommendations for best practice in flood management activities taking climate change into account are possible. The quantification of the impacts of climate change in a river catchment is the basis for the determination of the most efficient and at the same time sustainable adaptation measures.

Many floods in recent years have demonstrated that technical infrastructure has its limitations. As flood risk management focused on natural retention effects plays a subordinate role in more extreme events, the optimisation of the measures to be deployed concentrates on the substitution problem between built infrastructure and further self-provision, i.e. up to what degree of protection can infrastructure measures be justified? As this essentially depends on the respective circumstances, situation-specific individual decisions are required. The economic efficiency facts play a key role in this. However, cost-benefit analysis is sometimes far away from good practice. Besides quality, the planning content should be based on the information required in the decision making and communication process in a transparent and clear way.

In the existing protection schemes the main emphasis is on the reliability of the constructions and their emergency facilities. The need for adjustment measures results from the requirements of recognised rules of sound engineering practice. A whole range of different situations can occur. The most favourable circumstances exist when the scheme can be retrofitted in an economically efficient way, both with respect to the safety standards to be guaranteed and with respect to the degree of protection that follows the principle of proportionality of means. The case in which the safety adjustment occurs at the expense of the degree of protection is substantially more unfavourable. The worst case is given, when, for reasons of efficiency, it is not possible to compensate for the increase in load due to climate change.

On the basis of practical case studies it has been proven that a consideration of the effects of climate change for technical flood protection measures would only have led to a relatively moderate cost increase in most cases, if this load case had already been taken into consideration during planning and if at least appropriate precautions had been taken into account during construction for later adaptation. Later adaptations, by contrast, mostly involve very high costs.

For this reason, when planning new flood protection measures in the future, the load case climate change should be examined. This should include the demonstration of the consequences of the load case for the measures being planned and which additional costs are expected as a result. Thus decisions have to be made, on the basis of available findings, to what extent the adaptation necessary for future climate change should already be taken into consideration in current work. The possibilities for additional later adaptation should also be taken into account.

The following examples should explain how existing and future civil works as well as plans can be implemented in case of increased design values, i.e. taking into consideration the impacts of climate change.

- Planning of a flood protection dam: The dam is built according to current guidelines. However, additional measures are taken, which would not be required according to previous planning regulations. For example, an additional strip of ground on the valley side is reserved, enabling a future increase of the dam, if necessary, without additional problems.
- New constructions where a future alteration or adaptation is not possible or is only very expensive (e.g. bridges), should immediately be planned to take account of future increased calculation parameters relating to the level of water, if necessary.
- New constructions where a future adaptation is less difficult (e.g. river walls) should in view of the construction features (e.g. statics) be planned to a higher specification than currently required so that any further adaptation necessary (e.g. increase in height using stationary or mobile elements) would be possible without high costs.

As the principle of proportionality of means ranks higher and higher there will be a strong need to foster further self-provision in future in addition to a foresighted technical adaptation as mentioned above exemplary. This goes along with the communication of the residual risk which has been badly neglected in the past.

The case study "Fränkische Saale" is an ideal example of how to deal with the challenges of climate change and how to develop comprehensive options and strategies in a transnational context. It was greatly enriched by the transnational cooperation within the ESPACE project.

References

ESPACE (2006d), ESPACE Decision Making Framework and Tools – Phase 3 International Comparison. Final Report. www.espace-project.org.

– (2006f), ESPACE Project: Analysis of Partner actions. Final Report. October 2006. www.espace-project.org.

– (2007b), ESPACE Project: Final Common Transnational Strategy. April 2007.

Belau, M., Kleinhans, A., Weber, H. and Rosentritt, L. (2007), Berücksichtigung des Klimawandels im Flusseinzugsgebietsmanagement: Klimarisiken in der Raumplanung, Ökologisches Wirtschaften, März 2007.

IPCC (2001), Climate change 2001, Summary for Policymakers. Intergovernmental Panel on Climate Change, Cambridge University Press, Cambridge.

Landesamt für Umwelt, Referat 81 (2007), Final Outcomes of the case.study Fränkische Saale: Policies, guidance, tools and case-study, Referat 81, LfU, not published.

– (2007), Mainbericht I: Klimawandel und Auswirkungen auf den Wasserhaushalt im Maingebiet, Referat 81, LfU, not published.

– (2007), Mainbericht II: Klimawandel und Auswirkungen an der Fränkischen Saale, Referat 81, LfU, not published.

Stern, N. (2006), The Economics of Climate Change – The Stern Review, Cambridge University Cabinet Office – HM Treasury, Cambridge University Press.

Chapter 15

The Impact of Municipal Solid Waste Management on Greenhouse Gas Emissions in Estonia

Harri Moora, Viktoria Voronova, Alvina Reihan

Abstract

Waste management has an influence on the greenhouse gas (GHG) formation. The emissions of greenhouse gases vary between the EU countries depending on waste treatment practices and other regional factors such us composition of waste. The aim of this paper was to examine, from a life-cycle perspective, Municipal Solid Waste (MSW) management in the context of greenhouse gas formation and to evaluate the possible reduction of climate change potential of alternative waste management options in Estonia. Air temperature as the main climatic parameter is analysed in the context of climate change. The paper summarises the results of a case study in Estonia, assessing the climate change impact by 2020 in terms of net greenhouse gas emissions from two possible management scenarios. The paper also provides updated information on the composition and generation of MSW in Estonia. As a result it can be concluded that better management of municipal waste and diversion of municipal waste away from landfills could significantly reduce the emissions of GHG and, if high rates of recycling and incineration with energy recovery are attained, the net greenhouse gas emissions may even become negative.

1. Introduction

It is emphasised worldwide that the main climatic parameters such as air temperature, precipitation, etc. have changed and the changes are related to the increase of greenhouse gases (IPCC 2007). Estimation of climate change impacts on different ecosystems is based on different rates of GHG emissions. Therefore, it is necessary to minimise the GHG emissions resulting from waste management, in order to enhance the use of low carbon economy and adaptation measures to ensure ecosystem sustainability. GHG emissions represent a considerable environmental effect from the management of municipal solid waste. The GHG emissions from the waste sector depend on a number of factors including waste generation, waste composition and used waste management practices.

The generation of MSW in the European Union (EU) has increased steadily. The MSW generation and treatment options vary significantly in the old Member

States (EU-15) and the new Member States (EU-12). The EU-15 landfilled less than 60% of the municipal waste in 2005, while the majority of the EU-12 landfilled most of the MSW (more than 80%) (ETC/RWM 2008). The new EU Member States undergo currently a rapid economic development, resulting in a significant increase of waste quantities, while their waste management systems still require much effort to be adjusted to the European state-of-the-art. Therefore, in the context of climate change the diversion of municipal waste away from landfills and choosing optimal waste recovery practices is especially important in the new Member States including Estonia.

Through iterative examination of various waste management technologies and treatment alternatives, life cycle based methodologies can help to identify optimal environmental solutions for managing the waste. Life-cycle information can also help to identify the benefits and trade-offs of different waste management options in terms of direct GHG emissions (from landfills, incineration plants, recycling and collection of the waste) and indirect GHG emissions that are associated with the extraction and processing of primary resources or fossil fuels versus those associated with recycling or incineration operations. In the new Member States the life-cycle assessment (LCA) models are not widely used for waste management planning and calculation of GHG emissions. The major limitation of using LCA in waste management planning in these countries is the lack of relevant data and knowledge of the analysed systems. Also it is difficult to compare specific local waste management information and data with the data used by LCA models developed for other countries.

The main aim of this paper is to examine the climate change impact in Estonia in terms of GHG emissions from MSW management between years 2000 and 2020. Two most feasible waste management options for Estonia, material recycling with biological recycling in terms of composting and material recycling with intensive incineration, were compared in terms of their possible contribution to climate change. The paper also provides information on MSW quantities and composition in Estonia.

Most of the earlier studies have focused on the calculation of direct GHG releases associated with municipal waste management. In this paper calculations are based on life-cycle information and the calculated emission levels include not only direct emissions from MSW management, but also the 'avoided' emissions from material recycling and energy recovery. The study is based on the LCA model for waste management planning – WAMPS (Moora et al. 2006).

As such the results of the study also provide additional country specific background information and data for an ongoing discussion about the implementation of waste hierarchy and provisions of the EU Waste Framework Directive.

2. Climate change aspects in Estonia

Investigations of climate features in Estonia have a long tradition (Kant 1927; Kirde 1940). While earlier investigations were focused mainly on climate impacts on agricultural production, the present climatological studies are focused on the environmental and human impact aspects. Recent studies show different variations in climate parameters, such as increase in air temperature and precipitation in autumn and winter, decrease of snow and ice duration (Jaagus 1998) and a decrease in reflected radiation in spring (Russak 1998), increase of frequency of extreme events such as storms, inundations, droughts, etc. The most intensive warming (figure 1) took place in the winter and spring seasons during the last forty years with significant growth in the continental part of Estonia (Reihan 2008). All these changes are in good agreement with those reported by the Intergovernmental Panel on Climate Change (IPCC 2007) and are of a great importance for the development of any ecological management plans including waste management.

Fig. 1: Variation of the long-term air temperature with 5 years moving average in Pärnu MS (South-West Estonia); the straight line indicates the long-term mean value

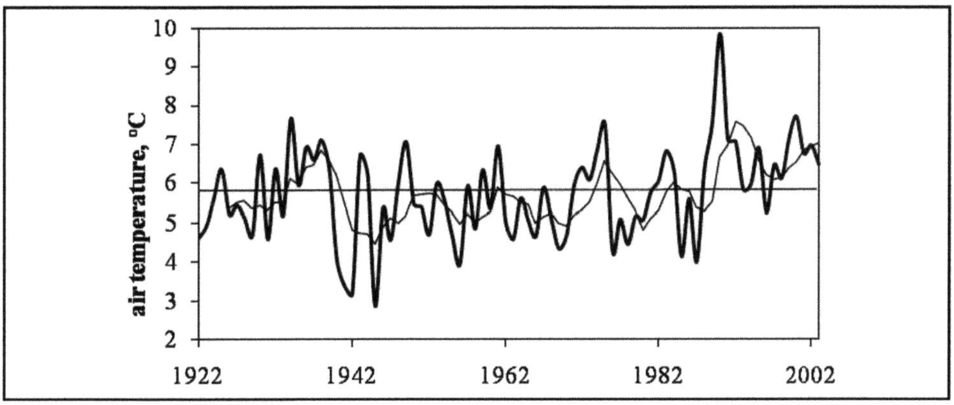

Estonia has signed and ratified the Kyoto Protocol, according to which emissions of GHG must be reduced by 8% in 2008-2012, compared to the baseline year 1990. All EU Member States report regularly their direct and projected GHG emissions in order to apply policies and measures to reach this target. The methodology used for the estimation of GHG emissions follows so called IPCC guidelines (IPPC 2006) produced by international expert groups for the IPCC and are followed also by Estonia to calculate the national yearly GHG emissions. These calculations focus mainly on direct GHG emissions of different sectors. According to the latest GHG emission calculations, which were made for the 4th National Communication under the United Nations Framework Convention on Climate Change (UNFCCC) in 2006, the total emission level in 2005 was 20.7 Mt of CO_2 equivalents. The main contributors to GHG emissions in Estonia are energy supply and use together with transport (88%) (figure 2). The electricity production in Estonia is mainly based on an Estonian specific fossil fuel – oil shale. The production of electricity from oil shale entails higher emissions of CO_2 than most of the other fossil fuels used in EU (OSELCA 2006). The share of waste management (mainly CH_4 emissions from landfills and waste water sludge treatment) is only 3% of the total GHG emissions. However, waste management is an important source of GHG reductions. Modern waste management is very closely linked to energy production, therefore, when planning waste management systems the context of climate change should be taken into account, since it contributes to the meeting of the Kyoto targets.

Fig. 2: Greenhouse gas emission by sectors in Estonia

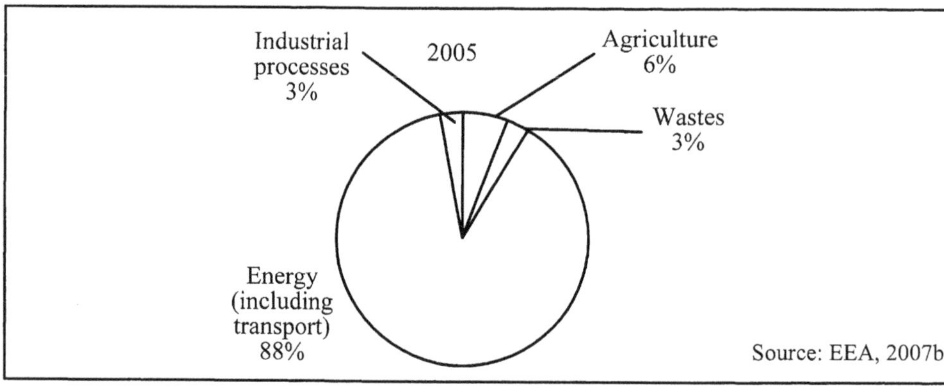

3. Greenhouse gas emissions from waste management

The GHG emissions from the MSW management are sum of the direct emissions (from landfills, incineration plants, recycling operations and collection of the waste) and indirect emissions. Indirect emissions arise from the energy and secondary products produced when incinerating and recycling waste replace energy production from fossil fuels and the use of raw materials for plastic, paper, metals, etc

Direct emissions from the waste management sector in the EU-15 contributed by 2,6% of the total GHG emissions in 2005 (EEA 2007a). The key sources of waste-related greenhouse gas emissions are illustrated in Figure 3.

As shown in Figure 3, the main sources of direct GHG emissions from waste management are *landfills*. It could be estimated that GHG emissions from landfills are larger in new Member States including Estonia due to a bigger share of landfilling in these countries. The landfilling of biodegradable waste results in the formation of landfill gas which contains mainly methane (CH_4). In addition to CH_4, landfills also produce carbon dioxide (CO_2), non-methane volatile organic compounds (NMVOC) and a small amount of nitrous oxide N_2O, nitrogen oxide NO_x and carbon monoxide CO (IPPC 2006). The origin of CH_4 in landfill gas is the degradation of organic waste under anaerobic conditions. Compared with CO_2, the global warming potential of CH_4 is 21 times higher over a time horizon of 100 years. In the modern landfills the landfill gas is partly (usually 10-50%) collected and either disposed by flaring or used as a fuel for energy production. Contrary to GHG emissions from other waste management practices (e.g incineration or recycling), landfill GHG emissions are characterised by the large time lag of emissions (Sundqvist 1999).

Fig. 3: Main sources of GHG emissions from the waste management sector in EU-15, 2005

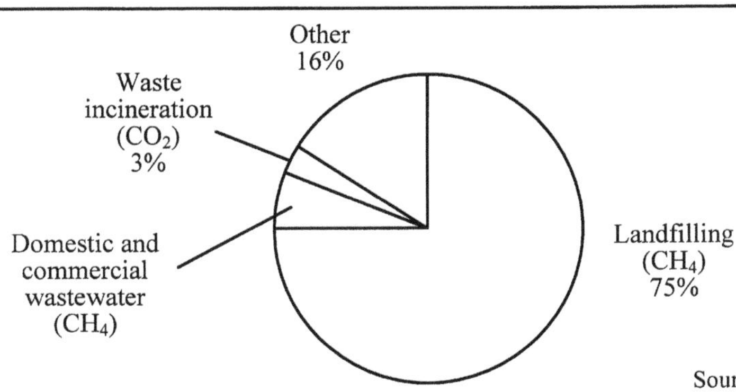

Source: EEA, 2007a

The most widely practised alternative to landfilling is *mass-burn incineration* where MSW is burnt with little or no pre-treatment. The modern MSW incinerators are required to recover energy released by the combustion process. Incineration is a source of GHG emissions like other types of combustion process. GHG emissions are estimated by the carbon content of the incinerated waste material. The carbon content contributes mainly to CO_2 emissions and less to CO, CH_4 and NMVOC emissions. Exhaust gas cleaning or incineration technology does not influence CO_2 emissions. Emissions of CO_2 from incineration of biological waste material do not contribute to net GHG emissions and should therefore not be taken into account. Calculation of net CO_2 emissions from waste incineration is based on the fossil carbon content of the MSW. The net climate change impact of incineration depends on how much fossil carbon CO_2 is released – both at the incinerator itself (direct emissions) and in savings of fossil fuel from marginal energy sources displaced by incineration.

Recycling diverts components of the waste stream for reusing the materials. If the GHG emissions resulting from the separating and processing of the recycled material into new products are less than those generated while manufacturing the products from primary material, net savings of GHG emissions results.

Composting is an aerobic process and a large fraction of the degradable organic carbon (DOC) in the waste material is converted into CO_2. CH_4 is formed in anaerobic sections of the compost, but it is oxidised to a large extent in the aerobic sections of the compost. The estimated CH_4 released into the atmosphere ranges from less than 1 percent to a few per cent of the initial carbon content in the material (Beck-Friis 2001; Detzel et al. 2003; Arnold 2005). Composting can also produce emissions of N_2O. The range of the estimated emissions varies from less than 0.5 percent to 5 percent of the initial nitrogen content of the material (Petersen et al. 1998; Hellebrand 1998; Vesterinen 1996; Beck-Friis 2001; Detzel et al. 2003). Poorly working composts are likely to produce more both of CH_4 and N_2O (e.g., Vesterinen 1996). This is the reason why the so-called home composting could have a relatively high climate change impact in terms of CH_4 emissions.

There are several other waste management options available. One of the most common practices is *mechanical-biological treatment* (MBT), which is a combination of mechanical and biological steps to reduce the amount and biological activity of the processed MSW. Pre-treatment of MBT prior to landfilling significantly reduces CH_4 emissions from the landfilled waste, compared with untreated MSW.

A common link between different waste management practices is the need for *collection and transport* from the source of the waste to the waste treatment/disposal facilities. It all has GHG impacts, mostly through the use of fossil fuels and associated emissions of CO_2. N_2O is also emitted from vehicle engines, but this has a minor impact (EC 2001).

4. Methodology

To calculate the GHG emissions from MSW management the important input data in terms of waste composition and generation in Estonia were studied. Also a review of possible future waste management technologies and practices was conducted. Based on collected information two possible waste management scenarios by 2020 were developed. As a basis for determination of GHG emissions from municipal waste management the LCA model for waste management planning WAMPS was applied. WAMPS model is intended to be applied during the waste management planning process to find optimal solutions and alternatives for waste management systems (Moora et al. 2006). WAMPS presents the environmental and economic consequences of different waste management scenarios in a life cycle perspective. WAMPS was developed by the Swedish Environmental Research Institute and is based on a more in-depth LCA model ORWARE (Sundqvist et al. 2002; Björklund 2000; Eriksson 2000).

WAMPS compares a waste management system with a background system. The waste management system can produce different products depending on the choice of treatment and recycling: heat, steam, electricity, vehicle fuel (biogas), compost, paper, plastic, metals, etc. In background system similar products are produced from virgin origin. When a product is produced from waste, it substitutes a product in the background system. Each waste product has an alternative in the background system with a virgin raw material source and a production process that is included in the model. In WAMPS different recovery options are compared with the background system and potentially 'saved emissions' are assessed. The background system consists of heat production (alternative to waste incineration and combustion of biogas and landfill gas), electricity production (alternative to waste incineration and combustion of biogas and landfill gas), vehicle fuel production (alternative to biogas), fertilizer production (alternative to compost and digestate) and production of materials (plastic, newsprint, paper packages, glass, steel, aluminium, etc). All these products from the background system can also be produced by different waste recovery methods.

The net emissions from the studied system are calculated according to:

$$E_{net} = E_{waste} - E_{Background}$$

E_{net}: Net emission (tonnes/year or kg/year)

E_{waste}: Emission from a waste process that produces a certain amount of product (tonnes/year or kg/year).

$E_{Background}$: Emission from the same amount of alternative virgin production in the background system (tonnes/year or kg/year).

This calculation can give negative net emissions. This means that the recycling method will give lower emissions or energy consumption than the corresponding virgin production. The global warming impact is calculated as CO_2-eqivivalent emissions. The basic functional unit in WAMPS is the waste generated within a specific region.

5. Case study

5.1 Waste composition and generation

The contribution made by the waste management sector to climate change is primarily determined by the volume and composition of municipal solid waste as well as the waste management options chosen.

The amount of MSW generated per person and its composition varies significantly in different Member States. Figures from different data sources cover different time periods and geographical locations and are rarely comparable. Most of the waste studies rely on official databases such us Eurostat. However, in most cases, especially in the EU-12, these data are notoriously unreliable. Since the lack of information on actual waste composition is one of the main barriers for waste management planning in Estonia, the composition of mixed municipal waste was explored by a countrywide sorting analysis of MSW in 2007/2008. The waste analysis was carried out as a part of the current study. Based on the results of the sorting analysis and the corrected statistical data, the most recent and accurate composition of MSW in Estonia was compiled (figure 4). The biodegradable fraction (organic waste, paper and cardboard, wood and textiles), which is the main source of GHG emissions, makes up a considerable share of municipal waste in Estonia (63%). The packaging waste amounts to 27% and the share of combustible waste to 80% of total MSW.

Fig. 4: The average composition of MSW for Estonia

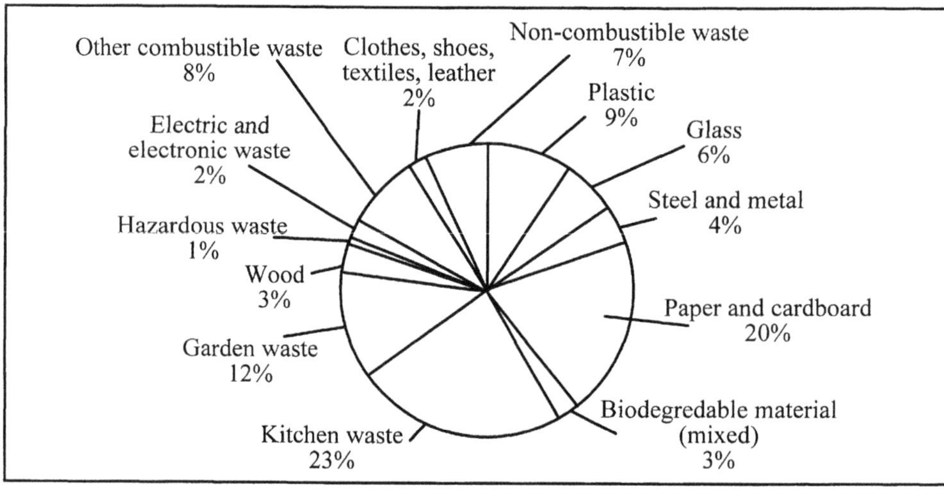

The generation of MSW relates mainly to a number of economic activities and the size of the population. In general there is a strong link between GDP and waste generation. The quantity of municipal waste in Estonia has rapidly risen in line with the economic growth and growing consumption. According to specified statistical data, approximately 435000 tonnes of municipal waste (302 kg per person) were generated in 2000. In recent years the amount of municipal waste has been growing by 6% per year and reached 542000 tonnes in 2007 (402 kg per person). MSW volumes are expected to grow even more. Since the number of population is expected to remain roughly the same, economic growth or specifically private final consumption will be the key driving force behind the growing waste volumes in Estonia. The forecast of the municipal waste generation in this study is based on the future estimations of final consumption by households expressed in Purchasing Power Standard (PPS). The growth rate of municipal waste generation is expected to decrease in the coming years (average growth 3% per year) due to the slowdown of economic growth. It is expected that waste generation will stabilise after 2015 (average growth 1% per year). During the period 2000 to 2020 the generation of municipal waste is projected to increase by approximately 58% (figure 5). In 2020, the generation of municipal waste per person is estimated to be 509 kg (690000 tonnes). In general, this is in line with the projections made by the EEA

– in the new Member States, the generation of MSW is projected to increase by approximately 50% from 2005 to 2020 (ETC/RWM 2007).

Fig. 5: Projected growth in MSW generation between years 2000 and 2020

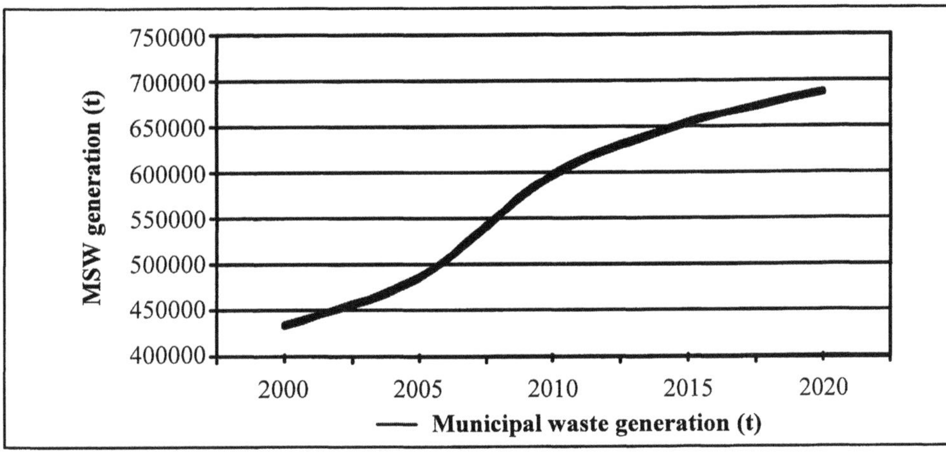

5.2 Scenarios

For the GHG emission calculation the waste management situation in 2000 was taken as a starting point or a base scenario. Two waste management scenarios were developed to analyse possible future alternative waste treatment options and their climate protection potential by 2020. Estonia, similarly to other new EU Member States, has to comply with the EU legal requirements and recovery targets for waste management. Since the pros and cons of waste incineration as a possible MSW management option were recently discussed in Estonia, the incineration-based scenario was compared with the scenario where legal targets are achieved with intensive biological recycling (composting) (see also table 1). Both alternative future scenarios are in compliance with the requirements and recycling targets of the following legal acts:

1. European Packaging Directive 2004/12/EC – minimum packaging recovery target 60% for packaging waste.
2. Landfill Directive 1999/31/EC – target amount of biodegradable municipal waste going to landfills must be reduced to 35% by 2020 (baseline 1995).

Base scenario (scenario 0)

In 2000, waste management in Estonia primarily involved landfilling of MSW (92% of the total MSW). There was no landfill gas collection in landfills at that time. Only a small amount of packaging waste (mainly PET-bottles and cardboard) was collected separately and sent to recycling. There was no centralised collection system for biodegradable waste. Approximately 17000 tonnes of biodegradable waste (mainly garden waste) were composted in the households (4% of the total MSW). It is assumed that the share of home composting will remain the same till 2020.

Material recycling with intensive incineration (scenario I)

Scenario I is a projection for 2020, where the dominant option of MSW management in Estonia is incineration. 45% of MSW is incinerated in the mass-burn incineration plant. This assumption is based on the plans to build an incinerator close to the capital of Estonia, Tallinn. The incineration plant is expected to start its operation in 2012. The incinerator complies with all the EU requirements and it is assumed that the gross efficiency of energy recovery from the incineration process will be relatively high (80%). A large amount of the heat could be utilised since Tallinn has large dwelling areas with district heating system. In this scenario increased amounts of recyclable materials (mainly packaging, paper, cardboard and metals) are separately collected and recycled to fulfil the recycling targets of the EU Packaging Directive. The recycling of material is expected to be 30%. As incineration is already contributing to the reduction of biodegradable waste, the share of biological recycling is not expected to exceed 15% of the total MSW. The centrally collected kitchen waste is composted using static composting method with forced aeration. Collected garden waste is composted in open windrows. Intensive material recycling and incineration leads to a relatively small amount of rest waste, which is landfilled (13% of the total MSW).

Material recycling with biological recycling in terms of composting (scenario II)

Scenario II is a projection for 2020, where the legal targets are archived by material and biological recycling. Also in this scenario material recycling is expected to amount to up to 30% of the total MSW. The Landfill Directive requirement to divert biodegradable waste away from landfilling, is fulfilled by increasing composting to 37% of the total MSW. An increased amount of wet biodegradable waste is composted using centralised reactor-composting method (without gas collection and energy recovery). It is assumed that the remaining waste will be deposited in a landfill.

The predictions about the future MSW generation presented in the earlier chapter were considered when developing the alternative future scenarios. It is assumed that waste composition remains the same during the studied period.

For both future scenarios it is also assumed that all landfills will be equipped with a landfill gas collection system at the latest by 2010 and the landfill gas recovery rate will increase up to 50% by 2020. Before 2010 the collected gas is flared and after 2010 it is used for electricity and heat production, which is substituting oil shale based electricity and natural gas based heat used for district heating. The energy produced in waste incineration is also substituting the electricity produced from oil shale and heat from natural gas.

Tab. 1: Municipal Solid Waste management scenarios

Scenario	Material recycling	Biological recycling (composting)	Incineration	Rest waste (landfilling)
2000 Base scenario	4%	4%	0	92%
2020 Scenario I Recycl + Incin	27%	15%	45%	13%
2020 Scenario II Recycl + Comp	27%	37%*	0	36%

*Landfill Directive: compliance with the target for 2020

6. Results

The results of the scenario analysis in terms of net GHG emissions are shown in Figures 6 and 7. The diagrams show the net GHG emissions from the waste management system minus saved emissions in the background system. When the emissions from the studied waste management scenario or waste management practice are less than the saved emissions in the background system then net result is negative.

Fig. 6: Emissions of net GHG from studied waste management scenarios, 2000-2020 (tonnes CO_2-equivalents)

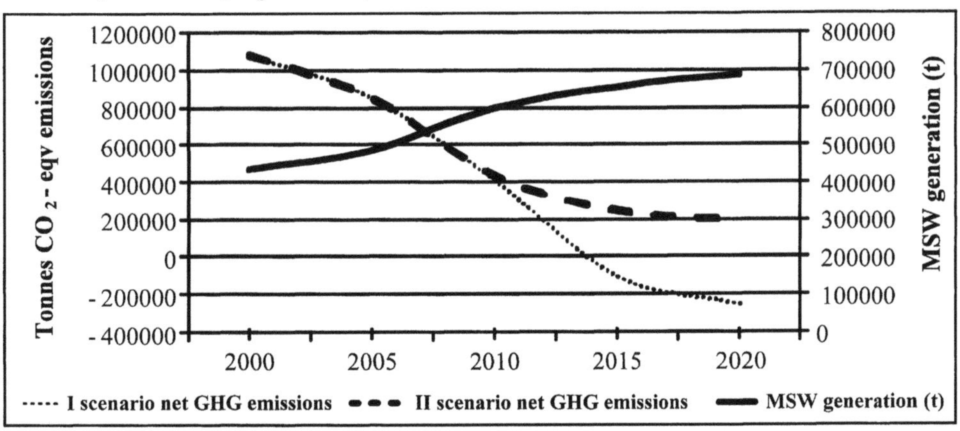

The results of the case study indicate that net GHG emissions from the management of municipal waste in Estonia are projected to decline significantly by 2020 from a peak of around 1.1 million tonnes CO_2-equivivalents per year in 2000, largely because of increased recovery of MSW and the diversion of waste away from landfills.

Fig. 7: Emissions of GHG from studied waste management practices and scenarios (tonnes CO_2-equivalents)

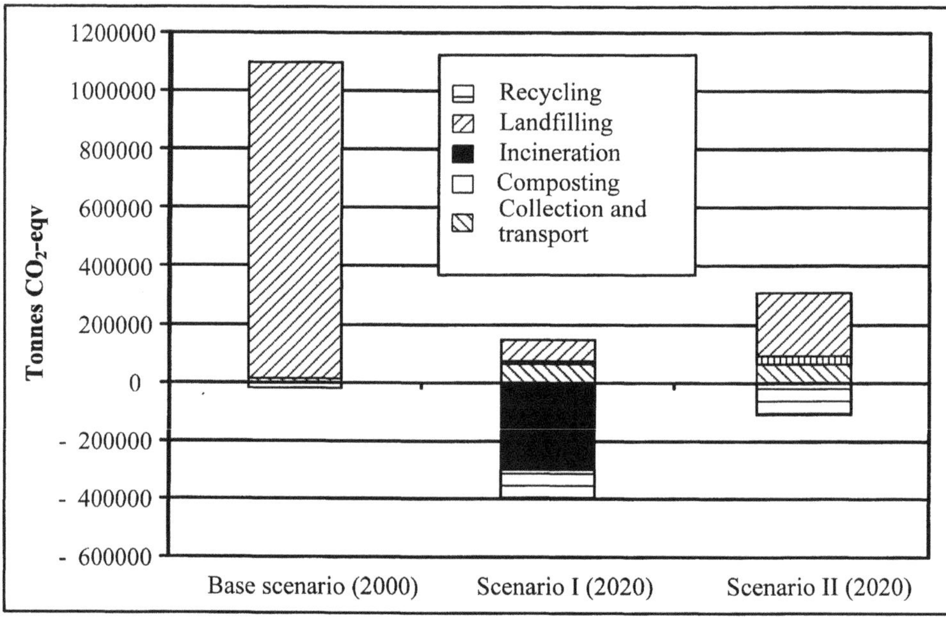

When comparing the two studied scenarios we can see that the incineration scenario (scenario I) has a higher climate protection potential than the alternative composting scenario (scenario II). In scenario I where high rates of recycling and incineration with energy recovery are attained, the net emissions of CO_2-equivivalents are even negative. The reason for the negative net GHG emissions is a relatively low amount of waste sent to landfills as well as a high share of material recycling (avoided primary production of materials) and recovered energy in incineration plants (avoided emissions as a result of substituting heat and electricity produced from natural gas and oil shale in the background system). Incineration gives approx. 75% and recycling almost 25% of the total avoided emissions. In scenario II sources of GHG savings are mainly material recycling and the avoided emissions from landfilling. As in this scenario composting without energy recovery is applied, the net GHG emissions are higher than in scenario II.

Direct emissions from landfills continue to be a major source of GHG emissions till 2020 despite of the fact that the landfilling rate will degrease significantly and a relatively high share of landfill gas is recovered in both studied scenarios. The GHG emissions from waste collection and transport will increase by 2020 due to increased recycling. In scenario II a higher collection rate of biodegradable waste causes slightly more emissions of CO_2-equivivalents. In spite of that, the collection and transport of waste accounts for a relatively small amount of estimated net GHG emissions in both future scenarios.

7. Conclusions

As a result of the case study in Estonia it can be concluded that better management of municipal waste and especially diversion of municipal waste away from landfills could significantly reduce the emissions of GHG despite an almost 60% increase in waste generation by 2020. This is valid even if landfill gas is recovered at a high rate. Material recycling and incineration with high rates of energy recovery should be favoured compared to other waste management options where energy is not recovered (e.g. composting). It is important to stress that if high rates of recycling and incineration with energy recovery are attained, the net GHG emissions may even become negative, which means that these waste management options can partly offset the emissions that occurred when the products were manufactured from virgin materials and energy was produced from fossil fuel/oil shale. Intensive recycling and incineration of MSW usually also lead to a lower landfilling rate of MSW compared to other possible waste management scenarios. Overall, emissions of GHG associated with the collection and transport of the waste and recovered materials are small in comparison with other waste management practices. However, along the increased recycling rate (especially additional collection of biodegradables) the collection system and transport distances could have considerable impact on the overall GHG emissions of the waste management options.

In general, the conclusions of the study concur with other recent LCA studies (German Federal Environmental Agency 2005; ETC/RWM 2008; Sander 2008), but due to the fact that in Estonia energy produced from waste substitutes oil shale based electricity which has high climate change impact in terms of CO_2 emissions, incineration and other waste management options (e.g anaerobic digestion) where energy is produced, should be preferred to other waste management practices.

The aim of this study was to evaluate the climate change impact of the possible future waste management options in Estonia and not to predict the exact GHG emissions generated in the waste sector. The total emissions of GHG from MSW management depend on several factors. The waste composition was assumed to be the same throughout the studied period. In reality, it may be expected that certain waste fractions (e.g plastic packaging) will probably grow compared to other fractions. It would be good to study the possible change in the composition of waste and its possible impact on the results. Also the future energy source in Estonia is assumed to change including more renewable and nuclear energy while the share of oil shale is assumed to decline. Thus, it would be interesting to analyse how a possible change of the marginal energy source would affect the net GHG emissions, especially the relative effects of waste incineration. Waste management has a wide variety of impacts on the environment apart from those associated with climate change. Therefore, in the future also life cycle data and other parameters should be collected in order to evaluate such environmental impacts as acidification, eutrophication, toxicity, social impacts, etc.

References

Beck-Friis, B.G. (2001), "Emissions of ammonia, nitrous oxide and methane during composting of organic household waste." Uppsala: Swedish University of Agricultural Sciences, 331 p. (Doctoral Thesis).

Björklund A. (2000) "Environmental systems analysis of waste management – experiences from applications of the ORWARE model." PhD thesis TRITA-KEM-IM 2000:15, ISSN 1402-7615, AFR report 30. Stockholm, Sweden: Royal Institute of Technology.

Detzel, A., Vogt, R., Fehrenbach, H., Knappe, F. and Gromke, U. (2003), "Anpassung der deutschen Methodik zur rechnerischen Emissionsermittlung und internationale Richtlinien: Teilbericht Abfall/Abwasser." IFEU Institut – Öko-Institut e.V., 77 p.

EEA (2007a), "Annual European Community greenhouse gas inventory 1990-2005 and inventory report 2007." technical report no. 7/2007.

– (2007b), "Greenhouse gas emission trends and projections in Europe 2007 – country profile, Estonia." European Environment Agency, EEA Report, Copenhagen.

Eriksson, O. (2000), "A systems perspective of waste and energy." Licentiate thesis TRITA-KEM-IM 200:16, ISSN 1402-7615, AFR report 30. Stockholm, Sweden: Royal Institute of Technology.

ETC/RWM (2007), "Environmental outlooks: municipal wastes", working paper 2007/1, Copenhagen, available http://waste.eionet.europa.eu/publications.

– (2008), "Municipal waste management and greenhouse gases", working paper 2008/1, Copenhagen, available http://waste.eionet.europa.eu/ publications.

German Federal Environmental Agency (2005), "Waste Sector's Contribution to Climate Protection." Research Report 205 33 314 UBA-FB III, August 2005, available www.bmu.bund.de/files/pdfs/allgemein/application/pdf/klima_abfall_en.pdf.

Hellebrand, H.J. (1998), "Emissions of nitrous oxide and other trace gases during composting of grass and green waste." J. agric, Engng Res. 69, pp. 365-375.

IPCC (2006), 2006 IPCC Guidelines for National Greenhouse Gas Inventories, Prepared by the National Greenhouse Gas Inventories Programme, Eggleston, H.S., Buendia, L., Miwa, K., Ngara, T. and Tanabe, K. (eds.). Published: IGES, Japan, available www.ipcc-nggip.iges.or.jp/public/2006gl/index.html.

– (2007), "Summary for Policymakers." In: Climate Change 2007: The Physical Science Basis. Contribution of Working Group I to the Fourth Assessment Report of the Intergovernmental Panel on Climate Change [Solomon, S., Qin, D., Manning, M., Chen, Z., Marquis, M., Averyt, K.B., Tignor, M. and Miller, H.L. (eds.)]. Cambridge University Press, Cambridge, United Kingdom and New York, NY, USA.

Jaagus, J. (1998), "Climatic fluctuations and trends in Estonia in the 20th century and possible climate change scenarios." In: Kallaste, T. and Kuldna, P. (eds.), Climate Change Studies in Estonia. SEI, Tallinn, pp. 7-12.

Kant, E. (1927), Päikesepaiste kestus, Tartu, pp. 324-325.

Kirde, K. (1940), Andmeid Eesti kliimast, TÜ Toimetised, A.XXXIV, Nr. 8.

Moora, H., Stenmarck A. and Sundqvist, J.O. (2006), "Use of life cycle assessment as decision – support tool in waste management planning – optimal waste management scenarios for the Baltic States." Environmental Engineering and Management vol. 5 no. 3, pp. 445-455.

OSELCA (2006), "Introduction and implementation of Life Cycle methodology in Estonia: Effects of oil shale electricity on the environmental performance of products." Layman's report, LIFE03 ENV/EE/000194, pp. 2-3, available www.energia.ee/OSELCA/eng/index.php.

Petersen, S.O., Lind, A.M. and Sommer, S.G. (1998), "Nitrogen and organic matter losses during storage of cattle and pig manure." J. Agric. Sci. 130, pp. 69-79.

Reihan, A. (2008), "Analysis of Long-term river runoff trends and climate change impact on water resources in Estonia." Doctoral thesis, TUT Press, p. 147.

Russak, V. (1998), "Seasonal peculiarities of long-term changes in radiation regime in Estonia". In: Kallaste, T. and Kuldna, P. (eds.), Climate Change Studies in Estonia. SEI, Tallinn, p. 13-20.

Sander, K. (2008), "Climate protection potentials of EU recycling targets." Ökopol GmbH, Germany, available www.eeb.org/publication/documents/RecyclingClimateChangePotentials.pdf.

Sundqvist, J.O. (1999), Life cycle assessments and solid waste – Guidelines for solid waste treatment and disposal in LCA. Stockholm, AFR-Report 279.

Sundqvist, J.O., Granath, J., Reich Carlsson, M. (2002), Hur skall hushållsavfallet tas om hand? Utvärdering av olika behandlingsmetoder. Stockholm, IVL Report B1462, available www.ivl.se.

Vesterinen, R. (1996), Impact of waste management alternatives on greenhouse gas emissions: Greenhouse gas emissions from composting. Jyväskylä: VTT Energy. Research report ENE38/T0018/96. (In Finnish), 30p.

Chapter 16

Conducting an Interdisciplinary Workshop on Climate Change: Facilitating Awareness and Adaptation in Sajama National Park, Bolivia

Karina Yager, Daniela Ulloa, Stephan Halloy

Abstract

Understanding how climate change affects both biological and human systems requires interdisciplinary collaborations. In addition, researchers and policy-makers must be informed of the local knowledge base of communities that are currently experiencing the impacts of rapid climate change. How do we bridge this knowledge gap and encourage collaborations that increase the understanding of climate change and facilitate adaptation at the local level? This paper describes recent efforts to generate a dialogue among scientists, conservationists, park administration and community members on the subject of climate change in Sajama National Park, Bolivia. Within the paper, a discussion of the methods applied to conduct an interdisciplinary and multi-agency workshop that is focused on understanding climate change impacts and adaptation at the local level is presented. The methodology applied considers the multi-dimensional nature of understanding climate and climate change. The interdisciplinary workshop model presented may be useful for application in other communities that are similarly coping with abrupt climate change.

Introduction: Climate Change and Interdisciplinary Research

Climate change is not a new phenomenon. However, climate change is occurring at accelerating rates (IPCC 2007a). Although there is some debate on whether such rapid rates of climate change occurred in the past, it is certain that the present situation is unique due to the combination of rapid change with high human population and habitat modification. The magnitude of change is creating new pressures on biological and human systems to rapidly adapt for their survival. Biological systems, such as plant communities, are by nature passive systems and therefore highly vulnerable to climate change. While human societies are recognized as having the adaptive capacity[1] to be proactive in the face of climate change, this does

1 The concept of adaptive capacity is taken from the fourth assessment of the IPCC working group II report, and is defined as "the ability or potential of a system to respond successfully to climate variability and change, and includes adjustments in both behavior and in resources and technologies" (Adger et al. 2007: 727).

not necessarily translate into individual or institutional action, which is dependant on the awareness of risks and the analytical capacity to act (Moser &Luers 2008).[2] An initial step in strengthening the adaptive capacity of society is to identify the local perceptions of climate change and to increase the awareness of climate change risks for decision makers. In this paper, we present the development of a participatory workshop on climate change aimed at increasing the adaptive capacity of Andean communities in Sajama National Park, Bolivia.

Throughout human history, societies have undergone social change, encompassing a broad range of challenging developments including increasing population, resource scarcities, and environmental change and degradation.[3] Yet in today's global milieu, it is increasingly recognized that the current rates of rapid global and climate change will challenge the ability of societies of multiple scales, from community to nation, to sustain their current livelihoods. In particular, the poorest and indigenous (often one and the same) people in society are recognized as the most vulnerable to current and future global climate change scenarios (Macchi et al. 2008; Watkins 2007). Therefore, increasing the adaptive capacity of society, and in particular that of communities that are already highly marginalized, is urgent.

Many efforts are underway to understand climate change; its causes, its effects, how to mitigate the causes, and how to adapt to the effects (IPCC 2007a; 2007b; UNFCCC 2007). Current efforts are aimed at avoiding reductionist science or simplistic solutions by embracing interdisciplinary and network approaches. As an example, the Global Observation Research Initiative on Alpine Environments (GLORIA) has initiated global monitoring of the climate change impacts on vegetation in alpine (above timberline) environments through the creation of research networks that emphasize open collaboration and apply a standardized methodology that facilitates data generation and comparison (Grabherr et al. 2000; Bayfield et al. 2005). Recently a developing network of researchers across the Andes has built multidisciplinary and multiscale modules around the GLORIA system (Cuesta et al. 2007; Halloy et al. 2005) to assist in a more integrated understanding of global change and hence capability for adaptation. Stud-

---

2 Moser and Luers identify three critical dimensions of adaptation: awareness, analytical capacity, and action. Analytical capacity is the ability to develop policy initiatives, strategies and plans that lead to concrete actions (2008: 312).

3 Historically, when traditionally effective management plans were impacted by severe or abrupt climate change during a socio-political period already rife with other stresses, such as political conflict, social unrest or top-heavy bureaucracy, collapse has followed in some cases (e.g. the Maya (Huag et al. 2003)). While societal change is understandably the result of multiple factors, some researchers have carefully considered the ways in which past recorded climate change events may have impacted local decision-making and resource management (e.g. for the Chimu (Moore 1991), Moche (Shimada et al. 1991) and Manchay (Burger 2003)).

ies in the Andes are now underway to understand the dynamics of glacial retreat; vegetation, fauna and soil bacterial succession and upward migration; changes in productive systems (tuber cultivation at higher levels, upward migration of livestock, pests and diseases, droughts and floods); cultural perceptions of change; and participatory adaptation (e.g. see Seimon et al. 2007; Halloy et al. 2005; Ulloa & Yager 2007). Interdisciplinary approaches and research networks are recognized as necessary to deal with the multiple scales and interactions of ecological and socio-economic systems in view of the complex interactions of climate change.

In this chapter, we report on the development of a participatory workshop aimed at increasing the awareness among various stakeholders on climate change and its impacts on the indigenous communities of Sajama National Park, Bolivia. The workshop was used as a pilot project to test and develop a more integrated approach to climate change adaptation at the local community level. The applied methodology may be replicated more widely, particularly in other Andean regions. We address the broader questions of the how and why of the overall situation, rather than present a description and quantification of a particular compartment of local reality.

People and Environment of Sajama

Sajama National Park (PNS) is located in the Cordillera Occidental of Bolivia. The park was created in 1939 to protect the endangered forests of Queñua (*Polylepis tarapacana*) that are the highest growing forests in the world (up to 5200 m). Located on the nation's frontier, the western park boundary of PNS is shared with Lauca National Park, Chile. There are five indigenous Aymaran communities living in the park, at a base altitude of 4200 m, and in the surrounds of the highest mountain in Bolivia, Nevado Sajama (6542 m). The flora and fauna in the park include many Andean endemic species, including the threatened Andean albatross, or suri (*Pterocnemia pennata*), and the endangered vicuña (*Vicugna vicugna*), the wild ancestor of the domesticated alpaca (*Vicugna pacos*). Nearly 200 species of vascular plants and some 110 vertebrates live in PNS. Although it is the first national park in Bolivia, it remained a "paper park" until the late 1990's when the Bolivian National Park Service (SERNAP) was established and park management began (Hoffmann 2007).

Applying Andean ecological classifications, the environment of PNS is described as *puna* (Troll 1968) or semi-humid to dry *altoandino* (Ribera 1992).[4] Annual precipitation is low and temperatures are relatively cold. According to one study, the mean annual temperature is 3.5° C and the mean annual precipita-

---

4 Puna refers to the Andean tropical belt of vegetation above treeline, and is the high altitude vegetation zone, typically dominated by grasses, below the nival (snowline).

tion is 347 mm (Liberman Cruz 1986). PNS's geographic location (S 17°55' to 18° 15', W 68°41' to 69° 10') and altitudinal range (4200 m to 6542 m) contributes to a climate characterized by extreme diurnal temperature variations that exceed seasonal variations. Glacial water melt, fed into underground aqueducts and other hydrological veins, is the primary source of water. This is seasonally augmented with low amounts of precipitation occurring primarily during the wet season, between December and March. The climate is considered harsh by production standards, with nightly frosts year round.

The indigenous people of the area are descendants of the Carangas culture, a pre-hispanic Aymaran polity (Klein 2003). A majority of the local population is fluent in both Spanish and the indigenous language, Aymara.[5] Archaeological sites in the park are evidence of a long and historical human occupation in the area, spanning several centuries to include formative, Inca and colonial period sites. Today, the five indigenous *ayllus* (communities) include: Sajama, Lagunas, Caripe, Manasaya and Papelpampa. The human population in PNS is around 1,700 and includes approximately 300 families.

The communities living in the park depend on the herding of domesticated animals including llama (*Lama glama*), alpaca (*Vicugna pacos*) and sheep (*Ovis aries*) for subsistence and production (Alzérreca 2001; Espinoza Terán 2001). A current estimation of the total grazing population includes 5,000 vicuñas and 49,000 heads of livestock (49% llama, 43% alpaca and 8% sheep) within a total park area of 100,000 hectares (Guzman 2007, pers.com.). There is no agricultural production in the park.

A trans-international highway was constructed in 1994 through the southern end of the park, linking the Bolivian capital of La Paz to the Chilean port of Arica. This development has contributed to many local changes, including a facilitation of out-travel to regional markets and urban centers for work. Alpine tourism has also grown and contributes to a small but growing amount of local revenue, with an estimated average of 4,000 visitors per year. Seasonal out-migration for temporary work and education has increased in recent years, along with permanent re-settlement.

Sajama as a case-study and its relation to other Andes research areas

Many communities throughout the Andes rely on small-scale agro-pastoral production, including planting tubers and grains, and the herding of camelids, cattle and sheep, to meet their subsistence needs and produce nominal economic re-

---

5 There is a gradual loss of Aymara among the younger generation. The local education system teaches in Spanish and while many of the youth can understand Aymara, some are not fluent in the indigenous language.

turns. These communities face and witness a series of current environmental challenges including the degradation of pastures, rapid deglaciation, temperature warming, unpredictable weather cycles and out of season climate events that threaten the survival of both crops and animals. These environmental issues are superimposed on major social changes driven by improved physical infrastructure, access to markets, education, health, and consequent out-migration, loss of traditional language and knowledge, and fragmentation of management systems. A few researchers have addressed several of these complex causes and effects at different scales, and have identified declining agrobiodiversity and traditional methods of conservation (Halloy et al. 2005; Ortega Dueñas 1989; Ortega et al. 2002). Along with these kinds of studies, it is also necessary to integrate scales by focusing on the ongoing development of change in a particular region. Methods arising from such focalized research can then be replicated across a wider spectrum of communities, adapting them to each particular case. Here we study Sajama communities because of their characteristics that are widely applicable to other regions of the Andes. Sajama shares many of the same broader issues facing Andean communities including rapid climate change, environmental degradation, population and management pressures, rapid social change, increasing connectivity to large urban centers and to the global market.

One of the most evident impacts of global warming for many communities in the Andes is deglaciation. The pan-Andean experience of shrinking glaciers is witnessed locally in the visibly rapid retreat of Nevado Sajama. *Tatay* (honorable) Sajama is an important *apu* (mountain deity) for local indigenous people and others throughout the region. Along with the glacier's cosmological, cultural and spiritual significance, it is the primary water source for local rivers and streams, on which ecosystem integrity and human livelihoods depend. Glacial retreat is the result of an increase in global temperature warming that is directly and adversely impacting the inhabitants of Sajama National Park. There are several baseline and continuing studies of the park's ecological parameters, including pastures, glaciers and forest, from which change can be monitored and measured now and in the coming decades.

In a broader sense, it is living in a national park that is a fundamental difference between communities of Sajama and many other Andean communities. Since 1997, the community members have worked with SERNAP park administration to develop a PDM (Resource Management Plan), not without a certain degree of expected constructive disagreement, which has ultimately led to the development of a "model protected area" (Hoffmann 2007). There is an overarching goal of sustainable management that fosters livelihoods, but also natural resource conservation. A monthly *co-gestión* (cooperative management) meeting has been organized where community members and park officials meet to discuss concerns, issues, news and projects relevant to park management.[6] Since the development of park administration, conservationists and non-governmental organizations have worked in the park promoting tourism infrastructure building. Locals are accustomed to hosting outsiders, through community rotation of room and board; and many are welcoming to outside visitors on a more personal level. Despite the congenial atmosphere and relationship between the park and local communities, local resource management decisions are monitored and critiqued to a certain degree by national park officials, environmental scientists and conservationists, with the intention of resource protection.

Objectives of the climate change workshop in PNS

A principal goal of the workshop was to create a dialogue between local community members and outside researchers whose studies are relevant to the climate change discussion in Sajama. Our GLORIA research team began work in the park in March of 2006 to establish vegetation-monitoring sites to be evaluated in relation to climate change in the coming years and decades (Halloy et al. 2008).[7] Stemming from this work and other continuing ecological studies in the park (e.g. Yager et al. 2007), we wanted to share with the local communities in greater depth the broader significance and relevance of ongoing research in the park on ecological and climate themes. While this had already been communicated with the park administration, local authorities, and a select amount of local individuals who were assisting in our research, there were many other community members that we had not spoken with directly on these subjects. We also understood that there were abundant studies on various climate change themes on a regional, national and international scale that would be beneficial to share and discuss with the local community given their current predicament in facing and managing the impacts of rapid climate change. Equally important, we wanted to learn more about the local perspectives and knowledge of ecological and climate change topics. Local and traditional knowledge is of great value, especially in understanding how environmental change factors into local beliefs and livelihoods. In essence, we wanted to create a discussion forum wherein multiple actors could share their personal knowledge, create an opportunity to recognize the important issues relevant to climate change, and to identify collaborative measures that could be taken to support and improve adaptive capacity. The workshop was held in PNS on September 19 and 20, 2007, and included the participation of various stakeholders invested in the impacts of climate change at the local level.

---

6 The monthly *co-gestión* meets in addition to the traditional intra-community meetings that also occur on a monthly basis.

7 For more on GLORIA see www.gloria.ac.at.

The stated primary objectives of the workshop were to:

1. Identify the local perspectives of climate change experienced by local communities in PNS;
2. Present to the communities in PNS current scientific and social research on themes related to climate change, including local and regional studies;
3. Identify possible strategies of adaptation to climate change and to other social and environmental changes;
4. Raise youth awareness in PNS on the subject of climate change and promote a discussion on the specific theme of water (Ulloa & Yager 2007).

Methodology of climate change workshop

The workshop was approached through three primary steps: first, the identification of information networks at the local level; second, an identification of key stakeholders in local climate change; and third, the creation of an arena and means for cross-cultural, interdisciplinary and multi-agency participation and communication on the subject of climate change in PNS.

In PNS, community information is primarily shared at the monthly held community gatherings, as previously mentioned, but also includes radio communication, informal conversation, meetings and social gatherings. Information about the workshop was shared through all of the aforementioned means, up to two months prior to the event, in order to generate awareness of the workshop and to increase the expected number of total participants. Following local customary norms, formal invitations were also presented to local authorities.

Invited participants from outside the local community were identified as stakeholders in climate change at the local level, including policy-maker, local and regional authorities, scientist and conservationist. Each stakeholder is recognized to have particular experience and knowledge relevant to climate change that can inform and raise awareness that affects management decision-making at the local level. Because the interactions between human and environmental systems are numerous and complex, it is necessary to develop participatory analyses of these interactions to reach a shared understanding and develop pathways towards actions that produce more sustainable results with less unexpected outcomes (see Box 1).

Box 1: Management decisions require information

Management decisions require that decision makers be informed and understand potential consequences of their decisions under different scenarios. This implies

1. basic information (e.g. what is the situation, where can we find what, monitoring and indicators appropriate to scale, including scientific and traditional indicators),
2. assumptions on future trends (models, scenarios, conceptual, mathematical, computational),
3. requires several options and potential pathways given one or the other decisions and alternative future trends in environment and social systems (e.g. 'what'if' scenarios),
4. have a broad understanding of the context of variability and diversity (physical and biological, landscape and human heterogeneity),
5. must consider and reflect ethical and cultural preferences (including also the use of traditional knowledge),
6. must be possible (opportunity costs, cost-benefit, capability, political will),
7. must be compatible with the decisions of other interacting groups,
8. flexible feedback loops of information and monitoring to adjust targets and actions as the situation changes.

Support for the workshop, including financial contributions and organizational support, was provided by some of the identified stakeholders, including Conservation International (CI), the Institute of Ecology (IE), the National Bolivian Herbarium (HBN), the National Program on Climate Change (PNCC), Sajama National Park (PNS) and SERNAP (National Park Service). Generalizing the stakeholders and participants, actors included local indigenous community members, local authorities, conservationists, development organization representatives, departmental and national institutional agents, and research providers, both national and international (see table 1). The total number of participants was 75.

Participants	Local (PNS)	56%
	Outside	44%
Gender	Women	45.9%
	Men	54.1%
Age of Participants (PNS)*	20 to 30	26.2%
	31 to 40	26.2%
	41 to 50	16.7%
	51 to 6	7.1%
	61 to 75	11.9%
	No Response	11.9%
Place of Residence	Sajama	35.4%
	Caripe	6.7%
	Lagunas	5.4%
	Papelpampa	4%
	Manasaya	2.7%
	La Paz	39.1%
	Oruro	2.7%
	Other	4%

Institutional Participants**	IE (Institute of Ecology)	24.7%
	UMSA (University Mayor San Andres)	12.1%
	PNCC (National Climate Change Program)	9%
	Sajama National Park	9%
	DDRNMA (Bolivian Department of Natural Resources and Environment)	9%
	Conservation International	6%
	CBF (Bolivian Faunal Collections)	6%
	Other	21.2%
	No response	3%
Occupation	Herder	16%
	Homestay	8%
	Park Guard	10.6%
	Community Authority	5.3%
	Tourism	5.3%
	Artisan	2.6%
	Researcher	17.3%
	Consultant	10.6%
	Director	6.6%
	Student	4%
	Other (teacher, farmer, vendor, medicine practitioner, etc.)	13.3%
	No Response	4%

* Only of local community members, ** Only non-local participants

The workshop model applied places emphasis on facilitating learning through intercultural communication and participatory methods. While many of the park's inhabitants are bi-lingual, formal community gatherings are conducted in Spanish, following park protocol. We elected to use Spanish as the official language of the workshop, understanding that this was the shared language among all participants. However, we also provided a translator of Aymara to summarize discussion topics and questions throughout the meeting. In some cases, people chose to share their remarks, question or concerns in Aymara and were welcome and able to do so with the assistance of a translator. The specific activities of the workshop included:

1. Introductory formal presentations. This activity created an opportunity for non-local presenters to share with the local community an overview of their investigations or relevant experience in relation to climate change, some of which had been realized in PNS or a similar region (e.g. Chipaya, Lake Poopo, Bolivia). Following each presentation, there was a period of open discussion wherein all participants could further question and share knowledge in relation to the presented topic. Presentations were scheduled throughout the duration of the two-day workshop. Through this kind of presentation, knowledge of one indi-

vidual was transmitted to the entire assembly. In some cases, key terminology on climate was identified and discussed as a group in order to facilitate further discussions during thematic work exercises.

2. Thematic work. Two dynamic participatory activities were organized around specific themes related to climate change. The activity involved splitting into groups, with each group assigned to discuss a particular theme and provide feedback to the larger assembly. This was an opportunity for individuals to reflect on specific themes in smaller groups wherein ideas could be more intimately shared and individuals could discuss their personal perspectives. The main points, including recognized commonalities and differences, were then shared with the group for others to consider. The first thematic session was aimed at identifying perspectives of environmental change (e.g. in pastures, water, sun, animals, and plants) and the second session, held at the end of the workshop, focused on identifying adaptation strategies.
3. Group Maps and Models. This tactile and visual activity provided an opportunity for local participants to create maps and models that reflected processes of environmental change in the park. Maps were made to reflect the different categories of pasture quality (i.e. good, drying and poor). Models, made of clay, were constructed to reflect the extent of the glacier, Nevado Sajama, during a particular time span (five year interval periods, spanning between twenty years in the past to five years from the present). After group map and model construction, an elected team member described their work to the assembly.
4. Climate Timeline. Local community members were asked to recall specific climate events in the past and their impacts. Results of this discussion were added to a timeline of climate events in PNS, from 1950 to the present.
5. Field presentations. This activity was a group visitation to local field-plots (i.e. pasture parcels) of researchers currently working in the park. Workshop participants examined pastures, assessed their quality, and discussed different pasture management strategies. It was an excellent opportunity for investigators and locals to talk freely in the field.
6. Student activity. A parallel activity on the theme of climate change, and in particular water, was presented at the local elementary school to include youth participation. The objective of this exercise was to facilitate reflection on environmental change and to express their ideas of change through the drawing of their local landscape.

Summary of Local Perceptions of Climate Change

The results of the workshop indicate that the community is most concerned with the need for community management of water and pastures to optimize pastoral production in the face of changing climate. In particular, they recognize a decrease in available pastures, linked to grazing pressures and a decrease of pasture production. Community members also recognize that the typical patterns of climate have changed in recent years and that this may play an increasing role in reducing pasture productivity.

The climate change variables identified by community members include rapid deglaciation, an increase and difference in timing of wind (leading to increased erosion), more extreme and unpredictable weather events, and increased extremes of weather patterns (including greater frost occurrence and increased heat (both air and solar temperatures)). In general, community members perceive a decrease in overall precipitation and increase in drought frequency, especially over the last five years. The soils are perceived as drier and their ability to retain moisture has decreased. Unseasonal weather patterns have increased. For example, many pointed out that on any given day of the year, one can experience weather that exemplifies all of the seasonal patterns.

Both animal and human health has been compromised with changing climate. People recognize an increased vulnerability to sunburns and many reported skin and eye problems as a result of increased solar intensities. Animal malnutrition and an increased susceptibility to illness and death due to rapid temperature fluctuations and pasture depletion were also reported.

The community members of PNS were able to identify local changes in climate over recent decades, primarily in the past fifty years. In particular, some recall the occurrence of extreme climate events that can be correlated with regional Andean climate events on an annual scale. For example, participants remembered extreme climate events occurring in 1974, 1982 and 2002 that wrought sickness and death for many animals. The July 2002 winter blizzard (local accounts of over 50 to 100 cm of snow on the ground and up to 200 cm in gullies) resulted in a total loss of over 1,500 animals and severe malnutrition of an estimated 8,676 animals (SERNAP 2002: 23). Climate records confirm the occurrence of these climate events on a pan-Andean scale, and are further associated with the occurrence of El Niño or La Niña events.

Discussion of Workshop Results

The identified climate change processes were understood to be a result of complex social and environmental interactions. While many of the identified patterns of climate change correlate with general regional trends of deglaciation and temperature warming, further research is needed to identify the social and environmental linkages to local trends and to quantify the degree of change that is being perceived at the local level.

Overall, the climate change workshop permitted a discussion of important topics related to pre-existing resource management issues. For example, there are different perspectives on how to manage climate change in relation to current production patterns. Individuals disagreed on some aspects of pasture management, including whether or not burning should be performed, and if so, to what degree and in what season. These questions of management strategy were identified during the workshop as an important topic that would need further consideration at the local level in relation to planning of climate change adaptation. It was recognized that climate change is an added stressor to a variety of other existing issues; an umbrella under which many of these can be reconsidered and their increasing severity addressed in a more holistic way.

In addition, adaptation was recognized as an ongoing process. This falls in line with current research on past climate events and its social management that can offer relevant knowledge and experience to inform current strategies of adaptation (e.g. McIntosh et al. 2000; Orlove 2005). Throughout our human past, societies all over the globe have been exposed to climate change and "adaptation" to climate change has resulted in multiple responses including migration, resettlement, change in social networks, technological development and production shifts. Current adaptation to global climate change is driving similar adaptations at the local level.

In Sajama, there is a high percentage of out-migration for work and a growing dependency on outside production to meet economic needs. This is related not only to climate change but to multiple other drivers including population growth, cultural changes, market changes, road access, etc. Another strategy of continued adaptation identified during the workshop, is to strengthen the tourism sector in PNS. This was perceived as a way in which to decrease the local dependence on pastoral production and supplement economic needs. However, much of the tourism to PNS is alpine tourism, dependant on glacial mass, which is undergoing rapid retreat. Nonetheless, local community members are already seeking to diversify economic livelihoods in the face of climate change. Thus, it is understood that adaptation is identified as an ongoing process incorporating multiple strategies to cope with multiple changes (not just climate), and not a static one-time action with simple solutions.

Conclusions

The workshop brought together people from multiple communities, from local to international, with a range of backgrounds, including scientists from multiple disciplines, allowing for an exchange of experiences and learning. The workshop facilitated awareness and communication of the multiple aspects of local climate change and in part raised the community expectations that something can be done to strengthen local adaptive capacity. Further planning is in progress to develop follow-up projects on adaptive community actions.[8] In particular, the communities of Sajama identified that an improvement in livestock quality and pasture productivity through improved irrigation and herd management is a way in which to increase the local adaptive capacity to manage current and future climate change. Despite rapid change and the lure of out-migration, the community's greatest concern was for the protection and maintenance of the traditional practice of herding, which is an important aspect of their current livelihood and cultural identity.

The features that Sajama has in common with other Andean communities ensure that such interdisciplinary and multi-agent workshops can be widely replicated. Clearly Sajama also has its unique characteristics, and allowance has to be made for adaptation of the workshop techniques in each particular area. In a generic way, methods can be replicated in other continents with communities transitioning in a similar way from small scale, and often traditional, resource management to the impacts of the modern market economy and simultaneous global environmental change, and to identify the pertinent climate change issues in relation to the local social and environmental context.

Establishing processes like the one described here, replicated in different areas, while adapting to the particular local conditions, provides information and empowerment to local communities to follow their own paths toward adaptation to local changes in an integrated fashion. Ensuring a broad participation at all levels (gender, age, professions, interests, stakeholders, urban, rural, social and political) helps ensure people take the findings as their own and participate in the further ongoing processes of putting words into actions. Replicating can be achieved through informal networks sharing knowledge through training or workshops, and moreover through living together and sharing experiences in a community of knowledge.

Acknowledgements

We would like to especially thank the people of Sajama National Park, without whose interest and openness the workshop would not have been possible. A very special thanks is also extended to the local park guards and director of PNS. We are grateful for the support provided by Conservation International (CI), National Program on Climate Change (PNCC), National Bolivian Herbarium (HBN) and the Institute of Ecology (IE), La Paz, Bolivia.

---

8 Pending financial support.

References

Adger, W.N., Agrawala, S., Mirza, M.M.Q., Conde, C., O'Brien, K., Pulhin, J., et al. (2007), "Assessment of adaptation practices, options, constraints and capacity." In: Parry, M.L., Canziani, O.F., Palutikof, J.P., van der Linden, P.J. and Hanson, C.E. (eds.), Climate Change 2007: Impacts, Adaptation and Vulnerability. Contribution of Working Group II to the Fourth Assessment Report of the Intergovernmental Panel on Climate Change, pp. 717-743. Cambridge: Cambridge University Press.

Alzérreca, H. (2001), Los Campos Nativos de Pastoreo en el Área del Parque Nacional Sajama y su Capacidad de Carga, MAPZA-GTZ-Parque Nacional Sajama-SERNAP, La Paz, Bolivia, 151 pp.

Bayfield, N., Brooker, R. and Turner, L. (2005), Some lessons from the ECN, GLORIA and SCANNET networks for international environmental monitoring Mountains of Europe: Conservation, Management, People and Nature, p. 213-222.

Burger, R. (2003), "El Niño, Early Peruvian Civilization, and Human Agency: Some Thought from the Lurin Valley." In: Hass, J. and Dillon, O. (eds.), El Niño in Peru: Biology and Culture Over 10,000 Years, vol. 43. n.s.: Fieldiana Botany.

Cuesta, F., García, C. and Halloy, S. (2007), Memorias del curso: Monitoreo del impacto del cambio climático en la biodiversidad de los ecosistemas de alta-montaña en los Andes, GLORIA, CONDESAN, CI, La Paz, Bolivia, 21 pp. Noviembre 2007.

Espinoza Terán, C.W. (2001), Caracterización de los Sistemas de Producción del Parque Nacional Sajama y la Zona de Amortiguación, MAPZA-GTZ-Parque Nacional Sajama-SERNAP, La Paz, Bolivia, 117 pp.

Grabherr, G., Gottfried, M. & Pauli, H. (2000), "GLORIA: A Global Oberservation Research Initiative in Alpine Environments." Mountain Research and Development 20(2), p. 190-191.

Guzmán, F. (2007), Ecologia de PNS. Taller de Cambios Climaticos en PNS. Sept. 19, 2007.

Halloy, S.R.P., Seimon, A., Yager, K. and Tupayachi Herrera, A. (2005), "Multi-dimensional (climate, biodiversity, socio-economics, agriculture) context of changes in land use in the Vilcanota watershed, Peru." In: Spehn, E.M., Liberman Cruz, M. and Körner, C. (eds.), Land Use Changes and Mountain Biodiversity, 2005. CRC Press LLC, Boca Raton FL, USA, pp. 323-337.

Halloy, S.R.P., Yager, K., García, C. and Beck, S. (2008), "South America: Climate Monitoring and Adaptation Integrated Across Regions and Disciplines." In: Settele, J. (ed.), Atlas of Biodiversity Risks – from Europe to the globe, from stories to maps. Pensoft (www.pensoftonline.net/alarm-atlas-info), Sofia & Moscow, in press.

Haug, G.H., Gunther, D., Peterson, L.C., Sigman, D.M., Hughen, K.A. and Aeschlimann, B. (2003), "Climate and the collapse of Maya civilization." Science 299, pp. 1731-1735.

Hoffmann, D. (2007), "The Sajama National Park in Bolivia: A model for cooperation among state and local authorities and the indigenous population". Mountain Research and Development 27, pp. 11-14.

IPCC (2007a), Climate Change 2007: The Physical Science Basis – Summary for Policymakers. www.ipcc.ch/SPM2feb07.pdf, 20070215.

– (2007b). Working Group II Contribution to the Intergovernmental Panel on Climate Change Fourth Assessment Report. Climate Change 2007: Climate Change Impacts, Adaptation and Vulnerability – Summary for Policymakers, 23 pp.

Klein, H.S. (2003), A Concise History of Bolivia. Cambridge: Cambridge University Press.

Liberman Cruz, M. (1986), Microclima y distribucion de *Polylepis tarapacana* en el Parque Nacional Sajama. Documents Phytosociologiques N.S., X, pp. 235-272.

Macchi, M. et al. (2008), Indigenous and Traditional Peoples and Climate Change. IUCN, 66 pp.

McIntosh, R.J., Tainter, J.A., & McIntosh, S.K. (eds.) (2000). The Way the Wind Blows: Climate, History, and Human Action. New York: Columbia University Press.

Moser, S.C. and Luers, A.L. (2008), "Managing climate risks in California: the need to engage resource managers for successful adaptation to change." Climate Change 87, pp. 309-322.

Moore, J.D. (1991), "Cultural Response to Environmental Catastrophes: Post-El Niño Subsistence on the Prehistoric North Coast of Peru." Latin American Antiquity 2(1), pp. 27-47.

Orlove, B. (2005), "Human adaptation to climate change: a review of three historical cases and some general perspectives." Environmental Science & Policy 8, pp. 589-600.

Ortega Dueñas, R. (1989), Contribución campesina en la conservación y ampliación de la variabilidad genética de papa, Manejo Campesino de Semilla en los Andes. PPEA-PRATEC, Cusco, p. 9.

Ortega, R., Quiros, C. and Arbizu, C. (2002), Strengthening the on-farm conservation and food security of Andean tubers in the fragile ecosystems of the Southern Peruvian highlands, Universidad Nacional de San Antonio Abad del Cusco, Perú, Cusco, 22 pp. December 1, 2001-November 30, 2002.

Ribera, M.O. (1992), "Regiones Ecológicas." In: Conservación de la Diversidad Biológica en Bolivia. La Paz: CDC (Centro de Datos para la Conservación).

Seimon, T.A., Seimon, A., Daszak, P., Halloy, S., Schloegel, L., Aguilar, C. et al. (2007), "Upward range extension of Andean anurans and chytridiomycosis to extreme elevations in response to tropical deglaciation." Global Change Biology 13, pp. 288-299.

Shimada, I., Schaaf, C.B., Thompson, L.G. and Mosley-Thompson, E. (1991), "Cultural impacts of severe droughts in the prehistoric Andes: application of a 1,500-year ice core precipitation record." World Archaeology 22(3), pp. 247-270.

SERNAP (Servicio Nacional de Areas Protegidas) (2002). Informe de Gestion (1995-2002) no. 68, Parque Nacional Sajama (pp. 28). Sajama, Department of Oruro, Bolivia: SERNAP.

Troll, C. (1968), "The Cordilleras of the Tropical Americas – Aspects of Climatic, Phytogeographical and Agrarian Ecology." In: Troll, C. (ed.), Geo-ecology of the Mountainous Regions of the Tropical Americas. Ferd. Dümmlers Verlag, Bonn, pp. 15-56.

Ulloa, D. and Yager, K. (eds.) (2007), Memorias del Taller "Cambio Climático: Percepción local y adaptación en el Parque Nacional Sajama". Conservación Internacional – Bolivia, Comunidad de Sajama, Bolivia, 42 pp.

UNFCCC (2007), Climate Change: Impacts, vulnerabilities and adaptation in developing countries. United Nations Framework Convention on Climate Change.

Watkins, K. (ed.) (2007), Informe sobre Desarrollo Humano 2007-2008 – La lucha contra el cambio climático: Solidaridad frente a un mundo dividido. Programa de la Naciones Unidas para el Desarrollo (PNUD, UNDP), New York, 402 pp.

Yager, K., Resnikowski, H. and Halloy, S. (2007), Pastoreo y Variabilidad Climática en el Parque Nacional Sajama, Bolivia. Paper presented at the Segundo Congreso Latinoamericano de Parques Nacionales y otros Áreas Protegidas, Bariloche, Argentina.

Chapter 17

Differences of Degree?
–
Public Opinion on Climate Change in a Cross-National Context

Sandra T. Marquart-Pyatt

Abstract

The issue of climate change is uniquely positioned at the nexus of the physical and social sciences. Its expansive audience ranges from scientists, policy-makers, governments, communities, stakeholders, to citizens, underscoring the importance of multiple perspectives for understanding how views about climate change are being created and sustained, and outlining possible implications for policy discussions. A number of important questions remain about the global reach of climate change and how to explain its influences, especially as reflected in public opinion data. This research chronicles recent cross-national datasets with available information on public opinion about climate change, focusing on the International Social Survey Programme survey on the environment. An investigation of the driving forces of views of climate change describes two sets of influences: individual-level and country level forces. Results demonstrate more similarities than differences across countries with regard to factors affecting views of climate change.

Introduction

What does the general public know about climate change[1], and to what extent can we identify similarities across countries in this regard? Climate change is an issue uniquely positioned at the nexus of academic fields spanning the physical and social sciences. In brief, its expansive audience includes scientists, policy-makers, governments, communities, stakeholders and citizens, thus underscoring the importance of multiple perspectives for understanding how views about climate change are being created and sustained. A number of important questions

---

1 The terms global warming, climate change, and the greenhouse effect have been to some extent used interchangeably over the years in media accounts (Dispensa & Brulle 2003) as well as in surveys querying about various attitudes. This chapter will utilize climate change and global warming to convey similar notions, though noting the growing preference for the term climate change to more accurately portray the complexity of the issue and expansive potential effects.

remain about the global reach of climate change as reflected in public opinion data and how to explain its influences.

This research chronicles public opinion on climate change across countries including three recent cross-national surveys, highlighting the International Social Survey Programme survey on the environment from 2000. Empirical investigations of the driving forces of views of climate change focuses on two sets of influences drawn from previous scholarship: individual-level forces (e.g., socio-demographics, education, environmental knowledge and attitudes) and aggregate level indicators (e.g., CO_2 emissions and industrialization). Results show some similarities with regard to influences on climate change views, especially the role that knowledge and environmental attitudes play in shaping perceptions of the impending threat of climate change. Exploratory analyses from multilevel modeling show some support for contextual factors, though pointing to additional possible layers of influences. Although scant attention has been played to public opinion on climate change globally in recent years from the perspective of social science scholarship, the issue has been addressed using other lenses. Given this recent attention, and especially its rise to prominence in global affairs, a comprehensive description of the current state of public opinion on climate change is indeed timely and provides an important backdrop for future research.

Previous Research on Public Opinion about Climate Change

Interest in environmental issues like climate change has been strong recently as environmental topics continue to be recognized as pressing social problems. Compared with other issues, climate change has unique aspects, especially with regard to uncertainty about its primary causes and in determining levels of public knowledge and awareness about it. Due in part to the absence of comparative, cross-national research, our understanding of factors influencing opinions about climate change is currently underspecified. Climate change has recently gained prominence as one of the most pressing issues of our times. Documenting the issue of climate change in public opinion surveys has not, however, achieved similar notoriety for any number of reasons. Some scholarship points to inaccuracies regarding what members of the public know, and other studies suggest the possibility of regional differences regarding the urgency of attention to climate change. Indeed, the causes of climate change are complicated, and given the incomplete picture provided in some media accounts, it is not necessarily surprising to see mixed views about climate change reflected in public opinions surveys.

In what ways have there been significant inroads made with respect to what the general public knows and thinks about climate change? Answering this question is more complex than it seems on the surface, as research documenting public opinion on climate change varies widely on a number of dimensions. For instance, some research focuses on a single country, others study a small number of countries comparatively, and a few include a relatively large number of countries[2]. Sociological research addressing what influences attitudes about global warming is to some degree limited, and cross-national studies are indeed quite rare. Generally, research on the extent of climate change information and knowledge among the general publicsuggests they are limited and often intertwined with other global environmental issues like the greenhouse effect and ozone depletion. Some research points to differences in knowledge of and opinions about climate change across nations. For instance, in comparing the state of awareness in the mid to late 1980s and into the 1990s, when climate change was emerging as an issue on the public agenda, Dunlap (1998) argues that the public (in six countries) generally seemed to be poorly informed about global warming, but that this was not unanticipated given the limited scientific and technological knowledge among members of the lay public. From its introduction to the public arena in the late 1980s [3] up to the late 1990s, the public's understanding of climate change did not appear to have changed. Brechin (2003) expanded Dunlap's analysis to encompass nineteen countries, only to reach a similar conclusion: the existence of a knowledge gap with regard to factors influencing climate change. For instance, with the exception of the Finnish case, less than one-tenth of general publics surveyed were able to correctly identify greenhouse gases as a major source of climate change (Brechin 2003). According to Brechin (2003: 120), "Perhaps the most striking finding from these cross-cultural comparisons is how poorly people from a wide-ranging number of countries understand the anthropogenic causes of global warming." It is only in recent years that this appears to be changing.

Scholarship documenting public opinion varies widely in measures of views of climate change, including knowledge, awareness, risk perception, level of concern, and policy views, indicators that highlight a variety of distinct issues and topics related to climate change overall. These studies utilize different datasets and analytical techniques, and consist of different sets of countries, all of which have consequences for knowledge accumulation. In summary, existing research on public opinion regarding climate change differs broadly in conceptual focus and empirical models, making comparisons across studies difficult and in some respects further hampering our understanding of the current state of affairs related to what the public thinks and knows about climate change. Of pertinent interest for this chapter is how general publics compare with one another with regard to views

---

2 Some research investigates the US general public or a representative sample thereof (Bostrom, Morgan, Fischoff & Read 1994; Brewer 2005; Dietz, Dan & Shwom 2007; Kempton 1991; Nisbet & Myers 2007; O'Connor, Bord, Yarnal & Wiefek 2002; PIPA 1998). Others adopt a comparative approach (Bord, Fisher & O'Connor 1998; Brechin 2003; Dunlap 1998; PIPA 2007).

3 1988 is a pivotal date regarding expanded awareness of global warming in the United States, as it was the year James Hansen, a climatologist for NASA, first testified before the US Senate, thus placing the issue in the public arena.

about climate change, and whether similarities can be identified in terms of what factors influence these views. This investigation is important because understanding what factors drive public opinion has implications for policy creation.

What inroads have been made with regard to public understanding of climate change since the 1990s? Answering this question is deceptively simple, given the current state of public opinion research on climate change. A recent synthesis piece describes a number of key dimensions of public opinion research on climate change including issue awareness, understanding global warming's causes and related policy debates, perceptions regarding scientific certainty and expert agreement, concern about the effects of global warming, and various policy aspects (i.e. costs of proposed policies and international agreements like Kyoto) (Nisbet & Myers 2007). This review only included data from the US general public. While some assert that the US general public indeed differs from much of the global public (Brechin 2003), recent research modifies this somewhat whereby levels of concern are not necessarily that different (Brewer 2005; PIPA 2007). Citing a number of public opinion surveys conducted in 2006 and 2007, a recent report from the Program on International Policy Attitudes' (PIPA) project suggests an awareness of climate change across the world that appears to be growing (PIPA 2007). Comparison data are, however, largely absent from this research.

To briefly review, majorities express that global warming is a problem or threat, that climate change results from human activity, and that action needs to be taken with regard to climate change (PIPA 2007). Yet variation remains on a number of fronts in these survey results. For instance, there are issues related to information dissemination, as majorities of publics had heard of climate change in most of the countries surveyed, though it was still the case that publics in some countries had not heard of it. A majority of survey respondents also expressed support for taking action regarding climate change soon, although views were not as uniform as other measures with regard to the urgency of necessary responses to deal with potential effects of climate change. Views regarding policy options for addressing the issue of climate change were among those exhibiting the greatest variation among countries. Thus, though gains appear to have been made, a number of issues remain to be explored. For instance, on the face of it scientific evidence about climate change would suggest that small island states and those countries with long stretches of coastline that are close to sea level would evince greater concern with the potential implications of the rise in sea level proposed in some global warming circles, yet this hypothesis has not been explored in social scientific research as it intersects with what publics think about climate change. Public opinion data might suggest that this explanation, while plausible, may not necessarily be the case. In order to bridge the gap between description and explanation, it is necessary to delve into such results in more depth. This topic is addressed in the next section.

Cross-national Data on Public Opinion about Climate Change: Select Recent Datasets

Table 1 provides a select overview of cross-national public opinion polls compiled from a recent, comprehensive search of data archives of the Interuniversity Consortium for Political and Social Research (ICPSR) at the University of Michigan. As shown in the questions included in these surveys, cross-national public opinion surveys vary widely in the measures of climate change included. Indeed, some surveys only have one question regarding climate change/global warming (Gallup 2007), and others generally have two to four items. Overall, the questions tap a number of different aspects of climate change, including knowledge and awareness (ISSP 1993; 2000), risk or threat perception (GMF 2005; 2006; ISSP 1993; 2000), level of concern (Eurobarometer 1999), and policy views. The surveys vary, moreover, with regard to geographic coverage and the number of countries included. The datasets are all cross-sectional, disallowing in some respects comparison over time on opinions about climate change.

In selecting a dataset, there are thus trade-offs regarding the span of countries and questions that can be investigated. If our interest is in whether members of the general public perceive global warming to be an important threat, GMF Transatlantic Trends data from 2005 or 2006 would be used. If our interest is in having the most recent data available from the ICPSR archive with the broadest geographic coverage, however, the Gallup (2007) dataset would be utilized. In order to determine the factors affecting opinion about climate change, including how it intersects with general environmental attitudes, and to gauge change over time, the ISSP Survey is ideal. Given limitations of current scholarship noted earlier, each of the datasets is briefly explored, although the empirical analyses use the ISSP Environment data.

Tab. 1: A Select Overview of Cross-National Public Opinion Data

Survey Organization	Year of Survey	Number of countries	Number of cases	Survey Question (s)	Pooled Response	Range
International Social Survey Programme	2000	26*	30604	The greenhouse effect is caused by a hole in the earth's atmosphere	28% (16%)	Ireland (10%) to Japan (46%)
				Every time we use coal or oil or gas, we contribute to the greenhouse effect.	84%(36%)	Philippines (66%) to Sweden (90%)
				In general, do you think the rise in the world's temperature caused by the 'greenhouse effect' is… Dangerous to you and your family	63% (26%)	Finland (40%) to Portugal (87%)
	1993	22*	28301	In general, do you think the rise in the world's temperature caused by the 'greenhouse effect' is… Dangerous to the environment	60% (24%)	40% in Netherlands to 80% in West Germany
Transatlantic Trends GMF	2005	11	11080	I am going to read you a list of possible international threats to Europe/US in the next 10 years. Please tell me if you think each one on the list is an extremely important threat, an important threat, or not an important threat at all. The effects of global warming	71% (35%)	79% in the United States to 96% in Portugal and Spain
				In the next ten years, how likely are you to be personally affected by the following threat? How likely are you to be personally affected by the effects of global warming?	89% (52%)	26% in Slovakia to 90% in Portugal
Transatlantic Trends GMF	2006	13	13044	I am going to read you a list of possible international threats to Europe/US in the next 10 years. Please tell me if you think each one on the list is an extremely important threat, an important threat, or not an important threat at all. The effects of global warming	90% (55%)	82% in Netherlands to 97% in Portugal
Gallup	2007	60	60182	Please tell me whether you agree or disagree with the following statements: global warming is having a serious impact now on the area where I live.	75% (37%)	Albania and Romania (98%) to Russia and Iceland (50%; 33%)

Table 2 provides information about public opinion on climate change from two recent surveys: GMF (2005) and Gallup (2007). It presents the percentage of respondents in each country for samples including 11 and 54 countries, respectively, and responses to three questions. A quick perusal of the table indicates substantial majorities across 11 countries feel that global warming is an important threat (i.e. nearly 80% of the population in each country), majorities in ten countries responded that they were likely to be personally affected by global warming in the next ten years (i.e. over two-thirds responding somewhat or very likely in ten of eleven countries), and substantial majorities across 56 countries agreed that global warming has serious impacts now on where they live (i.e. roughly 2/3 of the population in 42 countries). To answer the question posed in previous research regarding whether populations of some countries are more or less likely than others to find that climate change poses a potential danger, responses to the individual questions suggest somewhat different answers. For instance, nontrivial percentages of populations in 6 countries do not view global warming as an important threat in the next ten years, ranging from 10 percent of the population in Germany to 21 percent in the United States. A similar situation can be seen for responses regarding the likelihood of being personally affected by global warming in the next ten years, ranging from 11 percent of the population in Portugal to 75 percent of the population in Slovakia responding not too likely and/or not likely at all. Collapsing categories for the third question, furthermore, presents a similar picture, as in Iceland, Norway, and Russia, roughly half of the population agree and roughly half disagree about whether global warming is presently having a serious impact where they live.

The GMF data are for instance regionally concentrated, including European countries and the United States, while the Gallup data includes a wider geographic span (Europe, North America, Latin America, Africa and Asia/Pacific). Both datasets are limited, however, in available socio-demographic and attitudinal measures. In the analyses that follow, therefore, this research focuses on the ISSP 1993 and 2000 Environment data given the range of questions and scope of countries available, and because of their special focus on environmental issues and topics, enabling the examination of hypotheses from previous research on environmental concern cross-nationally.

Tab. 2: Public Opinion Surve Results for Two Recent, Cross-National Datasets

a. Results from GMF 2005 , Transatlantic Trends Data

	threat of GW[a]			personally affected[b]			
	not important threat	important threat	extremely important threat	not likely at all	not too likely	somewhat likely	very likely
Pooled	11	38	52	10	18	37	35
Germany	10	45	45	8	20	36	35
Spain	3	28	68	5	15	46	34
France	4	36	60	3	13	33	51
Italy	7	36	57	4	17	42	37
Netherlands	13	41	47	8	24	37	30
Portugal	4	19	77	3	8	37	53
UK	7	36	56	11	13	36	40
Poland	15	50	35	9	22	47	21
Slovakia	19	42	39	38	37	19	7
Turkey	16	42	42	13	14	35	38
USA	21	39	40	10	18	36	36

[a] The effects of global warming.

[b] How likely are you to be personally affected by the effects of global warming?

b. Results from Gallup 2007, Voice of the People Survey

	Agree strongly & slightly	Disagree strongly & slightly		Agree strongly & slightly	Disagree strongly & slightly
Pooled	75	25	Luxembour[	76	24
Albania	99	1	Macedonia	92	8
Argentina	87	13	Malaysia	82	18
Austria	57	43	Moldova	86	14
Bolivia	87	13	Netherlands	64	36
Bosnia-Herzegc	57	43	Nigeria	62	38
Bulgaria	87	13	Norway	52	48
Cameroon	83	17	Pakistan	78	22
Canada	77	23	Panama	80	20
Colombia	79	21	Peru	82	18
Croatia	91	9	Philippines	77	23
Czech Republic	68	32	Poland	65	35
Denmark	58	42	Portugal	82	18
Dominican Rep	81	19	Romania	98	2
Ecuador	91	9	Russia	51	49
Finland	65	35	Senegal	74	26
France	75	25	Serbia	83	17
Germany	56	44	Singapore	81	19
Ghana	83	17	South Africa	70	30
Greece	94	6	Spain	85	15
Hong Kong	94	6	Sweden	77	23
Iceland	34	66	Switzerland	60	40
India	71	29	Turkey	75	25
Indonesia	74	26	Ukraine	75	25
Ireland	61	39	UK	57	43
Italy	81	19	USA	60	40
Japan	87	13	Venezuela	70	30
South Korea	90	10	Vietnam, S	75	25
Kosovo	92	8			

"Global Warning is having a serious impact now on where I live."

Tab. 3: Frequencies for Knowledge Measures from the ISSP Environment Datasets

	Sample Size	Sample Size	Greenhouse Effect Fossil Fuels [a] 1993				2000								Greenhouse Effect Hole [b] 1993				2000							
	1993	2000	Def. not true	Prob. not true	Prob. true	Def. true	Def. not true	Prob. not true	Prob. true	Def. true	% Chg not true	% Chg prob not true	% Chg prob true	% Chg true	Def. true	Prob. true	Prob. not true	Def. not true	Def. true	Prob. true	Prob. not true	Def. not true	% Chg	% Chg	% Chg	% Chg
Pooled	28301	30604	4	14	49	33	4	12	48	36	0	-2	-1	3	31	43	11	15	29	43	12	16	-2	0	1	1
Australia	1779	.	2	12	59	28	.	.	.	.	.	.	.	.	18	46	17	18	.	.	.	.	.	.	.	.
Austria	.	1011	.	.	.	.	3	8	51	38	.	.	.	.	.	.	.	.	38	42	8	13	.	.	.	.
Britain	1261	971	2	11	51	36	3	11	50	37	1	0	-2	1	36	43	8	13	33	45	8	14	-4	2	0	1
Bulgaria	1183	1008	7	18	35	40	6	10	50	34	-1	-8	15	-6	49	34	8	9	45	40	8	8	-4	6	-1	-2
Canada	1467	1047	4	12	47	38	3	11	43	43	-1	0	-4	6	27	37	10	27	28	39	10	22	2	2	0	-4
Chile	.	1503	.	.	.	.	2	7	77	14	.	.	.	.	.	.	.	.	16	73	7	5	.	.	.	.
Czech Republic	1005	1244	5	13	38	44	6	14	42	38	1	1	4	-6	40	35	11	15	35	39	11	15	-5	4	1	0
Denmark	.	1069	.	.	.	.	2	8	51	40	.	.	.	.	.	.	.	.	23	42	11	25	.	.	.	.
East Germany	1092	527	3	12	45	39	3	12	47	39	-1	0	2	-1	44	35	9	13	42	33	8	17	-2	-2	-1	4
Finland	.	1430	.	.	.	.	3	20	55	22	.	.	.	.	.	.	.	.	14	41	29	16	.	.	.	.
Hungary	1167	.	5	17	44	34	.	.	.	.	.	.	.	.	32	42	11	15	.	.	.	.	.	.	.	.
Ireland	957	1184	3	12	51	34	3	8	50	39	0	-4	0	5	45	43	5	7	38	51	4	6	-7	8	-1	0
Israel	1198	1197	8	6	31	54	4	12	48	36	-4	6	16	-18	47	27	9	17	32	46	13	8	-15	19	5	-9
Italy	1000	.	5	11	45	39	.	.	.	.	.	.	.	.	46	36	7	11	.	.	.	.	.	.	.	.
Japan	1305	1144	3	11	53	33	4	13	53	31	0	3	-1	-2	21	38	23	19	16	38	25	21	-4	0	2	3
Latvia	.	1000	.	.	.	.	12	21	44	23	.	.	.	.	.	.	.	.	.	.	.	.	.	.	.	.
Mexico	.	1259	.	.	.	.	7	5	35	53	.	.	.	.	.	.	.	.	52	33	5	11	.	.	.	.
Netherlands	1852	1548	5	15	48	32	4	14	51	30	-1	-1	3	-2	25	36	11	28	16	40	15	30	-9	4	4	1
New Zealand	1271	1102	3	12	50	35	3	11	50	36	0	-1	0	1	35	40	8	16	26	45	9	20	-10	5	0	4
Northern Ireland	767	741	2	12	57	29	1	10	59	30	-1	-3	3	1	37	49	5	9	32	57	7	4	-5	8	2	-5
Norway	1414	1452	2	8	46	44	3	8	49	40	1	0	2	-4	26	38	9	27	21	39	11	28	-4	1	2	1
Philippines	1200	1200	15	23	46	16	13	21	43	24	-2	-3	-3	7	20	48	22	10	26	44	20	9	7	-4	-2	0
Poland	1641	.	5	22	51	22	.	.	.	.	.	.	.	.	31	59	7	2	.	.	.	.	.	.	.	.
Portugal	.	997	.	.	.	.	2	10	57	32	.	.	.	.	.	.	.	.	29	55	13	3	.	.	.	.
Russia	1931	1633	5	13	51	31	4	10	53	33	-1	-3	2	2	25	53	13	9	23	56	12	10	-3	3	-1	1
Slovenia	1032	1077	8	11	50	31	7	10	40	43	-1	-1	-9	12	33	46	10	11	36	41	9	14	3	-5	-2	4
Spain	1208	957	2	13	69	17	5	7	51	37	3	-6	-18	21	15	65	15	5	32	52	8	9	16	-13	-7	3
Sweden	.	1067	.	.	.	.	2	7	49	42	.	.	.	.	.	.	.	.	17	40	15	28	.	.	.	.
Switzerland	.	987	.	.	.	.	2	8	39	51	.	.	.	.	.	.	.	.	35	32	7	26	.	.	.	.
United States	1557	1275	4	22	53	21	5	22	51	21	1	0	-2	0	21	49	15	15	20	44	22	15	-1	-5	7	0
W. Germ.	1014	974	3	12	39	47	3	11	46	40	0	-1	7	-6	49	35	7	9	43	39	8	11	-6	4	1	1

[a] Every time we use coal or oil or gas, we contribute to the greenhouse effect.
[b] The greenhouse effect is caused by a hole in the earth's atmosphere.

The ISSP 1993 and 2000 datasets on the Environment includes data for 22 and 26 countries and four and three specific questions about global warming, respectively, as shown in Table 3. Examining frequencies for responses to the individual items shows interesting patterns regarding the cross-national distribution of climate change views. With regard first to knowledge, two questions were asked in each survey: 'Every time we use coal or oil or gas, we contribute to the greenhouse effect' and 'The greenhouse effect is caused by a hole in the earth's atmosphere' Correct responses for survey respondents pooled and by country show widespread knowledge and some confusion in 1993 and 2000. For instance, responses in Table 3 show an awareness of anthropogenic sources of greenhouse gas emissions, as the pooled responses indicate 82% in 1993 and 84% in 2000 of publics providing correct responses. These values range in 1993 from 62% in Philippines to 90% in Norway and in 2000 from two-thirds of the Philippines' public to 91% of the general public in Sweden and Chile. In 2000, 8 in 10 survey respondents in all but 4 countries responded correctly to this query. The second question evinces some possible confusion related to the greenhouse effect, where minorities of publics correctly responded to the query, as pooled responses indicate only 26% in 1993 and 28% in 2000 of publics providing correct responses. The values range in 1993 from 9% in Poland to 42% in Japan and in 2000 from about 11% of the Irish public to 46% of the Japanese public. These results are not entirely surprising (Brechin, 2003; Smith, 1996). Indeed, as noted earlier, the terms global warming and the greenhouse effect have been used interchangeably in some respects with climate change. In summary, we see mixed evidence of knowledge gains where the link between fossil fuel use and climate change has existed for some time while the greenhouse effect and the hole in the ozone layer of the atmosphere remain linked.

The third question taps a different aspect of climate change, that of perception of risk: "In general, do you think the rise in the world's temperature caused by the 'greenhouse effect' is ...", extremely dangerous for you and your family, very dangerous for you and your family, somewhat dangerous for you and your family, not very dangerous for you and your family, or not dangerous at all for you and your family? Pooled results in Table 4 show 60% of publics in 1993 and 63% in 2000 indicating a very or extremely dangerous threat posed by global warming. These results range in 1993 from 26% in Netherlands to 73% in Philippines and in 2000 from 40% of the Finnish public to 87% of the Portuguese public. A perusal of the percentages from 1993 and 2000 in some respects shows variation across countries with regard to perceived danger associated with climate change. Yet some consistency is also apparent. There is some movement toward identifying global warming as a danger. For instance, while more than one-fifth of survey respondents in twelve countries indicated that the rise in world's temperature was extremely dangerous for themselves and their families, by 2000 this had expanded to eighteen countries. Interestingly, in 2000 in only a few countries did somewhat large proportions of respondents not perceive global warming to be a particularly substantial personal threat, all of which are located in northern Europe (Denmark, Finland, and Norway). This represents a considerable change from 1993, where nontrivial percentages (i.e. greater than 10 percent) were shown in all but five countries. In sum, no discernable patterning linked with region in terms of geographic location is apparent, and a meaningful division linked with coastline length or an attribute like sea level also does not correspond with perceptions of danger linked with climate change.

Moving beyond description to explanation, this research fills an important gap in the literature. Drawing from previous research on environmental attitudes, this study investigates factors influencing views about climate change in order to determine whether there are similarities across contexts. Previous research suggests two sets of influences. On the one hand, it is possible that individual-level characteristics are likely to be driving attitudes of mass publics regarding climate change. This approach has characterized much of previous scholarship. On the other hand, it is possible that country-level attributes may in fact be influencing mass publics' attitudes over and above individual-level influences. In concert with these explanations, this research includes both sets of factors in empirical analyses, to be described in the next section.

Individual-level sources of views regarding climate change rely extensively on research examining environmental attitudes. Previous research suggests sociodemographic correlates of concern for the environment such as age, sex, urban residence, education, and income. Research suggests that younger individuals are more likely to express pro-environmental attitudes (Dunlap et al. 2000; Jones &Dunlap 1992; Kemmelmeier et al. 2002; Klineberg et al. 1998; Weaver 2002), differences between men and women exist (Bord & O'Connor 1997; Dietz, Stern & Kalof 2002; Kemmelmeier et al. 2002; Stern et al. 1993), and urban residents tend to exhibit higher levels of environmental concern (Jones & Dunlap 1992). Previous research confirms a positive effect of education on environmental concern (Jones & Dunlap 1992; Kemmelmeier et al. 2002; Klineberg et al. 1998; Weaver 2002). Prior scholarship also points to individual attributes such as knowledge and attitudes that may influence opinions about environmental issues. For instance, knowledge-based explanations suggest that correct information may lead to higher levels of environmental concern, though research generally is mixed (Blocker & Eckberg 1997; Dunlap 1998; Hayes 2001) especially across nations (Brechin 2003; Smith 1996). In a similar vein, attitudes about environmental issues generally may have an influence on those of a specific topic like climate change.

Contextual explanations suggest that various country-level attributes may be influential in shaping attitudes about climate change. For instance, greenhouse gas emissions may be linked with opinions about global warming. It is also possible that geographic factors indicative of vulnerability to the potential effects of a rise in the world's temperature may play a role (Roberts & Parks 2007). In summary, previous studies demonstrate that concern about environmental issues differs across countries, and it is also possible that the sources underlying its expression may also differ across countries and regions.

Tab. 4: Frequencies for Potential Danger Posed by Climate Change from ISSP Environment Datasets (1993 and 2000)

	1993					2000									
	not at all	not very	some-what	very	extreme-ly	not at all	not very	some-what	very	extreme-ly	% chg 'not at all'	% chg 'not very'	% chg 'some-what'	% chg 'very'	% chg 'extreme-ly'
Pooled	3	*12*	31	36	24	1	*7*	29	37	26	-2	-5	-2	1	2
Australia	2	*14*	32	37	15	.	.	.	.	.	.	.	.	.	.
Austria	.	.	.	.	.	1	*4*	19	49	28	.	.	.	.	.
Britain	2	*13*	37	25	22	0	*7*	39	29	25	-2	-6	1	4	3
Bulgaria	2	*7*	28	30	32	0	*4*	29	41	25	-2	-3	1	11	-7
Canada	2	*11*	33	31	23	1	*7*	32	34	26	-1	-4	-1	4	2
Chile	.	.	.	.	.	0	*1*	12	49	37	.	.	.	.	.
Czech Repul	2	*11*	27	40	20	1	*4*	21	47	27	-1	-7	-6	7	7
Denmark	.	.	.	.	.	2	*18*	35	27	18	.	.	.	.	.
East Germar	2	*10*	25	37	25	0	*2*	15	40	42	-2	-8	-11	3	17
Finland	.	.	.	.	.	2	*20*	38	26	14	.	.	.	.	.
Hungary	4	*20*	40	23	13	.	.	.	.	.	.	.	.	.	.
Ireland	2	*11*	34	29	24	0	*7*	39	35	19	-2	-4	4	5	-5
Israel	6	*11*	34	29	21	1	*9*	30	33	27	-5	-2	-4	5	7
Italy	2	*7*	27	38	26	.	.	.	.	.	.	.	.	.	.
Japan	2	*13*	34	33	19	0	*3*	23	41	32	-2	-10	-10	9	13
Latvia	.	.	.	.	.	1	*9*	35	32	23	.	.	.	.	.
Mexico	.	.	.	.	.	2	*6*	20	45	28	.	.	.	.	.
Netherlands	5	*25*	44	21	5	1	*11*	43	35	9	-4	-14	-1	15	4
New Zealanc	2	*15*	39	21	23	1	*8*	31	31	29	-1	-7	-8	11	6
Northern Irel:	2	*15*	42	21	20	0	*7*	29	36	29	-2	-8	-13	14	8
Norway	5	*24*	37	23	11	1	*16*	41	28	13	-4	-8	4	5	2
Philippines	1	*6*	20	53	20	0	*3*	16	33	48	-1	-3	-4	-20	28
Poland	2	*8*	49	27	14	.	.	.	.	.	.	.	.	.	.
Portugal	.	.	.	.	.	0	*1*	11	47	41	.	.	.	.	.
Russia	3	*12*	36	30	19	2	*6*	34	35	23	-1	-6	-1	5	4
Slovenia	2	*9*	27	37	25	1	*5*	27	39	28	-1	-4	0	2	3
Spain	2	*5*	28	50	14	0	*1*	17	54	27	-2	-4	-12	4	13
Sweden	.	.	.	.	.	2	*12*	40	31	15	.	.	.	.	.
Switzerland	.	.	.	.	.	0	*4*	25	42	29	.	.	.	.	.
United State:	3	*16*	38	26	16	2	*11*	39	29	19	-1	-5	2	3	3
W. Germ.	1	*7*	24	39	29	0	*3*	21	47	30	-1	-4	-3	8	1

Data and Measures

In the following analyses, data from the *International Social Survey Programme (ISSP) 1993 and 2000: Environment* are used (International Social Survey Program, 1996; 2003). For regression analyses in 2000, twenty-six cases are included: Austria, Britain, Bulgaria, Canada, Chile, Czech Republic, Denmark, former East Germany, Finland, Ireland, Israel, Japan, Latvia, Mexico, Netherlands, New Zealand, Northern Ireland, Norway, Philippines, Portugal, Russia, Slovenia, Spain, Sweden, Switzerland, the United States, and former West Germany. For analyses investigating change between 1993 and 2000, eighteen countries are included[4]. In this section, the measurement of the variables, methodological approach, and estimation strategy are discussed for the first set of analyses.

The dependent variable in the analysis gauges the degree of danger that the survey respondent perceives that global warming/climate change poses. To account for socio-demographics, five indicators are included. Respondent's sex (female=1) and place of residence (urban=1) are dummy variables. Age is measured in years. Education is measured as the number of years of schooling completed, and income is measured as total family income per capita (logged and standardized). Two additional measures are included to capture arguments that suggest cognitive antecedents to expressing concern about climate change. The first, linked with cognition, is a measure of scientific knowledge. The knowledge scale sums responses to five items and ranges from 5 to 20, with higher scores demonstrating greater knowledge (Smith 1996). The second measure is a scale of general environmental attitudes that sums responses to five questionnaire items selected based on their broad scope and from previous research[5]. These questions gauge broad ideas about interrelationships among societies and the environment. Taken together, they recognize that threats to the environment exist, addressing environmental problems is important and collective effort is necessary for resolving these issues, and that environmental issues are often entangled with broader processes of social change like economic issues and progress.

---

4 Britain, Bulgaria, Canada, Czech Republic, former East Germany, Ireland, Israel, Japan, Netherlands, New Zealand, Northern Ireland, Norway, Philippines, Russia, Slovenia, Spain, the United States, and former West Germany.

5 The first indicator is 'Many of the claims about environmental threats are exaggerated'. The second indicator is 'People worry too much about human progress harming the environment'. The third indicator is 'We worry too much about the future of the environment and not enough about prices and jobs today'. The fourth statement is 'There are more important things to do in life than protect the environment'. The fifth item is 'There is no point in doing what I can for the environment unless others do the same'. All of these items were rescaled such that a higher score indicates a pro-environmental response.

Results

The first set of empirical analyses regarding individual-level sources of opinion about climate change explores whether similarities can be identified across countries. Table 5 presents results of models assessing the sources of the perceived danger posed to survey respondents from the rise in the world's temperatures across all countries in the dataset. Dummy variables for each country (not shown) are also included in pooled analyses, representing fixed effects that account for unobserved differences across countries. Pooling the cases provides a preliminary assessment of whether there are representative patterns across national boundaries with regard to opinions about climate change.

Tab. 5: Regression Results for Climate Change Danger (Pooled)

	2000		Change between 2000 and 1993	
Intercept	2.853	**	-1.408	**
	(.076)		(.136)	
Female	0.062	**	0.206	**
	(.015)		(.030)	
Age	-0.001		0.000	
	(.001)		(.001)	
Urban	0.028		-0.041	
	(.016)		(.031)	
Education	0.007	**	-0.009	
	(.003)		(.005)	
Income	-0.024	**	-0.078	**
	(.008)		(.017)	
Knowledge	-0.023	**	0.001	
	(.003)		(.007)	
General Envtal Attitudes	0.054	**	0.075	**
	(.002)		(.004)	
R-sq	0.15		0.11	

* p < .05, ** p < .01

notes: models contain dummy variables (not shown)

In looking first at the general patterns, the variables tend to display the anticipated effects from previous research in the column of results from 2000. Since the sample sizes are quite large, the direction of the effect is of more interest than the variables' significance levels. Considered broadly, the relationships point to similarities across this set of countries. Briefly, being female, younger, having more education, and individuals having greater degrees of environmental concern express higher degrees of personal threat (i.e. to themselves and their families) from global warming. Having lower income and less scientific knowledge leads to expressing a higher degree of danger from global warming. Overall, the model explains 15 percent of the variation in perceived threat regarding climate change. For the results dealing with change between 2000 and 1993, results show some similarities for influences of being female, income, and general environmental attitudes. The differences in the coefficients across Table 5 could be caused by differences across countries in the regression slopes, thus leading to analyses that examine the countries separately.

To address the potential explanation of differences in regression slopes, Table 6 presents results for the regression models of opinions about the danger of climate change for each of the twenty six countries. Generally, the socio-demographic variables represent control variables in these analyses, and support for these indicators as driving forces of opinions about climate change is largely absent in these results. Beliefs about the dangers climate change poses appear to be largely a function of knowledge levels and general environmental attitudes, though these are not without exception. For instance, knowledge significantly affects opinions about the danger posed by climate change in half of the countries, where lower levels of knowledge affect higher levels of perception of threat due to climate change, all else being constant. Environmental attitudes have the strongest and most consistent effect on climate change opinions, with significant effects in twenty countries. Pursuing the coefficients across the countries in the models suggests striking similarities in patterning and in effect sizes to some degree, notably for the effect of environmental attitudes. For example, environmental attitudes significantly increase perception of danger posed by climate change in twenty cases, and the impact of unit increase of environmental attitudes on opinions about climate change is between .025 and .100 units across the whole sample. The adjusted r-squared values further suggest that these models share some similarities, ranging from two to fifteen percent of the variation in opinions about the threat of climate change (excluding four countries with values below one percent). They range widely however with regard to providing an overall picture of what individual-level measures influence views of climate change. They do, however, point to links between climate change views, knowledge, and general environmental attitudes, showing remarkable consistency in this respect. With regard to the second set of analyses examining change in views between 2000 and 1993 (not shown), these demonstrate that environmental attitudes are important influences on change in views of climate change.

Tab. 6: Results for Climate Change by Country (unstd coeffs; std errs in paren.)

	Female	Age	Urban	Educ	Income	Know.	Envtal Att	R-sq
Austria	-0.019	0.000	0.011	0.013	-0.042	-0.027	0.034 **	0.004
	(.105)	(.003)	(.113)	(.018)	(.056)	(.022)	(.014)	
Britain	0.158 **	-0.011 ***	-0.166 **	-0.012	-0.097 ***	-0.088 ***	0.123 ***	0.333
	(.065)	(.002)	(.069)	(.014)	(.035)	(.015)	(.009)	
Bulgari	0.021	-0.003	-0.119	-0.009	-0.111 **	-0.004	0.026 **	0.023
	(.104)	(.004)	(.115)	(.018)	(.055)	(.026)	(.014)	
Canad	0.031	0.001	-0.041	0.004	-0.025	0.006	0.091 ***	0.146
	(.072)	(.003)	(.069)	(.012)	(.038)	(.015)	(.014)	
Chile	-0.021	0.002	-0.016	0.012	0.011	-0.022 **	0.013	0.008
	(.052)	(.002)	(.053)	(.008)	(.032)	(.010)	(.009)	
Czech I	-0.006	0.000	0.050	-0.003	0.093 **	-0.041 **	0.047 ***	0.051
	(.082)	(.003)	(.089)	(.017)	(.043)	(.017)	(.010)	
Denma	0.122	-0.007	0.119	-0.020	-0.045	0.006	0.073 ***	0.096
	(.117)	(.005)	(.115)	(.015)	(.063)	(.025)	(.014)	
E Germ	0.171	0.003	0.131	0.024	-0.043	0.020	0.025 **	0.021
	(.109)	(.004)	(.112)	(.023)	(.062)	(.023)	(.013)	
Finland	0.349 ***	-0.012 ***	0.074	0.006	0.031	0.001	0.090 ***	0.188
	(.080)	(.003)	(.081)	(.010)	(.042)	(.019)	(.012)	
Ireland	0.104	0.002	0.052	0.010	-0.063	-0.020	0.078 ***	0.083
	(.072)	(.002)	(.079)	(.013)	(.042)	(.017)	(.011)	
Israel	-0.042	-0.003	-0.069	0.049 ***	0.086 *	-0.038 *	0.013	0.020
	(.084)	(.004)	(.090)	(.016)	(.048)	(.020)	(.013)	
Japan	0.032	0.001	0.054	-0.001	0.051	-0.030	0.039 ***	0.021
	(.096)	(.004)	(.098)	(.023)	(.061)	(.022)	(.012)	
Latvia	0.072	-0.003	0.022	0.009	-0.015	0.005	0.037 **	0.003
	(.092)	(.003)	(.096)	(.018)	(.049)	(.027)	(.017)	
Mexico	0.103	0.008 **	0.016	0.023 *	-0.015	-0.065 ***	0.003	0.029
	(.084)	(.003)	(.091)	(.012)	(.052)	(.019)	(.012)	
Neth.	-0.024	-0.002	0.159 ***	-0.011	0.025	-0.002	0.100 ***	0.158
	(.059)	(.002)	(.058)	(.009)	(.032)	(.013)	(.009)	
New Z	0.083	-0.002	0.005	0.000	-0.036	-0.044 ***	0.078 ***	0.087
	(.070)	(.002)	(.070)	(.016)	(.038)	(.017)	(.010)	
N Irel	0.118	0.004	-0.196	-0.009	-0.017	-0.083 **	0.104 ***	0.156
	(.119)	(.004)	(.217)	(.019)	(.068)	(.034)	(.058)	
Norway	-0.036	-0.004	0.087	0.002	-0.052	0.008	0.097 ***	0.150
	(.067)	(.002)	(.066)	(.012)	(.034)	(.014)	(.009)	
Philippi	-0.004	0.000	0.124 *	0.000	0.044	-0.037 **	-0.006	0.010
	(.059)	(.002)	(.066)	(.010)	(.035)	(.016)	(.010)	
Portug	0.158 ***	0.000	-0.090	0.024 **	-0.053	-0.012	-0.030 ***	0.049
	(.061)	(.002)	(.070)	(.010)	(.036)	(.021)	(.008)	
Russia	0.140 *	0.002	0.010	0.014	-0.058	-0.062 ***	-0.012	0.024
	(.084)	(.003)	(.090)	(.014)	(.046)	(.021)	(.012)	
Sloveni	0.044	0.001	-0.114	0.018	0.035	-0.058 **	0.051 ***	0.037
	(.103)	(.004)	(.134)	(.021)	(.067)	(.023)	(.016)	
Spain	0.021	-0.015 ***	0.013	-0.007	0.002	-0.045 *	0.024	0.082
	(.099)	(.004)	(.100)	(.014)	(.051)	(.025)	(.015)	
Sweden	0.038	-0.001	-0.079	0.026 **	-0.031	-0.068 ***	0.094 ***	0.149
	(.082)	(.003)	(.089)	(.013)	(.043)	(.018)	(.011)	
Switz.	-0.115	0.000	-0.038	0.004	-0.083 *	-0.034 **	0.073 ***	0.110
	(.088)	(.003)	(.086)	(.015)	(.047)	(.016)	(.011)	
US	0.033	-0.001	0.123 *	0.004	-0.136 ***	-0.054 ***	0.091 ***	0.144
	(.071)	(.002)	(.071)	(.014)	(.038)	(.018)	(.009)	
WGerm	0.019	-0.006 **	-0.021	0.010	-0.034	-0.050 ***	0.060 ***	0.098
	(.078)	(.003)	(.079)	(.013)	(.048)	(.016)	(.010)	

* p < .10, ** p < .05, *** p < .01

Some differences remain, however. For instance, though the models were generally quite consistent, it is not possible to completely rule out contextual explanations. In other words, how do levels of development, carbon emissions, or length of coastline matter with regard to shaping perceptions of climate change's imminent effects? In some respects, current literature, and especially popular media depictions, suggests that contextual factors linked with social, political, economic, and environmental conditions may be factors shaping views of climate change. These are not meant to be exhaustive, nor necessarily causal, but rather are considered here to be exploratory explanations that may account for how popular representations of the potential threat of climate change are realized. For instance, living along a coastline may be indicative of vulnerability to climate change's proposed effects (Roberts & Parks 2006), thus also affecting opinions regarding how dangerous climate change may be perceived to be. It may also be the case that having a higher level of wealth may insulate various populations from potentially harmful effects of a rise in the world's temperatures.

As an exploration of such 'context matters' approaches, a series of nested models are explored here. For the second set of analyses, three country-level variables are included in addition to the measures from the ISSP 2000 Environment data: gross domestic product (GDP) (World Bank 2002), CO_2 emissions (UNDP 2007), and coastline length (CIA Worldfactbook 2007). A series of steps for discerning differences between and within countries regarding opinions about climate change were devised from multilevel modeling to facilitate examination of aggregate-level effects to build on models previously described and present a preliminary assessment of within and between country variation in opinions about climate change. In other words, do features of countries influence how members of general publics react to an issue like climate change?

Hierarchical linear modeling allows researchers to consider nested models or multiple levels of data in accounting for relationships among contextual factors and individual-level outcomes (Raudenbush & Bryk 2002; Snijders & Bosker 1999). Multilevel modeling results show that 10.2% of the variance in opinions about climate change's potential danger is between countries. Next, results from a series of one-way ANCOVAs with random effects models show whether differences in average levels of the individual-level variables may partially explain some of the variation in countries with regard to climate change opinions. That is, if a particular variable like education has a strong effect on climate change opinions, and differences in average levels of education are apparent across countries, then these compositional differences in average levels of education can explain some of the country differences in opinions about climate change. Results (not shown) suggest potential compositional effects for five of the seven level-one variables included in the analyses (being female, education, income, knowledge, and environmental attitudes), or that about 5.6% of the variance in views regarding climate change is explained by the model including all of the individual-level predictors. Most of the compositional differences or between-country variation for the individual-level model is the result of the social psychological variables (knowledge and environmental attitudes).

While there is some evidence of compositional differences, it appears as though a non-trivial amount of variation rests at the country-level and may be due to contextual variables rather than individual-level variables. Though the aforementioned country-level factors (GDP per capita, CO_2 emissions, and coastline) do not have significant effects on country differences in opinions about climate change (results not shown), these factors may however influence the slope coefficients of the individual-level predictors.

Results from random-coefficient models (RCRMs) show that in a country, education, being female, and environmental attitudes are all positively related to views of climate change, while scientific knowledge is negatively related to perceptions of threat regarding climate change. In summary, results show that four of the individual-level variables have significantly different effects on environmental concern across the twenty four countries included in this analysis. Taking into account level one predictors, 8.6% of the variation in the model is explained. Next, building on the RCRM results, four slopes-as-outcomes models are specified. In each case, the slopes of sex, education, knowledge, and environmental attitudes were regressed on GDP, carbon dioxide emissions, and length of coastline separately (results not shown), then all in a combined model. Results for two of the analyses of slopes as outcomes (i.e. being female and environmental attitudes) suggest some support for contextual hypotheses, while those for education and knowledge suggest that none of these three country-level variables significantly affect the relationship between these variables and views of climate change.

Although presented here as exploratory analyses, this discussion is interesting as an extension of popular media accounts regarding climate change opinions. Indeed, the results suggest that searching for additional country-level variables could aid in accounting for remaining variability. Prior research regarding cross-national patterns in opinions of climate change does not include contextual effects as explanatory factors. This research therefore provided exploratory analyses aimed at discerning the extent to which perceptions of climate change's threat across countries can be explained by differences in within and between country variability. Results indicated that there is some variation among the twenty four countries in this sample, suggesting potential compositional effects. Though there is some evidence of compositional differences, it appears that a considerable amount of variation rests at the country-level and may be due to contextual variables rather than individual-level variables. While promising in an exploratory sense, the results also suggest that a search for additional country-level variables, particularly those that account for possible increased risk of being adversely affected by climate change, could aid in accounting for the remaining variability in slopes. Regional investigations would also make important contributions to understanding spatial patterns underlying environmental conditions like pollution.

Discussion and Conclusion

Are differences noted in previous studies documenting public opinion about climate change only a matter of degree? This research addresses this question through an examination of public opinions surveys that include measures of climate change. Contrary to previous studies (Brechin 2003; Dunlap 1998), knowledge and awareness of climate change appears to be growing among general publics across the globe. Indeed, more similarities than differences are apparent with regard to the role that knowledge and environmental attitudes play in shaping these views. Although climate change is an issue uniquely positioned at the nexus of academic fields spanning the physical and social sciences, scholarship from the social sciences documenting its distribution across various social groups and across contexts has not adequately addressed issues relevant for current policy discussions. Moreover, future scholarship needs to pay careful attention to the type of question explored and how this influences what drives particular responses. For instance, perceptions of danger related to climate change may play a role for some social categories (e.g. income) but not others, where economic or political factors might shape responses to some questions, especially those gauging policy views.

With regard to the global reach of climate change and how to explain its influences, as reflected in public opinion data, this research provides a partial explanation, and important questions remain regarding contexts. For instance, much of what we know about public opinion regarding climate change involves information sources that play a key role in disseminating materials to the general public. The mass media is a vital source for such information; thus a considerable amount of research centres on how the media has depicted the issue (Dispensa and Brulle 2003). How the mass media shapes public opinion within and potentially across contexts remains an issue to be investigated. The extent to which public opinion about climate change intersects with various policies regarding proposed strategies for mitigation, including especially the prominent role of industrialized countries in policy construction and implementation, remains an area to be explored in subsequent research. Such research will, in time, and with additional waves of data collection, serve as influential to the captive audience that includes scientists, policy-makers, governments, communities, stakeholders, and citizens, all as important actors in shaping what lies ahead.

References

Blocker T.J. and Eckberg, D. (1997), "Gender and Environmentalism: Results from the 1993 General Social Survey." Social Science Quarterly 78, pp. 841-858.

Bord, R. and O'Connor, R. (1997), "The Gender Gap in Environmental Attitudes: The Case of Perceived Vulnerability to Risk." Social Science Quarterly 78, pp. 830-840.

Bord, R.J., A. Fisher, and O'Connor, R.E. (1998), "Public Perceptions of Global Warming: United States and International Perspectives." Climate Research 11, pp. 75-84.

Bostrom, A.M., Morgan, G., Fischhoff, B. and Read, D. (1994), "What Do People Know About Global climate Change? 1. Mental Models." Risk Analysis 14, pp. 959-70.

Brechin, Steve (2003), "Comparative Public Opinion and Knowledge on Global Climate Change and the Kyoto Protocol: The US Versus the World?" International Journal of Sociology and Social Policy 23(10), pp. 106-134.

Brewer, T.L. (2005), "US Public Opinion on Climate Change Issues: Implications for Consensus Building and Policy Making." Climate Policy 4, pp. 359-376.

Dietz, T., Dan, A. and Schwom, R. (2007), "Support for climate change policy: social psychological and social structural influences." Rural Sociology 72(2), pp. 185-214.

Dietz, T., Kalof, L. and Stern, P. (2002), "Gender, Values and Environmentalism." Social Science Quarterly 83(1), pp. 353-364.

Dispensa, J.M. and Brulle, R.J. (2003), "Media's Social Construction of Environmental Issues: Focus on Global Warming – A Comparative Study." International Journal of Sociology and Social Policy 23, pp. 74-105.

Dunlap, R. (1998), "Lay Perceptions of Global Risk: Public Views of Global Warming in Cross-National Context." International Sociology 13(4), pp. 473-498.

Dunlap, R., van Liere, K., Mertig, A. and Jones, R. (2000), "Measuring Endorsement of the New Ecological Paradigm: A Revised NEP Scale." Journal of Social Issues 56(3), pp. 425-442.

Hayes, B. (2001), "Gender, Scientific Knowledge and Attitudes Toward the Environment: A Cross-National Analysis." Political Research Quarterly 54, pp. 657-671.

International Social Survey Program (ISSP) (1993/1996). INTERNATIONAL SOCIAL SURVEY PROGRAM: ENVIRONMENT, 1993 [Computer file]. ICPSR version. Koeln, Germany: Zentralarchiv fuer Empirische Sozialforschung [producer], 1993. Koeln, Germany: Zentralarchiv fuer Empirische Sozialforschung/Ann Arbor, MI: Inter-university Consortium for Political and Social Research [distributors], 1996.

– (2003) INTERNATIONAL SOCIAL SURVEY PROGRAM (ISSP), 1985-2000 [CD-ROM]. Cologne, Germany: Zentralarchiv fuer Empirische Sozialforschung an der Universitaet zu Koeln [producer], 2003. Cologne, Germany: Zentralarchiv fuer Empirische Sozialforschung/Ann Arbor, MI: Interuniversity Consortium for Political and Social Research [distributors], 2003.

Jones, R. and Dunlap, R. (1992), "The Social Bases of Environmental Concern: Have They Changed Over Time?" Rural Sociology 57, pp. 28-47.

Kemmelmeier, M., Krol, G. and Kim, Y.H. (2002), "Values, Economics, and Pro-Environmental Attitudes in 22 Societies." Cross-Cultural Research 36(3), pp. 256-285.

Kempton, W. (1991), "Public Understanding of Global Warming." Society and Natural Resources 4, pp. 331-335.

Klineberg, S., McKeever, M. and Rothenbach, B. (1998), "Demographic Predictors of Environmental Concern: It Does Make a Difference How It's Measured." Social Science Quarterly 79(4), pp. 734-753.

Nisbet, M. and Myers, T. (2007), "The Polls-Trends: Twenty Years of Public Opinion About Global Warming." Public Opinion Quarterly 71(3), pp. 444-470.

O'Connor, R.E., Bord, R.J., Yarnal, B. and Wiefek, N. (2002), "Who Wants to Reduce Greenhouse Gas Emissions?" Social Science Quarterly 83, pp. 1-17.

Program on International Policy Attitudes (PIPA) (1998). Americans on Global Warming. College Park, MD: University of Maryland.

– (2007), "International Polling on Climate Change." Available at www.worldpublicopinion.org. Electronic resources accessed April 4, 2008.

Raudenbush, S. and Bryk, A. (2002), Hierarchical Linear Models, Second Edition. Thousand Oaks: SAGE.

Roberts, J.T. and Parks, B. (2006), A Climate of Injustice: Global Inequality, North-South Politics and Climate Policy. The MIT Press.

Smith, T. (1996), "Global Environmental and Scientific Knowledge", pp. 105-127. In: Frizzell, Alan and Pammett, Jon H. (eds.), Shades of Green: Environmental Attitudes in Canada and Around the World. Canada: Carleton University Press.

Snijders, T.A.B. and Bosker, R. (1999), Multilevel Analysis: An Introduction to Basic and Advanced Multilevel Modeling. Thousand Oaks: SAGE.

Stern, P., Dietz, T. and Kalof, L. (1993), "Value Orientations, Gender and Environmental Concern." Environment and Behavior 25, pp. 322-348.

Weaver, A. (2002), "Determinants of Environmental Attitudes." International Journal of Sociology 32(1), pp. 77-108.

Chapter 18

The Campus Climate Challenge – Innovative Multi-Stakeholder Approaches to Reducing Greenhouse Gas Emissions at Canadian Colleges and Universities

Anjali Helferty, Amelia Clarke, Rosa Kouri

Abstract

The Campus Climate Challenge in Canada involves multi-stakeholder approaches to reducing greenhouse gas emissions on campuses, and transforming university and college campuses into leaders in addressing climate change. Coordinated by the Energy Action Coalition, it is a US-Canada campaign operated by the Sierra Youth Coalition (SYC) on Canadian campuses. Each campus project is based on a unique multi-stakeholder approach, bringing together campus operations, administration, faculty, students, and external organizations to implement sustainability initiatives. This process ensures that each climate change plan is specific to its local context, engages individual campus expertise, builds the capacity of the next generation of Canadian leaders, and has broad commitment from the campus community. Four cases are presented and discussed in this chapter, highlighting the multi-stakeholder aspect of each: Sustainable Ambassadors at Concordia University; a student-built wind turbine initiative at the University of Saskatchewan; Residence Reduction Challenges in southern Ontario; and Go Beyond in British Columbia. These cases provide a demonstration of initiatives at different scales for reducing GHG emissions at complex public institutions. The conclusions offer reflections on the implications of these different approaches for the stakeholders involved; student leaders, sustainability coordinators, faculty, staff, external NGOs and provincial governments.

Introduction

The Campus Climate Challenge in Canada involves multi-stakeholder approaches to reducing greenhouse gas emissions on campuses, and transforming university and college campuses into leaders in addressing climate change. Coordinated by the Energy Action Coalition, it is a US-Canada campaign which is operated by the Sierra Youth Coalition (SYC) on Canadian campuses. With SYC, over 65 campuses have worked through the Challenge to adopt emissions reductions plans; resulting in 35 applied projects, including 14 comprehensive greenhouse gas (GHG) emissions inventories as of April 2007. SYC, the youth arm of the non-profit organization Sierra Club Canada, provides training, networking, and common evaluation and tracking tools. Each campus project is based on a unique multi-stakeholder approach, bringing together campus operations, administration, faculty, students, and external organizations to implement sustainability initiatives. This process ensures that each climate change plan is specific to its local context, engages individual campus expertise, builds the capacity of the next generation of Canadian leaders, and has broad commitment from the campus community. The cases in this Chapter provide an excellent demonstration of best practices for reducing GHG emissions at complex public institutions, engaging a diversity of stakeholders and laying the foundation for deeper and long-lasting sustainability.

This chapter begins by looking at the campus sustainability literature, and the multi-stakeholder literature. Then there is an introduction of the overall Campus Climate Challenge, the Canadian component, and the multi-stakeholder approach. This is followed by a description of four Canadian campus initiatives, highlighting the multi-stakeholder aspect of each: Sustainable Ambassadors at Concordia University; a student-built wind turbine initiative at the University of Saskatchewan; Residence Reduction Challenges in southern Ontario; and Go Beyond in British Columbia. The discussion compares the campus projects and the conclusion offers implications for different stakeholders and for addressing climate change.

Campus Sustainability Management and Climate Change Initiatives

There is a burgeoning literature on campus sustainability management, which is particularly found in the *International Journal of Sustainability in Higher Education*, and numerous books on the topic (such as: Leal Filho 1999; Keniry 1995; Corcoran & Wals 2004). The Sustainable Campuses program of the Sierra Youth Coalition has been detailed by Beringer (2006) in a comprehensive article on the experience at the University of Prince Edward Island using the Campus Sustainability Assessment Framework (CSAF). The CSAF is a Sierra Youth Coalition tool which is used across Canadian campuses to provide a broad and detailed overview of the social, environmental, and economic sustainability of the institution (Cole & Wright 2005; Cole 2003). Canadian experiences in campus sustainability have been sporadically documented in the academic literature (For example: Clarke 2006; Moore et al. 2005; Beringer et al. 2008; Bardati 2006; Gudz 2004; Richardson & Lynes 2007; Conway et al. 2008), and, for the past several years, comprehensively documented by the Sierra Youth Coalition (Sierra Youth Coalition 2007b).

Within the campus sustainability literature is a limited focus on climate change-related initiatives. The first group of initiatives focus on energy reduction through building design, through energy efficient equipment, and through social marketing to students (Marcell et al. 2004; Kahler 2003; Pearce 2006). This is complemented by increased measuring of energy savings (Pearce 2006) and conducting greenhouse gas inventories (Marcell et al. 2004). The second group of initiatives involves the purchasing or generation of renewable energy. For example, the University of Vermont has installed solar panels, and Carnegie Mellon University has shifted its power supply portfolio to include wind power (Marcell et al. 2004).

> "According to a 2001 survey of 891 American higher education institutions undertaken by the National Wildlife Federation, nearly a quarter meet some of their energy needs from renewable energy, and more than half have developed energy efficiency design codes for new and existing buildings" (Marcell et al. 2004: 169).

In addition to direct energy reduction and renewable energy generation, transportation, offsets, refrigerants, waste, agriculture, and purchasing initiatives can also be used to reduce direct and indirect greenhouse gas emissions from university and college campuses (Sierra Youth Coalition 2008). Many of these initiatives require a multi-stakeholder approach.

Multi-Stakeholder Approach

At the organizational level, "a stakeholder in an organization is (by definition) any group or individual who can affect or is affected by the achievement of the organization's objectives" (Freeman 1984: 46). When considering issues-based collaborations, such as climate change initiatives, the stakeholders are any group or persons who can affect or are affected by the common agenda. For the purpose of this study, the stakeholders are the formal partners in the four initiatives discussed. These stakeholders may come from within the leading organization, or be external to the organization. Collaboration, which is another term for multi-stakeholder approach, is defined by Gray (1989: 5) as "a process through which parties who see different aspects of a problem can constructively explore their differences and search for solutions that go beyond their own limited vision of what is possible." It is a "process of joint decision-making among key stakeholders of a problem domain about the future of that domain" (Gray 1989: 11).

This chapter builds upon the existing campus sustainability management literature by considering four different Energy Action Coalition initiatives being undertaken in Canada; all of which involve multi-stakeholder approaches.

The Energy Action Coalition

The Energy Action Coalition was founded in June 2004 by youth climate leaders who came together to build a youth climate movement in the US and Canada after the first Fossil Fools Day on April 1, 2004. One year and three international days of action later, the reach of the Energy Action Coalition had skyrocketed, with the second annual Fossil Fools Day in April 2005 being comprised of more than 300 actions in Canada, the United States, England, Nigeria and Panama.

A unique aspect of the Energy Action Coalition is the broad diversity of groups involved. Members are from Canada and the US; they are environmental justice organizations and actions-based NGOs, faith groups and student associations. The Energy Action Coalition has seen its share of challenges, but has become stronger and stronger each year.

> The **Energy Action Coalition** unites a diversity of organizations in an alliance that supports and strengthens the student and youth clean energy movement in North America. The partners of the Energy Action Coalition work together to leverage our collective power and create change for a clean, efficient, just and renewable energy future. The work of the Energy Action Coalition is focused on four strategic areas: campuses, communities, corporate practices, and politics.

The three-year goals of the Energy Action Coalition are to:

1. Build a broad and diverse coalition.
2. Empower life-long leaders for the climate and a just energy future.
3. Pass comprehensive policies for climate neutrality, education, curriculum, research, and community relations on 250 campuses. Engage traditionally disenfranchised communities in climate work and prioritize support and solidarity for these communities.
4. Make 2008 the "Climate Election" [in the United States] and help pass bold, comprehensive, and just federal legislation.
5. Pressure corporations to become accountable to communities and the climate and empower our generation to build the green economy (Energy Action Coalition 2008).
6. In spring of 2008, at the end of the second year in this three-year timeline, many of the objectives set out with these goals have already been surpassed.

The Sierra Youth Coalition and the Energy Action Coalition

> The **Sierra Youth Coalition**, in each of our program areas, strives to achieve the following objectives:
> **Involve**: empower youth in Canada to become active members of their community;
> **Educate**: create a widespread understanding that all things are connected;
> **Change**: challenge unjust and unsustainable systems by using a solutions based approach – www.syc-cjs.org

The Sierra Youth Coalition (SYC) was one of the founding members and is currently the sole Canadian member of the Energy Action Coalition. In the years since its founding, the Energy Action Coalition has undertaken a number of initiatives ranging from Days of Action, to participating in UN negotiations, to the *Road to Detroit*, a hybrid car and biodiesel bus trip to organize activists in local communities on energy issues with the goal of building support for clean car legislation and asking individuals to pledge that their next car will be green. It was determined, however, that a more comprehensive campaign was needed to build a youth movement around climate change – it was from this that the Campus Climate Challenge was developed.

The Sierra Youth Coalition was highly involved in the development of the Campus Climate Challenge. Jeca Glor-Bell was the Ontario Sustainable Campuses Coordinator for SYC at this time, and was a significant participant in developing strategies for the campaign. In the following years, Geneva Guerin and Anjali Helferty, Sustainable Campuses National Coordinators from 2004-2006 and 2006-2008, respectively, were very involved with the ongoing progress of the Campus Climate Challenge.

> The Campus Climate Challenge is a project of more than 30 leading youth organizations throughout the U.S. and Canada. The Challenge leverages the power of young people to organize on college campuses and high schools across Canada and the U.S. to win 100% Clean Energy policies at their schools. The Challenge is growing a generation-wide movement to stop global warming, by reducing the pollution from our high schools and colleges down to zero, and leading our society to a clean energy future.
> – www.climatechallenge.org

The Sierra Youth Coalition, through its Sustainable Campuses program, engages students on colleges and universities across Canada to make their own campuses more sustainable in their operations, curriculum and culture. Energy-related work is an intrinsic element of campus sustainability, and students in Canada had been working on energy efficiency and conservation initiatives (both directly through reducing the use of electricity and indirectly through reductions in paper use and waste), as well as renewable energy purchasing for several years before the Campus Climate Challenge began. What the Campus Climate Challenge uniquely brought to Sustainable Campuses was the collective strength of a diversity of organizations that were coming together to work on the same issue in many different ways – on campuses and in communities, through clean energy purchasing and retrofits, through direct actions and through collaboration. Each organization was able to take its own strengths and experiences and apply them to running the program. In the case of the Sierra Youth Coalition, one of the main strengths of Sustainable Campuses was, and continues to be, the multi-stakeholder approach to change.

Why Multi-stakeholder?

When running programs and campaigns, it is important that the strategies adopted fit the needs of the campaign. By 2002, Sustainable Campuses was designed around a large research tool – the Campus Sustainability Assessment Framework (Cole 2003). Given that the project was primarily run by students who were taking it on in a volunteer capacity, there was a clear need for collaboration with a variety of other campus stakeholders if the assessment was going to be successful and if there were going to be changes implemented following the assessment. The SYC staff and volunteers who were working to develop Sustainable Campuses in 2002 and 2003 took the most strategic route available to achieve institutional change, and made the multi-stakeholder approach one of the key elements of the program. The SYC Multi-Stakeholder Guide, in a sentence that builds on the work of Sharp (2002), explains the importance of this multi-stakeholder approach:

> "Evidence has shown that the greatest successes in achieving institutional change (for sustainability or otherwise) exist when all the sub-cultures of the campus community (students, faculty, staff and administration) come together to form a shared vision and organizational alignment in their respective departments" (Sierra Youth Coalition 2007a: 3).

Snap Shot of Why a Multi-Stakeholder Process Works:

- Students will graduate! Efforts will only endure if they are supported institutionally.
- Multi-stakeholder process builds bridges to overcome polarization of departments.
- Information, resources, priorities and perspectives of different stakeholders can be shared.
- The power of the multi-stakeholder group can leverage action from within the university.
- Official recognition by many stakeholders can give campus sustainability a higher profile.

(Adapted from: Sierra Youth Coalition 2007a: 3)

In SYC's Multi-Stakeholder Guide, advice, which is adapted from Innes and Booher (1999), is offered on how to develop a multi-stakeholder group. They suggest that in order to be successful in the initial stages the following is needed:

- Include representatives from all relevant campus stakeholder groups;
- Ensure everyone involved is aware of the purpose and task of the group;
- Be self-organizing, allowing participants to decide by consensus on ground rules, objectives, and tasks;
- Ensure the process is engaging for participants and incorporates informal interaction (community building);
- Ensure the process fosters creative and "out-of-the-box" thinking;
- Seek consensus only after the issues have been fully explored and a significant effort has been made to address and respond to differences amongst the group (Adapted from: Sierra Youth Coalition 2007a: 10).

There are also a number of ways to determine if the process is achieving the desired outcomes. Here are some examples of outcome criteria that can indicate a successful multi-stakeholder process:

- It increases the sense of community, trust and unity of the group or committee;
- It has the support of management (even in principle) that can be substantiated with action;
- The face-to-face communication is well coordinated, facilitated and maximized;
- It produces information and decisions that all the committee stakeholders understand and accept;

- It produces agreements that recommend actions;
- The learning and change are shared beyond the individuals in the committee or group;
- It sets in motion a cascade of changes in attitudes, behaviours and actions, spin-off partnerships and new practices; and
- It improves the ability of the entire campus community to be more effectively responsive to change and conflict (Adapted from: Sierra Youth Coalition 2007a: 10).

Figure 1 displays a summary of the outcomes of multi-stakeholder initiatives over three time frames.

Fig. 1: Initial, Interim, and Long Term Multi-Stakeholder Process Outcomes

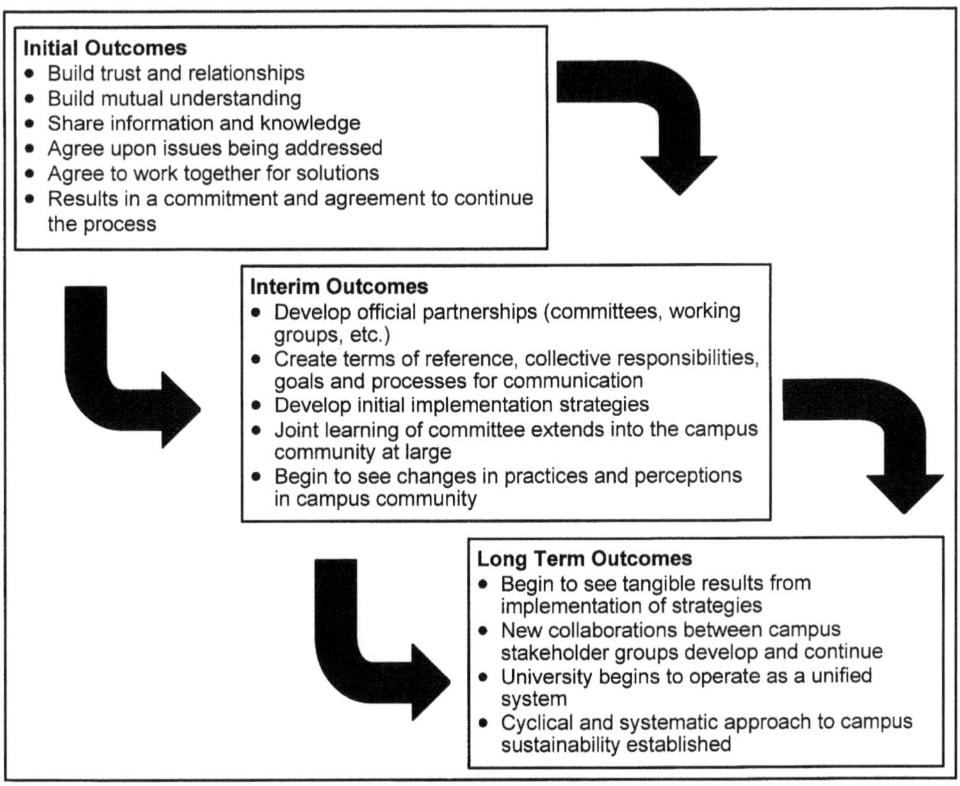

(Sierra Youth Coalition 2007a: 11)

The multi-stakeholder approach advocated by SYC for its Sustainable Campuses program has been used in different and innovative ways by the campuses involved in the Campus Climate Challenge.

Case Studies

In this chapter, we provide case studies of four different projects at Canadian campuses. Of these, two are located within a university, while the other two involve multiple universities at the regional and provincial level. The first example, the Sustainable Ambassadors program at Concordia University in Montreal, Quebec, focuses on energy use reductions directly through individual action of faculty and staff, as well as some minor types of institutional change. In the second case, the wind turbine initiative at the University of Saskatchewan in Saskatoon, Saskatchewan, focuses on local renewable energy production. The Residence Reduction Challenge addresses energy reductions from individual student action across three universities in southern Ontario. Our final example, the Go Beyond campaign, involves reductions across multiple universities in the province of British Columbia, and is coordinated by a formal coalition.

Table 1 displays the four cases in relation to their stakeholders involved, level of engagement (scale) and the type of climate change initiative. All of these cases involve what Rondinelli and London (2003) call moderately or highly intensive interactions.

Tab. 1: Introduction to Four 'Best Practice' Cases

		Level of Engagement			Climate Change Initiatives		
Case	Stakeholders Involved	Within One University	Between Universities	Provincial	Energy Reduction	Energy Production	Other GHG Reduction
Ambassadors	Students Faculty Facilities Mgmt. Employees	X			X		X
Wind Turbine	Students Department Facilities Mgmt. SYC	X				X	
Residence Reduction Challenge	Residences Universities Facilities Mgmt. SYC		X		X		X
Go Beyond	Universities Government Private Sector Other NGO SYC			X	X		

For each case, research was conducted by interviewing involved individuals and through the websites of the respective initiatives. Interviews were conducted for the Sustainable Ambassadors case with Jonathan De Luca, the student coordinator of Sustainable Ambassadors, and Jenn Davis, the Sustainability Coordinator at Concordia University; for the wind turbine initiative case with Jeh Custer, the SYC Sustainable Campuses Prairies Coordinator, and Margret Asmuss, the Sustainability Coordinator at the University of Saskatchewan; for the Residence Reduction Challenge case with Monique Woolnough, the SYC Sustainable Campuses Ontario Coordinator; and for the Go Beyond case with Maggie Baynham, the SYC Sustainable Campuses British Columbia Coordinator. All of these individuals were actively involved with the development and/or delivery of the cases.

Case I – Sustainable Ambassadors, Concordia University

Concordia University is a large university located in Montreal, Quebec, one of Canada's most dense urban centres. The university has a total student population of 31,640. Greenhouse gas emissions for the 2004-2005 academic year were calculated to be 13,471 metric tonnes of CO_2 equivalents (Sustainable Concordia 2006). Concordia University was one of the first campuses to be involved in SYC's program for Sustainable Campuses, and was the pilot campus for the Campus Sustainability Assessment Framework (CSAF). As a result, Concordia has a history of multi-stakeholder initiatives related to sustainability; these have resulted in the creation of a Sustainability Officer position in the university administration who works alongside student-led initiatives. The mandate of the sustainability office is to address a spectrum of environmental, economic, and social issues related to sustainability.

Influencing the consumption habits of campus community members is one piece of this work, and is addressed through an initiative called Sustainable Ambassadors. Their website explains the project:

> "A sustainable ambassador is a member of Concordia's staff or faculty who strives to promote and encourage sustainable actions and attitudes in the workplace by leadership and example. Many people today recognize that, as part of an institution of higher learning, Concordia campus members need to employ and disseminate their knowledge and expertise in moving towards a more sustainable community. This program has been created to access such potential. In creating this change towards more sustainable thinking and action, the passion, knowledge and creativity of the Concordia community are called upon. In becoming an Ambassador, you commit to: make every attempt to lead by example; endeavoring to understand the consequences of your actions; rethink personal behaviours that lead to wasteful consumption patterns and work to reduce them; engage in activities that build a sustainable campus and community; and encourage friends and colleagues to do the same" (Sustainable Concordia 2008).

In total, 12 ambassadors were active at the time of study. They conduct energy audits of their departments and provide a recommendations report that identifies pertinent areas for improvement. They also help fellow staff members and faculty identify any new needs or opportunities related to sustainability. In such a large university, communication can be weak about what programs and options are available. Sustainable Ambassadors are a liaison between what the university itself can offer, and how faculty and staff can take advantage of these options. This is relevant in the case of energy in particular, where much control over energy consumption is centralized under facilities management. Sustainable Ambassadors are given the information needed to, for example, know where motion-sensors for indoor lighting could be installed and are able to get in contact with facilities management who can then install the detectors. While the energy reduction of the overall program has not been monitored, one initiative within the program, a two-week lights out campaign, resulted in a 20% reduction in lights left on.

In the context of Concordia University, where a large-scale student initiative had been successful in putting sustainability on the forefront of campus discourse, and particularly amongst the student population, this program is designed to reach staff and faculty and tailor specific programs to their needs. The approach is intentionally bottom-up, in order to communicate effectively with the multiple stakeholders and foster ownership about implementing changes into every day work functions. Some departments are particularly important to work with in this program, such as information technology, as they can set up systems (such as default double-sided printing, or energy efficient computer power usage), which are then used throughout the campus.

In this case, the students have the advantage of working within a well-established and respected office on campus and are able to leverage the pre-existing administrative support present at Concordia University to establish their legitimacy as a student group running a program for faculty and staff. Given that the program has only just finished its second year, it is still quite small; however, there is a great opportunity for growth within the institution for this unique model of multi-stakeholder engagement.

Case II – Student-Built Wind Turbine, University of Saskatchewan

The University of Saskatchewan is a mid-sized university, located in a major city of a sparsely populated and largely rural province. The campus covers a large land base, and includes one of only four veterinary colleges in Canada, as well as one of the most comprehensive agricultural colleges in North America. It has a student population of 18,620. University of Saskatchewan C02 emissions for 2006 were calculated at approximately 197,805 tonnes of CO_2 equivalents. This is about 4.5% above the 1990 levels, much lower than the provincial increase of emissions of approximately 56% since 1990. A wind turbine is being constructed by students to be placed on campus and produce energy for campus use. The projected energy production of the wind turbine is unknown at this time.

Students have spearheaded the process, negotiating the regulatory approvals and fundraising through small-scale events. When interviewed, Footprint Designs, the student group leading the initiative, stressed that the turbine is a symbolic step, showing that student ideas can be implemented with the support of different stakeholders at the university. It can also serve as an educational tool for the university and wider community when completed and placed in a public space on campus. Furthermore, the operation and maintenance of the turbine is an opportunity for specific technical instruction – the project proponents are currently exploring having students manage the turbine as part of course work. According to Footprint Designs, it is a challenge that the university lacks a formalized commitment to sustainability – the multi-stakeholder process within the university could be formalized to ensure that the project can go further and that sustainability becomes the norm.

Although the project is student-driven, they have brought on board different stakeholders in facilities management and beyond. The completed project will bring the student-made turbine onto public university land, producing energy for facilities management. According to Margret Asmuss, the Sustainability Coordinator at the University of Saskatchewan, "We are trying to intersect our operational issues with student research projects and student education in general. This is the sort of thing we need to be doing, working with students, and how they can help us address sustainability issues".

In this case, there is no specific program being delivered and, as a result, there is less of a need for a formal multi-stakeholder approach. In order for the turbine to be operational on campus, there will certainly be a need to work with the facilities management; however, there is not a clear need to have a buy-in from many other campus stakeholders. Where the multi-stakeholder approach is necessary in this case is in looking beyond the operational aspect of the turbine to the learning and development elements; for example, the possibility of students working on the turbine as part of their coursework. There are many opportunities for this type of innovation on campuses, and it is essential that a diversity of stakeholders be able to participate in the initiative in order to maximize the potential for innovation.

Case III – The Residence Reduction Challenge, Southern Ontario

The Sierra Youth Coalition operated the first Residence Reduction Challenge in southern Ontario during the academic year 2007-2008. It ran on three campuses, the University of Waterloo, the University of Guelph, and Queen's University. The three universities are all mid-size, with student populations ranging from 20,550 to 24,160. The competition was both intra-university, held between residences at each university, and inter-university, between the three universities. It involved a number of challenges:

1. Which institution could encourage the most students to sign a sustainability pledge as a proportion of their population,

2. Which institution could provide evidence of the most behavioural changes related to energy use, through before and after surveys, and
3. Which institution could reduce the most energy and other consumption during the course of the challenge, according to metered records.

A solar panel was provided as a prize for the winning residence at each university. The combined totals of the proportion of students who had signed the sustainability pledge for each university determined the overall inter-university winner.

The two month long competition achieved many tangible results alongside building important relationships and heightening the dialogue on energy use. The winning university had 46% of students in residence signing the sustainability pledge, out of a residence population of 729. One of the residences at the University of Waterloo reduced their water consumption by 62%, with an average reduction across the residences of 21.5%, and electricity consumption at the university was reduced overall by 4%. The Sierra Youth Coalition estimates that around 10,000 students were exposed to the Residence Reduction Challenge through posters in residences, events and programming, and on-campus media attention. Of these 10,000 students, approximately 2500 students signed a pledge committing them to taking specific actions to reduce their ecological footprint. The Residence Reduction Challenge resulted in signed commitments at each university from residence staff to work with Facilities Management to reduce consumption in residences, and included an action plan for implementing the Residence Reduction Challenge on each campus. In several residences, energy and water were not metered at the time of the competition – there is now a concerted effort at these residences to have meters installed.

Major proponents of the initiative varied at different universities, from student union coordinators to staff positions in Facilities Management. At the residence level, it was the student-staff residence advisors who took on much of the responsibility. Different stakeholders were involved at each of the universities, primarily in facilities management and student relations. The Challenge provides both environmental gains as well as a financial incentive when consumption is reduced. As a result of the competition, reducing energy consumption has become part of the job training for residence assistants. The project strengthened relationships between facilities management, student groups, and students in the residence.

In the case of the Residence Reduction Challenge in southern Ontario, the entire program was dependent on buy-in from at a minimum the residence administrators and student staff and the facilities management, and received additional support from student union staff or environmental committees and clubs on some of the campuses. The three schools who participated in the Challenge were selected by the Sierra Youth Coalition explicitly because this high level of buy-in existed and the program was going to be supported. In this case, the multi-stakeholder approach was known to be an absolutely key component of the success of the program in a short timeline, and with that support from each campus the program was successful. In the case of campuses where this support did not previously exist, it would have been much more difficult to successfully run the program and a longer timeline would have had to have been utilised to establish time for the establishment of on-campus support.

Case IV – Go Beyond – British Columbia

Common Energy, the Sierra Youth Coalition, and the University of British Columbia Sustainability Office have come together to create a university initiative spanning the province of British Columbia. The Go Beyond project is set to launch on three campuses in September 2009, with twelve additional campuses involved and targeted for the coming years. The campuses involved range from large universities in urban centres to small colleges in rural settings. The largest university, the University of British Columbia, has a student population of 44,190. In comparison, the University of Victoria is under half that size, at 18,890 students. At 7480 students, Thompson Rivers University is the smallest of the three launch campuses. At the time of the release of their 2006 report, emissions at the University of Victoria were 35,612 T of CO_2 equivalents according to a scope of activities that includes direct and indirect emissions. For the University of British Columbia, they were 154,586.00 T of CO_2 equivalents counting direct and indirect emissions. Thompson Rivers University has not yet conducted a GHG inventory.

Go Beyond was started by a student network from across the province in order to facilitate collaborative work at different universities to address climate change. The project aims to see the universities in British Columbia not only achieve carbon neutrality, but go beyond by marshalling all their assets (land base, research capabilities, etc) to create regional solutions. The project originally went by the tagline 'Get Evolved, Be Involved' where 'Get Evolved' was the first phase targeting individual lifestyle changes, and 'Be Involved' was the second phase which encouraged participation in institutional change. Recently, the name was changed to Go Beyond.

A key window of opportunity emerged at the provincial level, with a new set of climate change policies introduced by the provincial government mandating that universities, as public institutions, would need to achieve carbon neutrality by 2010. Go Beyond was then developed as a project that would help students lead their institutions on the path towards creating a low-carbon future, and the formal partnership between Common Energy, the UBC Sustainability Office, and the Sierra Youth Coalition was established. Given the goal of carbon neutrality by 2010, the universities understand that capacity building is crucial and dialogue and planning have started immediately.

The scope of the project is broad and the campuses are diverse in their size, type, and location. The context at each school will be different and each stakeholder group will be able to contribute its unique experience with the campuses. Sierra Youth Coalition provides the grassroots networks, training student leaders and providing the broader perspective on sustainability management systems; the Sustainability Office at the University of British Columbia provides technical expertise from their history of cutting-edge sustainability initiatives; and Common Energy has led the way in bringing youth together to drive collaborative climate change planning. Common Energy marries both the top-down policy approach with bottom-up student initiatives, creating a space for both to work together. Faculty have an important role in Go Beyond both in developing curriculum that integrates practical and innovative steps towards carbon neutrality, and as champions within each specific university. Strategic support is provided by the formal partnership created by all stakeholders.

In addition to the main project partners, the British Columbia provincial government and BC Hydro, the power utility in British Columbia, have become involved as funding partners in the initiative and are also involved to a lesser extent in the program development. SYC, Common Energy, and the UBC Sustainability Office had been working together for several months prior to the involvement of the funding partners and were well poised to propose that this program be the youth outreach component of the climate change programming that the provincial government and BC Hydro were planning to undertake. In addition to funding central staff positions, students on the three pilot campuses will be paid to coordinate Go Beyond on their own campus through social marketing campaigns and encouraging involvement with campus-wide energy initiatives in collaboration with other campus stakeholders.

In this program, the pilot schools were selected by the group because of the level of on-campus support and administrative buy-in. Both the University of British Columbia and the University of Victoria have active sustainability offices and have the capacity to support large-scale campus sustainability projects. At Thompson Rivers University, the capacity was primarily present in the student group and in strong relationships with key members of the administration, faculty and staff. With the new provincial mandates around carbon neutrality in public institutions, support of the province for the initiative was essential and the additional support from the energy provider, BC Hydro, was also very beneficial. While this program has not rolled out yet on campuses, it has all the potential to be a successful initiative. While carbon neutrality is the ultimate goal, the first phase of student awareness initiatives alone are estimated to create a 4-10% reduction in GHG emissions, and the remaining reductions will come during the second phase from institutional changes.

Discussion

The case studies outlined in this chapter are at varying stages of development in their multi-stakeholder processes, and have therefore resulted in different types of outcomes. Each of the cases is summarized in Table 2 based on five factors: 1) the participants involved; 2) the structure; 3) the processes; 4) the leadership; and 5) their greenhouse gas emission reduction initiatives. Participants, structure and processes are common dimensions on which to explain collaborations (Huxham & Vangen 2000) though they tend to focus on the tangible aspects, so a category of leadership was added.

Tab.2: Cross-Case Comparison

	Sustainable Ambassadors	**Wind Turbine**	**Residence Reduction Challenge**	**Go Beyond**
Participants	Students Faculty Employees	Students Departments Facilities Mgmt. SYC	Students Residences Universities SYC	Students Universities Government Private Sector Other NGO SYC
Structure	Coordinated in student offices (Sustainable Concordia); involves student outreach to faculty, staff, admin. Implemented by faculty, staff, admin.	Coordinated by students involved in campus sustainability working with facilities management and faculty. Implemented by coordinating students and facilities.	Coordinated by student project team from three universities and SYC. Implemented by staff and students in residences.	Coordinated by SYC, UBC Sust Office, Common Energy; supported by BC Hydro and BC Climate Change Secretariat. Implemented by students.
Processes	Leaders embedded in multiple departments on campus; supported and initiated by Sustainable Concordia	Students spearheaded initiative and engaged facilities support	SYC engaged students and residence staff	Engaging in university planning process alongside broad communications campaign; initiated by multi-stakeholder partnership
Leadership	Individual champions, Sustainable Concordia	Students	Students, SYC	Universities, SYC
GHG Reduction Initiatives	– Individual reductions – Direct GHG emissions reductions as well as indirect reductions through water, transportation, waste – Institutional changes that reduce GHG consumption (e.g. light sensors)	Production of sustainable and local energy displaces coal, natural gas and hydro electric power in the grid	Individual actions by many students – measurable energy use reductions Direct GHG emissions reductions as well as indirect reductions through water Element of institutional change with increase in metering	Individual reductions Cross-sectoral approach and partnership with energy provider (BC Hydro) enables monitoring

In terms of participants, in all the cases, it was essential that students were involved in the implementation of the initiatives. The vast majority of initiatives associated with the Sierra Youth Coalition are student-led, and all SYC initiatives have a substantial student component. In analyzing initiatives associated with SYC, there is therefore going to be a substantial student influence and many students and young people in leadership roles. All of the cases required the support of Facilities Management, emphasizing the importance of the role that the Facilities staff have in energy-related initiatives on campuses. All of the cases also received direct or casual support from campus sustainability offices, with the highest level of involvement shown in the Sustainable Ambassadors, where the initiative is housed in Sustainable Concordia; and Go Beyond, where the University of British Columbia Sustainability Office is one of the key project partners. The Sierra Youth Coalition is linked to the creation of many of the sustainability offices mentioned in these cases – most directly to Sustainable Concordia, where the office was created as a result of the process of piloting the Campus Sustainability Assessment Framework in 2003 – but also to the office at the University of Guelph, the office in the student government at Queen's University, and the office at the University of Victoria.

In all the initiatives, a lead agency was required; this could be one of the stakeholders involved, or the partnership itself, as is the case of Go Beyond. This is in accordance with Waddell and Brown's (1997) findings on the two different kinds of partnerships. However, the stakeholders leading the coordination of the initiatives and those implementing the initiatives were not necessarily the same. The Sustainable Ambassadors program demonstrates this the most – the initiative is coordinated by students and implemented by staff and faculty. It is typically more difficult to have a voluntary program in which the group of stakeholders for which the program was intended did not create the program and are not involved in its delivery. This type of initiative would likely be easier to run had it been started by and was coordinated by faculty and staff. The same pattern is found in student-led initiatives that target students – it is easier to engage students on a project if it is student-led and student-developed. While the Sustainable Ambassadors program has been successful, it has not, at this point, had the widespread success that has been witnessed in the many other initiatives undertaken by Sustainable Concordia that are directed and implemented by students.

The wind turbine initiative is clearly-student led in the inception and building of the actual turbine, but its implementation as a successful campus initiative is entirely dependent on Facilities Management for installation and ongoing maintenance, and on faculty for integration into curricula. The wind turbine did not engage faculty members from the beginning, and this has presented the most challenges in implementing the project. It was also expressed by the students leading the initiative that it would have preferable had they been supported by an institutionalised commitment to sustainability from the University of Saskatchewan and through a formalized multi-stakeholder process.

While it has proved difficult to engage faculty in these student-led initiatives, this has not been the case with Facilities Management. In the three cases where Facilities Management was approached, there have been very successful partnerships. It is possible that Facilities Management is seen by the students as more essential to the programs than the faculty, and that more effort has been directed at this engagement. It is also possible that a different type of strategy is needed for engaging faculty, and that faculty are more difficult to engage with a bottom-up approach. It is also notable that no senior administrators have been key partners in these initiatives, and it may similarly be the bottom-up approach that has made this engagement difficult (and perhaps unnecessary).

Students have had certain measures of success independently leading the on-campus initiatives, such as in the cases of the wind turbine and Sustainable Ambassadors. When there was a cross-campus or provincial element, having an external organization, such as SYC, participating in a networking and coordinating role was determined to be very beneficial. The main need from external groups engaging in campus initiatives was a high level of buy-in from the campus stakeholders to the initiative.

In the case of the Residence Reduction Challenge, the initiative was successful. It was originally developed by SYC, coordinated by a small team of student volunteers from each of the universities involved, and implemented primarily by student staff in residences with the support of Facilities Management. This success could be due to SYC being more closely aligned with student perspectives and needs than students are with faculty or staff needs as in the case of the Sustainable Ambassadors. It could also be related to an existing power dynamic between students and faculty or staff that is not present between SYC and students. It was essential in the Residence Reduction Challenge to work with campuses that had the capacity and interest to run the initiative and had previously established relationships between on-campus stakeholders; in fact, one of the three universities where the Residence Reduction Challenge was originally planned to occur was replaced by another university very early on in the process when it was determined that the capacity, relationships, and interest were not present at that university.

In general, stakeholders implementing the project should be involved early on in the design process; however, it is possible to still achieve success when those designing the initiative are closely in line with the thinking of the stakeholders implementing the project, or these additional stakeholders are brought in following original development but before the project launch.

It is interesting to note the involvement of provincial governments in the two larger partnerships – the Ontario Ministry of Energy funded the Residence Reduction Challenge through their Community Conservation Initiatives program, and the British Columbia provincial government has committed to funding Go Beyond through their Climate Change Secretariat. It is likely that the partnership and multi-stakeholder element of both of these initiatives was appealing to the governments when they were making their funding decisions. An important difference, however, is that the Ontario funding came from an official funding program and only SYC was involved in writing the proposal; whereas the British Columbia funding came through a process of collaboration involving the provincial government. This is related to the scale of the initiatives – it is less likely that a provincial government would be involved in an SYC-driven initiative involving student residences at three universities than that it would be interested in a collaborative program targeting all the universities and colleges in British Columbia.

The role of the provincial governments, in particular the government of British Columbia, is interesting in analysing from where the leadership for these initiatives is stemming. Certainly, Go Beyond would not be developed to its existing level, nor would it have the funding it currently does, without the support and involvement of the provincial government through the Climate Change Secretariat. However, it is important to acknowledge the high level of involvement of SYC, Common Energy, and the University of British Columbia Sustainability Office in the program for many months before any funding was involved. The University of Victoria and the University of British Columbia were already leaders in addressing climate change on campuses, and Thompson Rivers University was also involved at a relatively high level, although with less institutionalised capacity to take on major initiatives. Since the provincial government has mandated that all public institutions are required to become carbon neutral, the many other colleges and universities that have not been as active will need to draw from the store of best practices available to succeed in making the on-campus reductions that are required. This government leadership has helped to raise the minimum level of involvement in energy issues on campuses.

From analysis of the structure of the relationships of all the cases, it appears that the larger the scale, the more formal the partnership. Go Beyond has the greatest level of formal partnership, then the Residence Reduction Challenge, then the Sustainable Ambassadors and finally the wind turbine initiative. This correlates directly with the scale of the initiatives – ranging from province-wide to regional to within an office on campus to a student group. The level of formality is also related to the ease of the project; students from Footprint Designs expressed that, had there been a stronger multi-stakeholder group at the University of Saskatchewan before the project began, the process of getting permission to erect the turbine on campus would have been simpler. In addition, key leaders from Go Beyond have also indicated that, despite a number of formal agreements, a more official structure would have been helpful in facilitating the complicated decision-making process of large partnerships.

Three of these four cases built on existing sustainability history on the campuses, particularly with the sustainability offices and multi-stakeholder groups. Without these existing conditions, it would have been much more difficult to launch the Sustainable Ambassadors program or the Residence Reduction Challenge, or to have the expertise required to develop Go Beyond. The only initiative that did not require existing sustainability infrastructure was the wind turbine;

however, it was clearly expressed that they would have appreciated more institutionalization of a commitment to sustainability and a more advanced multi-stakeholder process. In the case of the wind turbine, some of the challenges could have stemmed from being in the first stage of the outcomes of a multi-stakeholder collaboration – "Initial Outcomes" in Figure 1 – and that the more advanced outcomes had not yet been realised at the University of Saskatchewan.

An interesting finding regarding the scale of the initiatives is that there was no clear indication of which scale was the most effective for university and colleges to deal with the issue of climate change. In the three initiatives aimed at creating individual behavioural change, the level of the partnership varied significantly, as did the stakeholders involved. In addition, the complexity of the issue did not influence which stakeholders were leading the initiative – in both of the larger scale cases, students implement the program. This is in contradiction to what is predicted by Waddock (1991) in her article on types of social partnership organizations. She proposed that the level that the problem (such as climate change) is salient will determine the appropriate partnership type. Her other dimension, degree of interdependence, still holds true (Waddock 1991). Addressing climate change seems salient at all these levels.

A unique aspect in the within-campus cases of Sustainable Ambassadors and the Residence Reduction Challenge, which came directly from the students leading the initiatives, is the inclusion of different types of greenhouse gas reduction initiatives other than energy, as indicated in Table 1. Sustainable Ambassadors also includes water, transportation, and waste initiatives; and the Residence Reduction Challenge included a water use reduction element. In SYC's greenhouse gas inventory calculator tool, which is adapted from the Clean Air-Cool Planet tool in the United States, these indirect sources of emissions are also calculated in measuring the total emissions of the campus. This more holistic approach to addressing on-campus sources of greenhouse gas emissions is a unique aspect often brought by SYC or students. Whereas most campuses are addressing indirect sources through, for example, paper purchasing and use policies, it is often the students who contribute a more holistic framing, including emissions reductions, to these types of initiatives.

In terms of actual greenhouse gas emissions reductions, there were considerable differences between the cases. For the Residence Reduction Challenge, the average energy reduction over the two-month period in which the program took place was 4% at Queen's University and the University of Waterloo. Data from the University of Guelph was unavailable because the buildings were not individually metered, although Facilities Management expressed surprise at the level of engagement of the students in energy reductions and indicated that project provided an additional impetus for installing meters for each building. The impact of the program was constrained in this case because of the limited control that students have over their energy consumption – the 4% reduction therefore came from reduced electricity from lighting, the plug load, kitchen appliances and building systems but not heating for the buildings. A much more substantial decrease was in the water consumption which was only included in the program at the University of Waterloo, and was measured at an average of 21.5% since students have a much greater control of their water consumption. Overall, the impacts of the Residence Reduction Challenge were about institutional change as much as GHG reduction. The outcomes included engaging students in dialogue with administrators and facilities staff and integrating this type of challenge into the residence programming on an annual basis at all three schools. As a result of the success of this first year of the program, it is likely that there will also be longer-term impacts that have a larger impact on the actual energy use in the buildings.

In the case of the Sustainable Ambassadors program, while the overall GHG reduction has not been monitored, one initiative has shown concrete results. The Flick Off project, which was conducted as part of student-initiated coursework, was over a two-week period and involved tabling, posters and stickers on light switches on two floors of a classroom building. It was focused on getting students to turn off classroom lights when they left the room. While this was a short period of time, the results were quite dramatic with a 20% reduction in lights left on measured.

As both Go Beyond and the wind turbine initiative have not yet taken place, it is not possible to determine the level of climate change impact of the programs. An estimate for Go Beyond, based on individual lifestyle changes, is between 4-10%. The impact of the wind turbine remains to be seen at this point.

Conclusion

From analysis of these four cases, it is clear that students are able to provide the momentum and vision to spearhead campus energy initiatives; however, they need to have the support of key stakeholders for the initiative to be successful, and the need to engage these stakeholders early in the development of the initiative to ensure buy-in. In cases where there is not a significant level of institutional support, it is better to start with initiatives within a campus that targets a peer group rather than a multi-campus initiative or one that targets staff or faculty. In the case that an initiative is going to target staff or faculty, the likelihood of success is higher if the staff or faculty are brought into the decision-making for the initiative.

Sustainability coordinators, however, are in a unique position to support student initiatives, engage staff or faculty, and work collaboratively on regional or provincial activities. The student support element is particularly strong for many Canadian coordinators, who were previously involved as leaders in the student sustainability movement and often created the office on their campus with the support of SYC. They are also positioned within the administration and are therefore perceived to have a higher level of legitimacy among faculty and staff, and also are able to take on initiatives year after year on a full-time basis. This permanence can strengthen any initiative, and enables a larger scale of collaboration.

All of these four cases are complementary – one campus could be doing them all at the same time. In fact, initiatives similar to these are concurrently in progress at the University of British Columbia and are coordinated by the sustainability office on that campus. This speaks to the ability of well-supported sustainability coordinators to work in a variety of contexts and on a variety of projects whereas students without the institutional support find this very challenging. The Sierra Youth Coalition also plays a key role, particularly in the multi-campus initiatives, in terms of providing support and networking. While students are often not able to coordinate a variety of initiatives on their own campuses, the driving force, motivation, energy, and creativity for these initiatives often comes from the students, so the importance of their leadership cannot be minimized. What is less clear is which approach comes first – whether initiatives should be created by more structured and formal organizations, like SYC and sustainability offices, and implemented and directed by students; or whether students should be the drivers of the initiatives and be supported by SYC and the sustainability offices. A possible conclusion is that it depends on the capacity of the student group, the campus and regional context, and the relationships previously developed on the campus.

Another element of note is the emergent nature of these initiatives, and how little they resemble each other. This is very much like the early days of sustainability management systems in Canada, where students were trying to engage in all the steps of the process. Eventually, students found their leadership niche in the assessment phase, leading SYC to adopt the CSAF tool for the Sustainable Campuses program. Engaging through assessments enabled students to initiate institutional change rather than just undertake individual projects. For campus climate initiates, it is unclear whether conducting greenhouse gas inventories, which is what SYC was originally promoting with the Campus Climate Challenge, is the most strategic and effective role for students to commence with when addressing climate change on their campus.

What is certain is that a multi-stakeholder approach is essential to a successful energy-related initiative, whether on individual campuses, regionally between campuses, or provincially. In each of the cases, there is a clear correlation between the strength of the multi-stakeholder process and the success and ease of the initiative. However, whether the specific model that SYC has been promoting for Sustainable Campuses, creating campus-wide multi-stakeholder groups, is the best option for addressing climate change is not obvious. In these cases, there were more bilateral partnerships and a wider diversity in levels of scale of the multi-stakeholder approaches than has been observed with the broader Sustainable Campuses approach. These changes in scale, which stakeholders are involved, and in which ways they are involved, have created types of multi-stakeholder engagement which are specific to the needs of campus climate initiatives.

References

Bardati, D.R. (2006), "The Integrative Role of the Campus Environmental Audit: Experiences at Bishop's University, Canada." International Journal of Sustainability in Higher Education vol. 7 no. 1, pp. 1467-6370.

Beringer, A. (2006), "Campus Sustainability Audit Research in Atlantic Canada: Pioneering the Campus Sustainability Assessment Framework." International Journal of Sustainability in Higher Education vol. 7 no. 4, pp. 437-455.

Beringer, A., Wright, T. and Malone, L. (2008), "Sustainability in Higher Education in Atlantic Canada." International Journal of Sustainability in Higher Education vol. 9 no. 1, pp. 1467-6370.

Clarke, A. (2006), "The Campus Environmental Management System Cycle in Practice: 15 Years of Environmental Management, Education and Research at Dalhousie University." International Journal of Sustainability in Higher Education vol. 7 no. 4, pp. 374-389.

Cole, L. (2003), Campus Sustainability Assessment Framework, CSAF, Ottawa, Sierra Youth Coalition.

Cole, L. and Wright, T. (2005), "Assessing Sustainability on Canadian University Campuses: the Development of a Campus Sustainability Assessment Framework." In: Leal Filho, W. (ed.), Handbook of Sustainability Research, Peter Lang, Frankfurt, pp. 705-725.

Conway, T.M., Dalton, C., Loo, J. and Benakoun, L. (2008), "Developing Ecological Footprint Scenarios on University Campuses: A Case Study of the University of Toronto at Mississauga." International Journal of Sustainability in Higher Education vol. 9 no. 1, pp. 1467-6370.

Corcoran, P.B. and Wals, A.E.J. (2004), Higher Education and the Challenge of Sustainability: Problematics, Promise and Practice, Springer, New York.

Energy Action Coalition (2008), Three Year Goals and Objectives, Washington, DC; available from: www.energyactioncoalition.org/whatwerefor/3yeargoals.

Freeman (1984), Strategic Management: A Stakeholder Approach, Pitman, Boston.

Gray, B. (1989), Collaborating: Finding Common Ground for Multiparty Problems, Jossey-Bass, San Francisco.

Gudz, N.A. (2004), "Implementing the Sustainable Development Policy at the University of British Columbia: An Analysis of the Implications for Organisational Learning." International Journal of Sustainability in Higher Education vol. 5 no. 2, pp. 1467-6370.

Huxham, C. and Vangen, S. (2000), "Leadership in the Shaping and Implementation of Collaboration Agendas: How Things Happen in a (Not Quite) Joined-Up World." Academy of Management Journal vol. 43 no. 6, pp. 1159-1175.

Innes, J. and Booher, D. (1999), "Consensus Building and Complex Adaptive Systems, A Framework for Evaluating Collaborative Planning." APA Journal vol. 65 no. 4, pp. 412-423.

Kahler, S. (2003), “The Ripple Effect: How one Dorm Room can Affect a University’s Energy Use.” International Journal of Sustainability in Higher Education vol. 4 no. 3, pp. 230-238.

Keniry, J. (1995), EcoDemia: Campus Environmental Stewardship at the Turn of the 21st Century, National Wildlife Federation, Washington.

Leal Filho, W. (1999), Sustainability and University Life: Environmental Education, Communication and Sustainability, Peter Lang, Frankfurt.

Marcell, K., Agyeman, J. and Rappaport, A. (2004), “Cooling the Campus: Experiences from a Pilot Study to Reduce Electricity Use at Tufts University, USA, using Social Marking Methods.” International Journal of Sustainability in Higher Education vol. 5 no. 2, pp. 169-189.

Moore, J., Pagani, F., Quale, M., Robinson, J., Sawada, B., Spiegelman, G. and Van Wynsberghe, R. (2005), “Recreating the University from Within: Collaborative Reflections on the University of British Columbia’s Engagement with Sustainability.” International Journal of Sustainability in Higher Education vol. 6 no. 1, pp. 65-80.

Pearce, J.M. (2006), “Energy Service Companies as a Component of a Comprehensive University Sustainability Strategy.” International Journal of Sustainability in Higher Education vol. 7 no. 1, pp. 16-33.

Richardson, G.R.A. and Lynes, J.K. (2007), “Institutional Motivations and Barriers to the Construction of Green Buildings on Campus: A Case Study of the University of Waterloo, Ontario.” International Journal of Sustainability in Higher Education vol. 8 no. 3, pp. 1467-6370.

Rondinelli, D.A. and London, T. (2003), “How Corporations and Environmental Groups Cooperate: Assessing Cross-Sector Alliances and Collaborations.” Academy of Management Executive vol. 17 no. 1, pp. 61-76.

Sharp, L. (2002), “Green Campuses: The Road for Little Victories to Systematic Transformation.” International Journal of Sustainability in Higher Education vol. 3 no. 2, pp. 128-145.

Sierra Youth Coalition (2007a), Sustainable Campuses: Multi-Stakeholder Guide, Ottawa, Sierra Youth Coalition; available from: http://syc-cjs.org/sustainable/tiki-index.php?page=SYC+Guides.

– (2007b), Sustainable Campuses: Regional Campus Reports for 2006/2007, Ottawa, Sierra Youth Coalition.

– (2008), Campus GHG Emission Reduction Planning Guide, Ottawa, Sierra Youth Coalition; available from: http://syc-cjs.org/ sustainable/tiki-index.php? page=SYC+Guides.

Sustainable Concordia (2006), Campus Sustainability Assessment Framework – Chapter 7 – Energy, Montreal, Concordia University; available from: http://sustainable.concordia.ca/ourinitiatives/assessment/chapter7.

– (2008), Sustainable Ambassadors, Montreal, Concordia University; available from: http://sustainable.concordia.ca/ourinitiatives/ ambassadors/index.php.

Waddell, S. and Brown, L.D. (1997), Fostering Intersectoral Partnering: A Guide to Promoting Cooperation Among Government, Business, and Civil Society Actors. IDR Reports, Boston, MA, Institute for Development Research.

Waddock, S. (1991), "A Typology of Social Partnership Organizations." Administration & Society vol. 22 no. 4, pp. 480-515.

Chapter 19

Fostering Interdisciplinary Inquiry, Communication, and Action Among Diverse Populations

Susan R. Kephart, Catherine A. Collins

Abstract

Global climate change (GCC) is a pervasive, complex issue whose effects once hotly debated are now clearly evident to scientists, policy-makers, and the media. We initiated inquiry-based curricular innovations to foster greater understanding of the scientific underpinnings of GCC and to mobilize action among diverse college populations and the urban communities in which they interact. Among the educational projects implemented to date are: (1) interdisciplinary exchanges among science (Biodiversity) and humanities courses (Media & Environment) that interrogate the difference between scientific debate and mediated public argument about GCC; (2) surveys of student populations for scientific understanding and attitudes toward aspects as diverse as glacier melt, global dimming, the effects of climate change on the biota, the potential amelioration of carbon emissions in natural ecosystems, how GCC differs from paleoclimatic change, and actions that reduce or eliminate the impact of human-induced climate changes; (3) student-authored briefs on the effect of GCC on diverse ecosystems (biology students) and stories by media students that transform the scientific discourse into narratives accessible to all ages. Students presented these contrasting approaches to communication at the Focus the Nation Symposium as a white paper.

Global Climate Change represents one of the most critical issues garnering both scientific and media attention in the 21st century, yet despite over a century of scientific understanding of the relationship between carbon dioxide levels and atmospheric warming, international concern over global climate change (GCC) did not visibly emerge in the scientific and public consciousness until the late 1970's and 1980's. At a landmark climate conference sponsored by the World Meterological Organization (WMO) in 1979, participants urged cooperation among nations to reduce the impact of anthropogenic changes in climate, an act that culminated in the creation of the Intergovernmental Panel on Climate Change (IPCC) in 1988 from two United Nations based organizations, WMO and UNEP (Jarraud & Topfer 2004; UN Foundation 2008). The four IPCC major assessments thus far (1990, 1995, 2001, 2007) represent the efforts of three working groups, who evaluate the available scientific knowledge of global climate change (GCC), its environmental and socio-economic impacts, and the development of informed response strategies (IPCC 2007). Recent high-profile conferences and events (e.g.

United Nations Climate Change Conference in Bali, December 2007; Focus the Nation Teach-in, January 2008) further highlight the importance of global cooperation, interdisciplinary approaches, and locally supported initiatives at all levels, yet a significant portion of the world's citizens remain uninformed or unconvinced about the human-induced contribution to GCC and their roles in compounding or ameliorating the problem. Real solutions require comprehensive solutions that will depend on public as well as government policies. A key question remains: *How can we best link scientific knowledge to public understanding and action in diverse contexts across the globe*? We argue that public decisions on global climate change will be as much about accepting scientific evidence as about relying on value-based narratives. Educational initiatives can help frame both the scientific and the narrative forms of discourse that shape how the world's citizens understand and respond to climate change. However, if either scientists or politicians do not communicate effectively or avail themselves of the best, most current information, change will not occur at the individual or policy level.

This paper focuses on interdisciplinary, inquiry-based approaches to understanding global climate change. The goal was to challenge students to 1) integrate traditional narrative and scientific modes of understanding, 2) evaluate the differences between scientific and news media coverage of GCC, and 3) transfer their learning to the types of individual behaviors, community service and long-term scientific endeavors that enable societies to understand and ameliorate the effects of GCC for future generations. Surveys and self-evaluation tools provide a means of assessing student learning and attitudes over time and comparing these across institutions. We also urge students and citizens to consider their own ecological footprints, and what actions they might take to lessen anthropogenically induced elements of climate change.

Paradigms that Limit Global Understanding of Climate Change

As a result of scientific discourse and the most recent IPCC summary, current discussions of global climate change among scientists no longer question the existence of climate change, nor the presence of a clear anthropogenic component. Moreover, scientists have expanded their efforts to address the more serious causes and manifestations of GCC, to educate citizens and policymakers that reasonable opportunities for action can in fact ameliorate the effects of climate change, and to urge us all to find ways to help slow its exponential increase and thereby improve the quality of life in the future. Unfortunately, the same is not true of the way popular novels and the media cover the issue of global climate change. Nor can we claim that the public understands or accedes to the inevitability of serious environmental risk as a consequence of GCC, or the plausibility and potential role of individual citizens in reducing its impact even though the term or its more familiar referent, global warming, is part of the popular vernacular. The difference between what is commonsensical to these scientific and popular constituencies leads us to ask: *What are the barriers to global understanding of GCC, and how can educational efforts help marshal the transformative power of potential public and governmental actions?* Our short conclusion is that different communication norms have prevented a shared understanding of the problems, consequences, and necessary actions toward global climate change. In this essay we hope to identify the structural barriers to global understanding created by these varied communicative norms and to offer examples of how they might be overcome at the academic level and in the larger public sphere. More broadly, our goal in joint research and teaching has been to help scholars, students, and community members understand how the scientific and narrative forms of discourse are related, and when combined, offer compelling arguments on which to base public policy and personal action.

Narrative theorist Walter Fisher argues that in its inception in ancient Greek culture *logos* "meant story, reason, rationale, conception, discourse, thought" (1987: 5); later *logos* became associated with rational truth while *mythos* was less tied to reason and knowledge. Hierarchically, the logical treatment of a subject came to be afforded greater status than the poetic treatment. Today technical, scientific discourse is valued above the poetic, value-based claims that circulate in the public sphere. Fisher calls us to reconsider the original meaning of *logos* that celebrated both the logical and narrative responses to knowledge and truth. His theory of narrative asserts that comparable to logical proofs and scientific methods, narrative rationality employs a logic of good reasons that tests both the coherence and the fidelity of the stories that we tell to explain why we define situations as we do and act accordingly. Public understanding and action, from this perspective, is as much tied to value-based claims as technical logic-based claims. How the public decides about issues like global climate change will likely parallel the response to other crises such as nuclear energy development, family planning initiatives, or reducing global emissions of CO_2; the public response will require both accepting scientific evidence and responding to value-based narratives.

The first step in linking scientific and narrative discourse is to understand the criteria that define each type of legitimate discourse, as well as how competing voices in both domains define issues. In our classes and research efforts, we interrogate the "evidence" and the "good reasons" each discourse uses to support its claims, and the limits that prevent their universal acceptance. In the scientific domain, values are deemphasized in favor of evidential truth generated exclusively by scientific "experts," and clear replicable procedures. In this domain, clashes between academic and industry-funded research add to the cacophony of claims for the best science. In the narrative domain the debate revolves around the choice of story to tell and the type of characters that choice demands. For example, in editorial cartoons and news stories about dwindling arctic ice, carnivorous polar bears become the cuddly, tragic victims: that characterization is strategic and shapes public response to the issue. Narratives are tested by the values they promote and their fidelity to other stories that we tell. In the narrative paradigm it is familiarity with the culture, value hierarchies, and audience that determines one's expertise and hence credibility. Politicians, activists and media practitioners offer narrative rationales in the public sphere, and conventions for political and media discourse may be as much contested as the competing paradigms of scientists. But whether scientists, politicians, or citizens offer arguments about global climate change, all use extensively not only the conventions of their disciplines but the mass media to make their arguments heard.

News Coverage: Media Conventions

Mass media practitioners have developed conventions that define newsworthiness and effective news writing. These conventions highlight and hide particular kinds and sources of information, and topics that will be printed and aired in the public sphere. In a classic essay, Schudson argued, "News is not fictional, but it is conventional. Conventions help make news readable" (1982: 98), and include a preference for single events over ongoing processes, events involving important people, what is new and influences large numbers of people, and unique entities. Good news stories are designed to tell *both* sides of an issue (i.e., only two sides, with balanced coverage), and are seldom linked to earlier issues or events. Most importantly, good news stories are *objective* reports eschewing editorializing; they report rather than argue or advocate. These media conventions can distort the way issues are presented. If, for example, scientists all agree that global climate change is occurring, news accounts will still balance scientists' quotations against someone who argues that global climate change is not occurring, even if this perspective is less credible or representative of the consensus on the issue.

Established conventions in news reporting yield stories in the mass media that are simple, dramatic, and conflict-based. Because news is less about what happened, and more about what an authority says happened (Sigal 1973), reporters routinely turn to highly visible personalities to influence events. Hence news stories typically quote politicians, public officials, or prominent scientists and are largely political in content. These media conventions "perform for the reporter the function that rules of evidence do for the scientist," forming the criteria that guide newsworthiness (Sigal 1973: 3). Media conventions, despite touted objectivity in reporting, frame events and how they will be covered; they highlight and hide, and may distort in their desire to capture public attention. In contrast, scientists purport greater objectivity as they investigate the natural world and communicate their findings to colleagues.

The problems inherent in media conventions are exacerbated when the subject of the news story is science. Science is vast, encompassing diverse areas, all subject to political manipulation during competition for the top funding associated with scientific prominence. Moreover, science writers often interpret the rather specialized jargon of a given field of research for the larger scientific community, as evident in the generalized, "reader-friendly" sections of prestigious journals like *Science*, *Nature*, and the *Proceedings of the National Academy of Sciences*. So how much journalistic convention (drama, simplicity, conflict, and politicization) accompanies scientific articles on important topics that are visible to science and the public? Does the framing of popular paradigms by scientists/science writers change as awareness of the concept increases within the scientific community? We want our students to be asking these questions, especially with respect to the public debate over GCC.

Communicating key scientific concepts: Biodiversity and globalclimate change

Our joint research on two topics—how biodiversity as a concept entered the public sphere, and how narrative and media constructions of mythic concepts of nature or wilderness privilege particular frames of reference—made us acutely aware that the best scientific evidence does not always inform public opinion (Collins & Kephart 1995; 2007). Whenever values embedded in narrative arguments clash with scientific evidence, it is not surprising that the public may reject the experts in favor of what scientists might see as *mythos*. Part of what we need to understand, then, is how media conventions limit how all voices – scientific or narrative – are filtered. Our contention is that the best understanding of any issue of public debate comes with an appreciation of both forms of discourse. If we want public support of scientific issues like global climate change, we need to find value-based as well as scientific appeals; we need to find common ground from which to engage in productive public debate.

Also crucial is that students and the public know how our understanding of the natural world and GCC is constructed using the lens of rigorous scientific investigation, debate, and writing. Because the scientific issues associated with GCC are complex, we sought to develop teaching, survey, and educational tools that would help students and ideally local citizens better understand several key components of global climate change:

1. latest scientific information as to magnitude of climate change,
2. the key contributing factors (e.g. greenhouse gases from both anthropogenic and natural causes), and
3. the impact of the changes on the biota and natural ecosystems.

To this end, our curricular efforts link natural science and media based courses, narrative stories and scientific evidence, and academic and community initiatives.

Student populations and courses

We selected two biology courses taught annually for initiating units on global climate change and for testing a class questionnaire designed to assess temporal shifts in student comprehension of GCC. In *Plant Ecology and Conservation* (Bio257), biology majors use investigative approaches to link predicted changes in climate with the likely effects on plant populations. In contrast *Biodiversity: Discovering Life,* is a non-majors introduction to science as a mode of inquiry. Students in Bio210 often co-enroll in *Media and Environment* (Rhet210); they learn the most important scientific ramifications of GCC as well as how to analyze arguments and interpret environmental media.

Understanding the Science, Linking it to Narratives

To engage student learning on global climate change, we use texts, current journal articles, video/DVD media, investigative analyses, and community-based projects; all help biology students understand how GCC is related to important ecological concepts and the role scientists can play in influencing policy and community action. Initially, students complete a self-test on GCC (Appendix 1) prior to curricular activities (e.g., lectures and discussions of atmospheric and biotic effects, content analyses of scientific and news articles, student talks, and written reports). In small groups, students brainstorm important questions, then organize into teams focused on subtopics of special interest for further investigation. All teams co-author a visual presentation summarizing the content of their investigative efforts. Students also critique what they read, interact and learn from others, and suggest new future directions for scientific study, *i.e., What do scientists and policymakers still need to know to make good decisions related to climate change*?

In the majors ecology course, investigative studies of a particular biotic effect of GCC in Fall 2007 ranged from past climatic changes to the effects of shifting temperatures on the synchrony of leafing and flowering in plants and the first appearance of hungry, hibernating animals. These studies culminated in an edited volume made available to the public on the Willamette University campus during a "Focus the Nation Teach-in," a grass-roots effort throughout the United States instigated by Eban Goldstein and Chungin Chung. The briefs covered the climatic effects on the biota that each student perceived as important, focusing on five subtopics: paleoclimatic change (Appendix 2), arctic climate, Northwest climate, changing plant and animal emergence, and climate-induced range shifts in natural populations, including pests. Several key scientific points were that:

- Despite inherent uncertainties in projections, temperatures are rising at rates that well exceed paleoclimatic changes subsequent to the last great ice age (Bush et al. 2004);

- Thawing of arctic permafrost creates feedback loops of methane and carbon release that exacerbate climatic changes (Zimov et al. 2006);
- In the Pacific Northwest, a warming of up to 3°C coupled to high levels of precipitation as rain will likely lead to more disturbance (e.g., summer fire, winter flooding); the warmer water and low summer flows affect salmon and other populations (Parson et al. 2001).
- Climatic change results in asynchrony in the timing of bud-burst and flowering in plants relative to the emergence of overwintering vertebrate animals, insect pollinators, and pests; "heat-islands" in urban environments with absorptive surfaces and low circulation enhance such shifts (Inouye & Barr 2000; Inouye et al. 2002; Dixon 2003; Neil & Wu 2006)

Moreover, in an interdisciplinary collaboration, students in an advanced *Narrative Theory* course (Rhet341) used the briefs created by plant ecology students to craft fictional narratives, as in this excerpt of a story focused on the arctic by Jeffrey Collins.

> "Dr. Turcotte muttered as he shivered and drove through the Arctic summer's bog-like tundra .... As he reached the town where his storyteller – the widow of a native artist – lived, he noticed that the old church was torn down. Sad, he thought ... and noticed the hole. The hard permafrost melted beneath the church and the pews, the altar, the crucifix, and the great brass bell had all collapsed. Turcotte felt water on his cheek. He reached to brush away the tear when another one fell. Rain in the land of ice and snow."

Scientific elements embedded in the story include the loss of permafrost, and greater prevalence of rain over ice and snow in a warmer arctic region. Linking these ideas to the tear-stained loss of culturally important memories and artifacts might impress upon readers the importance of learning about global climate change and acting accordingly.

Narrative Voices in Communicating Global Climate Change

Here we offer a series of short examples of curricular units that examine the way narrative voice and visual media introduce GCC, including:

1. novels in which GCC prompts the plot and character development and
2. editorial cartoons. Combined with a focus on media framing of the issue in news coverage, these units introduce a broad range of messages that contribute to the debate over GCC in the public sphere. We plan to include these in 2009 in paired courses, Bio210-Rhet210.

Novels and Popular Narratives

Narratives can be powerful instruments influencing public understanding. Moreover, even in reading what is clearly fiction, we can become entangled in the arguments and characterizations, suspending temporarily, or entirely, our distinctions between fact and fiction. Scholars of popular media narratives repeatedly discover mistaken audience assumptions about real issues and people as treated in fictional accounts. Oliver Stone's films on JF Kennedy and Richard Nixon were controversial because his blend of historical fact and fiction was lost on viewers who thought the plot elements in the film depicted actual historic events, not because audiences are inherently stupid or filmmakers and television writers craftily manipulate their viewers, but because when narratives have a fidelity to other experiences we know to be true, their status as fiction may be ignored.

Global climate change is a premise for a numerous novels, television dramas, and Hollywood films. One novel is particularly worthy because it is the basis of a senator's arguments to Congress about why GCC is not real. The novel is *State of Fear* by Michael Crichton, an international bestselling fiction writer. Written in 2004, it became the basis for Senator James Inhofe's invitation for Crichton to address the Senate Committee on the Environment and Public Works on global climate change. Inhofe's belief that the novel should be required reading for the committee, conservative political columnist George Will's use of the novel's argument in one of his columns, and references to the arguments of the novel in conservative web sites all point to the novel's impact on public discourse among conservative doubters about the reality of global climate change.

In next year's Media and Environment class students will read the novel as a representative text challenging GCC. Already senior Bridget Sutherland (2008) noted Crichton's strategic blurring of two distinct genres, scientific and fictional. With fictional narratives we allow exaggerated plotlines, characterizations, and motivations. Drama allows science fiction writers to stretch the credulity of the reader, if enough plot exists to give potentiality to the exaggerated claims. The class will explore how Crichton manipulates narrative form to enhance the credibility of an argument that scientists universally dismiss. For example, the stylistic trappings of conventional scientific discourse allow the reader to attribute the authority of science to the fictive arguments. The novel demonstrates the "priestly voice," which Lessl argues is not "merely human," but "the voice of God or Nature, a mouthpiece for what is wholly real, ultimate and insoluble" (1989: 188-189). Combine these credentials with an invitation to discuss GCC with senators and a media propensity for statements from GCC skeptics in news articles and it is easier to understand how a fictive work gains adherents to its premise that human-induced GCC is a fabricated, scientific fallacy.

Crichton's unsupported contention that most scientists actually disregard global warming provides a springboard for a discussion of how science and the rational world paradigm regulate debate in the public sphere. Because science relies on expertise and reproducible data, a public unschooled in GCC research is functionally silenced in the debate. Imagine, then, the appeal of a narrative in which the plot asserts scientists only say what they are forced to say to maintain funding. Thus, the independent thinker who reasons through issues, even without specific evidence, may be seen as more believable then credentialed scientists, enabling conservative skeptics of GCC to challenge scientific authority. Even the reader who recognizes the work as fictive may retain part of the argument against the reality of GCC. The 1.2 million copies of the book sold in the U.S. in its first year of publication suggests that even a small portion of the readership taking part of the argument at face value has the potential of significantly impacting public discourse about GCC.

Despite the call to move the GCC debate beyond the question of if it really exists, popular writers and filmmakers only return us to issues of its impact on our common environment and our ability to discern truth from fiction. Students will compare assertions in Crichton's novel with the latest scientific evidence garnered in the Biodiversity class in which they are co-enrolled. They will also evaluate a second fictive piece whose implicit or explicit claims about GCC may influence public willingness to act on important environmental issues. Films like *The Day After Tomorrow* or novels such as Kim Robinson's *Forty Signs of* Rain, Robert Morris' *The Owl and the Oak*, or Mark Tushingham's *Hotter than Hell* give students an opportunity to explore politically charged issues of science and policy that arise from fictive treatments of controversial international problems.

Cartoons

This unit reflects an analysis of editorial cartoons by student Phil Handke to discover the shared assumptions about GCC that permeate public discussion and action on this issue. Cartoons represent visual metaphors that conceptually highlight and hide from view our understanding of a subject. Edwards (1993: 66) argues the visual dimension gives "metaphorical definitions a concreteness that affirms the 'reality' of their meaning," reducing a complex issue to a concrete, simplified form that forces the reader to attention and to engage the issue. Medhurst concurs when he concludes that cartoons are successful as "readers … are able to recognize that shared locus of meaning as expressed by caricature" (1981: 220).

Following a discussion of how cartoons are visual metaphors that argue enthymematically, students will examine a cartoon set from international news sources to identify arguments, metaphors, and visual narratives that transcend national and local commonalities. The more widespread the enthymematic understanding and archetypal the narrative argument, the more likely the cartoon is to be accepted as *true* rather than just one author's opinion. The more *taken for granted* the premises, the more readily the audience will accept unquestioned the premises offered. Student analysis of cartoons from other countries only reinforces the global nature of assumptions about climate change. The images are easily retrievable from a Google search of cartoons and GCC, Gagle Cartoons, the *New York Times*, the *Washington Post* and other online sources (Appendix 3).

GCC cartoons reflect four visual metaphors: Earth as human, Earth as boiling water, pollution as smoking, and climate change as a bomb. The Earth as human metaphor anthropomorphizes Earth and thereby associates our own experiences, and fears with the fate of the Earth under conditions of global warming. The Earth sweats, wants, and experiences pain and exhaustion in these cartoons: "Cognitive interplay is triggered by the outreaching arm and grimace" in Johannsen's 2007 cartoon (Handke 2008: 14). Handke concludes that the ontological metaphor allows the viewer to relate his/her own bodily experience with the Earth subject to GCC giving the visual enthymemes a strong experiential foundation" (2008: 14). The cartoons highlight pain but hide the causes of GCC. The Earth as boiling water employs images of cauldrons or teapots where the inhabitants of the planet are in the pot coming to a boil or the Earth itself is ready to explode because of GCC. The argument thus is that the kettle needs to be removed from the source of heat before permanent damage is done. In one of the cartoons how we use the teapots is linked to the G8's actions and how international policy contributed to GCC. In the pollution as smoking cartoons the person smoking the cigar (often figured to look like a smoke stack) is the cause of the problem, and he is usually Big Oil or other corporate polluters who ignore the science of GCC in favor of profits. The United States is also figured as the villainous smoking polluter whose emissions exacerbate the greenhouse effect. A less frequent, but still powerful image is the earth as a bomb because of global climate change. "The metaphor contends that climate change is an issue that, if not controlled, will end the world" (Handke 2008: 22).

Students explore these four visual metaphors in the selected cartoons to discover how the editorial arguments highlight and hide, rely on shared cultural values and assumptions about GCC, and might speak to the assumptions and experiences of the general public. We ask them to think about what is absent in the arguments advanced by cartoonists. What scientific evidence is excluded or included in the enthymemes? Who is blamed or victimized? Is any concrete action called for? Do the cartoons add anything new to our understanding of GCC or do they merely reinforce what we already assume to be true? How likely would the viewer be to change any specific behavior because he or she was exposed to the argument of the cartoon? Does the reinforcement of an association such as Earth as suffering or Earth as boiling water influence how we think about the issue of GCC?

Science and Media Framing of GCC

Previous research reveals significant differences in the way scientists, politicians, and journalists view environmental concepts and hence frame their messages (Collins & Kephart 1995; 2007). Frames are organizational, classificatory, and interpretive acts through which individuals, institutions, and the media "locate, perceive, identify, and label" events and information (Goffman 1974: 21). The frame choice reflects selection (and exclusion), offering salience to some ideas, voices, or organizational schema over others. To help students in paired Bio210-Rhet210 courses see how media conventions influence coverage of scientific issues, we discuss how the media and scientists differently frame GCC, *e.g.,* "How often do news stories use the same referential terms for GCC as scientists?" Are particular concepts treated differentially in scientific articles and news stories? We use the effect of GCC on biodiversity as a prevailing theme given the interconnectedness of climate, diversity (*i.e.,* genes to ecosystems), and the ability to sustain life on earth.

In Project 1, using Lexis-Nexis, students identify all news stories containing "Global Climate Change" *and* "Biodiversity" as key terms. Students form teams to discover how journalists employ or fail to employ a set of 20-30 terms common in the scientific literature. Teams then conduct two broad searches: first they identify media coverage of a specific term, e.g. "phenology" or "ice core," that appears concurrently in stories with primary term "Global Climate Change." A second search compares how often these terms appear in stories with "Global Warming" (GW), allowing students to map how the popular media frames GCC *vs* global warming. Recently, in a 5-year search of major newspapers, three biodiversity-related indicators of climate change – polar bears, robins, and migration – co-occurred frequently with GCC (99% of news stories) but appeared with GW in only 10% of stories. Prior research on media coverage of science shows that choosing species with whom the public will have a positive response is a popular way to explain scientific concepts. Both robins and polar bears personalize the consequences of GCC. Plant or animal migration stories similarly involve emotional appeal – indicator species are depicted as being harmed by increases in global climate change. Interestingly, of 27 scientific terms searched, only four (15%) appear in more than 5% of the news stories with GW *vs* 18 (67%) with GCC. Two popular depictions of weather changes – El Niño and La Niña--not surprisingly appear with GCC in 71-98% of stories, and tundra and ice cores are linked to GCC in ~50% of news stories but only 3-7% of stories with GW. Thus, what the public receives in association with popular terms like global warming from news media is very unlike what scientists consider to be important indicators of global climate change.

Students further query the database to answer: Do news media report GCC differentially over time? By plotting the number of GCC stories monthly, they discover that 15% of stories ran during Earth Day week. Other political events

(GCC conferences, presidential campaigns, court case) show corresponding increases in GCC stories appearing in the popular media.

In 2008, Biodiversity students conducted investigative analyses comparing keyword searches of newspaper media via Lexis Nexis with web engines that broadly access science articles, including BioOne and Academic Search Premiere. All students searched on "global climate change" in combination with several contrasting terms (e.g. "biodiversity" vs "politics" vs "economic") to compare frames in media stories relative to science articles, then read a selection of these articles for content. As part of the lab, pairs of students developed and presented independent investigative projects to the class. For example, one team evaluated media *versus* scientific framing of "greenwashing" whereby corporations use earth-friendly language to appear more environmentally conscious and thus more progressive. Discussions delved aspects of these topics that included examining implicit or explicit hypotheses and variables, the accuracy of portrayal of scientific concepts, the presence of underlying assumptions and value appeals, and the challenges of factoring in the often marginalized environmental costs of ignoring global environmental problems.

Similarly, Media and Environment students not only identify overall patterns in GCC coverage, but read stories in depth for the space allocation to both sides of an issue, examples of drama, or conflict, and how media conventions shape the messages the public receives. They look closely at arguments, values, metaphors based on rhetorical analysis; using their background in the biodiversity course they critique the scientific data and interpretations included in the news stories for their chosen terms. In each case they must consider how media conventions and scientific arguments privilege some understandings of GCC while hiding other important perspectives.

Finally, students are asked how they might reshape the news story to better reflect the scientific research understood from the Biodiversity class. Creating messages that improve the scientific information available to readers while acknowledging values and story lines that are powerful motivators for public action, helps students understand the consequence of drawing distinctions between scientific and narrative discourse on GCC. The final unit assignment is to design a message for the public that will use the best of these two domains of discourse.

Community Service and Web-based Educational Initiatives

We seek to couple the varied visual and textual media, as well as scientific evidence, with an educational approach that incorporates the worldwide web, service-learning, and hands-on investigations. Web active learning, community-based action, and long-term restoration of important local environments have been cornerstones of the interdisciplinary pedagogical and research efforts we undertook since the inception of our two courses in 1999.

Web-based activities will continue to form part of the joint Bio210 and Rhet210 courses and are designed to engage student learning on a topic of interest related to biodiversity. In teams, students develop a website offering new insights into one particular issue related to global climate change. The goal of the project is to produce an informative and provocative discussion that melds scientific, rhetorical and media knowledge they have gained throughout the semester. The web page integrates a science and media components and compares journalistic and scientific treatments of an important topic. Students examine other web pages and ask themselves which ones appealed most and why? Which demonstrate the most sophisticated understanding of how the media frames issues and the consequences of those choices? Which pages employ the medium of the worldwide web, and which are just "papers on the web" instead of on paper?

Student web pages reflect an informed understanding of the scientific issues – something missing in most news stories. They work through a careful analysis of the science missing in popular media stories and how scientific stories in journals like *Science* or *Nature* may be missing successful persuasive appeal. What can scientific coverage learn from rhetorical principles and vice versa?

Students critique both the scientific and media articles – the extent of their coverage, if scientists consider alternative hypotheses in their papers, and whether scientists and/or journalists recognize conflicting viewpoints. Students summarize the relevant scientific conclusions (including conflicting evidence) in tables or figures, and how the media portray those data or conclusions.

The result of this exercise is a more coherent narrative about GCC that takes advantage of the best of the scientific and the narrative domains of discourse. Classroom information and analysis is taken to the public in a way that should improve on discourse from either domain. Science becomes accessible to the public, not just to experts, and concomitantly shared cultural values and narrative patterns inform scientific debate about the complex issues surrounding global change.

Service-learning initiatives in biodiversity and plant ecology courses will link GCC to pollinator services, which have become an important international and national focus in relation to global climate change (NAPPC 2007). The design is relatively simple, yet has been successful thus far:

"*GCC, Flowers, & Pollinators*" is a popular two-lab collaboration between university students and both elementary and secondary school urban classes with at-risk student populations. In Lab 1, high school and college students jointly study the diverse forms of living flowers and roles of putative pollinators (e.g. bees, butterflies). Students use search engines to examine the:

1. services that pollinators provide in natural ecosystems and for humans,
2. potential effect of GCC on plant-pollinator interactions, and
3. potential actions that may help curb pollinator decline.

Prior to meeting all students will have guided access to DVD/video and web resources in their classrooms (suggestions available on request). Second, teams develop hypotheses to test in natural populations of lilies and buttercups the next week when university and high school students mentor elementary students in an investigative lab at a local park with diverse wildflowers. This field experience begins with an art-science project, *i.e.,* sketches of study flowers, after which teams test their hypotheses by noting insect visits to 1x1m manipulated and natural observation plots, adding to a long term database tracking flowering time and insect abundance with GCC. Activities and results are logged in student journals and shared during discussion among all groups with prompts like: what was most intriguing; what you would you do differently next time? These investigative labs bring new friendships and mentoring opportunities and reinforce understanding of GCC and plant-pollinator mutualisms. Students also learn how to communicate and teach ideas in a diverse environment and become aware of the challenges and complexities of GCC and at-risk learning environments.

Long-Term Community Based Action: Linking Environment and Media

Even without climate change, habitat fragmentation presents an imposing challenge for scientists, conservation managers, and policy-makers; it reduces global biodiversity, influencing the richness of species interactions and the services provided by healthy ecosystems. In Oregon's Willamette Valley, most grasslands and wetlands have been converted or filled for agriculture and urban expansion. Natural remnants persist largely on rocky, shallow soils or in degraded areas under restoration. Thus, our service-learning efforts now extend to Fairview Wetlands, the largest mitigated wetland and restoration site in Salem, Oregon (18.6 ha with ~2 km of enhanced riparian corridors). Mitigated sites are no substitute for pristine remnants of wet prairie and oak savanna, but in this case once degraded farmlands are enhanced functionally and in species richness. Since 2004, students, local employees, and citizen volunteers have reintroduced successfully over 3000 individuals of two formerly abundant plants: the endangered Nelson's checkermallow and the common camas lilies that once sustained both indigenous populations and early explorers.

The wetlands enhancement is a collaborative effort at ecological restoration where university and community partnerships enable long-term monitoring related to global change. Students track temporal changes in biodiversity, including plant bud-break and animal abundances, climatic variables, water levels, and restoration efforts. They learn long-term field assessment techniques, analyze the effect of plant reintroductions on species richness/diversity of animals, and work side-by-side with citizens. The wetlands also form an outdoor laboratory for media and biology students to document visual landscapes and test ecological and evolutionary hypotheses, including the possible effects of GCC and biodiversity enhancement. Moreover, the project is a venue for on-site public education, enabling local employees and citizens to learn the ecological value of diverse landscapes and the risks of invasive exotics. Investing in public education and restoration protects the ecological functions of the site (e.g., water supply, filtering capacity) while creating a visual pleasing, functional landscape for recreational activities. Project partners include Willamette University, Pringle Creek Watershed Council, the City of Salem Parks and Public Works, on-site industries, local K-12 schools, Rotary Club, Oregon Community Foundation, local Native Plant Society and Audubon societies, Earthwatch Institute, and citizen volunteers.

Surveys as Learning Tools for Sustainability

We use assessment tools to gauge student attitudes and learning on GCC-related topics. To track conceptual understanding, we administer open-ended questionnaires that students complete before and after units on GCC and biodiversity; these will also guide the design of future surveys with more objective, discriminatory responses. Also, to provide a broad context for evaluating student knowledge of GCC, we began surveying environmental attitudes of incoming and upper-level Willamette University students in 2007, at the inception of a new Center for Sustainable Communities; we augmented the NEP and ECO instruments to compare our students with national populations via the Ecological Society of America's TIEE site (Humston & Ortiz-Barney 2007).

Global Climate Change Assessment

The biology classes participating in pilot assessments were matriculating to senior students, including 27 science-biology majors (Bio257), and 31 students from diverse majors, ranging from art to politics and economics (Bio210). Prior to discussion, lectures, and readings on climate change, we invited students to respond in class to seven questions (Appendix 1) and indicate if they had seen "*Inconvenient Truth*" narrated by Al Gore, or had taken courses covering GCC. Students knew they would reevaluate their knowledge at the end of the course using the same questions (all responses coded for anonymity, entered by a third party). The open response design of the form maximized opportunities for students to explain concepts with which they had some familiarity; these questions focused on the magnitude and causes of climate change and its effects on the biota.

The assessment questions provided starting points for discussion; subsequent activities included were an introduction to global climate change, assigned readings, and invitations to campus events and lectures. Students also viewed and evaluated the scientific assertions, arguments, and value appeals in two DVDs, "*Inconvenient Truth*" (Appendix 4) and "*Dimming the Sun*". We assigned separate subsets of concepts and assertions to students who independently evaluated those aspects, met in small groups, and later shared their perspectives on specific segments of the DVD with the class. Students used Kellert's (1996) topology of values as we challenged them to consider images as well as narration, to explore how the creators of the DVDs depicted and interpreted scientific data, and to reflect on what they might do differently if given unlimited time, expertise, and resources.

Responses to GCC questions varied with learning and with student background. Initially, 41-42% of students left blank the predicted temperature rise with increasing atmospheric CO_2 whereas post-tests revealed 34.5% and 20% of questions unanswered by Bio210 and Bio257 students respectively. Initial estimates of global warming were also higher and more variable (0-25°C; mean=5.9°C) than post-test values (0.2-15°C; 3.7°C) that more closely matched climate models (Stouffer & Manabe 1999; Kerr 2004; IPPC 2007). Marked increases occurred in student recognition that global warming rates are higher than in paleoclimatic history (14.7-15.2% increase in student understanding) and that CO_2 is a major source of GCC (8-9% rise in accuracy). The question least effectively addressed initially was the nature of global dimming, i.e., the reduction in radiation from the sun, largely due to particulate pollutants. From "*Dimming the Sun*" students learn the value of independent discovery and careful documentation of initially unrelated scientific data, and of garnering new information from unexpected, and even tragic events, as during the fallout of "911" in New York City.

Not all respondents commented explicitly on anthropogenic *versus* natural causes of global warming, but most often cited the Kyoto protocol, emissions reduction, and hybrid cars as mechanisms for curbing human-induced climatic shifts. Some noted the importance of life-style changes, retaining and replanting forests, carbon credits, pursuit of 'green' energy alternatives, local consumption, education, recycling, and legal controls. By the post-GCC unit, students understood that oceans, rainforests, and tundra might function as CO_2 sinks and/or represent areas heavily impacted by melting glaciers, northward migration, and permafrost thaw. We did not assess directly how students evaluate assertions or appeals yet discussions revealed that asking them to categorize values and critique graphs in sources improved their awareness of the critical role of evaluating scientific assumptions. We concur that such assessments provide a powerful learning tool for students (Humston & Ortiz-Barney 2007).

Environmental Attitudes, Ecological Concepts, and Sustainability

We further sought to place student understanding of global change in the context of attitudinal perspectives on environment and conceptual knowledge. Biodiversity was one of the 'hottest" 1900s topics (Collins & Kephart 1995) whereas 'sustainability' and 'global warming' emerge as the newer, though variably interpreted "buzz words" of the 21st century. In 1999-2000, prior to a campus sustainability program, students in over 20 classes responded anonymously to questions about ecology, biodiversity, and habitat fragmentation. In 2007, the first full year of funding for the Center for Sustainable Communities, we further surveyed student populations for ecological concepts and environmental attitudes. The large data sets for these NEP surveys and the ECO case studies are still under analysis, the responses foreshadow how students may react to GCC. For instance, despite a split in whether humans have the right to modify environment, students uniformly agreed that such interference often has disastrous consequences (72%), that environmental health profoundly influences human health (94%), and most importantly that it is "still possible to solve critical environmental problems" (91%) and that national policy should yield to international censure (57%). Not surprisingly, when allowed to comment about environment, one student noted:

> "As for the environment, sea levels are rising, ice caps are melting, animals are running out habitats. The weather is haywire – I can't remember the last time Oregon or Washington has had such hot, long summers …."

However, 11-12% of respondents thought the "ecological crisis" greatly exaggerated as in:

> "I do not think global warming is as severe as it is made out to be, but something should be done about pollutant production. I do not think our production of CO_2 is causing global warming. I believe it is a natural process … which warms our earth …."

That education may make a difference is reflected in another comment:

> "In high school I disputed/was skeptical about global warming. After taking this and my 'Oh Canada!' Colloquium … there is no doubt whatsoever in my mind that global warming exists and is the most important issue of my generation, much more important than terrorism."

Conclusions

Creating a global climate that sustains our natural biosphere and its associated diversity, from genes to ecosystems, will require international collaboration of unprecedented proportions, yet the potential benefit and ability to set change in motion are within our grasp. The four hallmarks of many sustainability programs, including at Willamette University are the 4E's: Environment, Education, Equity, and Economy. Each element must be brought to bear on global and local responses to the present climate trends worldwide. Moreover, all participants in change must consider the diverse audiences they seek to reach. Certainly all constituent groups miss opportunities to craft sound decisions: the public often sidesteps the best scientific evidence, the media is too focused on balancing views to acknowledge virtual consensus on GCC, and few scientists communicate effectively to policymakers, the media, or the public. We sought to develop interdisciplinary exchanges among science and humanities students that would increase their understanding of both the scientific and media-based elements needed to marshal public awareness and action. Students responded with stories, scientific papers, initiatives for long-term community restoration, and a heightened sense of their role in facilitating the initiatives that will be required of present and future generations to address the causes and consequences of global change. The potential ramifications of inaction versus action are perhaps best summarized in the comments of two anonymous students at the end of the unit on climate change: "Humans beware," and, crediting musician Louis Armstrong, "Yes I think to myself, what a wonderful world." Let's hope societies of the world choose the latter.

Acknowledgments

We thank the many students who contributed their ideas or time, especially Jennifer Bufford, Emily McNeilly, Adam Kotaich, Molly Sultany, Jeffrey Collins. Our analyses reflect support from Willamette University Sustainability Grants program, the K. Bisbee II Fund of Oregon Community Foundation, the City of Salem Watershed Grants Protection Program, and NSF-IL-IP award 9751313 on inquiry-based learning to S. Kephart, J. Koprowski, and N. Perigo.

References

An Inconvenient Truth [DVD] (2006), Paramount Classics and Participant Productions, directed, Davis Guggenheim; produced, Laurie David, Lawrence Bender, Scott Z. Burns; executive producers, Jeff Skoll et al.; Paramount, Hollywood, CA.

Bush, M., Silman, M. and Urrego, D. (2004), "48,000 Years of Climate and Forest Change in a Biodiversity Hot Spot." Science 303, pp. 827-829.

Collins, C. and Kephart, S. (1995), "Science as news: the emergence and framing of biodiversity." Mass Communications Review 22, pp. 20-44.

– (2007), "Wilderness in the American imaginary: Media constructions of scientific and political knowledge", pp. 249-259. In: Wilderness, Advocacy, and the Media, Proceedings, 8th Conference on Communication and Environment, University of Georgia.

Dimming the sun [DVD] (2006), Dox production for Nova/WGBH and BBC; directed, Duncan Copp; producer, David Sington. WGBH Boston Video, Boston.

Dixon, A.F.G. (2003), "Climate change and phenological asynchrony." Ecological Entomology 28, pp. 380-381.

Edwards, Janis L. (1993), "Metaphors of enmity in gulf war political cartoons." Ohio Speech Journal 30, pp. 62-75.

Fisher, W.R. (1987), Human communication as narration: Toward a philosophy of reason, value, and action. University of South Carolina Press, Columbia, SC.

Goffman, E. (1974), Frame analysis, Free Press, New York.

Handke, P. (9 April 2008), "The pen is mightier than the metaphor." Top undergraduate paper at the Northwest Communication Association Conference, Coeur d'Alene, ID.

Humston, R. and Ortiz-Barney, E. (June 2007), "Evaluating course impact on student environmental values in undergraduate ecology with a novel survey instrument." Teaching Issues and Experiments in Ecology, p. 5, available http://tiee.ecoed.net/vol/v5/research/humston/abstract.html.

Inouye, D. and Barr, B. (2000), "Climate change is affecting altitudinal migrants and hibernating species." Proceedings National Academy Sciences 97, p. 1630.

Inouye, D., Morales, M.M. and Dodge, G.J. (2002), "Variation in timing and abundance of flowering by *Delphinium barbeyi* Huth (Ranunculaceae): the roles of snowpack, frost, and La Niña, in the context of climate change." Oecologia 130, pp. 543-550.

IPPC (2007), Intergovernmental Panel on Climate Change, available (IPCC, April 2008) www.ipcc.ch/index.htm.

Jarraud, M. and Topfer, K. (2004), Sixteen years of scientific assessment in support of the climate Convention. Intergovernmental Panel on Climate Change, available (IPCC, April 2008) www.ipcc.ch/.

Kellert, S.R. (1996), The value of life: Biological diversity and human society, Island Press/Shearwater Books, Washington, D.C.

Kerr, R. (2003), "Three degrees of consensus." Science 305, pp. 932-934.

Lessl, T.M. (1989), "The priestly voice." Quarterly Journal of Speech 75, pp. 183-197.

Medhusrt, Martin J. and DeSousa, Michael A. (1981), "Political cartoons as rhetorical form: A taxonomy of graphic discourse." Communication Monographs 48(3), pp. 197-236.

North American Pollinator Protection Campaign (2007), available (NAPPC, May 2008), www.nappc.org.

Neil, K. and Wu, J. (2006), "Effects of urbanization on plant flowering phenology: a review." Urban Ecosystems 9, pp. 243-257.

Parson, E., Mote, P., Hamlet, A., Mantua, N., Snover, A., Keeton, W., Miles, E., Canning, D. and Ideker, K.G. (2001), "Potential Consequences of Climate Variability and Change for the Pacific Northwest", pp. 247-280. In: Climate Change Impacts on the United States. New York: Cambridge University Press.

Schudson, M. (1982), "The politics of narrative form: The emergence of news conventions in print and television." Daedalus 3, pp. 97-112.

Sigal, L. (1973), Reporters and officials. Lexington, Mass: D.C. Heath Co.

Stouffer, R. and Manabe, S. (1999), "Response of a coupled ocean-atmosphere model to increasing atmospheric carbon dioxide: Sensitivity to the rate of increase." Journal of Climate 12, pp. 2224-2237.

Sutherland, B. (11 September 2008), State of controversy: A narrative analysis of *State of fear*' presented at Northwest Communication Association Conference, Coeur d'Alene, ID.

United Nations Foundation (2008), Available (INF, April) www.unfoundation.org/ipccinfo/.

Zimov, S., Schuur, E. and Chapin III, F. (2006), "Permafrost and the global carbon budget." Science 312, S. 1612-1613.

Appendix 1

Questions used for pre and post-assessment of student knowledge of global climate change; final responses of one exemplary student shared with permission.

Respond to each of the following ... (learning assessment, not for grade; use code on back)

1. Is global warming occurring, and if so, how much temperature rise is predicted over time for a doubling of CO_2 in our atmosphere?

   Global warming is a serious environmental and societal problem with severe ramifications for natural ecosystems and human communities worldwide. Evidence that global climate change is occurring include the melting of glaciers and the polar ice cap, rise in seal level with the risk of flooding low-lying coastal regions, changes in the flowering time and phenology of plants and animal migrations, and warmer air temperatures. The temperature rise by 2100 is predicted to be 3-8 degrees F. Carbon dioxide is now at about 380ppm and is much higher than a century ago.

2. Global warming is occurring, but paleoclimates have changed markedly over the earth's history. So how is the climate change that is occurring today different from the past or is it?

   Global warming today is remarkably different from the past as it is occurring at a rapid pace, much faster than the normal warming phenomenon of the greenhouse effect. Our earth is now warmer than in past paleoclimates. Human activity may prevent our climate from cycling back into the next ice age.

3. What are the major sources of global warming? Consider all aspects you can think of.

   Major sources include: combustion of fossil fuels, cars, trucks, industrial development, agriculture, electrical appliances, electricity dependent on burning coal and natural gas, globalization (with its dependence on oil resources), plane travel, deforestation, land clearing, rice cultivation, methane gas emissions from termites and livestock (such as cows) and driving SUVs.

4. How do natural ecosystems or atmospheric effects either worsen or help alleviate global warming? What ecosystems or biomes are most likely to influence climate and why?

   Natural ecosystems help alleviate global warming as primary vegetation and the oceans are a carbon sink and absorb carbon. Deforestation releases carbon back into the atmosphere worsening global warming. The greenhouse effect and the ability of our atmosphere to increase in carbon dioxide concentration and to trap additional heat will exacerbate global warming. Global dimming helps lessen and offset global warming. Ecosystems most influential on climate include large forested areas such as boreal, coniferous, and evergreen forests that act as a carbon sink. Tropical rainforests are also very influential on climate and deforestation results in a negative feedback effect that changes the climate dramatically with less humidity, less rainfall, and consequences for flora and fauna.

5. What are the predicted outcomes of global warming for:
   - plants, animals, humans, etc?

     changing plant phenology, changing animal migrations, mass extinctions, shifting biome distribution, loss of tundra habitat, extinction of large predators in arctic such as polar bears due to melting of ice, warmer temperatures may lead to increased frequency of hurricanes, greater intensity/size of storms in Pacific Northwest, less snow melt in Pacific Northwest, more severe forest fires, greater invasive species and insect outbreaks due to warmer temperatures, lower snow melt and a smaller water supply. The Pacific Northwest is one of the regions in the U.S. that is expected to be hit hard by the impacts of global warming.
   - for changes in other abiotic factors, land masses, and oceans

     mass bleaching of coral reefs, loss of global biodiversity, expansion of warm air masses over equator toward higher latitudes, eustatic and isostatic sea level rise (inundations or low lying areas and coastal regions such as Florida and Cape Cod). Warmer ocean temperature may impact sea life and lead to die off of temperature sensitive tidepool creatures or loss in salmon populations.

6. What is global dimming and what is it's primary cause?

   Global dimming is the reduction in solar radiation that reaches the earth's surface that is caused by human pollution that decreases air quality with soot particulates.

7. What mechanisms are currently in place to reduce the level or impact of human-induced changes?

regulations on carbon dioxide levels and country goals for reducing emissions.
creation of hybrid cars to offset fossil fuel emissions
few regulations for large corporations or companies
U.S. does not endorse Kyoto treaty or support international regulations.

Appendix 2

Student 'brief' by Jennifer Bufford: Paleoclimatic Change since the Holocene – A Context for Modern Global Climate Change.

The most important and most hotly disputed question in the current debate on global climate change is whether current warming trends are actually due to anthropogenic effects. One important way to address this question is by examining paleoclimatic changes and comparing them to modern changes. In this comparison, the causes, global extent, and rate of climatic change are all important.

It is appropriate to begin with a discussion of the methods used to determine the extent of paleoclimatic changes. To measure changes up to a million years ago, deep sea sediment cores, which contain the calcium carbonate remains of marine organisms, are used to estimate temperature. Temperature is estimated using O^{16}; the concentration of O^{16} in the sediments varies proportionally with temperature in a predictable way (University of Michigan 2008). Ice core samples, both from the poles and from glaciers, are used in a similar way and provide localized climate data for the last few millennia (Jones &Mann 2004). Climatologists can measure the carbon dioxide levels of air bubbles trapped in the ice and estimate temperature using O^{16} levels. Tree ring data, fossilized pollen and other plant fossils provide information on plant communities, which gives indirect estimates of temperature and moisture levels around 10,000 years ago (University of Michigan 2008). For the last 1000 years, agricultural data and historical records can give further information about climate changes; actual temperature measurements exist for the last 150 years for most of the world (Jones and Mann 2004), while carbon dioxide measurements have been made for the last 50 years.

Using these methods, scientists have tracked significant changes in global climate since the last ice age in the Pleistocene, 50 to 100,000 years ago. Significant warming occurred during the Holocene (about 6,000 years ago), ending the ice age and bringing temperatures close to present-day levels. A study of plant community changes in the Andes during this transition found that, though the plant community changed directionally toward lower-elevation, drought tolerant species, this change was very slow, with rates of change approximately equal to the slow, random variation that occurred before and after the transition (Bush et al. 2004).

In historical times, several smaller changes in global climate have been proposed, including the Medieval Warming Period and the Little Ice Age. The Medieval Warming Period, from about 800 to 1300 AD, is characterized by changes in plant communities and historical records, particularly the Viking explorations of the Arctic and Greenland, which suggest warmer temperatures and less ice than the present (University of Michigan). However, this "warming period" is hotly debated. Ice core data is equivocal and there is little data from the southern hemisphere, suggesting that, at best, this may have been a regional phenomenon (NOAA 2007). There have also been suggestions that this warming period was also anthropogenic, caused by changes in land use in Europe at the time (Goossel et al. 2006). The Little Ice Age, which lasted from the mid-1500s to the mid-1700s AD, was marked by lower temperatures in the Northern Hemisphere (NOAA 2007). Some studies suggest that volcanic forcing may have been responsible for this decrease in temperature (Crowley 2000), but others have disputed the existence of a global cooling (NOAA 2007).

It is important to recognize the uncertainties inherent in these temperature estimates. Naturally, the farther back in time the climate change occurred, the greater the uncertainty involved. Furthermore, data are often unavailable for many regions of the world, especially the southern hemisphere, making it difficult to distinguish global and regional climate changes (NOAA). A vast number of potential factors influence climate change, all of which occur on different scales and interact in complex ways to result in the observed climatic patterns. This makes interpretation very difficult and has resulted in extensive debate about the nature and causes of paleoclimatic change. Jones and Mann (2004) provide an excellent discussion of the methods used to determine paleoclimate and their inherent uncertainties.

Many different causes have been attributed to global climate change. These include the Milankovitch cycles, which describe periodic changes in the tilt of the Earth, the elliptical character of the Earth's orbit around the sun and the inherent wobble of the Earth's rotational axis. Solar forcing is caused by changes in the sun's solar output. This is decreasing over time, very slowly, as the sun dies, but it is also characterized by cyclical changes in the frequency and intensity of sunspots. Non-anthropogenic changes in the concentration of greenhouse gases in the atmosphere are caused primarily by volcanic activity. This is correlated with the movement of tectonic plates, since most volcanoes form along subduction zones. The formation of Pangaea, for example, is correlated with global ice ages, because there was little subduction occurring when all the plates were in one landmass, thus resulting in little volcanic activity and little carbon dioxide output (University of Michigan 2008).

The most dramatic forcing of today's climate appears to be caused by human production of greenhouse gases, especially carbon dioxide. The United States National Oceanographic and Atmospheric Administration estimates that 50% of the current global climate change is anthropogenic in origin (NOAA 2007). Crowley (2000) found that only 25% of the global warming seen in the last 150 years could be attributed to natural causes, when a 1000 year model was used.

Increases in greenhouse gases accounted for the majority of the warming in his model. He also noted that the current climate changes are on a magnitude unprecedented in the last 1000 years (Crowley 2000). What alarms climatologists the most about current changes is the rapidity with which they are occurring. In previous centuries, temperatures have risen 0.2°C per century. Now, temperatures are rising at a rate of 0.6-0.9°C per century, much faster than the paleoclimatic changes documented since the last great ice age (Bush et al. 2004). This is why climatologists are concerned and urge greater action be taken to reduce emissions and slow the rate of climate change.

Works Cited

Bush, M.B., Silman, M.R. and Urrego, D.H. (2004), "48,000 Years of Climate and Forest Change in a Biodiversity Hot Spot." Science 303, pp. 827-829.

Crowley T. "Causes of climate change over the past 1000 years." Science 289, pp. 270-277.

Goossel, H., Arzel, O., Luterbacher, J., Mann, M., Renssen, H., Reidwyl, N., Timmermann, A., Xoplaki, E. and Wanner, H. (2006), "The origin of European 'Medieval Warm Period'." Clim. Past 2. pp. 99-113.

Jones, P.D. and Mann, M.E. (2004), "Climate over past millennia." Rev. Geophys 42, p. RG2002.

NOAA (2007). NOAA Paleoclimatology. http://lwf.ncdc.noaa.gov/paleo, accessed 2008 Jan 15.

University of Michigan (2008), Climate models and predictions for the future. www.globalchange.umich.edu/globalchange1/current/lectures/samson/climate_patterns, accessed 2008 Jan 15.

Appendix 3

Sources of cartoons that focus on global climate change

Bagely, P. (5 June 2007), "Environmental president." Salt Lake Tribune, available: CagleCartoons.com, http://caglecartoons.com/ viewimage.asp?ID = {4A109F2E- 9DD9-4706-8F06-7D298CA46DBF [accessed 4 October 2007].

Bertrams (2007), "Montreal-climate change summit Bush. Het parool. Amsterdam, Netherlands." The New York Times, available: http://cartoons. nytimages.com/ wieck_preview_page_025036 [accessed 4 October 2007].

Heng (4 October 2007), "Global warming pot tea environment." The Straits Times. Singapore, Singapore;
The New York Times, available: http://cartoons. nytimages.com/ portal/ wieck_preview_page_110790 [accessed 4 October 2007].

Johansson, O. (26 June 2007), "The Mirage." Sweden, available: CagleCartoons.com http://caglecartoons.com/viewimage.asp?ID={FED26C2F-6411-48DC-99C0-5B519142F155} [accessed 4 October 2007] .

Komarnitski, C. (18 February 2005), "Bush and Kyoto Protocol." Bulgaria, available CagleCartoons.com http://caglecartoons.com/viewimage.asp?ID={C02A4270-EC75-455B-B46D-66F0C15EC61F} [accessed 4 October 2007].

Luojie (2007), "Climate Change Conference – USA Polluter." China Daily. Beijing, China.
The New York Times, available: http://cartoons.nytimages.com/portal/wieck_preview_page_000037 [accessed 4 October 2007].

Nath, P. (15 May 2007), "Global Warming issue." National Herald, India, available CagleCartoons.com, www.caglecartoons.com/viewimage.asp?ID={73F961D5-FBE4-4E1E-B5FE-200E56E77D95} [accessed 4 October 2007].

O'Farrell, V. (16 September 2007), "Mines bigger than yours." The Illawarra Mercury, Australia, available CagleCartoons.com www.caglecartoons.com/viewimage.asp?ID={D0A9A69C-E7DC-4E1D-964-48904484EA02} [accessed 4 October 2007].

Pismestrovic (2007), Kleine Zeitung, Graz, Austria.
The New York Times, available: http://cartoons.nytimages.com/ wieck_preview_page_112362 [accessed 4 October 2007].

Pismestrovic, P. (19 July 2007), "Sommer 2007." Kleine Zeitung, Austria, available: CagleCartoons.com www.caglecartoons.com/viewimage.asp?ID={E0C006D1-652C-4F77-A4EE-1F1EBAC85277)> [accessed 4 October 2007].

Toles, T. (29 May 2007), "Global frog." Washington Post [Internet], available: www.washingtonpost.com/wpsrv/opinions/cartoonsandvideos/toles_main.html?name=Toles&date=05292007 [accessed 4 October 2007].

Zapiro (2007), "G8 summit Africa global warming." Independent. Johannesburg, South Africa.
The New York Times, available: http://cartoons.nytimages.com/ portal/wieck_preview_page_113140 [accessed 4 October 2007].

Appendix 4

Assignment used for class viewing and discussion of DVD, *Inconvenient Truth*

As we view the DVD, we will take a break periodically, to discuss the information that you will gather in teams while watching the movie. Enjoy the show!

Aspects to consider generally as you view the DVD

- what is presented and what you record below
- how what you saw or heard compares with what you know

- scientific or value-based questions that viewing this raises for future research
- actions that individuals, organizations, or governments might take

TEAMS: Each team will record information for each DVD segment analyzed. Please note:

Team 1) key assertions

Team 2) evidence for assertions

Team 3) hidden or explicit value appeals

Team 4) audience targeted in that segment and credibility of the narrator

ALL: Topics that you would like to know more about, questions to consider

Note: The **audience** might be anything from citizens to uninformed voters, teachers, scientists, media, politicians, etc.

OTHER QUESTIONS FOR GENERAL DISCUSSION AFTERWARD:

What do we have confidence about in relation to global climate change?

What do we still need to know? What questions did viewing the raise in your mind?

Is this movie or segment informative, i.e., does the information presented contribute to our understanding of global climate change?

Will viewing the movie help motivate citizens, organizations, or governments toward effective actions?

What can each of us do as citizens to help reduce the severity of global climate change for ourselves and for future generations?

Chapter 20

An Assessment of India's Efforts on Education, Awareness and Capacity Building on Climate Change

Nitin Chaudhary

Abstract

There is a general consensus that the climate change threat is real. Education, awareness generation and capacity building are among proposed strategies to mitigate its predicted ill-effects. It has also been prioritized in Article 6 of UNFCCC. This study identified the Indian initiatives undertaken through education, awareness generation and capacity building. Keeping in mind the multitude of stakeholders and work done on the subject in the country, the State of Delhi was selected. Students and the general public were selected as two broad target communities to examine the effectiveness of awareness generation mechanisms. The study examined the impacts of the contents of the CBSE syllabus and outreach programs relevant to the issue on students. The impacts of programs by government and other institutions aimed at raising awareness about climate change on the general public were also assessed. Around 83% of students and only 47% of the sampled general public were found to be aware of climate change issues. The study found that only school children and teachers have been the major target groups. It is recommended that the message should reach wider sections of the population and here, the role of NGOs becomes important. The role of vernacular media is also important to spread the message at the grassroots level.

Introduction

Climate change is recognized as today's greatest environment challenge. It threatens to leave a major adverse impact on the natural world as well as on human society. Recent events have emphatically demonstrated our growing vulnerability to unpredictable climate. The only way to mitigate the adverse impacts is to reduce and stabilize total world green house gas emissions. It means big cuts right away. Education, awareness generation and capacity building are among the proposed strategies to mitigate climate change caused ill-effects. To achieve this objective, the general public needs to be sensitized, especially the future generation; and government should widely broadcast information on the issue so that the concerned adjust their behavior in ways that limit emissions. With this as an objective, the UNFCCC under Article-6 have identified the importance of communicating this environmental concern to the community.

It is evident that communication and awareness programs have a proven record in bringing about behaviour change in health and environmental practices. Successful awareness programs in the fields of nutrition, AIDS, ozone-depleting CFCs and conservation are a paradigm. A well-conceived communications programme for addressing climate change issues can similarly be successful if government takes proactive and suitable steps. Realizing this as a significant move, many countries have already started working actively to strengthen their national Article-6 programme taking into consideration their common but differently appropriate responsibilities and other specific national and regional development priorities. The countries are Kenya, Georgia, Albania, Cambodia, Namibia, Russia, Uzbekistan, India etc.

Methodology

This report follows an exploratory research approach to understand the situation of India's initiatives undertaken on climate change through education, awareness generation and capacity building. It also examines how well India is following up its commitment and the impact of the outreach activities and education initiatives on climate change on the general public and students. The study has focused on assessing the awareness levelsof students and the general public on the issue.

For the purpose of this study, 120 Students and 95 members of the general public were selected as two broad target groups on the basis of judgmental sampling[1]. Keeping in mind the preponderance of stakeholders and work done on the subject in the country, the State of Delhi was selected. After discussion with experts, scientists, NGO personnel and intense assessment of CBSE 8th to 12th standard syllabus, questionnaires were developed.

The study refers to both secondary and primary sources of data. Initial discussions about the active and completed outreach programs and CBSE syllabus contents were done with government officials and NGO representatives working in the climate change field. This was followed by field visit where personal interviews with studentsand the general public, were conducted. Discussions were also held with educators and mass media representatives and some other key informants.

---

1 Judgemental sampling is a selection process where the surveyor selects a sample based on his own discretion. It involves a subjective selection of items for testing and it gives a fair indication of the prevailing situation on any subject matter. With limited time and available resources, this sampling technique was the best to get a broader picture of level of awareness among students and general public and work done in the country on climate change.

Results and Discussion

India has already taken some remarkable steps in the field of climate change awareness. As per its commitment to UNFCCC, India has initiated a project titled 'Enabling Activities for the Preparation of India's Initial National Communication to the UNFCCC' or the NATCOM project through MoEF in 2001. The project aims at developing the National Communication on climate change involving research institutions, technical institutions, universities, government departments and NGOs. To facilitate the process, under the aegis of the project, several seminars, workshops and outreach programs have been conducted all over India for planning the work, developing linkages and developmental and economic processes pertaining to different components of the National Communication. The process has been initiated to identify areas of future research, to strengthen India's Initial National Communication experience, gaps and future needs have been identified for the development and strengthening of activities for creating public awareness, ensuring meaningful inputs into education, and enabling access to information. A website (www.natcomindia.org) has been developed for dissemination of information and publications arising out of the project.

Many non-governmental organizations (NGOs) namely WWF-India, Winrock International India, TERI etc. are also working actively to sensitize the public to climate change issues and have targeted students, teachers, youths and women. It is found that very few NGOs and organizations have targeted farmers and rural folk. The message could spread extensively if NGOs targeted a wide cross-section of the population because 70% of India's population live in rural areas. According to the NGOs the most common problems identified are:

- Funds from national and international sources/donors are inadequate
- Follow-up in developmental programs is feasible but difficult in the case of awareness programs.
- Motivating people to participate in program is a difficult task and very few turned up again to attend the workshops after participating once. (There is considerable initial enthusiasm; but very little support for serious follow up action.)

Learning from the field

Before conducting the survey on student awareness levels, the contents of textbooks were assessed to find out what topics on the subject have been covered by the C.B.S.E. syllabus committee. After examining 8 to 12 standard school books, it was noticed that Science and Social Studies textbooks comprise a considerable amount of subject matter according to student's level of understanding about various other environmental issues.

When the understanding about climate change was assessed and students were asked whether it's a global problem, the survey indicated that 83% of students were well informed and agreed that it's a global problem and many of them were quite clear about the basic facts of climate change. On being asked about the reasons, 67% stated that it is greenhouse gases in general and CO_2 in particular, but very few of them could specify other GHGs. Approximately 8 out of 10 respondents (i.e 80%) felt that it's humans and their increasing needs that are primarily responsible for this problem and some stated that because of the escalating use of fossil fuels by developed countries. Some suggestions to tackle the problem of climate change have been given such as by planting trees; lessen use of high voltage equipment, reducing carbon dioxide emissions and use of cleaner fuels like C.N.G etc. Almost 74% indicated that a change in behaviour brought about through, awareness creation would be an incisive strategy to tackle the problem. Students opinions were also sought on modern day lifestyles and the ways to reverse this process. About 79% of students agreed that accepting an environmently friendly lifestyle would benefit the environment as well as help in combating this problem and 72% agreed to practicing an environment friendly lifestyle by using a bicycle instead of motor vehicles when going to nearby places, switching off the lights and fans when leaving the room and not in use, using public transport and planting trees etc.

In all 95 respondents were surveyed, to assess the awareness level of the general public on climate change through personal interviews. Tthe questionnaire was designed based on previous studies on this subject. Only 47% of respondents were well informed and this is not a good sign for the future. Those who were aware also agreed this problem would increase if present emission rates of greenhouse gases are not checked specifically of carbon dioxide. Only 43% of respondents could specify which GHG is much harmful (CO_2) and very few specify other GHGs. 29% revealed that human beings and their increasing use of fossil fuels are responsible for this global problem. Approximately 3 out of 10 were in favour of changing lifestyle. As far as the source of information is concerned, 54% respondents indicated Television, 11.2% mentioned radio while 22.5% said newspaper. However, very few respondents indicated other sources such as exhibitions and seminars.

Conclusion

The surveys results showed that students are well informed about the issue; however more effort is required to raise general public awareness. This majority is mainly composed of adults who have little or no formal education, especially in the area of science. The issue of climate change has only recently been a topic within the high school science curriculum, and therefore the younger generation is more aware of the associated issues. Unfortunately, most people do not readily see the effects of climate change in their day to day activities so they do not care much about it. Here media has to play a significant role and government should also encourage them in creating documentaries, publishing articles, stories and organizing events on climate change so that message would spread across society.

The study evaluated the effectiveness of the programs in creating awareness among stakeholders and it has been found that only 47% of the general public and 83% of students are aware of the issue. But the fact is that despite being aware, people are not practicing the principles in their daily life and it is not showing in their behaviour. The need is to provide the information in a simple and usable manner that generates more action. It is also recommended that there is greater coordination of all current and future programs run by civil society or any other department with media. Through a vernacular media the message could reach a wider population as around 70% of India's population reside in rural areas. India as a party to UNFCCC has taken some initiatives to achieve the objectives of Article-6 but still there is a long way to go. It would be in the nation's best interest to move forward and strengthen the campaign and educate its people about the effects that climate change would have on society; of course education is persuasion.

Recommendations

- *Information Dissemination*: Information should be available in a simple and applicable manner to generate action-oriented results.
- *Participatory Effort*: Bring out books and literature documenting best participatory practices, involving local community and government, which would provide linkages to their day to day life, their lifestyle and how traditional knowledge of the common people can be integrated to some simple responses to climate change.
- *Easy and Effective Communication*: Such books and literature must be written in the language and that is localized with familiar illustrations, making it easy to understand.
- *Media's Role*: The Media is playing a significant role in awareness creation. Through vernacular media the message could reach a wider population as around 70% reside in rural areas. It is also recommended that collective action through coordination of all current and future programs is run by civil society or any other department with media.

References

Achanta, A.N. (1993), The Climate Change agenda: an Indian perspective. TERI, New Delhi.

Anil, M. and Kristen, H. (2002), Climate change and sustainabledevelopment, London: Earthscan.

Chambers, F.M. (1993), Climate change & human impact on landscape: studies in palaeoecology & environmental archeology, London: Chapman & Hall.

Chatterjee, Kalipada, Education, Training and Public Awareness on Climate Change, Development Alternative, New Delhi.

Faure, M. (2002), Climate change & the Kyoto protocol: the role of institutions to control global change, London: Earthscan.

Gates, D.M. (1993), Climate change and its biologicalconsequences, Sunderland: Sinaver Associates.

Hardy, John T. (2003), Climate Change: causes, effects & solutions, New Jersey.

Houghton, J.T. et al. (1992), Climate Change: the supplementary report to the IPCC scientific assessment, Cambridge: Cambridge University Press.

– (2001), Climate change 2001: scientific basis: Contribution of working group I to III assessment report of IPCC, Cambridge: Cambridge University Press.

Jager, J. (1983), Climate and energy system: – a review of their interactions, Chichester: Joh Wiley.

Jit, S. Chandan (1994), Organizational Behaviour, New Delhi.

Kates, R.W., Ausubel, J.H. and Berberi, M. (1985), Climate impact assessment studies of the interaction of climate & society, New York: John Wiley.

Long, Douglas (2004), Global Warming, Cambridge: Cambridge University Press.

McCarthy, J. (2001), Climate change 2001: impacts, adaptation and vulnerability: Contribution of working groupr I to III assessment report of IPCC, Cambridge: Cambridge University Press.

Mehta, S.R. (1992), Communication & development, issues & perspectives, Jaipur: Rawat.

Mohan (1988), Environmental awareness and urban development, New Delhi.

Shukala et al. (2003), Climate change and India: Vulnerability assessment and adaptaion, Hyderabad: University Press.

Soz, Saifuddin (1998), Climate change and awareness impact, Srinagar: Peoples Empowerment Mission, Kashmir.

United Nations Environment Programme's Division of Environmental Law and Conventions (2006), Raising awareness of climate change: A handbook for government focal points, The Government of Norway.

Werner, Wolfgang L. (1993), Aspects of ecological problems and environmental awareness in South Asia, New Delhi: Manohar.

Annexure I

United Nations Framework Convention on Climate Change

Article 6 – Education, Training And Public Awareness

In carrying out their commitments under Article 4, paragraph 1(i), the Parties shall:

(a) Promote and facilitate at the national and, as appropriate, subregional and regional levels, and in accordance with national laws and regulations, and within their respective capacities:

  (i) The development and implementation of educational and public awareness programmes on climate change and its effects;

  (ii) Public access to information on climate change and its effects;

  (iii) Public participation in addressing climate change and its effects and developing adequate responses; and

  (iv) Training of scientific, technical and managerial personnel.

(b) Cooperate in and promote, at the international level, and, where appropriate, using existing bodies:

  (i) The development and exchange of educational and public awareness material on climate change and its effects; and

  (ii) The development and implementation of education and training programmes, including the strengthening of national institutions and the exchange or secondment of personnel to train experts in this field, in particular for developing countries.

Annexure II

Questionnaire (For Students)

Name of the Student:

Name of the School:

Standard:

Age:

- Do you think that Climate change is a global problem? Could you express it in your own words?
- What are the main reasons of climate change and how can we tackle this problem?
- Could you mention which gases are responsible for climate change?
- Who is responsible for this global problem?

- From which standard you have started studying about climate change? What all does it cover?
- Which subject is covering climate change issues?
- Name any Organization/Institute/ NGO working on this problem?
- Have you received any lecture/promotional material etc. on climate change by any individual/ NGO/organization.
- Do you think that change in our life styles can solve this problem? If yes,
- OR
- Are our lifestyles responsible for Climate change? If yes, then how we should change our life styles to reverse this process (of climate change).
- Would you consider changing the way you travel to a more climate-friendly mode?
  - Yes
  - No
  - No suitable alternative
  - Already travel in a climate-friendly way
- Please, use this space to tell us about what efforts you have done to help to tackle climate change problem?

Questionnaire (For General Public)

Respondent Name:

Age:

- Mention any top three environmental problems in India, e.g., threat to species, air pollution, water pollution, climate change, flooding, solid waste, etc which you think require immediate attention.
- Do you think that Climate change is a global problem; could you express it in your own words?
- Who is responsible for this global problem? How can we tackle this problem?
- Could you mention which gases are responsible for climate change?
- From where you get the information about climate change.
  - Newspaper
  - T.V.
  - Magazine
  - NGO/Organization
  - Children
  - Other
- Name any Organization/Institute/ NGO working on this problem?

- Have you get any promotional material on climate change by any NGO or organization. Could you mention the name?
- Do you think that change in our life styles can solve this problem? If yes, then how?
- Would you consider changing the way you travel to a more climate-friendly mode?
  - Yes
  - No
  - No suitable alternative
  - Already travel in a climate-friendly way
- Please, use this space to tell us about what efforts you have done to help to tackle climate change problem?

Chapter 21

Engaging Australians on Climate Change – The Role of Education and Communication in Social Change

Estelle Gaillard, Jo-Anne Ferreira

Abstract

This chapter reports on an Australian research project exploring social diffusion theory as a theory of social change to gain a better understanding of the role of education and communication in assisting this change process. The research examines Al Gore's Climate Change Leadership Program, in partnership with the Australian Conservation Foundation (ACF) and assesses the effectiveness of the social diffusion model that underpins this educational outreach in encouraging and empowering individuals to take individual and collective action to address climate change. Initial findings of this research show that education and communication initiatives are more successful when they promote processes of learning that engage individuals in their social contexts in order to assist both individual and social change.

Introduction

Addressing the major challenges posed by climate change[1] requires societal engagement on strategies for mitigation and adaptation. As researchers in the field of education for sustainability, we are interested in processes of learning for individual and societal transformation and how education and communication[2] can effectively engage people in sustainable solutions – through changing their own ways of thinking and behaviours as well as mobilising for broader social change.

This research explores how education and communication can promote individual and social learning in order to help bring about individual and social change towards sustainability. We look specifically at education and communication ef-

---

1 The term climate change is used throughout this paper, and refers to the current or projected changes in climate, whether due to natural variability or to human activities (IPCC 2001).

2 In this paper, education and communication are inseparable processes that impact upon people's thinking and actions. Since education and communication both deal with the transfer or exchange of ideas, information and skills in a two-way process, they inevitably shade into one another (Palmer 1998). Both communication and education are needed if we are to successfully address the challenges posed by climate change.

forts that address climate change because climate change is a fundamental and pressing challenge facing the sustainability transition (IPCC 2007). This research is motivated by the assumption that effective education and communication have an important role to play in addressing climate change and in facilitating long-term changes towards sustainability. While environmental education and communication practices and activities are often directed at young people in formal settings, we have chosen efforts aimed at adults as the focus of our work. As Clover (1996: 97) argues, "it is the adults of the world, as citizens, consumers, employers and parents, who make critical decisions on a daily basis that affect others and the biosphere everyday". It is adults who are "the force of social and political change in both domestic and global arenas" (Lipschutz 1996: 2).

In order to understand how an educational effort such as The Climate Project-Australia might influence individual and social change, we examine this program through the lens provided by social diffusion theory. Social diffusion theory explains how new ideas and practices spread through certain communication channels over time among the members of a social system (Rogers 2003: 5). This research focuses on the spread of new understandings of and solutions to climate change among individuals, communities and organisations; and the potential change outcomes of educational initiatives such as The Climate Project-Australia.

Learning for Climate Change

In order to understand the role of education and communication in encouraging and empowering people to respond to climate change, in this section we examine the literature around climate change; education, communication and social change; social diffusion theory; and the current context of climate change education and communication in Australia.

Sustainability and climate change

One of key challenges facing the achievement of a sustainable society is climate change (IPCC 2007). The Intergovernmental Panel on Climate Change (IPPC), the leading international body on climate change science, indicates that beyond natural variability in Earth's climate system, human activities are playing a significant role in causing present climate changes and those projected in the future (IPPC 2001; 2007).

In the past couple of years, the release of the 4th IPCC report, the Stern Report, Al Gore and the IPCC winning the Nobel Peace prize, the United Nations Climate Change Conference in Bali, intense media coverage of climate change, and responses to climate change from some state and local governments, businesses, industries, environmental and social advocacy groups as well as civil society, have placed the issue of climate change high on the global agenda. As a result, addressing the challenges posed by climate change is beginning to be recognised as urgent, not only in the environmental sphere but also in political, social and economic spheres (Lorenzi & Pidgeon 2006).

The IPCC (2001) indicates that patterns of social relationships, cultures, political practices and economic institutions are not only part of the cause of climate change; they are also subject to its potential effects. The IPCC (2001) points out that changes in governance, lifestyles, economic activity, technological advances, policies, pricing and regulation mechanisms are critical if we want to tackle climate change:

> Social learning and innovation, and changes in institutional structure could contribute to climate change mitigation. Changes in collective rules and individual behaviours may have significant effects on greenhouse gas emissions but take place within complex institutional, regulatory and legal settings. (p. 299)

As Gelbspan (2007) states, responding to the climate crisis must involve every segment of society working at every level of action. Stabilization of the Earth's climate requires large-scale government policies and programs along with international cooperation as well as major shifts in individual and collective thinking and behaviours (Leiserowitz 2007). While governments are crucial in setting priorities and policies (Moser 2006), education and communication efforts have a key role to play in empowering individuals to engage in individual, social and political actions to address the challenges posed by climate change.

Education, communication and social change

Some education and communication efforts tend to be individualistic (Padolsky 2006). They are based on the assumption that the transition towards sustainability can be made simply through individuals changing their personal practices. These education and communication efforts encourage individuals to take personal action on social and environmental issues by providing a list of suggestions on what individuals can do (Padolsky 2006). As Oskamp (2002) indicates, there are an increasing number of 'what you can do' lists in various locations such as internet sites, newspapers, and books. Such individualistic approaches assume that each individual can make a difference by taking small, day-to-day actions. While such actions appeal to the power of individuals to change, Maniates (2001) argues that these actions tend to lead to the individualization of responsibility for social and environmental issues. However, as Princen et al. (2002) state, the current socio-environmental crisis will not be countered by actions that are mainly individual and typically consumer-oriented such as planting trees, riding a bicycle, or recycling. Although these actions are commendable, they do not reflect the size and nature of the current global environmental and social issues (Clover & Hill 2003). Clover (2002) argues that a sole focus on 'individual' behaviours ignores the cultural, social, economic and political structures that underlie the current socio-environmental crisis. Effective education and communication should, therefore, take into account the dynamics of socio-environmental issues originating in social, political and environmental attitudes, behaviours, actions, decision-making and policies (Hill 2003). Effective education and communication should expose people to exactly how their society operates and why, thereby allowing them to become fully involved in its transformation towards sustainability (Huckle 1993). Consequently, while education and communication efforts should encourage and empower individuals to make changes in their lifestyles, it is crucial that they also address the necessary changes to social, economic and political systems (Pearse 2007).

According to Fien (1993: 4), "the current environmental crisis is first and foremost a social crisis". Thus, effective education and communication must be based upon the dynamic interdependence of individual and structural/social transformation (Fien 1993). It cannot be based upon changing individual values or lifestyles or through structural changes alone (Fien 1993). As Clover, Follen and Hall (1998) argue, in addition to individual change, education and communication must have a goal of collective social transformation. Clover (2003: 10) sees education as "an engaged and participatory process of political and social learning and not solely a matter of information transmission and individual change". In the context of climate change, it is all the more important that education and communication programs seek to empower for individual change within a social or collective context. This is for four reasons. Firstly, it is important to acknowledge that individuals are limited in their actions by external barriers such as the lack of feasible alternative technology, public infrastructure or transportation choices (Moser 2006). As Tribbia (2007) indicates, an individual might not want to drive to work but unless public transport or safe bicycle lanes are available, she/he might have no other option but to drive. Secondly, people are heavily influenced by their social context (Rogers 2003; Bandura 1977). For example, the lack of action being taken by others might discourage individuals to act (Lorenzi et al. 2007). Thirdly, individuals are often influenced by internal factors such a sense of isolation, lack of motivation or the lack of belief that their individual actions can make a difference (Kollmuss & Agyeman 2002). Fourthly, as Harriss (2007) states, climate change solutions will need to confront the dysfunctional practices in our social, governance and economic systems. Thus, education and communication programs seeking to address the challenges posed by climate change must move from the simple transmission of information and facts, or from a sole individual focus, to learning processes that seek to empower individuals to transform themselves as members of a social group and in the process work towards achieving broader social change. Social diffusion theory offers one such strategy.

Social diffusion theory as a social learning process towards social change

Social diffusion theory is a theory of social change that illuminates the change process and how learning leads to change. It sees individuals as having a role to play in social change while fostering a sense of community/collective and building pressure for broader societal changes.

The doyen of social diffusion theory is Everett Rogers, a social scientist who studied over 1,500 innovations and their diffusion over the last three decades (Rabkins & Gershon 2007). Social diffusion theory has been used to study the spread of new ideas and practices for over 50 years in a wide variety of settings. Social diffusion theorists attempt to identify how new ideas, behaviours, values, technologies and products spread through populations. According to Rogers (2003: 6), social diffusion theory explains social change or "the process by which alteration occurs in the structure and function of a social system". When new ideas, practices, or products are invented, diffused and adopted or rejected, social change occurs. However, Rogers acknowledges that change can also happen in other ways; for example, as result of a political revolution, a natural event or a government policy. Our interest in this research is, however, in the role of education and communication efforts in assisting 'planned' individual and social change.

However for Rogers (2003), how a social problem (such as climate change) is defined is an important determinant of how we go about solving it, and ultimately of the effectiveness of the attempted solutions. Some factors underlying a particular social problem may indeed be individual in nature. Effective solutions to the problem should, therefore, focus on changing these individual factors. In many other cases, however, the causes of a social problem lie in the larger context or system of which the individual is a part. Rogers (2003) warns against overstressing individual-blame in defining a social problem and underestimating system-blame. As Kaplan (2000) states, solely 'individual' strategies might be inappropriate if macro conditions exist which are responsible for contributing to the issue or limiting the effectiveness of individual actions. Thus, Rogers emphasises the need to simultaneously address individual change and social and structural change.

Social diffusion theorists argue that individual actions undertaken collectively help to create a 'culture of possibility for everyone' (Rabkins & Gerhson 2007, 302). This is because, as Milbrath (1989) argues, people's sense of the possibility of systemic change is influenced by the attitudes and behaviours of the people around them. Individual actions taken collectively can also build a 'grassroots groundswell' (Rabkins & Gerhson 2007: 302). Or, as Suzuki (1999: 218) puts it, "each person, group or organization working towards a different world may seem powerless and insignificant, but all of them can add up to a force that can become irresistible".

Definition of social diffusion

In his book *Diffusion of Innovations*, Rogers (2003: 5) defines social diffusion as:

> [t]he process by which an innovation is communicated through certain channels over time among the members of a social system.

This definition can be broken down into four key elements- innovation, communication channels, time, and social system. The following section offers a detailed account of each of these elements.

1. The innovation

An innovation is an idea, practice or object that is perceived as new by an individual or other unit of adoption (e.g. a group of individuals or an organisation). An innovation can take various forms. It may be tangible product, like an energy-saving compact fluorescent light bulb, or a new idea, practice or behavioural pattern. According to Rogers (2003), the characteristics or attributes of an innovation, as perceived by individuals, help explain their different rates of adoption. These attributes are: relative advantage, compatibility, complexity, triability and observability.

2. Communication channels

According to Rogers (2003), information exchange is at the heart of the social diffusion process. Such information exchanges occur, he argues, through communication channels. Communication channels are the means through which information moves from one individual to another. Communication channels include mass media channels, such as radio, television and newspapers; interpersonal channels that involve a face-to-face exchange between two or more individuals; and interactive communication channels, such as via the internet (Rogers 2003). Different communication channels are most effective at different times in the social diffusion process. As Moser and Dilling (2004) state, mass communication channels are the most efficient way for spreading information widely. However, more direct and personalised communication channels are required to persuade individuals to take specific actions.

3. Time: The innovation-decision process

According to Rogers (2003), the innovation-decision process is the process through which an individual (or other decision making unit) passes from first knowledge of an innovation (knowledge), to the formation of an attitude toward the innovation (persuasion), to a decision to adopt or reject the innovation (decision), to the implementation and use of the new idea (implementation), to confirmation of this decision (confirmation). A key insight in the innovation-decision process is that each stage of the innovation process requires different communication and education approaches in terms of the communication channels, how the communication is framed, who is communicating, to whom and the needs of targeted groups (Birney et al. 2006).

4. A social system

Social diffusion occurs within a social system, defined by Rogers (2003: 25) as "a set of interrelated units that are engaged in joint problem-solving to accomplish a common goal". The members of a social system may be individuals, informal groups, organizations or societies. A social system is also based upon a social structure. Lopez & Scott (2000) define a social structure as the patterns of causal interdependence and interconnection among the members of a system, their actions as well as their positions in the system. In addition to this formal structure, the communication flows between a system's members, tracing who interacts with whom and under what circumstances, creates a communication structure (Rogers 2003). Social diffusion theorists' understandings of the nature of information flows from one individual to another is enhanced by the concepts of homophily and heterophily. Homophily is the degree to which two or more individuals who interact perceive themselves as similar in certain attributes such as beliefs, education, socio-economic status or interests (Rogers 2003). Homophily accelerates the diffusion process but it limits the spread of an innovation to those individuals who are already connected in a close network. In contrast, heterophily is the degree to which individuals who interact are different in certain attributes (Rogers 2003). Heterophily allows for a broader learning and diffusion of new ideas or innovations across diverse and different groups (Moser 2007). Ultimately, the social diffusion process requires a certain degree of both heterophily and homophily. In addition to social and communication structures, social norms guide individual behaviour by setting standards for socially acceptable behaviours (Stern 2000; Cialdini 2000). Social norms often discourage risk taking and the willingness to be the first to make a change (Johnston 2007). According to Rogers (2003), the social structure, the communication structure and social norms can all impede or facilitate the social diffusion process.

Adopter Categories: The Bell Curve

Individuals in a social system do not all adopt an innovation at the same time. Rogers (2003) states that adopters of any new innovation or idea could be classified in five categories: innovators, early adopters, early majority, late majority and laggards, based on a Bell Curve. Each adopter's willingness and ability to adopt an innovation depends on their awareness, interest, evaluation, and trial of the innovation (Rogers 2003).

Fig. 1: The Bell Curve

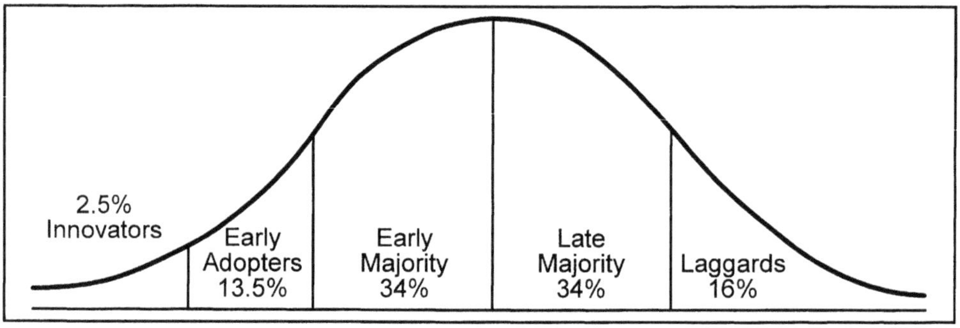

Source: based on Rogers 2003: 281

Moser (2006) indicates that it is the typical pattern for innovators and early adopters to create the conditions for a majority of actors to eventually adopt new ideas, practices or technologies. Early adopters are key individuals who are attracted to the innovation and who have a high tolerance for experimentation (Rabkins & Gerbson 2007). As Rabkins & Gershon (2007) state a key element in social diffusion is the idea that influencing a few people may be a precursor to influencing many.

In addition to these five adopter categories, Rogers (2003) identifies two additional categories in the diffusion process: 'opinion leaders' and 'change agents'. Opinion leaders are members of a social system within which they are able to exert some influence. Change agents, in contrast, are individuals who actively and effectively promote new ideas and behaviours (Atkisson 1999). Their role is to develop a need for change among early adopters. Once early adopters have recognised a need for change, change agents seek to motivate interest in the innovation and create an intent to change in early adopters. Once this intent to change has been identified, then change agents seek to assist early adopters to translate their intentions into actions (Rogers 2003; Atkisson 1999). In this study, educators and communicators are seen as change agents.

The S-shaped innovation adoption curve and critical mass

According to Rogers (2003), the adoption of an innovation in any given population follows a certain pattern. Gladwell (2002) refers to the diffusion process as a social epidemic as new ideas, behaviours and products move through a population very much as a disease does. An innovation starts with an innovator, often a single individual with an innovation. After its conception, an innovation spreads slowly at first – usually through the work of change agents, who actively promote it until it reaches the 'take-off' phase. At this stage, interpersonal networks become activated and start to spread an individual's subjective evaluations of an innovation from peer to peer in a system (Rogers 2003). As Ball (2005: 372) states, the key ingredient here is the effect that one person has on the other: 'the influence of interaction'. A critical mass (Rogers 2003) occurs when 10% to 20% of the population in a system has adopted an innovation. After reaching this point, sometimes referred to as 'tipping point' (Gladwell 2002), the innovation's further rate of adoption becomes self-sustaining. Critical mass bears on the relationship between individuals and the larger system of which they form a part. This stresses how individual's actions often depend on their perception of how many other individuals are thinking or behaving in a particular way (Cialdini 2000; Clayton & Brook 2005). After critical mass is reached, the norms of the social system encourage further adoption by remaining individuals within a system (Rogers 2003). Critical mass also puts pressure on decision makers to introduce policy and structural changes that will facilitate further diffusion and adoption. After reaching critical mass, the S-shaped curve eventually reaches a saturation level, where virtually everyone who is going to adopt the innovation has done so (Rogers 2003).

Fig. 2: The S-shaped innovation adoption curve

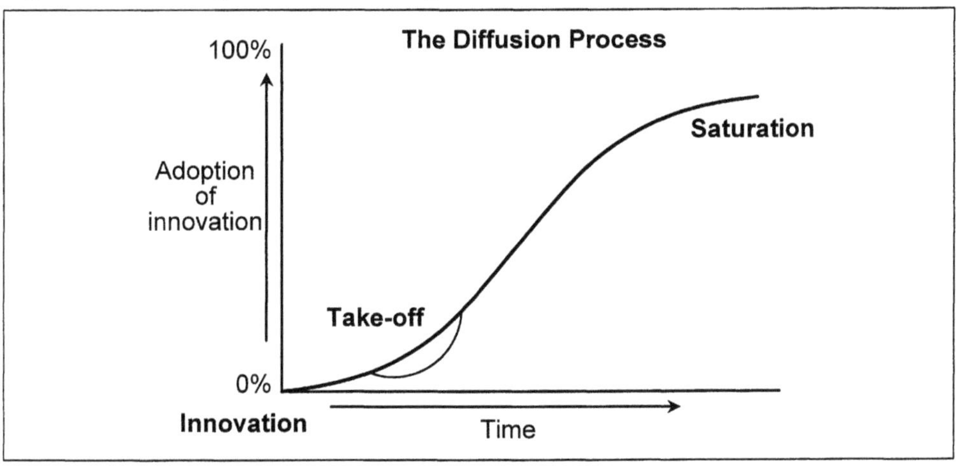

Source: based on Rogers 2003: 11

Social diffusion as a process of social learning

The Bell Curve, the S-shaped curve and the concept of critical mass are crucial in understanding the social nature of the diffusion process. Social diffusion is driven by an underlying social learning process. As people interact with each other, pass information on to one another, discuss the innovation and share their experiences, social learning occurs (Bandura 1977). Social learning theory (Bandura 1977) considers that people learn from one another through observational learning, imitation, modelling and communication. According to Bandura

(1977), individuals learn by observing and modelling the behaviours, attitudes, and emotional reactions of others. Allen (1988: 260) refers to this process as 'watching while being watched'. Social learning theory is concerned with how the social environment influences an individual's way of thinking and behaviours. As Serrano (1998) states, since individuals acquire knowledge and skills as well as form attitudes and values as they relate to one another and their environment, all learning is, therefore, social learning.

Climate change education and communication in Australia: The current context

In the past couple of years, public awareness of climate change has been growing quickly. The release of the Stern Report, Al Gore's documentary *An Inconvenient Truth*, the 4th IPPC report and intense media coverage coinciding with higher average temperatures, altered rainfall patterns, Australia's worst drought on record, intense bushfires, tropical cyclones Larry and Monica, have generated a broad national debate on climate change. Public opinion polls indicate a widespread and high level of awareness of and concern about climate change. For example, the latest BBC World Service poll, *Climate Change: Global Poll* (2007), indicated that 81% of Australians agreed that human activity was a significant cause of climate change. A vast majority (95%) endorsed taking action, including 70% who say major steps should be taken 'very soon' (BBC World Service 2007: 8-19).

Given the shift in public opinion, education and communication efforts should now move from an awareness-raising agenda (or trying to convince Australians that climate change is real) to encourage and empower Australians to change their own behaviours and to support collective action and policy changes. If we refer to social diffusion theory, this means that these efforts should now be moving into the persuasion stage. In order to help move along the diffusion S-shaped curve towards critical mass (Rogers 2003), the current challenge for educators and communicators is to persuade a large number of Australians that they should individually and collectively engage with the issue of climate change and take action. As Gladwell (2002: 69) states: "For a social epidemic to start, some people are actually going to have to be persuaded to do something."

To achieve this goal, Moser (2007) suggests that educators and communicators must make climate change visible to individuals by better educating them on the causes, impacts and implications of climate change as well as making the issue relevant to them personally. In the persuasion stage, educators and communicators, as change agents, should also concentrate on 'how-to-knowledge' (Rogers 2003: 173) to help individuals translate their awareness and knowledge into action to combat climate change. As Moser (2007) states, to facilitate engagement efforts, educators and communicators must persuade individuals that practical solutions are available and that they can make a difference by adopting them.

According to social diffusion theory, at the persuasion stage, an individual seeks information about the new idea, decides what messages he or she regards as credible and decides how s/he interprets the information (Rogers 2003). It is at the persuasion stage that an individual develops his or her perception of the innovation (Rogers 2003). The individual wants to know what the advantages and disadvantages of the innovation are for his or her particular situation (Rogers 2003). Interpersonal communication networks with peers are particularly likely to convey such evaluative information about an innovation. Individuals tend to turn to their peers to confirm, debate, or obtain further details. As people interact with each other, pass information on to one another, discuss climate change solutions, and share their experiences, social learning occurs (Bandura 1977). Rogers (2003) indicates that information obtained from peers located in social and organizational networks has more weight than information obtained from the media or scientists. Individuals want to know whether their thinking is aligned with the ways of thinking and behaving of their peers (Rogers 2003; Goldstein & Cialdini 2007).

As Cialdini (2000) indicates, individuals often seek what he calls 'social validation' or 'social proof'. If individuals see that many others, especially those perceived as similar, are taking action, they will be more willing to take action too (Cialdini 2000). This is especially important is the context of climate change. As studies on public views of climate change (e.g. Bulkeley 2000; Nicholson-Cole 2004; Lorenzi et al. 2007; Downing & Ballantyne 2007) indicate, individuals often lack motivation for taking action or discount the effectiveness of their actions in tackling climate change because they see few other people engaging in climate change solutions. Thus, education and communication efforts must give individuals a sense that they can play an essential part in tackling climate change. They must convince individuals that their personal actions will make a difference as part of a larger collective who can together successfully tackle the issue (Moser & Dilling 2004). This is accurately summarised by David Suzuki: "If you don't take your car to work one day, it is a drop in the bucket. But if 400,000 people leave their car at home for one day a week, that is huge" (Suzuki in Ambrose 2006: 12).

As Moser and Dilling (2007) indicate, although such activities might be limited in actual emission reductions they can spread a symbolic message that engages others. Such efforts might also help to increase individuals' sense of efficacy (the belief that one is capable of taking effective action) and create collective efficacy (the belief that together they are capable of taking effective action) (Bandura 2000). According to Bandura (2000) when people feel a strong sense of belief in their ability to act, they are more likely to try an innovation and adopt it long-term.

As social diffusion theory indicates, as more and more people decide to adopt climate change solutions, they develop common guidelines and create new social norms. Thus, the task of communication and education efforts is to show individuals that others are engaged and that new social norms are emerging

(Cialdini 2000; Moser 2007). For example, Bardwell (1991) advocates using success stories of ordinary people to encourage others to take action. Sharing examples of individuals, communities or organisations acting on climate change can show others the possibilities for effective action that might be available to them. As Milbrath (1989) states seeing that some people are able to bring about change, even a small one, can provide evidence that change is indeed possible. Therefore, it can be argued that education and communication efforts should create spaces for people to network, exchange information and ideas on climate change, discuss and evaluate potential solutions and actions available, and promote processes of learning that engage individuals in a social context.

In order to 'widen the circle of the engaged' (Moser 2007: 78) and penetrate every segment of society (Gelbspan 2007), it is also essential to widen the circle of educators and communicators beyond scientists and environmental advocates.

More business and religious leaders, politicians, health practitioners, lawyers, engineers, architects and community leaders are beginning to speak out on climate change (see Gaillard 2008). These 'new' educators and communicators are key individuals who act as entry points into specific groups. As 'peers', they are often seen as credible, trustworthy and persuasive (Rogers 2003). They can tailor their communication strategies to their audiences' specific needs, knowledge, concerns, motivation and sphere of influence. As Reardon (1991: 10), states "the greatest enemy of persuasion is assumption"; when motivating changes, "you must know what does matter to your audience, not what should matter to them". Sharing similar situational contexts, interests, values or concerns, educators and communicators can help their peers understand what climate change means to them and how they can effectively respond to the issue in their particular contexts (Moser & Dilling 2007).

Educators and communicators should also enlist the support of other individuals in the diffusion process. They should persuade and empower individuals, who are already engaged or receptive, to spread the word on climate change and climate change solutions, as the adoption of new ideas and behaviours frequently occurs as a result of friends, family members, colleagues or neighbours introducing them to their peers. The more people spread the word and talk to their peers and others, the more people will start to adopt new ideas and practices sending, clear messages to government and industry regarding citizen concerns and consumer preferences (Stern et al. 1999).

The study design

In order to understand how education and communication practitioners can use a social diffusion process in the current Australian context, this research study investigates the social diffusion model underpinning The Climate Project-Australia education program and its potential impact on audiences' understandings and perceptions of climate change as well as their willingness to undertake individual, social and political actions to tackle climate change. This section provides a brief overview of The Climate Project-Australia along with an outline of the research design for this study.

Brief overview of the Climate Project-Australia

The Climate Project-Australia seeks to spread the message of Al Gore's Oscar-winning documentary on climate change, *An Inconvenient Truth*. In late 2006, 84 Australians were personally trained by Al Gore to deliver versions of the documentary's slide show across Australia. In September 2007, an additional 170 Australians were trained using a more Australian version of the slide show called *An Australian Inconvenient Truth* (ACF 2008).

The Climate Project-Australia educational outreach draws heavily on social diffusion theory and peer education. It seeks to capitalise on presenters, from diverse social, professional and geographical backgrounds, as entry points in communities across Australia to spread the word on climate change (Gaillard 2008). After presenters learn how to utilise the Al Gore slide show, they return to their communities, workplaces and social groups and seek to engage as many people as possible in tackling climate change (Haag 2007). More than 70,000 Australians have attended The Climate Project-Australia slide show presentation since October 2006. By the end of September 2008, it is predicted than one in one hundred Australians will have attended a Climate Project-Australia presentation (ACF 2008). This is a cause for great optimism in terms of awareness and knowledge-raising. However, this research seeks to examine the outcomes of the program in light of the next stage in social diffusion theory, 'persuasion'; and to understand to what extent the program helps attendees to individually and/or collectively engage with climate change and decide to take action.

Research Design

This research utilises a multi-method approach of observations, questionnaires and semi-structured interviews to collect both quantitative and qualitative data from audience members both before and after attending a Climate Project-Australia presentation. This approach allows for a broader exploration of the research themes and the triangulation of findings (Lowe et al. 2006). The research design is flexible and intentionally leaves the study open to changing events inside and outside the study, new ideas and our own learning during the course of this research (Nicholson-Cole 2004). This research follows ten presenters from The Climate Project-Australia and surveys their audience members at selected presentations. The presenters were selected based on their occupational sector and geographical location to both reflect a wide spectrum of society in the research and mirror The Climate Project-Australia's social diffusion model (Gaillard 2008).

To date, 12 presentations have been attended, 140 questionnaires completed, and 60 semi-interviews conducted. In this research, the observation process is based on attending presentations and monitoring the extent to which presenters customise the messages to their audiences; the visuals they use and the solutions they promote; and the verbal and non verbal responses of audience members (Gaillard 2008). At each presentation, audience members are invited to complete a two-part questionnaire. The first part is completed immediately before attending the presentation and the second part completed right after attending the presentation. The first part solicits basic demographic information (gender, age and occupation) and asks eight questions on motivation, perceptions, understanding of the causes and impacts of climate change, level of personal concern and actions already taken to reduce climate change. In addition to the questions posed in part 1, part 2 contains a rating scale on several aspects of the presentation, self-reported changes in understandings of and level of concern about climate, intent to act as a result of attending the presentation, and a final open-ended question for respondents to share additional comments on the presentation and/or climate change (Gaillard 2008). Respondents who provide their contact details in the questionnaire are also invited to participate in semi-structured interviews about four weeks after attending the presentation to discuss in more depth their experience of the Climate Project-Australia presentation they attended and their views, opinions and concerns on climate change. At this stage of the research, it is expected that participants will be re-interviewed about six months after the initial interview in order to ascertain whether changes that might have been reported after the presentation and during the initial interview have continued over time (Leiserowitz 2004). The second interview will also monitor if intentions to act expressed by respondents after the presentation have translated into actions and what might have assisted or hindered their change process (Gaillard 2008).

Initial findings

To date, 12 presentations have been attended, 140 questionnaires completed, and 60 semi-interviews conducted. In this section, we present four main preliminary findings, based on the responses of presentation attendees on the questionnaire and during semi-structured interviews, in light of the two implications for climate change education and communication described earlier in this paper: 'widening the circle of the engaged' and 'widening the circle of educators and communicators'.

1. Widening the circle of the engaged

Providing solutions

Initial findings show that although a reasonable number of presentation attendees claim to attend a presentation to learn more about the science behind climate change, the majority attend in the hope of finding out more about what they can do to tackle climate change or "do my bit" (respondent). This shows a clear demand from audiences for available and practical solutions (Gaillard 2008). When asked to rate the usefulness and relevance of solutions promoted in the presentation to them, the average response from presentation attendees was between "fair" and "good". It is important to note, however, that audience members' level of engagement with climate change prior to attending a presentation as well as their interest and needs vary significantly. This has significant implications on communicating solutions (Gaillard 2008).

Results from the questionnaires suggest that the majority of respondents already take some actions to lead more sustainable lives prior to attending a presentation. Recycling, energy conservation at home, water saving and use of public transport are the most frequently mentioned (Gaillard 2008). Yet for some respondents "the presentation gave easy solutions to start the process of change" (respondent). This shows that some attendees are looking for additional practical ways in which they can make a difference or "to learn how to further make improvements to a climate aware lifestyle" (respondent). Some respondents indicated that information in the presentation on 'where do emissions come from?' helped them to identify further areas in their lives where they could reduce their emissions (Gaillard 2008). The standard slide in presentations on solutions or "What can you do?" includes four actions individuals can take: reduce your own emissions, buy Greenpower, offset the rest and spread the word". Findings from semi-structured interviews suggest that this short list of practical solutions for individuals seems to suit audience members looking to either get involved or to take additional actions in their daily lives.

However, for others respondents the presentation didn't suggest anything that they were not already doing. As one respondent indicated "I would have liked to have heard more in-depth ideas for ways an individual can contribute rather than the simplistic day-to-day activities that most of us do or try to do" (respondent). For these audience members, the 'four actions' slide appears to fall short. Some respondents would have liked a wider discussion on alternative solutions and for the presenter to touch upon the need for broader social transformations. As one respondent stated "solutions need to address not solely individual actions and technical fixes but also the social and political changes required to tackle the issue" (respondent). For audience members already pro-actively engaging with climate change and looking for further involvement, presentations could go beyond 'simple' steps and make a stronger case for citizen activism (Chase 2007) and bigger commitments.

These preliminary results show how crucial it is for presenters to customise the solution information to their specific audience by reflecting on who the audience is, what they know and do already, what they need and want, what their areas of interest are and what their capacity for action or further is. There is no 'one size fits all' list of solutions as the level of engagement on and interest about climate change seems to differ greatly among audiences.

Engaging people socially

Some respondents attend a presentation hoping to engage with other people interested and/or concerned about climate change, "I want to meet likeminded people who are interested in climate change" (respondent) or "other interested parties" (respondent). Although some presenters include interesting initiatives (such as using the slide-show as an introduction to a discussion forum among attendees or breaking up the slide show with interactive activities to allow more dialogue and interaction with audience members, most presentations attended to date follow the traditional slide-show and lecture format (Gaillard 2008). Some respondents would have liked some time either during or around the slide show to interact with other people, discuss the topic and "network" (respondent). They not only wish to learn from the presenter but also from other audience members. As one respondent stated in her questionnaire, "I am happy to see that there are more and more people passionate about climate change". However, this respondent later indicated during the interview that she was disappointed that she did not have the chance to interact at the presentation.

A few attendees indicated that they hoped to learn at the presentation how other people were responding to the climate change locally or "community projects and networks I could join" (respondent). Some attendees claimed that they would like to get involved but they did not always know how and where. This is in line with Kaplan (2000) who argues that individuals value opportunities to be involved in community activities and to take action. Based on observations, some presentations promote examples of successful local community projects such as climate actions groups to inspire audience members to take individual and collective action. These examples not only show that individuals, communities and/or organisations are already taking action on climate change but also how attendees can take action. As one respondent stated "to get people to act, you need to get them to see that other people are being proactive" (respondent). Indeed, some respondents claimed that the examples gave them ideas and increased their motivation to act, "If they can do it, we can do it" (respondent). Examples also can provide a way to make climate change relevant to people where they live and work. All these findings indicate a need for The Climate Project-Australia to shift from a mainly one-way 'message delivery' to a two-way communication process that encourages more dialogue among participants and promote active social learning, while still using the slide show.

2. Widening the circle of educators and communicators

Presenters' credibility

As Moser and Dilling (2007: 501) state in their book *Creating a Climate for Change*, effective communication and education has to achieve a match between the audience and what it perceives to be a credible and legitimate messenger. This raises the question of what makes someone persuasive (Gladwell 2002).

Findings from semi-structured interviews show that audience members have diverse ideas about what makes a presenter persuasive. These include:

- Similar professional background, "the presenter has a similar professional background than me so I could relate to his position" (respondent);
- Level of interest on the issue, enthusiasm and passion, "the presenter was very passionate and his passion was contagious";
- Good presentation skills, "the presenter was a very good and engaging speaker";
- Personal commitment to a sustainable lifestyle, "I know the presenter personally. She embraces a sustainable lifestyle, she is connected to this place and she is doing her part to tackle climate change. All these give her credibility";
- The fact that The Climate Project-Australia's presentations are based on Al Gore's slide show, as part of his leadership programme, added credibility for some respondents: "the fact that the presenter has been trained by Al Gore made him credible".

According to Chess and Johnson (2007: 23), the importance of credible educators and communicators has not yet been examined in the context of climate change. Further exploration in this study of what makes The Climate Project-Australia presenters credible or not offer some contribution to the field.

Spreading the word

When asked in the questionnaire, what actions audience members were likely to take to address climate change as a result of attending the presentation, only a few respondents stated that they would talk to their family, friends, neighbours and colleagues. However, when asked during semi-structured interviews who they spoke to after attending the presentation about either the presentation itself and/or climate change, a larger number of respondents reported having talked to colleagues, friends and families about either the presentation and/or climate change after attending the presentation. Interview findings indicate that some audience members consider climate change to be a risky issue to discuss with others. Respondents provided various reasons for this, including the complexity of the issue, the lack of opportunity to raise the issue, the lack of interest from others, the lack of confidence in own knowledge on the issue, and feeling like they were in a minority.

For some, "taking about climate change is not always easy due to the complexity of the issue" (respondent), "the opportunity does not always come up" (respondents) and "a lot of people are not interested or do not want to hear about it" (respondent). For others, they are afraid to be perceived as 'greenies" (respondent), as "messengers of doom and gloom" (respondent) or as 'unpopular" (respondent). Thus, audience members tended to only bring up the issue with people who they knew were receptive. A few respondents also hoped that by making changes in their own lives, they could encourage others to reflect and try out changes themselves.

Some attendees felt that they needed to know more on climate change (e.g. on climate change science, causes and impacts, implications for Australia and worldwide and alternative technological, financial, individual, social and political solutions) before engaging others on the topic, "I want to gain more knowledge of scientific facts related to climate change for use in discussions with friends and family members in this presentation" (respondent) or " how we can successfully tackle climate change in Australia" (respondent). A few attendees indicate that they hope to learn from the presentation "how to educate others on climate change" (respondent) and "some presentation skills" (respondent).

The findings indicate that a number of respondents still find it challenging to engage others on climate change and believe that they are a minority. Thus, it is key for The Climate Project-Australia to encourage and better equip attendees to reach out to others already engaged as well as not yet engaged with climate change; or as one presenter stated "in this presentation, I hope to provide you with ammunition when you are trying to engage with others and enthusiasm because 'spreading the word' is what it is all about". It is also crucial to encourage attendees to talk to one another, exchange stories, ideas and knowledge to build mutual support and feel as part of a growing collective.

Conclusion

This research explores social diffusion theory in the context of climate change and how education and communication initiatives, such as The Climate Project-Australia, can promote both individual and social change towards sustainability. The current challenge for education and communication in Australia is to persuade more Australians that they should engage with climate change and take action. A key finding from this research is that there is not one, but many Australian audiences and thus, it is crucial for education and communication efforts, such as The Climate Project- Australia, to be strategic in their approach to these audiences. Educators and communicators must reflect on: who the audience is, what they need or want, what they know and do already, what their capacity for further engagement is, what the key messages to be conveyed are, what solutions should be promoted and what the expected outcomes from the education and communication outreach are.

Education and communication initiatives such as The Climate Change Project Australia need to move from a one-way information delivery to a two-way learning process that engages individuals within their social context in order to assist social diffusion towards broader social change for a sustainable future. However this presents key challenges for educators and communicators.

References

ACF (2008), Australian Conservation Foundation, viewed 31 July 2008, http://www.acfonline.org.au.

Allen, D. (1988), "New Telecommunication Services: Network Externalities and Critical Mass." Telecommunication Policy vol. 12 no. 3, pp. 257-271.

Ambrose, M. (2006), David Suzuki: We can change the world in one Generation, Habitat Australia, pp. 12-13.

Atkisson, A. (1999), Believing Cassandra: an optimist look at a pessimist's world, Chelsea Green, White River Junction, Vermont.

Ball, P. (2005), Critical Mass: How one thing leads to another, Arrow Books, London, UK.

Bandura, A. (1977), Social Learning Theory, Prentice Hall, Englewoods Cliffs, N.J.

– (2000), "Exercise of Human Agency Through Collective Efficacy." Current Directions in Psychological Science vol. 9 no. 3, pp. 75-78.

Bardwell, L. (1991), "Success Stories: Imagery by Example." Journal of Environmental Education vol. 23 no. 1, pp. 5-10.

BBC World Service (2007), Climate Change: Global Poll, viewed 31 July 2008, http://news.bbc.co.uk/2/shared/bsp/hi/pdfs/25_09_07climatepoll.pdf.

Birney, A., Hren, B., Jackson, L. and Kendell, P. (2006), "Creating Pathways to Change." In: Leal Filho, W. (ed.), Innovation, education and communication for sustainable development, Peter Lang, Frankfurt am Main, New York, N.Y.

Bulkeley, H. (2000), "Common knowledge? Public understanding of climate change in Newcastle, Australia." Public Understanding of Science vol. 9, pp. 313-333.

Chase, S. (2007), The Other "Inconvenient Truth", Synthesis/Regeneration 42, Winter 2007, viewed 31 July 2008, http://www.greens.org/s-r/42/42-10.html.

Chess, C. and Johnson, B. (2007), "Information is not enough." In: Moser, S. and Dilling, L. (ed.), Creating a climate for change: Communicating climate change and facilitating social change, Cambridge University Press, Cambridge, UK, pp. 223-233.

Cialdini, R. (2000), Influence: Science and Practice (4th ed.), Allyn & Bacon, Inc., Boston, Massachusetts.

Clayton, S. and Brook, A. (2005), "Can psychology help save the world? A model for conservation psychology." Analyses of Social Issues and Public Policy vol. 5, no.1, pp. 87-102.

Clover, D.E. (1996), "Developing international environmental adult education: the challenge, theory and practice." In: Filho, L., Murphy, Z. and O'Loan, K. (ed.), A Sourcebook for Environmental Education: A Practical Review Based on the Belgrade Charter, The Parthenon Publishing Group, London, pp. 92-111.

– (2002), "Traversing the Gap: Concientizacion, educative-activism in environmental adult education." Environmental Education Research vol. 8 no. 3, pp. 315-323.

– (2003), "Environmental Adult Education: Critique and Creativity In a Globalizing World." In: Clover, D.E. and Hill, L.H. (ed.), Environmental adult education: ecological learning, theory, and practice for socio environmental change, San Francisco, pp. 5-15.

Clover, D.E., Follen, S. and Hall, B. (1998), The Nature of Transformation: Environmental, Adult and Popular Education, University of Toronto Press Inc., Toronto, Ontario, Canada.

Clover, D.E. and Hill, L.H. (2003), Environmental adult education: ecological learning, theory, and practice for socio environmental change, Jossey Bass, San Francisco.

Downing, P. and Ballantyne, J. (2007), "Turning Point or Tipping Point? Social marketing and climate change", Ipsos Mori Social Research Institute, UK, viewed 31 July, 2008, ipsosmori.co.uk/_assets/reports/turning-point-or-tipping-point.pdf.

Fien, J. (1993), Environmental Education: a pathway to sustainability, Deakin University, Geelong, Australia.

Gaillard, E. (2008), "Communicating climate change Down Under: an Australian case study." International Journal for Sustainability Communication vol. 3, pp. 133-151, viewed 31 July 2008, http://www.ijsc-online.org/en/details/gaillard_03.php.

Gelbspan, R. (2001), "Foreword." In: Dauncey, G. with Mazza, P. (ed.), Stormy Weather: 101 Solutions to Global Climate Change, New Society Publishers, Gabriola Island, British Columbia, pp. xi-xii.

Gladwell, M. (2002), The Tipping Point: How Little Things Can Make a Big Difference, Back Bay Books, New York, NY.

Goldstein, N.J. and Cialdini, R.B. (2007), "Using social norms as a lever of social influence." In: Pratkanis, A. (ed.), The Science of Social Influence: Advances and Future Progress, Psychology Press, Philadelphia, PA, pp. 167-192.

Haag, A. (2007), "Al' Army." Nature April 2007 vol. 12, pp. 723-724.

Harriss, R. (2007), "An ongoing dialogue on climate change: the Boulder Manifesto." In: Moser, S. and Dilling, L. (ed.), Creating a climate for change: Communicating climate change and facilitating social change, Cambridge University Press, Cambridge, UK, pp. 485-490.

Hill, R.J. (2003), "Environmental Justice: Environmental Adult Education at the Confluence of Oppressions." In: Clover, D.E. and Hill, L.H. (ed.), Environmental Adult Education: Ecological learning, theory and practice for socio environmental change, Jossey Bass, San Francisco.

Huckle, J. (1993), "Education and Sustainability: a view from critical theory." In: Fien, J. (ed.), Environmental Education: a pathway to sustainability, Deakin University Press, Australia, pp. 42-68.

IPCC (2001), Third Assessment Report 2001. Summary for Policymakers, viewed 31 July 2008, http://www.ipcc.ch/pdf/climate-changes2001/synthesis-spm/synthesisspm-en.pdf.

– (2007), Fourth Assessment Report 2007. Synthesis Report, viewed 31 July 2008, http://www.ipcc.ch/pdf/assessment-report/ar4/syr/ar4_syr.pdf.

Johnston, J. (2006), "Educating for Sustainable Development by Teaching about SD." In: Leal Filho, W. (ed.), Innovation, education and communication for sustainable development, Peter Lang, Frankfurt am Main; New York, N.Y.

Kaplan, S. (2000), "Human Nature and Environmentally Responsible Behavior." Journal of Social Issues vol. 56 no. 3, pp. 491-508.

Kollmuss, A. and Agyeman, J. (2002), "Mind the Gap: why do people act environmentally and what are the barriers to pro-environmental behavior?" Environmental Education Research vol. 8 no. 3, pp. 239-260.

Leiserowitz, A. (2004), "Before and after the Day After Tomorrow: A U.S. study of climate change risk perception." Environment vol. 46 no. 9, pp. 22-37.

– (2007), International Public Opinion, Perception, and Understanding of Global Climate Change, Human Development Report 2007/2008 "Fighting climate change: Human solidarity in a divided world", UNDP, viewed 31 July 2008, http://hdr.undp.org/en/reports/global/hdr2007-2008/papers/leiserowitz_anthony6.pdf.

Lipschutz, R.D. (1996), Global Civil Society and Global Environment Governance, State University of New York Press, Albany, New York.

Lopez, J. and Scott, J. (2000), Social Structure, Open University Press, Buckingham, UK.

Lorenzoni, I. and Pidgeon, N. (2006), "Public Views on Climate Change: European and USA Perspectives." Climatic Change vol. 77 no. 1-2, pp. 73-95.

Lorenzoni, I., Nicholson-Cole, S. and Whitmarsh, L. (2007), "Barriers perceived to engaging with climate change among the UK public and their policy implications." Global Environmental Change vol. 7, pp. 445-459.

Lowe, T., Brown, K., Suraje, D., De Franca Doria, M., Haynes, K. and Vincent, K. (2006), "Does tomorrow ever come? Disaster narrative and public perceptions of climate change." Public Understanding of Science vol. 15, pp. 435-457.

Maniates, M. (2001), "Individualization: Plant a Tree, Buy a Bike, Save the World?" Global Environmental Politics vol. 1 no. 3, pp. 31-51.

Milbrath, L. (1989), Envisioning a Sustainable Society: Learning Our Way Out, State University of New York Press, Albany, NY.

Moser, S. (2006), "Communicating Climate Change Motivating Civic Action: An Opportunity for Democratic Renewal?" In: Climate Change Politics in North America, Woodrow Wilson International Centre for Scholars, Canada Institute.

– (2007), "Communication Strategies." In: Isham, J. and Waage, S. (ed.), Ignition: What you can do to flight global warming and spark a Movement, Island Press, Washington, pp. 73-93.

– (2008), "Toward a deeper engagement of the U.S. public on climate change: the 44th president of the United States of America." In: International Journal for Sustainability Communication vol. 3, pp. 119-132, viewed 31 July 2008, http://www.ijsc-online.org/en/details/moser_03.php.

Moser, S. and Dilling, L. (2004), "Making Climate Hot: Communicating the urgency and challenge of global climate change." Environment vol. 45, pp. 32-46.

– (2007), Creating a climate for change: communicating climate and facilitating social change, Cambridge University Press, Cambridge, UK.

Nicholson-Cole, S. (2004), Imag(in)ing climate change: Exploring people's visual imagery, issue salience and personal efficacy, Unpublished PhD thesis, School of Environmental Sciences, University of East Anglia, Norwich.

Oskamp, S. (2002), "Environmentally responsible behavior: teaching and promoting it effectively." Analysis of Social Issues and Public Policy vol. 2 no. 1, pp. 173-82.

Padolsky, M.E. (2006), Bringing Climate Change Down to Earth: Science and Participation in Canadian and Australian Climate Change Campaigns, PhD Diss., University of California, San Diego.

Palmer, J.A. (1998), Environmental Education in the 21st Century: Theory, Practice, Progress and Promise, Routledge, London.

Pearse, G. (2007), High and Dry: John Howard, climate change and the selling of Australia's future, Penguin Group (Australia), Camberwell, Victoria.

Princen, T., Maniates, M. & Conca, K. (2002), Confronting Consumption, The MIT Press, Cambridge, Massachusetts.

Rabkin, S. and Gershon, D. (2007), "Changing the world one household at a time: Portland's 30's day program to lose 5,000 pounds." In: Moser, S. and Dilling, L. (ed.), Creating a climate for change: Communicating climate change and facilitating social change, Cambridge University Press, Cambridge, UK, pp. 292-302.

Reardon, K. (1991), Persuasion in practice, Sage, Newbury Park.

Rogers, E. (2003), Diffusion of Innovations, 5th ed., Free Press, New York, NY.

Serrano, I.R. (1998), "Social Learning: Which Way out?" Community and Habitat vol. 1, pp. 76-87.

Stern, P.C. (2000), "Toward a Coherent Theory of Environmentally Significant Behaviour." Journal of Social Issues vol. 56 no. 3, pp. 407-424.

Stern, P.C., Dietz, T., Gregory, T., Guagnano, A. and Kalof, L. (1999), A Value-Belief-Norm Theory of Support for Social Movements: The Case of Environmentalism, Research in Human Ecology vol. 6 no. 2, pp. 81-97.

Suzuki, D. (1999), The Sacred Balance: Rediscovering our place in nature, Allen & Unwin, Crows Nest, NSW, Australia.

Tribbia, J. (2007), "Stuck in the slow lane of behaviour change? A not-so-superman perspective on getting out of our cars." In: Moser, S. and Dilling, L. (ed.), Creating a climate for change: Communicating climate change and facilitating social change, Cambridge University Press, Cambridge, UK, pp. 237-250.

Chapter 22

Linking Climate Change and Sustainable Energy Technologies to Develop a 'People-Centred' Intervention Framework

Sam Wong

Abstract

This paper offers a critical examination of the role of sustainable technology in tackling climate change in developing countries. Drawing on solar home systems in Bangladesh as examples, it argues that institutional building, in the process of technological interventions, could reinforce political and gendered inequalities. Making new rules could provide opportunity for the local elites to gain political influence. Despite the positive impact, the solar home system could cause tension and undermine norms of cooperation within communities. This paper proposes a 'people-centred' framework which aims to be poverty-reducing, gender-sensitive and power-dynamic. It builds bridges between nature and society, and links the macro technical systems into the micro processes of people's everyday life and public acceptance.

1. Introduction

While sustainable technology is considered as an indispensable tool to tackle climate change, issues around it, such as human incentives, institutional arrangements and power dynamics, need close scrutiny. This chapter aims to use a communal solar home system in Bangladesh as an example in order to illustrate the complex relationships between climate change and sustainable energy technology in regard to governance, gender and the distribution of costs and benefits within communities.

Technology has long been placed high in the agenda of development. Amid the rising concerns over global warming and the recent energy 'crisis', the call for a more robust response by, applying sustainable technological interventions to mitigate the negative impact brought by climate change and help both the developed and developing countries to make adaptation to the changing climate, is pressing. The launch of the 'Lighting Africa' campaigns in 2008 by the World Bank has shown the optimistic views about the contributions of sustainable technology in poverty reduction and climate change (World Bank website).

The concept of technology in development and environmental studies has been evolved over the past 40 years. The shifting emphasis, from large-scale technology in the 1960s and 70s to more culture-specific, sustainable interventions in the 1990s and the early 2000s, has marked the changing perspectives about how (sustainable) technology is understood. Technology in the 1960s and 70s was considered as problem-solving in nature. Top-down, expert-led and the state-initiated engineering projects with large-scale, visible outcomes were the norms. Owing to the mixed results of the interventions, the neo-liberal agenda, the rising awareness of environmental sustainability and the increasing concerns of global warming, putting technology in the context of sustainability has entered the mainstream policy level since the 1990s (Chataway et al. 2007). It advocates smaller-scale, more context-specific interventions with participatory, community-based decision-making. It aims to use sustainable technology to achieve cost efficiency, policy effectiveness and environmental sustainability simultaneously. In short, the current thinking promotes the 6Cs principles: clean, cheap, capacity-building, context, convenient and collaborative.

This new perspective is useful in linking sustainable technology, poverty reduction and climate change. It has influenced the energy policies and social reforms by the World Bank and the Asian Development Bank (ADB 2000). The impact of this approach on the ground is, however, mixed – while some interventions help poor people tackle climate change (Paavola & Adger 2002), others are not so effective. In some cases, the technologies implemented need to be scrapped (Wong and Sharp, forthcoming). Literature offers three explanations for the limited success: firstly, the existing sustainable technological interventions are not all 'pro-poor' in nature. The distribution of costs and benefits by sustainable technology is so uneven that poor people often bear disproportionately high costs while local elites capture most benefits. Secondly, the interventions tend to under-estimate the structural constraints, such as caste, class, gender and race. All these can hinder poor people from gaining access to technology in their everyday lives (Feenberg 1999). Thirdly, the discourse of sustainable technology pays insufficient attention to power dynamics within communities which can reinforce social and environmental inequalities (Cleaver & Frank 2005).

This paper will build on these critiques and make two main arguments. Firstly, the complex relationships between governance and sustainable technology (specifically sustainable energy technology) require a critical re-examination. Good governance structures and effective technological interventions are necessary, but not sufficient, conditions to challenge existing social and power inequalities. This paper will argue with examples that institutional building and rule-making, related to sustainable energy policies, are not politically-neutral. They can, unwittingly, reproduce gender and social inequalities within communities. Secondly, this paper will shed light onto the complex socio-economic impact by sustainable technological interventions and the distribution of costs and benefits. It will show with examples that the interventions can undermine the sense of solidarity and harmony within communities, and worsen the poverty of the marginalised groups in the long-run.

This paper will draw on a communal solar home system in a rural Hindu village in South Bangladesh as a case study to develop the arguments. Using semi-structured interviews, focus-group discussion and social drawing techniques, it aims to build a 'people-centred' sustainable technological intervention framework in order to accommodate the competing needs between governance, people, technology and the nature within the climate change debate. This paper will start by examining the shifting paradigms about how (sustainable) technology is conceptualised and theorised over the past four decades. After an introduction to the case study and research methodology, it will focus the discussion on four issues: institutional building, gender, rule-making, and socio-economic changes. It will conclude by politicising the notion of sustainable technology and offering insights into building a 'people-centred' and 'pro-poor' framework for future technological interventions.

2. Theorising sustainable technology

The conceptualisation of technology has been changing over the past four decades. This can be roughly divided into three periods: 1960s-70s, 1980s-mid 1990s, and late 1990s-early 2000s. The dynamism is the result of competition between socio-economic, political and ecological drivers in our society. Table 1 summarises the paradigm shifts of understanding technology[1].

In the 1960s and 70s, the pursuit of economic growth and cost efficiency were the two main drivers. Technology was used to build big infrastructure and to achieve large-scale engineering projects, such as hydro-electric dams. The objectives of the interventions were to apply technologies to bring tangible benefits for affected communities. According to Mehta (2005), the colossal presence of the infrastructure is so visible that it has caught the public's imagination, and that has become part of the achievements of the technological interventions. The top-down, expert-led approach was the norm in the decision-making process. The cost-benefit analysis was a common appraisal tool to evaluate the success of the projects. The state and the big international development agencies, such as the World Bank, took an active role in promoting technological transfers and implementations. The understanding of technology in this time was considered 'positivistic' because technological interventions are problem-solving in nature, and the parallel interests of technical experts and policy makers were rarely challenged.

---

1 This section offers a general analysis about the shifting understanding of sustainable technology. It draws largely on the World Bank's policies and reforms in the developing countries from the 1960s. The author acknowledges that different countries may experience different paths of historical transformation in technology.

Tab. 1: Shifting paradigms of conceptualising (sustainable) technology over the past four decades

	1960s – 70s	1980s – mid-1990s	Late 1990s – early 2000s
Dominant drivers	Economic growth, cost efficiency	Socio-economic changes	climate change, sustainable development
Conceptualising technology	Positivism	Social constructionism	Social constructionism, post-modernism
Features	Large-scale engineering projects, aiming visible and tangible outcomes	Challenge the universal application of technology, promote appropriate and efficient use of technology	Sustainable technology
Objectives	Problem-solving	Context and culture sensitive, effectiveness, pursuit of quality	Good governance: equity, justice, legitimacy
Decision-making process	Top-down, expert-led	Public consultation	Bottom up, participatory, respect indigenous knowledge, aim at transparency and accountability
Decision-making criteria	Cost-benefit analysis	Multi-criteria assessment, considering social, economic impact	sustainability impact assessment, deliberation
Key players	Active roles of the state and big international agencies, such as World Bank	Market	Role of community, Public-private partnership

Source: My own table, inspired by Chataway et al. (2007) and Feenberg (1999).

The beginning of the 1980s marked drastic changes about how technology was perceived. The general weakness of the state, a rising level of global poverty, the emergence of the World Bank's neo-liberal agenda and the structural adjustment programmes in many poor countries had forced academics and practitioners to rethink the role of technology in development. The new thinking, such as appropriate use of technology and cultural- and context-specificity of technology, had gained more attention (Chataway et al. 2007). Public consultation had entered the policy making arena, suggesting that affected communities should have a voice in the development process. Multi-criteria assessment methods were introduced to complement and/or challenge the predominant cost-benefit analysis. Market forces were also encouraged in order to provide incentives for the private sectors to produce technological innovations. In this period, technology was conceptualised as 'social constructionistic' because social knowledge was co-produced between experts, policy makers as well as end users (Bijker et al. 1987).

Vocabularies, such as climate change, global warming and sustainable development, have gradually entered the mainstream policy thinking in the late 1990s and the early 2000s. The concept of sustainable technology has caught more attention. It aims to achieve not only environmental justice, but also social equity and policy legitimacy. In order to achieve good governance, the decision-making process has become more decentralised, democratic, transparent and accountable. Participatory approach is more often deployed to reach consensus amongst community members affected by the technological interventions (World Bank 2000). Sustainability impact assessments have become mandatory in most project designs. The role of public-private partnership is encouraged to provide sustainable technology while communities are mobilised to make decisions *ideally* in every process of intervention, from design to delivery. 'Post-modernistic' conceptualisation of sustainable technology is the feature of this time. Conflicting interests and values are appreciated, rather than compressed.

2.1 Building 'people-centred' sustainable technological intervention framework

The rising popularity of the notion of sustainable technology in the late 1990s and the early 2000s is constantly contested. It is generally analysed at three levels: macro, meso and micro. After examining the focus and assumptions of each level, this chapter will develop a 'people-centred' framework in order to build bridges between different levels of analysis. Our intention is to develop a new framework, so that we can have a deeper understanding of the complex relationships between sustainable technology, climate change, governance and poverty alleviation. Table 2 compares and contrasts the similarities and differences between the macro, meso and micro levels.

Tab. 2: Compare and contrast different levels of theorising sustainable technology

Levels	Perspectives	Assumptions	Key authors
Macro	Socio-technical systems, new logic of production and consumption	– focus on technological systems – assume that existing socio-technical systems are produced and reproduced through the interactions of many different social, technical and natural actor networks, institutions – take an evolutionary understanding of transitions to sustainable production and consumption – need to steer to achieve normative goals of sustainability	Smith and Stirling (2006); Spaargaren et al. (2006); Guy and Marvin (2001), Hughes (1987)

Levels	Perspectives	Assumptions	Key authors
Meso	'Technology-people-governance' nexus	– large technological systems and individuals' everyday practices are intertwined – (sustainable) technologies are socially constructed, mediated by people's perceptions and cultural contexts	Murphy (2006)
	'Subjectivity-institution-structure' analysis	offer a contextual understanding of sustainable technology – acceptance and effectiveness of sustainable technology lie in individuals' subjective values, institutional arrangements and structural conditions	Wong and Sharp (forthcoming)
Micro	Sustainable products	– assert that it is difficult to change human unsustainable behaviour – sustainable products are effective and efficient to reduce the impact of human unsustainable actions on the environment – market forces play a key role in creating demand and supply for sustainable products	Bessant and Tidd (2007)
	Practice-based perspective	– environmental innovations interact with experiences of individual users, patterns of routinised behaviour – refute the idea that sustainable products *alone* can achieve sustainability – need full engagement with the public for the whole process of technological interventions	Sofoulis (2005), Shove (2003)

Source: My own table

The macro process of environmental innovations focuses on technological systems and their organisational counterparts to oversee the process of different natural resource management, such as water collection and energy distribution. It has been extensively studied, for example, through ideas about large technical systems and processes of transition (Smith & Stirling 2006; Spaargaren et al. 2006; Hughes 1987). This perspective highlights the dynamics of system innovations and structural change and the complexity of governance. The basic assumptions of this level of analysis are that: the existing socio-technical systems are produced and reproduced through the interactions of many different social, technical and natural actor networks, institutions and ideas, cutting across multiple levels and scales. It takes an evolutionary understanding of transitions to sustainable production and consumption. It acknowledges the combined role of physical infrastructure, organisational structures and social norms in mediating the functioning of different technical systems. This school of thought also makes a significant assertion that the path-dependency of technological changes can be an obstacle to achieving sustainable development. For example, according to Guy and Marvin (2001), there has been a conventional supply side logic associated with large technical systems, which are based on the logic of network expansion to meet ever expanding consumer needs. This technological thinking, they argue, is unsustainable in the long run. In order to steer transitions along pathways that seek the normative goals of sustainability, there is a need to call for new management of network demand which aims to achieve the 'new logic of production and consumption'.

While the large technical systems literature provides a language to discuss changing environmental systems at a macro-level, the micro perspective offers the potential for a more detailed contextual analysis of the process of interactions between technology and users. There are two schools of thought in this level of analysis: the product perspective and the practice perspective.

The sustainable product perspective highlights the role of sustainable products in achieving sustainability. The inventions of smart water metering and water-saving washing machines are a few examples of sustainable products that aim to reduce the demand for water. The assumptions of this perspective are that: firstly, it is difficult, though not impossible, to change human unsustainable behaviour. As long as the sustainable products do not create inconvenience to the users, they should be efficient and effective in reducing the impact of human unsustainable actions on the environment (Bessant & Tidd 2007). Secondly, if sustainable technology is vitally needed, it will be found by market forces. This school of thought is keen to conduct a full-cost analysis, measuring the level of CO_2 emissions from production to delivery and implementation.

The practice perspective has a very different view on human rationality and on how human behaviour is interacted with technology and sustainable development. It highlights how environmental innovations interact with experiences of individual users, patterns of routinised behaviour and the transformation over time (Sofoulis 2005). Shove (2003) argues that there are three elements of practice: understanding (thinking), procedures (doing) and engagement (relating to ends or goals of a practice). Using dish washing as an example, Sharp (2006) shows the complexity of the interactions between different levels of practice. The 'understanding' includes the know-how that hot-water added to detergent will help in the removal of remnant food, and what elements (patterns of the dish) are expected to remain. The 'procedures' are actual processes of putting a dish in the soapy water. The 'engagement' concerns the shared cultural understanding that dishes are required to be 'clean' such that they can be used for further food consumption. This school of thought does not agree with the idea that sustainable products *alone* can achieve long-term sustainability. It stresses the need to engage the public for the whole process of technological interventions. It calls for recognition of wider socio-economic, cultural and political contexts that shape the interactions between technology, society and the nature.

The meso level of analysis offers an attempt to link the macro and micro understanding of sustainable technology and to show how large technological systems and individuals' everyday practices are intertwined. Joseph Murphy's 'people-technology-governance' nexus (2006) provides a good theoretical discussion on how technology can be governed for sustainability. The first component of the nexus is people and their behaviour and motivation. He suggests that we need to have a deeper understanding of the multiple identities of people and the complexity of their motivation in adopting sustainable innovations. He criticises the rational choice model for over-simplifying human actions. He suggests that most of our decisions are not made consciously, and we tend to follow our routines, habits and social conventions in our everyday lives. The second component is technology. Murphy suggests that technology can be seen as a means to achieve sustainability, but the process of selecting technology is not apolitical. He challenges technological determinism, and argues that technologies are socially constructed, mediated by people's perceptions and cultural contexts. There has been a paradigm shift, moving away from the top-down, hierarchical and expert-driven approach to a more local, decentralised and participatory direction (although there are many obstacles in achieving this). The third component is governance. He defines governance as the relationships between the state and people. The shifting emphasis from government to governance not only marks the changing role of the state, but also the roles of the market, NGOs and community. While the state is now generally expected to create an enabling environment for the public-private partnership, communities are encouraged to participate more actively in the decision-making process. The institutional arrangements shape, and are shaped by, how people and technology interact. The nexus has five distinct features:

1. it problematicises the notion of technology and opens up the implicit assumptions underlying the choice and application of specific technologies;
2. it politicises technology by highlighting the uneven distribution of costs, benefits and risks amongst stakeholders. It unravels the power dynamics, such as who makes what decisions on which technology;
3. it contextualises the interactions and the impact;
4. it can be used in various levels, ranging from a micro and community level to a regional and national level;
5. it works closely with inter-disciplinary research.

Wong and Sharp (forthcoming) argue that Murphy's nexus has three limitations: firstly, it plays down the role of the nature/ecosystem in the interactions with technology and people. Secondly, it pays inadequate attention to the detailed processes about how the interactions take place. Thirdly, it fails to highlight the structural constraints that hinder the marginalised groups from governing technology for sustainability. In addressing these limitations, Wong and Sharp propose a 'agency-institution-structure' framework in an attempt to offer a more contextual understanding of sustainable technology. They argue that the acceptance and effectiveness of sustainable technology lie in individuals' subjective values, institutional arrangements and structural conditions. Drawing on the UK's first-ever combined rain-water and grey-water system in an elderly social-housing estate in North-west England as an example, they expose a huge divergence between the professional stakeholders and the poor, elderly council housing tenants over the nature and the environment surrounding them. While the tenants perceived an improvement in pollution level in their area, the professional stakeholders had an opposite view. This gap affected how the new water innovation was implemented and accepted. Wong and Sharp blame the failure of the project for a lack of tenant participation in the decision-making process. Their framework also highlights the role of institutions in legitimising the sustainable interventions. Institutional factors connect the inner (subjective) and the outside (structural) worlds. The concepts of cleanliness and hygiene, for example, are culturally and collectively shaped. The water innovation, using rain-water to rinse clothes, has caused tenants' big concerns. They perceived rain-water as unclean, and that caused the tenants to feel that their right to access to clean water was denied. Their research also explored multiple social and psychological constraints, such as a strong sense of powerlessness and a lack of human capital, that hindered tenants from raising objections at the initial stage of the project.

As a matter of fact, Murphy's nexus and Wong and Sharp's analysis are not completely incompatible. They can help develop a 'people-centred' sustainable technological intervention framework. The emphasis on 'people-centred' does not mean 'anti-nature'. It calls for a need to have a better understanding of human rationality and motivation in understanding sustainable technology. While climate change and global warming are partially man-made phenomenon, the effectiveness of sustainable technology hinges on the linkages between the dynamism of the ecosystem and the actions of human beings at the macro and the micro levels. Institutions are the social values and cultural practices that build governance structures to manage technology for long-term sustainability. The 'people-centred' framework underlines power and gender dynamics as well as ecological equity. This framework will be developed in the coming sections.

3. Communal solar home systems in Bangladesh

Solar home systems play a significant role in the current renewable energy technology policies. They offer clean energy and helps reduce the reliance on fossil fuels. They exploit the natural resources (abundant and continuous solar supplies) that most developing countries, especially those lying in the tropics, can claim as their own. The systems also match the objectives of the Clean Development Mechanisms. The active advocacy by the World Bank and the UNDP has popularised solar power (Martinot et al. 2001). The launch of the World Bank's 'Lighting Africa' campaigns in 2008 shows the Bank's faith in the solar home systems in the sustainable energy policies. For example, 120,000 solar photovoltaic systems for household use have been sold in Kenya alone since 1990 (Hankins 2000).

This paper draws on our British Council-funded project. The overall aims were to challenge the notion of sustainable technology and issues around it, such as governance and institutional building. It raised questions about who controls technology and how technological interventions intertwine power (in)equalities. It compared and contrasted two different solar home systems – communal and individual, and examined which one was more cost-effective and simultaneously resulted in more equitable outcomes. Because of the space limitation, this paper will only focus on the communal solar home system.

3.1 Contexts and research methodology

This paper is based on research on a Hindu rural community in Chokoria, South Bangladesh (see Map 1).

Map 1: Location of Chokoria

It was a joint project of the United Nations Development Programme (UNDP), the Department of Fishery and the Local Government Engineering Department (LGED) in Bangladesh (2003-5) (Diagram 1). A community-based solar home system was built for 70-100 households (approximately 700 population) living in a Hindu fishing village in Cox's Bazer district. The project aimed to provide alternative energy supply to the village which was not covered by the grid-based electrification systems. Each household was required to pay 40 Taka per month, and it would receive two bulbs and lighting from 6pm to 11pm. Four to five households were also able to afford extra electricity for televisions. This project was also intended to improve some villagers' health by avoiding respiratory diseases from burning charcoal and cow dung (although most of them used gasoline). The lighting also enabled poor community members to prolong their income generation activities, such as making fishing-nets, in the evening at home.

Diagram 1: Key players in the solar home system in Chokoria

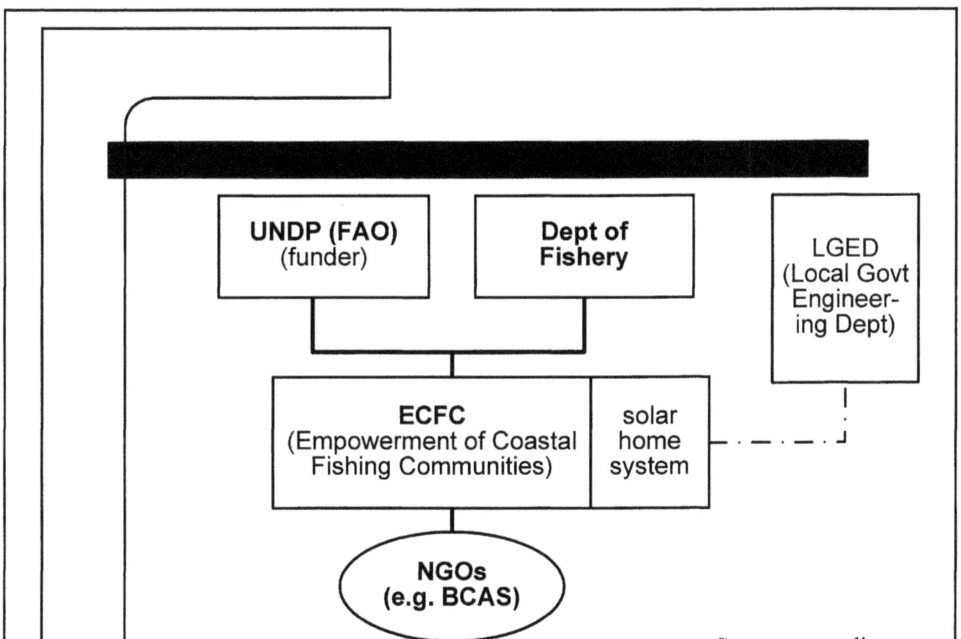

Source: own diagram

The project targeted a Hindu fishing community (Photo 1). These Hindu resettled in Chokoria in 1968 after their home village was severely flooded. The low level of boat ownership and the exploitation from rich Muslim boat owners mean that they are generally poorer than their Muslim counterparts. The Local Government Engineering Department was also keen to find a demonstration site in order to test the feasibility of the communal system. The selection of that village was also owing to its proximity to the main road.

Photo 1: The communal solar home system in Chokoria

Source: own photo

The author worked with a local NGO and we visited the village in August 2007. We adopted a semi-structured interview approach and talked to various social groups, including village leaders, local politicians, women, children, religious leaders and governmental officers. We conducted a 'poverty-sensitive' research by dividing local people as non-poor, poor and ultra-poor according to their income levels, ownership and size of land, availability of assets, such as the numbers of livestocks, and farm productivity. We assumed that the impact of the solar home system was unevenly distributed, and the level of poverty affects how local people share the benefits and costs in the sustainable technological interventions. We also used social drawing techniques in order to capture local people's complex and diverse understanding about the changes. We taped, transcribed and coded the interviews and used secondary data to make analysis.

4. Politicising good governance

The implementation of the communal solar home system in Chokoria was not simply a technical intervention. Apart from offering tangible benefits (ie cheap, clean and lighting) to poor villagers, it intended to re-engineer the governance structures of the village. In interviews, both officials of the UNDP and LGED suggested that the village faced poverty because of 'bad' governance structures.

From their perspective, the village had been dis-organised and poorly-managed. They criticised that indigenous leaders, called *Sardars*, captured most of the political and economic power. They were not democratically-elected, and thus not accountable to their people. The leadership was also male-dominant, and women were excluded in the decision-making process.

In order to address these governance issues, the government officials and NGOs decided to use the communal solar home system to give power back to local people and to undermine the elites' authority by democratising the village management.

Diagram 2: Governing structure of the solar home system in Chokoria

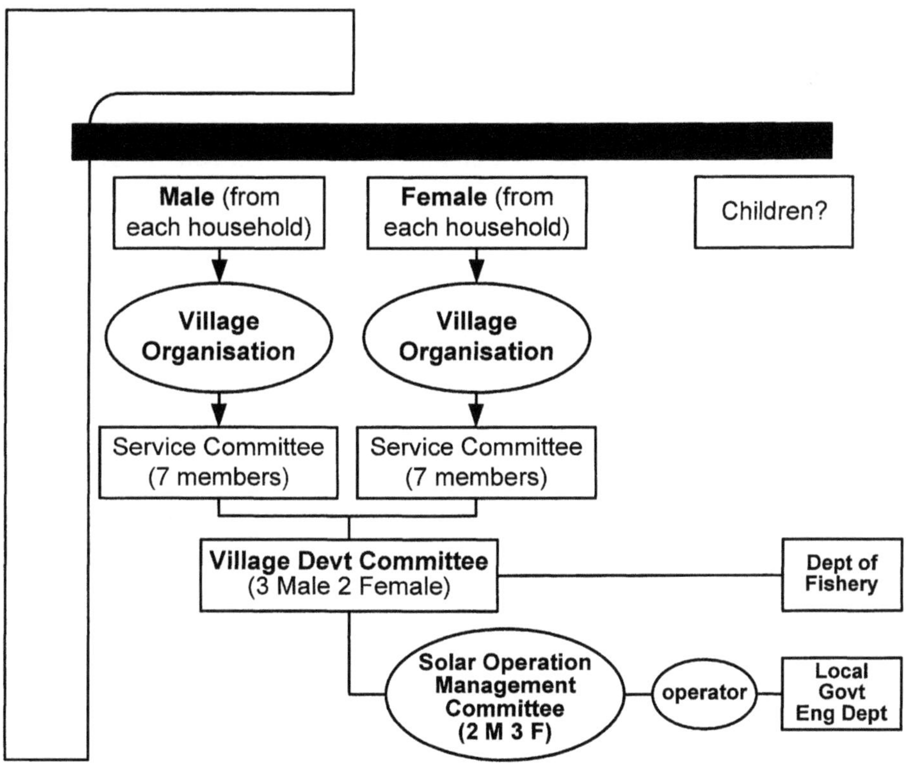

Source: own diagram

Diagram 2 summarises the structure of the institutional building. Any men and women above 18 years old from each household were encouraged to join the Village Organisations, one for men and another for women. The separation was intended to help women feel more comfortable about expressing their political opinions without worrying about men's interference. From each Village Organisation, seven members were elected to form a Service Committee. During the construction of the solar home system, a Village Development Committee was set up. It comprised three male and two female members, all elected by local villagers. It worked with the Department of Fishery to deal with local affairs. Under the Committee, a Solar Operation Management Committee was formed to take control of the solar home system. This Committee comprised another five members (two male and three female), also elected by the public.

4.1 Can new institutions challenge local power inequalities?

Ideally, the new institutions were expected to improve local governance by offering village members opportunities to make decisions and to build networks by working with the government officials. In reality, however, *Sardars* felt that their dominant positions had been undermined, so they exerted their influence in the new committees through their control of the finance – most of the committee members borrowed money from *Sardars*. On one occasion, *Sardars* received 40,000 Taka micro-credit funding from the Village Development Committees for investment. They refused to pay back on the due date. Other community members followed *Sardars*' examples. This caused the collapse of the micro-credit system. Some villagers made formal complaints to the LGED. The government officials intervened and tried to pressurise *Sardars*, but their interventions proved ineffective because of their limited political influence at the local level.

This case study shows that institutional building alone, without the political and financial support from outside, was not adequate to challenge local inequalities. Since the livelihoods of poor community members depended on the rich and the powerful, they were willing to bow to the pressure from the privileged in return for stable livelihoods and community harmony. This posed a dilemma for external interventions: should we seek cooperation from local elites by including them in the process of institutional building or should we confront them by excluding them in the process? The latter, from our case study, suggested that the local elites could find a way to re-gain their political capital and risk derailing the project.

4.2 Achieve gender equalities by institutional building?

Another dimension of institutional strengthening was to increase women's representation in the committees. Our case study, however, shows that raising women's visibility in the Village Organisations does not necessarily achieve gender equality or enhance their decision-making power in their village. Questions, as to who these women were, what interests they represented, what role they held in the committees, and why they were included, need to be examined.

In interviews, villagers considered that the women in the committees tended to possess certain characteristics: they had time to serve; they were educated and had relevant skills, such as accounting; they were 'honest'. Our investigation suggested that these women were not 'so poor' in their village. They had already been active in local affairs and tended to have influential husbands. On the contrary, the very poor women, such as the disabled and the widowed, were not included in the committees. Worse still, the discrimination was strong between women in the village. Women with many daughters and very few sons, for example, were looked down on by the less-poor women.

The reasons for including women in the committees were instrumental – they needed women to carry out daily administration while the men were out at sea for fishing for six months every year. Women were considered as possessing better social skills, so they, rather than men, were more suitable to collect electricity fees. The division of labour inside the committees was also gendered. Female committee members were not supposed to meet the government officials since they thought they were not educated enough. They tended to leave their male counterparts to make decisions while they carried out most of the administrative tasks.

4.3 How effective is rule-based management?

Another way to improve local governance was to promote rule-based management because rules facilitate clarity and transparency (Ostrom 2005). For instance, eleven new rules about how the solar home system was managed and maintained were laid down. The first rule suggested that the solar home system only worked for five hours a day while the second rule warned that villagers needed to ask the operator for help if bulbs do not work property.

While the rules sounded reasonable and flexible, the process of rule making was not politically-neutral. Our case study shows that the Local Government Engineering Department (LGED) and system operator gained tremendous power in the process. LGED devised the rules without consulting community members. According to the regulation, villagers could change the rules provided that they got the approval from LGED. This example showed that while a 'young' community needed support from external bodies in setting up rules, the non- collaborative relationships risked perpetuating a culture of dependence that the local community became over-reliant on external forces and failed to develop itself.

UNDP and LGED acknowledged the significance of capacity building in ensuring the sustainability of the solar energy project, they worked very hard to train a 20-year old college student in the village to maintain the system. He received 1,500 Taka a month and was responsible for the daily functioning of the system. He became influential since he practically controlled the times of the electricity supplies (although literally LGED had fixed the time). In interview, he admitted that some villagers attempted to bribe him in order to extend the opening hours of the electric supplies during wedding celebrations. Although local villagers could file complaints to LGED, they would not do so because this would upset inter-personal relationships in the community.

In a nutshell, the new institutional building process throughout the implementation of the communal solar home system failed to transfer power back to local people. The ideas of the governance re-engineering could turn out reinforcing, rather than challenging, existing political and gender inequalities. New rule making also provided an opportunity for a few people, such as the operator, to gain political influence.

5. Complexity of the socio-economic impact

Our research conducted a detailed examination of the socio-economic impact that the solar home system had on the village. In our participatory research exercise, we asked local children to do drawings about their community and the impact of the solar home system. Diagram A shows that electric lines and poles had become new symbols in their community. The Hindu temple remained on the maps, but its significance was not as much as the community hall (where the solar system was built on the top).

Drawing 1: Socio-economic impact of the solar home system in Chokoria

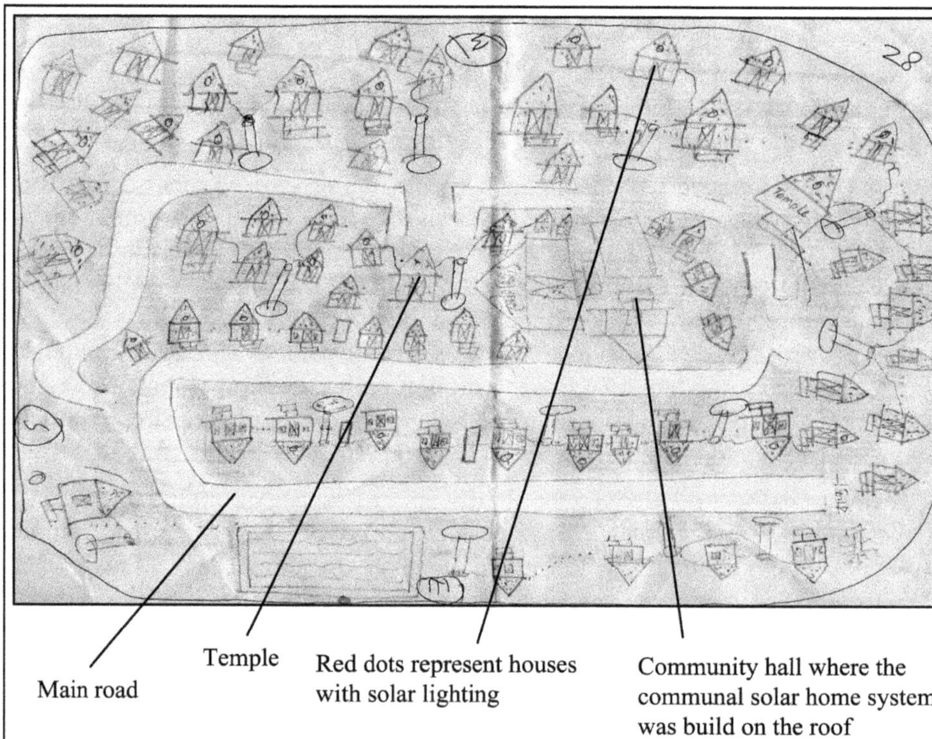

Source: drawn by a 19 years old local villager

On the positive side, the solar home system lowered the cost of lighting. Before the technological intervention, households on average needed to pay 60 Taka monthly for gasoline. Now, they paid 40 Taka for solar energy. The system had brought tremendous changes to their everyday lives. The income-generating activities, such as net making, were prolonged till 11:00pm. Children could finish their homework in a bright environment (when compared to kerosene). They also enjoyed entertainment provided by the television, and they felt that they knew what happened in the outside world. The religious leaders also welcomed the change because the lighting makes the temple brighter, and villagers were more reverential towards gods because they could see their fearsome facial expressions.

However, the solar home system has its downside. From the gender perspective, the provision of lighting extended the working hours of women. Before the solar home systems, villagers relied on kerosene for lighting. The lighting was, however, so dim that they could not continue their work after 8pm. The solar lighting enabled women to prolong their income-generating activities, such as net repair, at home till 11pm when the system was off. In interviews, women felt that they needed to work hard in the evening; otherwise, they felt that they lagged behind of their neighbours. Working extra hours also exerted burden on their already-tired bodies.

The sustainable interventions also generated tension within the community. The Solar Operation Management Committee (SOMC) went out to collect fees monthly, but the committee members told us that the experience was not pleasant. Owing to seasonal fluctuations of the fishing industry, many poor households had problems in paying the fees punctually. Quarrels easily arose about delayed payment. On one occasion, a SOMC member felt she was pushed by a local member, and she decided to quit the post. Other members followed. As a result, the SOMC was disbanded in December 2006. After the intervention of LGED, it was re-established and new committee members were elected. In order to avoid conflict, LGED passed a new rule, stressing that no more late payment would be tolerated; otherwise, the Committee would disconnect the electricity supplies. The unintended consequence of this measure was that 56 out of 70 households were disconnected when we visited the village in August 2007 since most poor villagers failed to pay the fees promptly.

The high disconnection rate has created a big gap between the haves and the have nots. Lighting became a visible symbol of status and wealth. This impact was more obvious at night time because only a few households could enjoy lighting and TVs. In interviews, those mothers whose electricity supplies were disconnected felt guilty because their children were deprived of the opportunity that the lighting could bring. Some children also felt inferior because they could no longer study after 8pm. The sustainable innovation has, unwittingly, widened, rather than narrowed, the gap of inequality. It risked undermining, rather than enhancing, the norms of cooperation amongst community members. In addition, the solar home system creates a crowding-out effect and disincentivises the national grids to extend their networks into the village. This will affect the long-term economic development of the village because electricity generated by the solar home system is limited and unstable when compared to the grid electrified systems.

6. Conclusions and policy implications

Technology plays different roles in the domains of climate change. While some technologies help detect climate change and reduce its variability, others are useful in mitigating the impact of global warming. However, whether sustainable technologies are effective in assisting developing countries and poor villages to adapt to changing climate, or whether the technologies are accepted by the public, are less certain. This chapter has drawn on the communal solar home system in Bangladesh and suggested that the process of institutional building could reinforce, rather than challenge, political and gendered inequalities. Making new rules could provide opportunity for the local elites to gain political influence. Despite the positive impact, the solar home system could cause tension and undermine norms of cooperation.

We can draw four lessons from the case study: firstly, green technologies and climate change policies are not necessarily 'pro-poor'. The uneven distribution of costs and benefits of the sustainable technological interventions can make the marginalised groups worse off. Secondly, the impact of sustainable technology is so context- and culture-specific that it requires a more flexible approach in the design of the interventions. Thirdly, institutional building and capacity building are time-consuming processes, and they need continual support in terms of resources and time. Fourthly, restructuring existing governance structures is a necessary, but not sufficient, condition to achieve social and ecological equity and full participation of the poor in the decision-making process.

These observations are useful to develop a 'people-centred' framework for a better understanding of the complex relationships between sustainable technology and climate change policies. Our framework suggests that any sustainable technological interventions need to link the macro technical system (such as solar home systems) and the micro process (such as users' daily practices) on the one hand and build bridges between nature and society on the other. It calls for a different approach to conceptualising human motivations (Jackson 2005). Before any interventions take place, we need to acquire a deeper understanding of local power and gender dynamics (Chaurey et al. 2004). We also need to examine how institutions affect public acceptance of the innovations. There is nothing automatic about the application of sustainable technology and finding the solutions to climate change. Good answers lie in a new kind of analysis, based on greater awareness of the limitations of institutional crafting and more confidence in drawing the poor end-users on board.

References

Asian Development Bank (2000), Energy 2000. Review of the Energy Policy. Manila.

Bessant, J. and Tidd, J. (2007), Innovation and Entrepreneurship. Chichester: John Wiley and Sons.

Bijker, W.E., Hughes, T. and Pinch, T. (eds.), The Social Construction of Technological System. Cambridge: MIT Press.

Chataway, J., Maat, H. and Waldman, L. (eds.) (2007), "Science, Participation and Development." IDS Bulletin 38(5) Nov 2007.

Chaurey, A., Ranganathan, M. and Mohanty, P. (2004), "Electricity Access for Geographically Disadvantaged Rural Communities – Technology and Policy Insights." Energy Policy 32, pp. 1693-1705.

Cleaver, F. and Frank, T. (2005), Water Governance and Poverty: What Works for the Poor? Bradford Centre for International Development [www.bradford/acad/bcid/research/wqp, accessed on 26th Sept 2006]

Feenberg, A. (1999), Questioning Technology. London: Routledge.

Guy, S. and Mervin, S. (2001), "Urban Environmental Flows: Towards a New Way of Seeing", pp. 22-40. In: Guy, S., Marvin, S. and Moss, T. (eds.), Infrastructure in Transition: Urban Networks, Buildings, Plans. London: Earthscan.

Hankins, M. (2000), "A Case Study on Private Provision of Photovoltaic Systems in Kenya", pp. 92-10. In: World Bank (2000), Energy and Development Report. Washington: Washington DC.

Hughes, T. (1987), "The Evolution of Large Technological Systems." In: Bijker, W.E., Hughes, T. and Pinch, T. (eds.), The Social Construction of Technological System. Cambridge: MIT Press.

Jackson, T. (2005), Motivating Sustainable Behaviour: a Review of Evidence on Consumer Behaviour and Behavioural Change. University of Surrey, Centre for Environmental Strategy.

Martinot, E., Cabraal, A. and Mathur, S. (2001), "World Bank/GEF Solar Home System Projects: Experiences and Lessons Learned 1993-2000." Renewal and Sustainable Energy Reviews 5, pp. 39-57.

Mehta, L. (2005), The Politics and Poetics of Water: Naturalising Scarcity in Western India. New Delhi: Orient Longman.

Murphy, J. (ed.) (2006), Governing Technology for Sustainability. London: Earthscan.

Ostrom, E. (2005), Understanding Institutional Diversity. Princeton: Princeton University Press.

Paavola, J. and Adger, N. (2002), Justice and Adaptation to Climate Change. Tyndall Centre for Climate Change Research. Working Paper 23.

Sharp, L. (2006), "Water Demand Management in England and Wales; Constructions of the Domestic Water User." Journal of Environmental and Planning Management 49(6), pp. 869-889.

Shove, E. (2003), "Converging Conventions of Comfort, Cleanliness and Convenience." Journal of Consumer Policy 26, pp. 395-418.

Smith, A. and Stirling, A. (2006), Moving Inside or Outside? Positioning the Governance of Sociotechnical Systems, SPRU Working Paper Series, Paper no. 148.

Sofoulis, Z. (2005), "Big Water, Everyday Water: a Sociotechnical Perspective." Journal of Media and Cultural Studies 19(4), pp. 445-463.

Spaargaren, G., Mol, A. and Bruyninckx, H. (2006), "Introduction: Governing Environmental Flows in Global Modernity." In: Spaargaren, G., Mol, A. and Buttel, F.H. (eds.), Governing Environmental Flows: Global Challenges to Social Theory. MA: MIT Press.

Wong, S. and Sharp, L. (forthcoming), "Making Power Explicit in Sustainable Water Innovation: Re-thinking Subjectivity, Institution and Structure through Environmental Citizenship." Environmental Politics.

World Bank (2000), Energy and Development Report. Washington: Washington DC.

World Bank website: http://web.worldbank.org accessed on 20th April 2008.

Chapter 23

Sensitivity of Bulk-Cargo Dependent Industries to Climate Change – First Results of a Case Study from the River Rhine

B. Rothstein, A. Scholten, E. Nilson, R. Baumhauer

Abstract

The River Rhine is one of the most important inland waterways in Europe. Bulk-cargo dependent industries of different sectors (e.g. energy, chemistry) use inland navigation on the River Rhine as an efficient way of transport. Therefore, the reliability of transport on the River Rhine represents an important location factor for different industrial sectors and therefore has great relevance for economy as a whole.

Water-levels have always changed due to climate variability. The debate on climate change has raised the question is there a risk that current strategies of authorities responsible for management of inland waterways as well as industrial managers might not be suitable to cope with future conditions anymore?

First, the current knowledge of observed and projected water-level change in the River Rhine is summarized. Second, the vulnerability of inland waterway transport and industries dependent on this transport mode is quantified. Finally, several adaptation options for bulk-cargo dependent industries are given taking coal-fired power plants as examples.

1. Introduction

The River Rhine is one of the most important inland waterways in Europe. Bulk-cargo dependent industries of different sectors (e.g. energy, chemistry) use inland navigation on the River Rhine as an efficient way of transport. Therefore, the transport on the River Rhine represents an important location factor for different industrial sectors and therefore has great relevance for economy as a whole.

Inland navigation strongly depends on water-level. Low-flow situations as well as floods can limit, slow down or even stop the transport on the River Rhine. For bulk-cargo dependent industries, low-flow situations are more relevant than floods, because they generally last longer and potentially lead to more severe limitations of transport.

Along the River Rhine, there are several bottlenecks with relatively shallow water that are critical for navigation at low-flow conditions. The relevant bottleneck for all destinations passing the Middle Rhine is represented by the Kaub gauging station. For example, ships with destinations in the industrial area of Mannheim/Ludwigshafen (Upper Rhine) have to pass Kaub. Even if the water depth in Mannheim is sufficient, the bearing capacity might be limited due to the water depth around Kaub. Thus, the gauge at Kaub can be regarded as a reference gauge for the calculation of the bearing capacity.

Water-levels have always changed due to climate variability. As a consequence, authorities responsible for the management of inland waterways as well as industrial managers always had to develop and apply strategies to adapt to these changes (Moser et al. 2008a). The debate on the current "additional" climate change, which is likely anthropogenic (IPCC 2007), has raised the question is there a risk that current strategies might not be suitable to cope with future conditions anymore? The key question is: Are the bulk-cargo dependent industries vulnerable to future climate change?

Here, the current knowledge of observed and projected water-level change in the River Rhine is summarized with a special focus on low-flow conditions. Second, the vulnerability of inland waterway transport and industries dependent on this transport mode is quantified. Finally, several adaptation options for bulk-cargo dependent industries are given, taking coal-fired power plants as examples.

2. Observed changes of discharge in the River Rhine during the 20th century

The efficiency and safety of inland waterways as transport routes depend on reliable navigation conditions in terms of sufficient water width and depth as well as moderate flow velocities. These conditions have always been subject to changes, which are caused by long- to short- term variations of hydrometeorological variables like temperature and precipitation. An exhaustive study about changing hydrometeorological conditions and discharge in the Rhine basin during the 20th century has recently been compiled by BELZ et al. (2007; also cf. Moser et al. 2008b). Key statements are:

1. The Rhine basin consists of two discharge regimes: A nival regime (influenced by the Alps) showing a meltwater-related discharge maximum in summer (defined as May to October) and low-flow occurring typically in winter (defined as November to April), and a pluvial regime (influenced by the mid-mountain ranges) with a precipitation-related discharge maximum in winter and low-flow occurring typically in late summer. Both regimes merge at the mouth of the River Main to a combined regime showing equalised intra-annual variability downstream. Figure 1 illustrates the standardized multi-annual monthly discharge (so-called Pardé coefficient (cp. Pardé 1933)) at representative gauging stations.

Fig. 1: Standardized multi-annual (1951-2000) monthly discharge (so called Pardé coefficient) at Ilanz (nival discharge regime), Treves (River Mosel, pluvial discharge regime), and Cologne (combined nival + pluvial discharge regime).

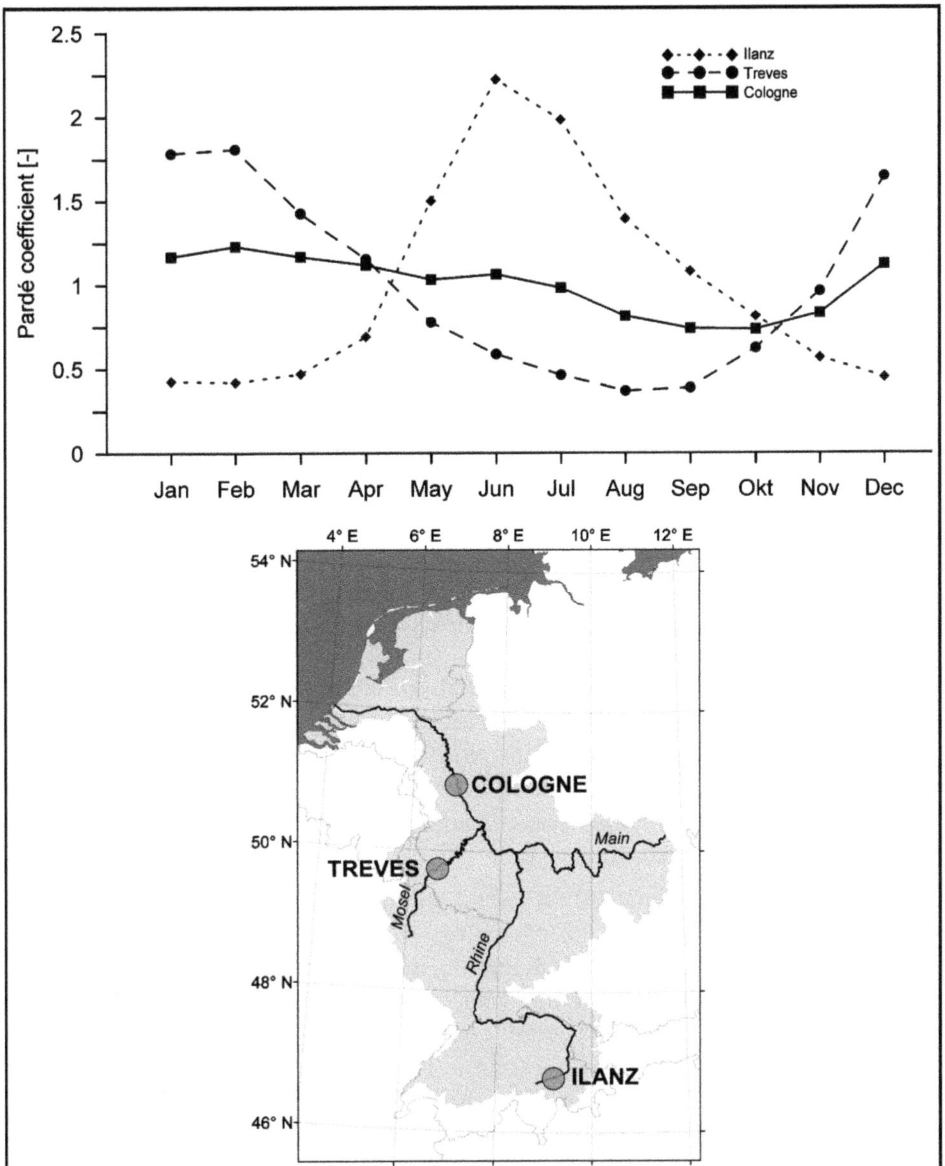

Source: modified after BELZ et al. 2007

2. In the 20th century significant increases of mean monthly discharge occurred during the winter in the whole Rhine basin. During summer a small increase

of mean monthly discharge can be observed in the pluvial part of the Rhine basin. In the nival part, the summer discharges decreased, leading to an equalisation of the mean hydrograph (i.e. differences between seasons decreased).

3. The increase in mean discharge in the pluvial part of the Rhine basin was caused by a rise in precipitation, which in turn was due to an increasing frequency of "wet" atmospheric circulation patterns over Central Europe. The winter discharge intensification in the nival part was initiated by a change in precipitation type. Due to a general increase of air temperature over the 20th century that was most pronounced after the 1970's, more precipitation fell as rain (direct runoff) instead of snow (storage). The summer discharge decrease in the nival part was due to reduced snow storage, resulting in less meltwater. Discharge from shrinking alpine glaciers has no significant effect on mean and low discharge conditions in the main stream.
4. Peak discharges have generally increased. This results from higher precipitation sums in parts of the Rhine basin and more frequent extreme precipitation events.
5. The frequency of low-flow situations has not changed in the pluvial part and tended to decrease in the nival part. Occasionally, a shift of the mean month of low-flow from October to September was observed (figure 2).

For industries dependent on river transportation, the severity of a low-flow situation is mainly determined by both, the duration of transport limitations and the absolute deficit. Low flow situations with water-levels below an important threshold for navigation have occurred throughout the 20th century. Current research shows that there is no general change in the number of low-water periods comparing the three 25 year intervals.

Obviously enterprises dependent on inland waterway transport and other sectors using water (e.g. electricity sector, agriculture) had to deal with changing natural (and non-natural) conditions for navigation over many decades. Even extreme low-flow situations like that of the year 2003 on the River Rhine were not unique in the 20th century. Other equally or even more prominent low-flow years were 1921, 1962 and 1971. The question is whether future discharge conditions will be modified to a degree that makes new or additional adaptation measures necessary in order to keep inland transport efficient and safe.

Fig. 2: Standardized multi-annual (25 year) monthly discharge (so called Pardé coefficient) at Cologne. Between the beginning and the end of the 20th century, the month with lowest mean discharge has shifted (arrow)

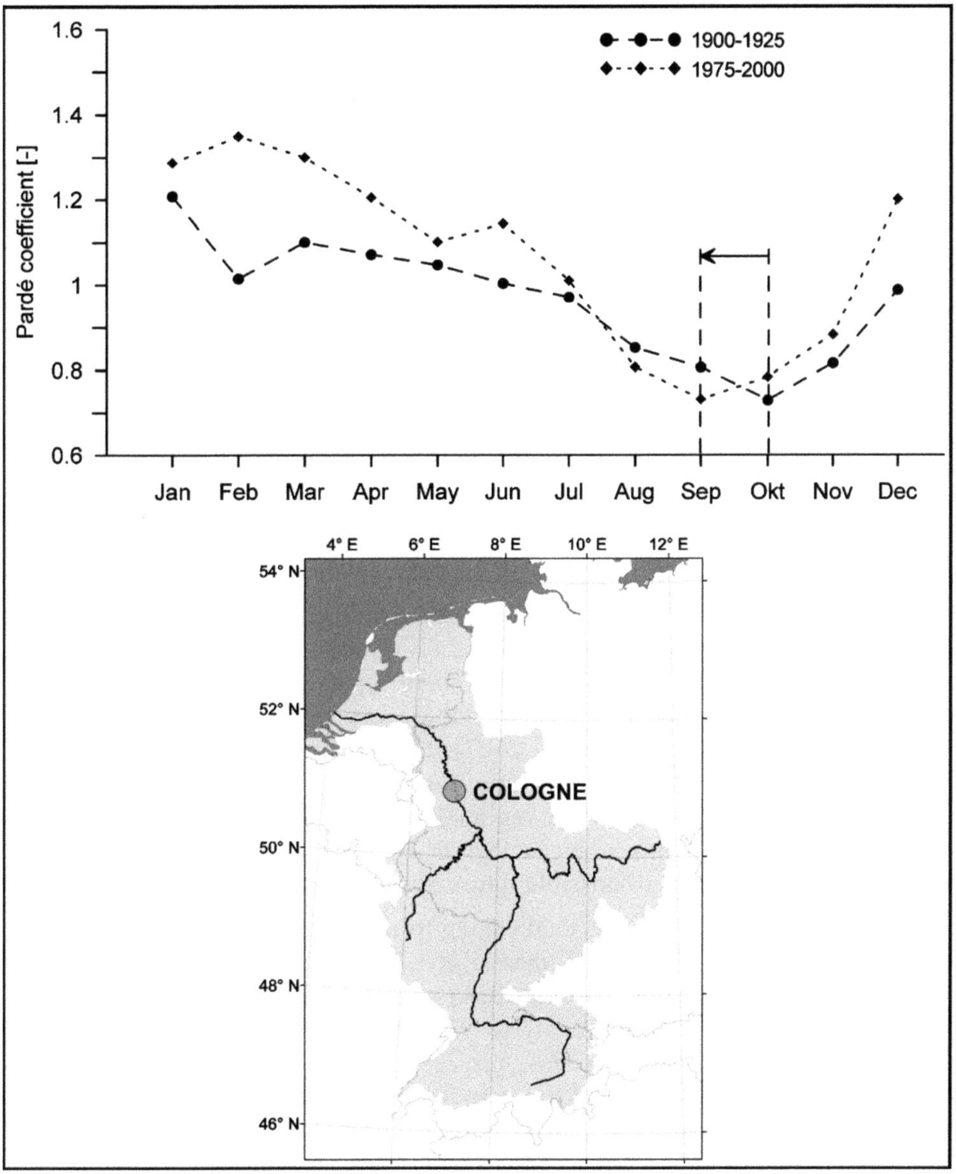

Source: modified after BELZ et al. 2007

3. Projections and scenarios of discharge for 21st century in the River Rhine

According to observational evidence (cf. Section 2) a clear trend in frequency and duration of low-flow situations on the River Rhine is not discernible in the 20th century. However, the 4th Assessment Report (AR4) of the Intergovernmental Panel on Climate Change (IPCC) (2007) states that the significant warming observed over the past 50 years in Europe (and other continents except Antarctica) will continue in the 21st century. This could lead to significant changes to important variables of the water balance. Against this background, research effort has increased to obtain more information of regional and sector-specific consequences of future climate change. Summaries for the navigation sector have been compiled recently by BMVBS (2007) and PIANC (2008).

The IPCC (2007) also points out, that "impact research is hampered by uncertainties surrounding regional projections of climate change, particularly precipitation". Impact research is often based on regional climate models, which downscale information from global climate models (GCMs, ~ 200 km grids) to the regional scale (meteorological stations or 10-50 km grids). These regional data serve as input to further models (e.g. hydrological or hydrodynamic models), which should give information about system-specific impacts. Each model has its own specific uncertainties and errors that accumulate on the pathway through the model cascade.

Within this cascade, the GCMs have been identified as a major source of uncertainty (FOWLER et al. 2007). With focus on downscaling methods applied to GCM data, large intercomparison studies were set up to obtain a better understanding of uncertainties. They concluded that there is no optimal method available so far (CHRISTENSEN et al. 2007). Up to now, the multi-model mean value resembles observational data of a control period better than any single model. It is therefore argued, that regional climate-impact analysis should be based on multi-model approaches (model ensembles). This is essential in order to represent the full range of possible future hydrometeorological developments.

Figure 3 demonstrates, that it is challenging to decide which pathway should be followed through the cascade of uncertainty. There is no clear benchmark by which a climate or hydrological projection can be regarded as reliable and can be taken as scenario for impact analyses. Note that IPCC DDC (2007) has made a clear distinction between the terms "projection" which means essentially "forward-looking model output" and the term "scenario", which refers to "a coherent, internally consistent and plausible description of a possible future state [...]". Thus, the selection of a scenario has to be made deliberately and on the basis of a scientific rationale taking additional information (e.g. about baseline conditions) into account. In many impact studies published so far, projections are chosen without any scientific argument, often simply because no alternative data sets are available.

Fig. 3: Cascade of uncertainties and errors, pathways of decisions in a model chain. Modified after VINER, 2002

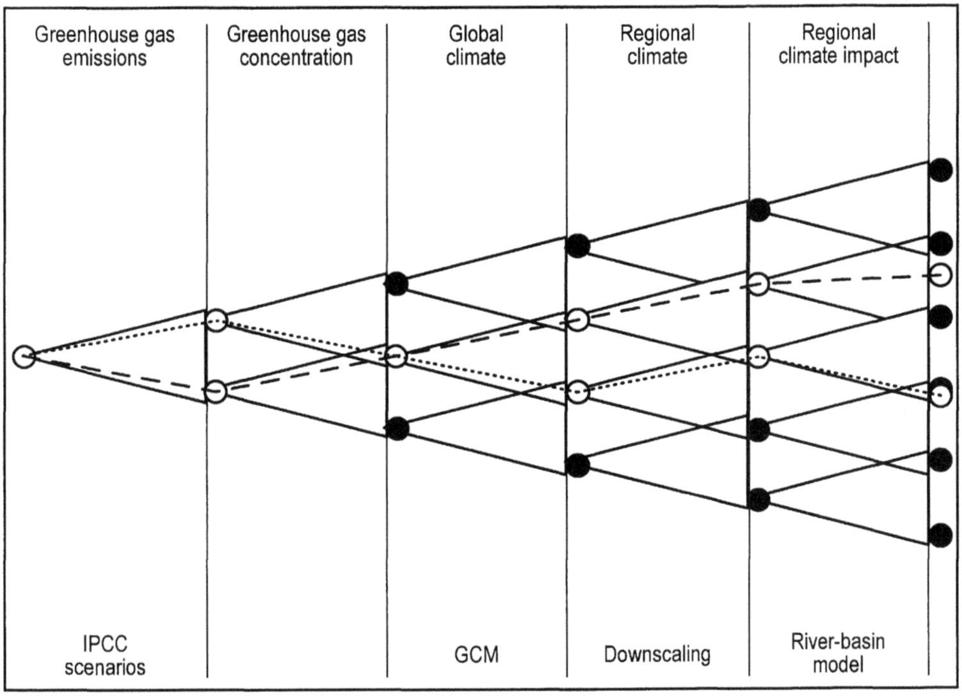

Source: modified after VINER 2002

For the Rhine basin, multi-model impact analyses using the latest generation of climate models and downscaling methods are not yet available. Studies published so far are based on one single GCM (e.g. ECHAM5-MPI-OM) and are only taking a few regional climate projections into account (table 1). Thus, they can be regarded only as "snapshots" into future and do not show the full band width of possible projections.

Some of these "snapshots" as well as studies which are based on earlier climate models – e.g. those underlying the third IPCC assessment report (IPCC 2001) concluded that there could be an increase in extreme discharges. According to Schabalova and Buishand (2002) the frequency of low-flow situations may raise in future e.g. in the River Rhine. Also Krahe (2000) assumes a higher risk of floods in winter and an intensification of low-flow in summer e.g. in the River Rhine. However, most of these statements remain qualitative.

Scenarios of future climate change and derived reliable quantitative discharge data, which can serve as a planning basis for adaptation strategies in inland navigation, still have to be developed. As adaptation measures in this sector have often great monetary consequences, scenarios have to be developed in the best possible way (BMVBS 2008; Moser et al. 2008b).

Tab. 1: Overview of regional climate data available in Germany

Method	Driving GCM	Concept	Modelling work by	Made by order of	Year of publication
REMO	MPI-ECHAM5-MPI-OM (run 1)	dynamic	Max-Planck-Institute for Meteorology, Hamburg	Federal Environment Agency, Dessau	2006
CosmoLM	MPI-ECHAM5-MPI-OM (run 1, 2, 3)	dynamic	Max-Planck-Institute for Meteorology, Hamburg (Group M&D), Hamburg	Federal Ministry of Education and Research	2007
WETTREG	MPI-ECHAM5-MPI-OM (run 1)	statistic	MeteoResearch, Potsdam, CEC Potsdam GmbH	Federal Environment Agency, Dessau	2006
STAR	(MPI-ECHAM5-MPI-OM)	statistic	Potsdam Institute for Climate Impact Research	Potsdam Institute for Climate Impact Research	2003

WETTREG and STAR are available for German territory only. REMO and CosmoLM also include neighbouring areas. The selection takes only the most recent approaches into account which were developed in the context of, or following IPCC AR4.

4. Sensitivity of inland navigation

For flood events, there is an absolute maximum in water-level, when all ships have to moor. On average, floods cause inland navigation to stop on four to five days per year on the River Rhine. Low-flow situations generally may persist longer. They are not related to general navigation stops, but can lead to critical limitations of transport. The amount of cargo transported depends on the available fairway depth and is the skippers' responsibility.

The load capacity of a vessel depends on the size of the ship (width, length) and on its draught. Figure 4 illustrates the bearing capacity of selected inland vessels in relation to their draught. The bearing capacity generally increases with growing draught. While the smaller vessels reach their maximum bearing capacity at a draught of 2.50 m, larger ships can make more use of higher water-levels. At a draught of 3.50 m the bearing capacity of the largest vessel is about 14 times higher than those of the smallest vessel. However, the large vessels lose much of this advantage if the navigable depth is limited. Relative to the smaller vessels the bearing capacity of larger vessels decreases much more when navigable depth is reduced. At extreme low water-levels, all vessels have a lower bearing capacity. However, smaller vessels are still able to carry some goods at low-flow, when larger ships are not able to navigate at all (ZKR 2005).

Without limitations in navigable depth (and width) of the fairway larger vessels are more cost-efficient than smaller ones (economies of scale) (Jonkeren et al. 2007). They have higher bearing capacities and usually can operate 24 h per day (ZKR 2005).

Fig. 4: Loading capacity of exemplary inland vessels in dependence on draught

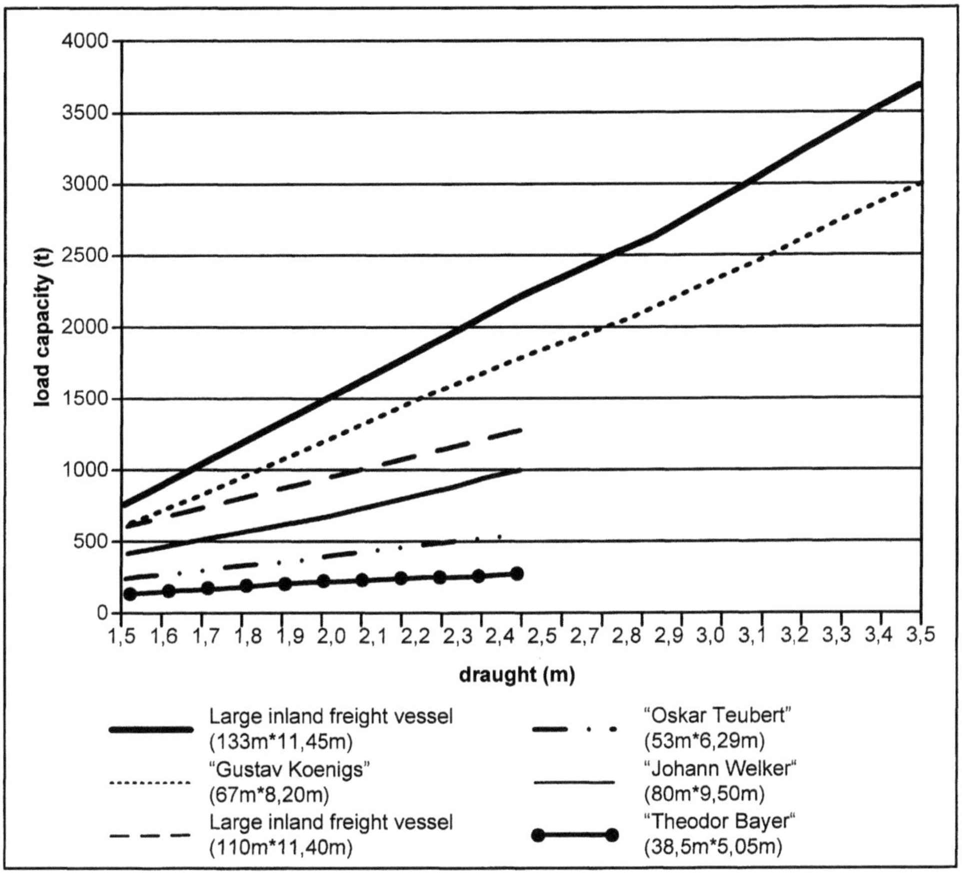

Source: own illustrations based on WESKA 2006

In the past the average size of new vessels augmented, smaller ships were not replaced by new units (ZKR 2005, 2006, 2007). While the capacity in dry transport rose, the number of vessels became smaller. This might be due to assemblies of two vessels or the disposal of smaller vessels to countries outside the River Rhine area (ZKR 2007). Obviously, the navigability of the River Rhine was regarded as comfortable and stable enough to allow for an adaptation of the fleet structure towards larger vessel sizes. As a consequence of this change, the sensitivity of inland navigation to low-flow situations increased, too.

In the summer of 2003, the bearing capacity of small vessels going from Duisburg (Lower Rhine) to Karlsruhe (Upper Rhine) was reduced to two thirds of their normal bearing capacity. Larger vessels could transport only one third of their normal capacity. Temporarily, only vessels with a draught of 1.45 m were able to navigate through the bottleneck at Kaub (BUWAL et al. 2004; Seidel 2003).

The sensitivity of the inland navigation sector to low-flow conditions is related to its fleet structure. The higher the share of large inland vessels, (1) the better is the efficiency when the navigability of the River Rhine is optimal and (2) the higher the loss can be in transport efficiency and reliability in case of extreme low-flow situations.

Magnitude of the correlation between transport quantity and water draught

Important questions are how strong the correlation between transport quantity and water-level is and how strong the impact of other influences might be.

Fig. 5: Cross-correlation of transport rate and water-level for an exemplary harbour on the Middle Rhine,

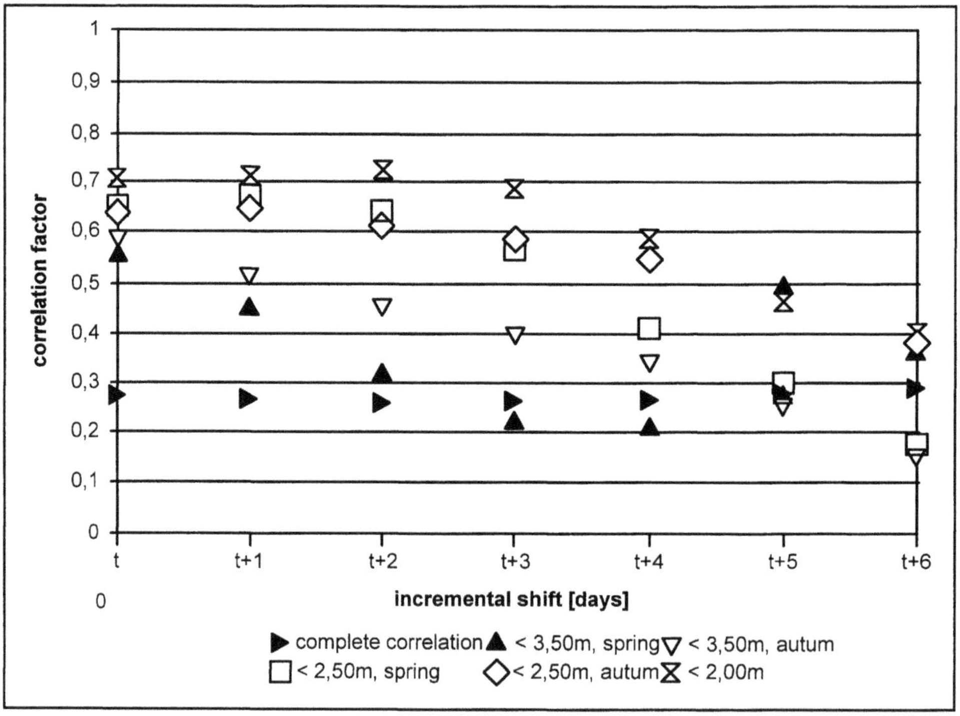

Source: calculations by the authors

To gain a basic understanding of these aspects, a case study at the Middle Rhine was done. Figure 5 demonstrates the cross-correlation (cp. Schönwiese 2006) between the transport quantity by inland vessels, transshipped at a harbour on the Middle Rhine, and the water-level at the same harbour for the year 2007. On the x-axis, the incremental shift of the correlation in days is indicated. The y-axis shows the strength of the connection (correlation factor). As can be seen, the cross-correlation of the total time series, named "complete correlation", is weak. The influences of other factors on the transport quantity (esp. demand) are stronger than the influence of the water depth. This is why in the following analyses the time series were split into several parts, containing time slices of more than 30 days with water levels < 3.50 m, < 2.50 m or < 2.00 m. For water levels < 3.50 m and < 2.50 m, two time periods were found in which the water-level dropped below these levels for more than 30 days. One of these periods was during the spring of 2007, the other one in late summer/autumn of the same year. At a water level < 2.00 m only the second period in summer/autumn had more than 30 days in a row, so only this period was analyzed. Since there were no persistent low-flow periods in 2007, no time period beyond 30 days with water-levels lower than 2.00 m could be found.

As can be seen, the strength and the course of the cross-correlations of the two seasons differ slightly. Of more importance is that the strength of the cross-correlation increases with dropping water level, reaching a maximum of 0.72 (strong connection) for the "< 2.00 m" level with a time delay of two days. While most of the cross-correlations show a peak without a time shift, this delay probably indicates that the pre-arrangements of a transport get longer when water-levels are extremely low. The rising strength of the correlation with falling water-level means, that the impact of the water-level on the transport quantity increases while other influences like demand decrease with dropping water-levels.

5. Sensitivity of bulk-cargo dependent industries

For industries dependent on transport of bulk-cargo like hard coal or mineral oil, inland navigation is generally the most efficient transport mode. However, the River Rhine has a natural discharge regime over most of its course. As a consequence, the amount of cargo that can be transported and the sizes of ships that can be used vary over time and space due to changes in the navigability of the waterway.

During a low-flow situation the bearing capacity of a given ship is limited (cf. above). As the companies have fulfilled their demand for a good anyway, more ships are needed to carry the required amount of cargo. Regardless of the amount of cargo transported, the expenses for a tour with a given ship are about the same (Jonkeren et al. 2007). As a result, transport expenses for companies are higher during low-flow periods.

Figure 6 illustrates the expenses of an exemplary power plant on the Upper Rhine. It compares transport costs assuming optimal navigability with maximum draught as compared with real navigability based on measured water-levels and related draught between 1998 and 2005. For low-flow years like 2003, the expenses assuming real conditions for navigation were up to twice as high as expenses for optimum conditions for transport. Thus, bulk-cargo dependent industries are sensitive to transport cost effects of low-flow conditions.

Fig. 6: Expenses of an exemplary power plant at the Upper Rhine for transport with 135 m inland vessels assuming optimal flow conditions ("maximum draught") and real flow conditions ("real draught")

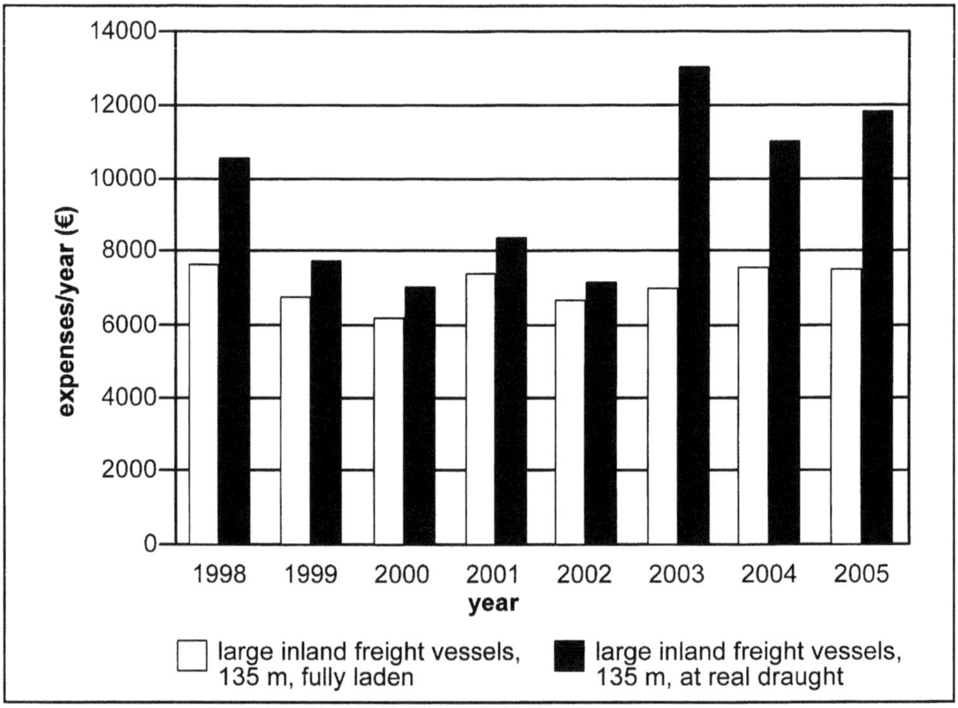

Source: Own calculations based on Jonkeren et al. (2007).

Some goods have seasonal cycles of transport volumes due to seasonality of demand (e.g. fertilizers) or seasonality of production (e.g. crops). The transport of most of these goods cannot be shifted in time easily. For example the storage of vegetables requires a refrigerated warehouse, which would be a rather expensive investment. Consequently, companies dependent on these goods are very sensitive to transport limitations in a specific season.

In the 20th century, a shift in the seasonality in the discharge of River Rhine has been observed at some river sections (figure 1). Currently, the effect of this shift on transport of seasonal goods is not recognizable. Ongoing and future research has to evaluate if the shift in low-flow seasons will continue in future.

Some industries are clustered in industrial parks along the River Rhine. Here, one company produces the basic material needed by the next company which carries out the next production step and so on. At the end of this production chain a final product results. Examples for this kind of cooperation between different chemical companies are the "Chempark" in Leverkusen (Lower Rhine; Chempark.de) or the "ChemSite" in Marl (at the Wesel-Datteln-Canal, an important waterway which is connected to the River Rhine) (Chemie.de). The company Evonik Degussa GmbH produces pyrogenic-silicic-acid using chlorine-silane produced by Momentive Performance Materials Inc. The latter company needs the pyrogenic-silicic-acid to produce silicone rubber. To close this small circle, hydrochloric-acid, a side-product of this process, is used to produce chlorine-silane (Chempark.de). But of course this is just a very small chain. A bigger chain for example can be found in Marl. In this zone the petrochemical propylen, produced by two crackers in Gelsenkirchen (approx. 30 km distance to Marl), is post-processed by several companies to basical, fine and special chemicals ranging from polystyrene to polyvenyl chloride (PVC) and tensides (Chemie.de).

Only the first and the last "link" in the production chain have an interface to "external", i.e. waterway transport, to bring raw material in or the final product out of the industrial park. Both links are sensitive to low-flow conditions. If the transport capacity for raw material is reduced, the processing company has to fall back on its stocks. Based on stock resources normal production generally can be kept up for a few days. If the low-flow period lasts too long, the production has to be reduced. As a consequence, the production of the whole production chain is affected. If the "export" of the final product is reduced due to transport limitations on the waterway, stocks can overflow. To avoid this, production has to be reduced, too. As a result, all other companies in the chain have to diminish their production volume.

Therefore, in industrial parks with close links between companies, not only is one company sensitive to transport limitations, but the whole chain of companies. Low-flow situations may be one reason for transport limitations. Others may be floods or ship accidents.

6. Possible adaptation options

Bulk-cargo dependent industries are sensitive to changing discharge conditions and have to react flexibly. Against the background of the climate-change discussion, adaptation to climate change becomes a more and more relevant topic for many companies. Using coal-fired power plants as case study industries, this section gives an overview of how bulk-cargo dependent industries can adapt. Overviews of possible adaptation strategies for other sectors like water management and navigation have been given by Moser et al. (2008a) and PIANC (2008).

6.1 General remarks

To operate economically coal-fired power plants have to be situated near a coal mining area. Alternatively, hard coal may be transported cost-effectively by inland vessels. For example, each year three million tons of coal are shipped to power plants along the River Rhine in the Federal State of Baden-Württemberg

(South-West Germany) (Hönemann 2005). Most of this coal is imported from overseas. For destinations along the River Rhine or its tributary streams, cargo is often handled in the ports of Amsterdam, Rotterdam or Antwerp (the so-called ARA-harbours). Here, hard coal is transhipped from seagoing vessels to inland navigation ships.

Compared with other German areas, for example the Ruhr region (North-West Germany), the Federal State of Baden-Württemberg has a serious disadvantage in location due to higher transport costs for coal needed for its power plants (WM 2004). Additionally this Federal State is economically affected by the risk of extreme water-levels which influences safety and reliability of inland navigation.

In the past, missing capacities due to low-flow periods at the River Rhine were only partially compensated by trains, as engines and wagons as well as drivers were not available (BUWAL et al. 2004). The summer of 2003 affirmed the importance of inland navigation for the transport of bulk-cargo, as no alternative transport options could be mobilized at short notice (ZKR 2003).

To reduce the impacts of climate change on waterways and inland navigation two general options exist: on one hand, investments, for example in hydro-engineering projects, can be made (e.g. barrages). On the other hand, operative measures can be taken, e.g. flexible transport chains or prognostic arrangements. It is of great importance that critical shortages of logistics are analyzed and eliminated. For potential alternatives in logistics good knowledge of the background conditions is essential. As mentioned above, during periods of low-flow and therefore reduced capacity of inland-navigation vessels, there is usually a simultaneous shortage in train capacities. Furthermore, alternative logistics require the corresponding infrastructure at the power plant.

It has to be emphasized that many other factors (e.g. oil prices, demand) are influencing the transport of goods as well. For that reason it cannot be answered finally if these reasons or in fact low-flow situations are the final motivation for the described adaptation efforts.

6.2 Examples of adaptation options

To depict some adaptation measures in more detail, the specific situation of the German electricity supplier EnBW is illustrated. To ensure the supply of hard coal for their power plants, EnBW has developed several logistic alternatives that retrospectively can be regarded as adaptation measures. Some of these options are exemplified: Creation/Enlargement of a second coal storage dump, paying higher freight prices, installing additional infrastructure, transport by train, construction and operation of power plants in cooperation with other power suppliers.

One important adaptation option for power plants is the creation of more storage capacities for coal. This goal can be achieved in three different ways: The easiest solution is an enlargement of the coal storage dump at the power plant, as it will be done at the coal-fired power plant in Karlsruhe. If the remaining space at this plant is not sufficient, a neighbouring plant could extend its coal dumps. At the power plant Stuttgart-Münster, the dump capacity is too small and additional unloading capacities are confined to one ship per day. If required (like in winter 2003/2004 and in winter 2005/2006), coal can be delivered from a neighbouring power plant in Gaisburg by trucks (10 km distance, around 20 to 30 truck loads per day). Thus, the coal dump in Gaisburg acts as buffer for the power plant Stuttgart-Münster. Today, a truckage company is assigned if needed, but binding contracts are getting more and more important.

The third possibility is the creation of an external dump, which can be easily reached by ships. For instance the coal dump of the power plant in Altbach/Deizisau on the River Neckar, is too small to ensure coal supply for several weeks. Therefore an external dump was created at Duisburg-Orsoy (on the Lower Rhine; considerably closer to the ARA-harbours). Adequate supply can almost always be guaranteed because the navigability from the ARA-harbours to Duisburg is very rarely hindered. In Duisburg-Orsoy coal can be transferred from ships to trains if necessary (Rothstein 2007).

Another adaptation option is paying higher freight prices depending on threshold values for new ship categories (instead of one overall threshold for older ships). This provides an incentive for the transport even at low-flow situations. Having experienced the summer of 2003 and the subsequent low-flow, EnBW now concludes three-year contracts with specified terms with the cooperating shipping companies. This results in a higher planning reliability for both partners (Rothstein 2007).

A wise adaptation measure is the installation of tri-modal infrastructure capacities at power plants for ships, trains and trucks. For example, at the coal-fired power plant Karlsruhe the infrastructure of the harbour will be supplemented by railways in order to be able to receive coal by train regularly (Rothstein 2007).

It also may be regarded as an adaptation method that coal-transport contracts for two years were concluded with the German train company Bahn AG following the summer 2003. EnBW does not possess its own train wagons yet. But, facing the planned annual need of two million tons of coal at the power plant in Karlsruhe (because of the construction of an additional plant), the company considers an acquisition or a transport alternative to the Bahn AG. Although the dump will be extended, the coal consumption of the enlarged power plant in Karlsruhe cannot be buffered sufficiently. Therefore, a train shuttle service, which secures basic supply from the ARA harbours to the plant, is planned. Additional coal transportation with ships is possible (Rothstein 2007).

Another possible adaptation measure is the construction and operation of power plants in cooperation with other power suppliers in regions with better transport possibilities. For example, the electricity company STEAG is going to build a coal-fired power plant in Walsum, close to Duisburg on the Lower Rhine (North-West Germany), where the possibilities for coal transport are favourable. EnBW is not the owner of the plant but is going to collaborate with STEAG in the construction and operation phases. Furthermore, EnBW placed a long-term contract with STEAG for the purchase of electricity. This electricity could be transported to South-West Germany (EnBW 2006; Rothstein 2007).

The next section gives a conclusion of the general aspects mentioned in sections 2 to 4 and the examples in section 5.

7. Conclusions

The influence of water-levels on inland-navigation vessels and bulk-cargo dependent industries like coal-fired power plants or industrial parks intensifies with decreasing water availability and growing vessel size. A decline of water-levels results in a smaller maximum loading capacity of vessels and therefore reduced availability and higher costs for transport. A trend towards larger vessels results in a larger dependence of inland navigation on sufficient water and a higher vulnerability potential to changes in water-levels.

The precipitation patterns and types in the Rhine basin changed during the 20th century, leading to higher winter discharge and lower discharge in summer in its nival part. In the pluvial part of the River Rhine the discharge increased throughout the whole seasonal cycle. Peak discharges generally increased, while in the frequency and duration of low-flow situations only small positive alterations were noticed in the nival part and no changes at all in the pluvial part. Occasionally a shift in the month of lowest-flow was observed, which could affect industries with a transport maximum at this time of the year.

In the recent past, some industries took measures, which may serve as adaptations to changing discharge conditions and an altered situation in transport business. An increase in storage quality and quantity, advanced links with other means of transport and shifts in transport modes are the most prominent strategies of adaptation to climate change.

Current knowledge about future effects of climate change on the discharge of the River Rhine is up to now not adequate to allow for an evaluation of specific adaptation measures. Some climate projections indicate that more extreme discharges might occur in the future with a trend to higher discharges in winter and more low-flow periods in summer. This could require additional adaptation efforts in waterway management, inland navigation, and bulk-cargo dependent industries along the River Rhine.

In order to quantify (1) effects of future climate change on water-levels, (2) the impacts of water-level changes on the efficiency and safety of inland waterway transport, and thus (3) the effects of climate change on inland navigation and bulk-cargo dependent industries, additional research is needed. Ongoing projects (e.g. the Project KLIWAS – Consequences of climate change for navigable waterways and options for the economy and inland navigation) will add to these topics.

References

Belz, J. et al. (2007), "Das Abflussregime des Rheins und seiner Nebenflüsse im 20. Jahrhundert. Analyse, Veränderungen, Trends." Bericht Nr. I-22 der KHR, 377p.

Bundesministerium für Verkehr, Bau und Stadtentwicklung, BMVBS (2008), "Schifffahrt und Wasserstraßen in Deutschland – Zukunft gestalten im Zeichen des Klimawandels." www.bmvbs.de/Anlage/original_1031087/ Schifffahrt-und-Wasserstrassen- in- Deutschland- Zukunft- gestalten- im-Zeichen-des-Klimawandels.pdf, called 21.5.2008.

Bundesamt für Umwelt, Wald und Landschaft (BUWAL); Bundesamt für Wasser und Geologie (BWG); Bundesamt für Meteorologie und Klimatologie (Meteo Schweiz) (Hrsg.) (2004), "Auswirkungen des Hitzesommers 2003 auf die Gewässer." Schriftenreihe Umwelt 369, Bundesamt für Umwelt, Wald und Landschaft, Bern.

Chemie.de, "Chemieparks bieten Wettbewerbsvorteile." www.chemie.de/news/d/230704/?sort=3, called 21.5.2008.

Chempark.de, www.chempark.de/index.php?page_id=80, called 21.5.2008.

Christensen, J.H. et al. (2007), "Prediction of Regional Scenarios and Uncertainties for Defining European Climate Change Risks and Effects: The PRUDENCE Project." Climatic Change 81, supp. 1.

European Communities (EC) (2004), "European Energy and Transport Scenarios on Key Drivers." www.eu.int/comm/dgs/energy_transport/figures/ scenarios/doc/2005_flyer_scenarios_on_key_drivers.pdf, called 9.5.2006.

Energie Baden-Württemberg AG (EnBW) (2006c), "STEAG und EnBW wollen beim Bau und Betrieb von thermischen Kraftwerken zusammenarbeiten." EnBW-Pressrelease 16.1.2006, www.enbw.com/content/de/presse/pressemitteilungen/2006/01/PM_20060116_cu_he01/index.jsp, called 9.12.2006.

Fowler, H.J. et al. (2007), "Linking climate change modelling to impacts studies: recent advances in downscaling techniques for hydrological modelling." International Journal of Climatology 27, pp. 1547-1578.

Hönemann, W. (2005), "Klimaschutz und Klimawirkung als Herausforderung für Unternehmen der Binnenschifffahrt." Presentation at the BMBF-Symposium "Rheinklima" held in Bonn 15.4.2005.

– (2006), "Climate Protection and Climate Change as a Challenge for Companies in the Inland Shipping Industry." Climate Change – Consequences for Water Resources Management, International DWA-Symposium for water management, pp. 101-104.

IPCC (2001), "Climate Change 2001." Synthesis Report, 408 p.

– (2007), "Climate Change 2007." Synthesis Report, 73 p.

Jonkeren, O., van Ommeren, J., Rietveld, P. (2007), "Climate change and inland waterway transport – welfare effects of low water-levels on the River Rhine." Journal of Transport Economics and Policy vol. 41, part 3, pp. 387-411.

Krahe, P. (2000), "EU-Projekte zu Auswirkungen von Klimaveränderungen." Arbeitskreis KLIWA (AK KLIWA) (Hrsg.), KLIWA-Symposium vom 29./30.11.2000 in Karlsruhe, KLIWA-Bericht 1, pp. 236-249.

Moser, H., Krahe, P., Maurer, T., Nilson E., Rothstein, B. and Scholten, A. (2008a, in press), "Wasserstraßen – Handlungsoptionen für Wirtschaft und Binnenschifffahrt." Forum für Hydrologie und Wasserbewirtschaftung 24.

Moser, H., Hawkes, P.J., Arntsen, Ø.A., Gaufes, P., Mai, S. and White, K.D. (2008b), "Impacts of Climate Change on Navigation." Task Group 3 of PIANC, Seminar Beijing 2008.

Middelkoop, H. and Kwadijk, J. (2000), "Towards integrated assessment of the implication of global change for Water management – The Rhine Experience." Physical, Chemical and Earth Sciences Research Reports 26, pp. 7-8, 553PIK, 2005

Pardé, M. (1933), "Le régime des cours d'eau de l'Europe orientale (en collaboration avec St.Kolupaila)." Revue Geographie Alpine, T.XXI (IV): pp. 651-748.

PIANC (2008), Waterborne transport, ports and waterways: A review of climate change drivers, impacts, responses and mitigation. Report EnviCom – Task Group 3: Climate Change and Navigation, PIANC General Secretariat, Brussels.

Rothstein, B. (2007), Elektrizitätswirtschaft als Betroffene des Klimawandels – Eine Identifikation von Betroffenheiten und Ansätze zur Anpassung an den Klimawandel dargestellt am Beispiel der Energieunternehmen EnBW und EDF. Habilitation, University of Wuerzburg, 398 p. (unpublished).

Schabalova, M. and Buishand, T. (2002), "Project SWURVE, Interim annual report 2002" www.ncl.ac.uk/swurve/downloads/2002team2knmi.pdf, called 14.12.2006.

Schönwiese, C.-D. (2006), Praktische Statistik für Meteorologen und Geowissenschaftler. Gebrüder Borntraeger Verlagsbuchhandlung, Stuttgart, 302 p.

Viner, D. (2002), "A qualitative assessment of the sources of uncertainty in climate change impacts assessment studies." In: Benislon, M. (ed.), Climatic Change-lmplications for the Hydrological Cycle and for Water Management, pp. 139-149.

WESKA (2006), Europäischer Schifffahrts- und Hafenkalender. Verein für europäische Binnenschifffahrt und Wasserstraßen e.V., Binnenschifffahrtsverlag, Duisburg.

Zebisch, M. et al. (2005), Klimawandel in Deutschland. Vulnerabilität und Anpassungsstrategien klimasensitiver Systeme. UBA-Forschungsbericht 20141253, Dessau, 203 p.

Zentralkommission für die Rheinschifffahrt (ZKR) (1997-2002), Jahresbericht der Zentralkommission für Rheinschifffahrt, volumes 1997-2002.

– (2005-2007), Marktbeobachtungen der europäischen Binnenschifffahrt, volumes 2005-2007.

About the Authors

Roland Baumhauer, Geography and Geology (Dipl.), Universities of Wuerzburg and Mainz. PhD in Geo-Sciences, University of Wuerzburg, Postdoctoral lecture qualification (Habilitation) in Geography, University of Wuerzburg Current position: Chair in Physical Geography, Dept. of Geography, University of Wuerzburg

Institute of Geography
University of Wuerzburg
Am Hubland
D-97074 Wuerzburg
Germany
baumhauer@uni-wuerzburg.de

Michael Belau, deputy head of unit Climate Change, Impacts of Climate Change and Water Balance in the Bavarian Environment Agency. As an civil engineer specialised in water resources management he is among other duties in the field of climate change and impact on the water budget responsible for the coordination of the german participation in the project ESPACE.

Bavarian Environment Agency
Climate Change
Impacts of Climate Change and Water Balance
Bayerisches Landesamt für Umwelt
Ref. 81: Klimawandel, Klimafolgen und Wasserhaushalt
Hans-Högn-Straße 12
D-95030 Hof
Germany
Michael.Belau@lfu.bayern.de

Zoltán Botta-Dukát, born in Hungary in 1970 he studied agriculture in Georgikon Faculty of PATE. He has taken PhD degree in Botany in 2003 at the University of Pécs. Currently he is a senior researcher at the Institute of Ecology and Botany of the Hungarian Academy of Sciences. His main research topics are invasibility of plant communities and application of statistical methods in the vegetation science.

Institute of Ecology and Botany of the Hungarian Academy of Sciences
Alkotmány u. 2-4
H-2163 Vácrátót
Hungary

Dr. *Lionel Boxer*, CD PhD MBA BTech(IE), is a corporate quality manager for a multinational civil engineering design business. Previously he has consulted to business and government since 1981 through Intergon – his private consultancy – and several multi-national professional service firms. At Australia's Royal Melbourne Institute of Technology (RMIT) he is an Adjunct with the Graduate School of Business and a sessional lecturer for masters of engineering programs. Previously he lectured at Centennial College in Toronto, Canada. He studied industrial engineering at the Royal Military College of Canada and Ryerson University in Toronto and business at RMIT. In 1993 he was listed in the International Who's Who of Quality and was made a Fellow of the Quality Society of Australasia. He contributes to a variety of community and voluntary groups and continues to serve as an engineer officer and bagpiper in the Australian Army Reserve after 20 years intermittent service in the Canadian Army Reserve.

Group Operations System Coordinator
Connell Wagner
Graduate School of Business
Royal Melbourne Institute of Technology
PO Box 2103
South Melbourne
3205 Victoria
Australia
lionel.boxer@rmit.edu.au

Nitin Chaudhary is currently working as a Program Associate in Natural Resource Management of Winrock International India (WII). At WII, he has been involved in research based projects and in addition, developing proposals, report writing, data collection, exploring collaboration avenues, tracking web-based new business opportunities are some of her other responsibilities. Prior to joining WII, he has done her M. Phil in Natural Resource Management, from the Indian Institute of Forest Management, Bhopal, MP, India and Maters in Environment Management from the Forest Research Institute, Dehradun, Uttaranchal, India. He has conducted two research dissertations in the area of Climate Change. He was also associated as an intern in People's Power Campaign with Climate Change and Energy Division, WWF-India, Delhi. Recently, she has attended a short term course on Sustainable Development in Slovakia. Her areas of interest are climate change, livelihood and community based natural resource management.

248-A Pocket J&K Dilshad Garden,
Delhi-110095
India
nitin.chj@gmail.com

Remigijus Ciegis is a professor at Vilnius University, Kaunas Faculty of Humanities. He holds a PhD in Economics and a doctor degree in Management. Research interest: sustainable development, climate change mitigation policies, cross-cultural management and implementation of sustainable development. The author of more than hundred scientific publications in foreign and Lithuanian scientific journals including monographs in the field of sustainable development.

Vilnius University
Kaunas Faculty of Humanities
Muitines 8
LT-44404 Kaunas
Lithuania

Amelia Clarke is a PhD Candidate in Strategy in the Desautels Faculty of Management at McGill University and the Past President of the Sierra Club Canada. She has taught in the School of the Environment at McGill University and the Faculty of Management at Dalhousie University. Her research interests include: sustainable development, corporate social and environmental responsibility, campus environmental management, and cross-sector partnerships. Amelia was also a member of the Canadian Advisory Committee to the World Peace Forum (2006), a member of the federal government's National Advisory Committee to the U.N. Habitat's World Urban Forum III (2006), and an advisor on the Canadian delegation during the United Nations World Summit on Sustainable Development (Johannesburg) negotiations (2002). Perhaps best known for founding and being the first National Director of the Sierra Youth Coalition, she is proud of having launched their climate change campaign in 1997 and their Sustainable Campuses program in 1998. Amelia has been working on environment and sustainability issues since 1989, and was recently recognized as one of 50 Canadian environmental leaders in the spring 2008 issue of the magazine Green Living.

Desautels Faculty of Management
McGill University
Montreal, Quebec H3A 1G5
Canada
Amelia.Clarke@mail.mcgill.ca

Catherine Ann Collins, Professor and Chair of Rhetoric and Media Studies. Ph.D. University of Minnesota, M.A. University of Arizona, B.F.A. Pacific Lutheran University, Magna cum laude. Areas of research interest and expertise: mass media and environmental rhetoric, narrative theory, war, visual, and commemorative rhetoric. Recent publications have appeared in Media International Australia (May 2008), International Dimensions of Mass Media Research (2007), Wilderness, Advocacy and the Media (2007), and Experiencing War: Trauma and Society in Ancient Greece and Today (2007).

Departments of Biology and Rhetoric and Media Studies
Willamette University
OR 97301 Salem, U.S.A.
ccollins@willamette.edu

Francesc Cots worked at the Institute of Environmental Science and Technology of the Autonomous University of Barcelona and now is a consultant at Eco-institut, Barcelona. His focus of research is on climate change and sustainable development institutions from a political science perspective.

Institute of Environmental Science and Technology
Autonomous University of Barcelona
Campus UAB E-08193 Cerdanyola del Vallès
Barcelona
Spain

Bálint Czúcz, born in Hungary in 1975 he studied meteorology at the Eötvös Loránd University, and horticulture at the Szent István University in Budapest. Currently he is a PhD student and a junior researcher at the Institute of Ecology and Botany of the Hungarian Academy of Sciences. His main research interests are landscape level evaluation of biodiversity and ecosystem services, and the potential impacts of climate change thereon.

Institute of Ecology and Botany of the Hungarian Academy of Sciences
Alkotmány u. 2-4
H-2163 Vácrátót
Hungary
czucz@botanika.hu

Dr. *Xingang Dai*, a senior scientist working at the Institute of Atmospheric Sciences of the Chinese Academy of Science. His interests are in global change, climate change adaptation and mitigation. He was a professor in Lanzhou university during 1989-2003 lecturing dynamic meteorology, climatology, air-sea interaction and computer modelling.

Institute of Atmospheric Physics
Chinese Academy of Sciences
P.O. Box 9804
100029 Beijing
P.R. China

Michael Dutschke, Master in Socio-Economics and in Political Science. Michael Dutschke has been a policy advisor on the design and implementation of flexible compliance mechanisms in international climate policy since 1997, being observer at all relevant meetings of the Climate Convention. He has been advising the European and developing country Governments and numerous UN institutions, including FAO, CIFOR, GEF, UNDP, UNEP, UNFCCC, environmental NGOs and private sector enterprises. In 2004 and 2005, he was a member of the CDM Executive Board's Afforestation & Reforestation Working Group. He is Lead Author of the 2007 Fourth Assessment Report of the Intergovernmental Panel on Climate Change, awarded with the Nobel Peace Price. Before founding his own company biocarbon consult in 2007, he worked with the Hamburg Institute of International Economics and at the University of Sao Paulo, Brazil. His focus is on climate and development issues in land use and biomass.

Badstr. 41
D-77652 Offenburg
Germany
michael@biocarbon.net

Dr. *Geraint Ellis* is a Senior Lecturer in the School of Planning, Architecture and Civil Engineering at Queen's University, Belfast, where he teaches and researches in the fields of sustainable development, planning governance and equality in the spatial planning process.

School of Planning Architecture and Civil Engineering (SPACE)
Queen's University Belfast
Level 3
David Keir Building
Stranmillis Road
Belfast, BT9 5AG
United Kingdom

Maria Falaleeva works at Institute for Environmental Studies (IVM), Amsterdam, The Netherlands on climate mainstreaming in local development strategies.

Institute for Environmental Studies IVM
VU University Amsterdam
De Boelelaan 1087
1081 HV Ámsterdam
The Netherlands

Jo-Anne Ferreira has taught in the Master of Environmental Education programme at Griffith University since 1997. Her background in the social sciences has led to her current research interest in the ways in which we as environmental educators encourage the adoption of environmentally friendly means of conduct.

Griffith School of Environment
Griffith University
Nathan QLD 4111
Australia

Tony Ferrier is a Director, Community Planning and Development, Kingborough Council, Tasmania, Australia. He is also a Board Member of Greening Australia (Tasmania).

Kingborough Council
Civic Centre
Locked Bag 1

Channel Hwy
Kingston
7050 Tasmania
Australia

Zsuzsanna Flachner is senior researcher at the Research Institute for Soil Science and Agricultural Chemistry, Hungarian Academy of Science supporting regional integration of climate change adaptation policies in Hungary.

RISSAC – Research Institute for Soil Science and Agricultural Chemistry
Hungarian Academy of Science
P.O. Box 35
H-1525 Budapest
Hungary

Aysha Fleming is a PhD Candidate with the Tasmanian Institute of Agricultural Research at the University of Tasmania. With a background in education, her PhD is a discourse analysis of farmers' views about climate change information.

Tasmanian Institute of Agricultural Research
University of Tasmania
Private Bag 54
Hobart TAS 7001
Australia
aysha.fleming@utas.edu.au

Estelle Gaillard is a PhD candidate at Griffith University, Brisbane, Australia. As a researcher in the field of education for sustainability, she is interested in processes of learning for personal and societal transformation and how education and communication can effectively engage people in sustainable solutions – through changing their own ways of thinking and behaviours as well as mobilizing for broader societal change.

Griffith School of Environment
Griffith University
Nathan QLD 4111
Australia

Prof. *Undine Giseke* is a certified engineer and landscape architect. In 2003 she was appointed at the TU Berlin. Since 1987 she has worked for her company bgmr – Becker Giseke Mohren Richard, Berlin/Leipzig. She is Head of Chair Landscape Architecture/ Open Space Planning at TU Berlin and project leader of the research project "Urban Agriculture as an Integrative Factor of Climate-Optimised Urban Development, Casablanca/ Morocco" funded by the BMBF programme on Future Megacities. She deals with the topics urban development and productive open space systems.

Technische Universität Berlin (TU Berlin)
Berlin Institute of Technology
Department of Landscape Architecture and Environmental Planning
Chair of Landscape Architecture/ Open Space Planning
Strasse des 17. Juni 145
D-10623 Berlin
Germany
undine.giseke@tu-berlin.de

Stefanie Greis, Dipl.-Geographer, University of Würzburg. Scientist at the European Institute for Energy Research (EIFER)

European Institute for Energy Research (EIFER)
Emmy-Noether-Str. 11
D-76131 Karlsruhe
Germany
greis@eifer.uka.de

Stephan Halloy, Ph.D., is currently Director of the Biodiversity Science Unit for the Andes Center for Biodiversity Conservation of Conservation International, coordinating research and teaching on conservation and sustainable development in the context of climate change. He is also working in environmental impact assessment, restoration, and monitoring of impacts of management and global change on biodiversity, with special thematic areas in Felids (particularly Felis jacobita), Amphibians and Vascular Plants, and regional emphasis in the Andes, Chaco and Amazonia. This work is linked to integrated social-environmental and economic development, scenarios of global change, and developing alternative diversified land uses for improved quality of life and wealth distribution. He has published over 70 papers and books and other reports and extension articles.

Conservation International
PO Box (Casilla Correo) No. 13593
La Paz
Bolivia

Dipl.-Ing. *Silvia Martin Han*, born in 1976, certified landscape planner and postgraduate "Centre for Advanced Training in Rural Development" (SLE). Since 2006 she is working as project coordinator and research assistant within the Moroccan-German project "Urban Agriculture as an Integrative Factor of Climate-Optimised Urban Development, Casablanca/ Morocco". Her focus is set on urban and rural development, structures of international cooperation in research and development and intercultural communication.

Technische Universität Berlin (TU Berlin)
Berlin Institute of Technology
Department of Landscape Architecture and Environmental Planning
Chair of Landscape Architecture/ Open Space Planning

Salzufer 12
D-10587 Berlin
Germany
silvia.martinhan@tu-berlin.de

Anjali Helferty is a graduate of Queen's University in Environmental Chemistry. She was involved with the Sierra Youth Coalition for almost three and a half years as a volunteer before being hired as National Coordinator for Sustainable Campuses in summer 2006. She also worked full-time in student government at Queen's University as Social Issues Commissioner in 2004-2005. She was profiled in the 2007 'Green Issue' of Vanity Fair along with other young people running the Campus Climate Challenge, and most recently featured in Alternatives Journal. She is passionate about campus sustainability and is particularly enthusiastic about integrating social and human rights issues into the environmental sector.

National Sustainable Campuses Coordinator
Sierra Youth Coalition
1 Nicholas St, Suite 406
Ottawa, Ontario K1N 7B7
Canada
national@syc-cjs.org

Dr. *Frank Helten*, born in 1942, graduated in social sciences and doctorate in philosophy. Long standing experience in social research in interdisciplinary and European research projects as coordinator, project manager and participant. Experience in urban and technology assessment studies as well as in urban and regional planning. He is working with TU Berlin as a freelancer within the Moroccan-German project "Urban Agriculture as an Integrative Factor of Climate-Optimised Urban Development, Casablanca/ Morocco".

Technische Universität Berlin (TU Berlin)
Berlin Institute of Technology
Department of Landscape Architecture and Environmebiontal Planning
Chair of Landscape Architecture/ Open Space Planning
Salzufer 12
D-10587 Berlin, Germany
Frank.Helten@t-online.de

Ferenc Horváth is an ecologist research fellow, working for the Institute of Ecology and Botany of the Hungarian Academy of Sciences. He is responsible for development and maintenance of MÉTA database and information tools and services.

Institute of Ecology and Botany of the Hungarian Academy of Sciences
Alkotmány u. 2-4
H-2163 Vácrátót
Hungary

Susan R. Kephart, Professor of Biology, Ph.D. Indiana University, BA Ohio Wesleyan University, Summa cum laude. Areas of research interest and expertise: Systematics, plant evolutionary biology, especially hybridization and speciation; pollination ecology and conservation biology. Kephart and students conduct research on native plant populations in coastal to montane habitats, focusing on plant-pollinator interactions, the events and processes affecting species distinctions and successful restoration efforts, and how scientists versus news reporters write about concepts like wilderness and biodiversity. She publishes broadly on plant genera as diverse as Asclepias, Camassia, and Silene, including two recent feature articles on the covers of American Journal of Botany (October 2005) and New Phytologist (February 2006). Among other professional activities, she has served or continues to assist as a National Science Foundation panelist (2007, 2008), an associate editor, a technical advisor for the Pringle Creek Watershed Council, the opponent on a PhD defense in Uppsala, Sweden, and past president of the Oregon Academy of Sciences. Her work has been supported by multiple grants from Earthwatch Institute, the M.J. Murdock Foundation, Oregon Community Foundation, the National Science Foundation, and Willamette University.

Departments of Biology and Rhetoric and Media Studies
Willamette University
900 State Street
Salem, Oregon 97301
U.S.A.
skephart@willamette.edu

Jane Korck is a qualified hydrologist who works at the Bavarian Environment Agency in the field of climate change quantification, adaptation to climate change and the implementation of the EU Water Framework Directive.

Bavarian Environment Agency
Climate Change
Impacts of Climate Change and Water Balance
Bayerisches Landesamt für Umwelt
Ref. 81: Klimawandel, Klimafolgen und Wasserhaushalt
Hans-Högn-Straße 12
D-95030 Hof
Germany

Rosa Kouri was the National Director of the Sierra Youth Coalition from 2005-2007. She is a tireless spokesperson and advocate for young people, as she firmly believes that youth deserve the power to shape a sustainable world. She founded the Sustainable McGill Project at McGill University, bicycled the west coast of California to raise awareness about industrial agriculture and poor working conditions for migrant farmers, sat on the board of the national Sierra Club Canada, and organized youth delegates at the 11th UN global climate summit. In September, 2006, she was a founding member of the Canadian Youth Climate Coalition. Raised in Saskatoon, Saskatchewan, Rosa completed a joint honors degree in economics and political science from McGill University in Montreal. She was recognized by the World Conservation Union for outstanding work on climate change issues.

111 Albert Ave
Saskatoon
Saskatchewan S7N 1E6
Canada
rosakouri@gmail.com

György Kröel-Dulay is a senior researcher at the Institute of Ecology and Botany of the Hungarian Academy of Sciences. He got an MSc in 1995 and PhD in 2002 in biology at the Eötvös Loránd University, Budapest. In 1996-97 he studied at the Colorado State University and worked in semiarid and arid grasslands in Colorado and New Mexico, US. Currently working mostly in the sand grasslands and forest steppes of central Hungary, his main research interest is the effect of changing land-use and climate on vegetation composition and dynamics, from local to landscape and regional scales.

Institute of Ecology and Botany of the Hungarian Academy of Sciences
Alkotmány u. 2-4
H-2163 Vácrátót
Hungary

Prof. *Walter Leal Filho* (BSc, PhD, DSc, DL), Head of the Research and Transfer Centre Applications of Life Sciences, Hamburg University of Applied Sciences, has been working in the field of Environmental Management since 1987. He deals with topics such as General Life Sciences, Environmental Management, Innovation and Sustainable Development (including Energy and Climate issues).

Research and Transfer Centre "Applications of Life Sciences"
Hamburg University of Applied Sciences
Faculty of Life Sciences
Lohbruegger Kirchstraße 65
D-21033 Hamburg
Germany
walter.leal@haw-hamburg.de

Peat Leith recently completed his PhD at the University of Tasmania, jointly between the Tasmanian Institute of Agricultural Research and the School of Geography and Environmental Studies. His research focussed on the integration of local and scientific knowledge to better manage for climate variability in the Australian rangelands.

Tasmanian Institute of Agricultural Research
University of Tasmania
Private Bag 54
Hobart TAS 7001
Australia
peat.leith@utas.edu.au

Franziska Mannke, Project Manager, Research and Transfer Centre Applications of Life Sciences, Hamburg University of Applied Sciences, has been co-ordinating EU projects on a national and an international level since 2005. Her work currently focuses on development economics, climate change and renewable energies.

Research and Transfer Centre "Applications of Life Sciences"
Hamburg University of Applied Sciences
Faculty of Life Sciences
Lohbruegger Kirchstraße 65
D-21033 Hamburg
Germany
franziska.mannke@haw-hamburg.de

Evangelos I. Manolas, BA, MA, PhD, Assistant Professor, was born in Naxos, Greece, in 1961. He has received a Bachelor of Arts in Sociology from the University of Essex (1983), a Master of Arts in International Relations from the University of Kent at Canterbury (1985) and a Philosophy Doctorate from the University of Aberdeen (1989). He is Assistant Professor of Sociology and Environmental-Forest Education in the Department of Forestry and Management of the Environment and Natural Resources at the Democritus University of Thrace.

Department of Forestry and Management of the Environment and Natural Resources
Democritus University of Thrace
193 Pantazidou Street
68200 Orestiada
Greece
email: emanolas@fmenr.duth.gr

Sandra T. Marquart-Pyatt is Assistant Professor of Sociology at Utah State University. She has authored papers on environmental concerns cross-nationally, political attitudes and behaviors across contexts, and co-edited a volume on social change in Poland over time. This research has appeared in such journals as *Society and Natural Resources*, *Social Science Quarterly*, *Rural Sociology*, and *Political Behavior*. Her current work investigates pathways to environmental activism across countries and contextual influences on environmental attitudes cross-nationally and in specific cases. She received her doctoral degree in sociology in 2004 from The Ohio State University in Columbus, OH, USA.

Department of Sociology
Social Work and Anthropology
Utah State University
0730 Old Main
Logan, UT 84322-0730
U.S.A.
s.marquartpyatt@usu.edu

Darryn McEvoy is senior researcher at the International Centre of Integrated Assessment and Sustainable Development of the University of Maastricht where he works on climate regional and local adaptation from a comparative perspective.

ICIS – University of Maastricht
P.O. Box 616
6200 MD Maastricht
The Netherlands

Solveig Mimler, M.A. Sociology, Political Science, Anthropology, University of Trier. PhD at the European Institute for Energy Research (EIFER)

European Institute for Energy Research (EIFER)
Emmy-Noether-Str. 11
D-76131 Karlsruhe
Germany
mimler@eifer.uka.de

Zsolt Molnár is a botanist and landscape historian born in 1966. He is the leader of the MÉTA program mapping the actual vegetation of Hungary. Currently he is a senior researcher at the Institute of Ecology and Botany of the Hungarian Academy of Sciences.

Institute of Ecology and Botany of the Hungarian Academy of Sciences
Alkotmány u. 2-4
H-2163 Vácrátót
Hungary

Harri Moora, MSc in Environmental Management and Policy, International Institute for Industrial Environmental Economics, Lund University, Sweden. The field of research: sustainable consumption and production, environmental systems analysis, life cycle assessment of waste management systems.

Tallinn University of Technology
Institute of Environmental Engineering
Stockholm Environmental Institute
SEI-Tallinn
Lai 34
10133 Tallinn
Estonia
harri.moora@seit.ee

Ulrike Müller, Dipl.-Geographer, University of Würzburg. PhD at the European Institute for Energy Research (EIFER)

European Institute for Energy Research (EIFER)
Emmy-Noether-Str. 11
D-76131 Karlsruhe
Germany
mueller@eifer.uka.de

Enno Nilson, Geography (Dipl.), University of Bonn. PhD in Geography, University of Aachen. Scientist, Department Water Balance, Forecasting and Predictions, Federal Institute of Hydrology, Germany.

Federal Institute of Hydrology
Department Water Balance
Forecasting and Predictions
Am Mainzer Tor 1
D-56068 Koblenz
Germany
nilson@bafg.de

Dr. *Melissa Nursey-Bray*, Lecturer, originally from Adelaide, Melissa comes from a history and politics background, with degrees from the University of Adelaide. She subsequently moved into Indigenous resource management and trained Indigenous Rangers in natural resource management across Northern Australia. In this field she then spent a number of years working for government, Indigenous groups, ENGOs and universities. Her core interest is the investigation of how communities can become involved in and contribute to environmental management in interdisciplinay ways. Most recently this has included consideration of how social science can help build community capacity to resolve global problems such as climate change, specifically local government. Currently a social sciences lecturer at the National Centre for Marine Conservation and Resource Sustainability, Australian Maritime College Melissa delivers courses in marine sustainability and management and the application of qualitative methodologies. She has a PhD in Indigenous resource management, and a Master degree majoring in water quality management. She is co-convenor of the Social Sciences Climate Change Research Network (SSCCRN) and currently working on developing courses to train professionals in local government, NRM and ports and shipping in how to adapt to climate change.

Social Sciences
National Centre for Marine Conservation and Resource Sustainability
Australian Maritime College
University of Tasmania
Locked Bag 1370
Launceston Tas

Australia
m.nursey-bray@amc.edu.au

Alvina Reihan, hydrologist, Doctoral degree in civil engineering. Specific research area: climate change impact on surface water resources and water quality, river runoff modelling, water balance, water resources estimation and management and scenarios for water management.

Tallinn University of Technology
Institute of Environmental Engineering
Ehitajate tee 5
19086 Tallinn
Estonia
alvina.reihan@ttu.ee

Prof. Dr. habil. *Benno Rothstein*, Head of the Bachelor course BioEnergy, University of Applied Forest Sciences Rottenburg, has been working in the field of Georisk Management in the Electricity Sector since 2002. He deals with the topics adaptation to Climate Change, inland navigation, bulk-cargo dependent industries, (Bio-)Energy.

Hochschule für Forstwirtschaft Rottenburg
University of Applied Forest Sciences
Schadenweilerhof
D-72108 Rottenburg am Neckar
rothstein@hs-rottenburg.de

Dr. *Manoj Roy* is Lecturer in the School of Planning, Architecture and Civil Engineering at Queen's University, Belfast, where he teaches and researches in the fields of sustainable development, spatial anaylsis and modelling, material flow analysis and urban climate change.

School of Planning Architecture and Civil Engineering (SPACE)
Queen's University Belfast
Level 3
David Keir Building,
Stranmillis Road
Belfast, BT9 5AG
United Kingdom
m.roy@qub.ac.uk

Anja Scholten, Meteorology (Dipl.), University of Bonn. Ph.D. student, University of Wuerzburg

Institute for Geography
University of Wuerzburg
Am Hubland
97074 Wuerzburg
Germany
anja.scholten@uni-wuerzburg.de

Dalia Streimikiene is a professor at Vilnius University Kaunas Faculty of Humanities. She graduated from Kaunas Technological University in 1985 and obtained a PhD in Vilnius Technical University in 1997 and passed habilitation procedures in 2005 in the same university. The main areas of research are energy and environmental economics and policy, development of economic tools for environmental regulation in energy sector, promotion of renewable energy sources and sustainable energy development. The author of more than 70 scientific publications in foreign and Lithuanian scientific journals.

Vilnius University
Kaunas Faculty of Humanities
Muitines 8
LT-44404 Kaunas
Lithuania
dalia@mail.lei.lt

J. David Tàbara is senior researcher at the Institute of Environmental Sciences and Technology of the Autonomous University of Barcelona where he also lectures participatory social science methods for Integrated Assessment and Sustainability Science.

Institute of Environmental Science and Technology
Autonomous University of Barcelona
Campus UAB E-08193 Cerdanyola del Vallès
Barcelona
Spain
joandavid.tabara@uab.cat

Gergely Torda obtained his M.Sc. in Applied Zoology at the Szent Istvan University, Hungary, and is working towards his Ph.D. in Marine Ecology, currently at the University of Las Palmas de Gran Canaria, Spain. He also holds a position of junior researcher at the Institute of Ecology and Botany of the Hungarian Academy of Sciences, collaborating in landscape ecology studies that require the use of GIS techniques.

Institute of Ecology and Botany of the Hungarian Academy of Sciences
Alkotmány u. 2-4
H-2163 Vácrátót
Hungary

Daniela Ulloa is a specialist on issues related to environmental education and communication as is active in various projects in Bolivia.

Instituto de Ecología
Universidad Mayor de San Andrés
La Paz
Bolivia

Frank Vanclay is professor of rural sociology and leads the Rural Social Research Group in the Tasmanian Institute of Agricultural Research at the University of Tasmania in Australia. He specialises in social research relating to farming, agriculture and natural resource management. His other interests include social impact assessment and the social meaning of place.

Tasmanian Institute of Agricultural Research
University of Tasmania
Private Bag 54
Hobart TAS 7001
Australia
frank.vanclay@utas.edu.au

Viktoria Voronova, MSc in Environmental management and cleaner production. Fields of research: environmental management and its systems, life cycle assessment in waste management, scenarios for waste management.

Tallinn University of Technology
Institute of Environmental Engineering
Ehitajate tee 5
19086 Tallinn
Estonia
viktoria.voronova@ttu.ee

Ceri Warnock, LLB, LLM (Auck) is a lawyer and practiced as a barrister in the UK for many years prior to joining the Faculty of Law at the University of Otago, New Zealand. She lectures in resource management law, energy law and the law relating to climate change. She has written widely on climate change related topics.

Faculty of Law
University of Otago
P.O. Box 56
Dunedin,
New Zealand
ceri.warnock@otago.ac.nz

Claire Warnock, BA DipArch (Sheffield) RIBA/ARB, is an architect with 11 years experience ranging from high-density housing schemes to restoration projects. Since joining Matthew Lloyd Architects in October 2002, Claire has been involved in a wide range of projects including master planning for mixed use, mixed tenure developments; strategic analysis of underused urban sites and regeneration of dilapidated buildings. She is particularly interested in sustainable design, materials and techniques and is currently responsible for the construction and completion of the Norfolk Park Green Homes project in Sheffield for the Tower Hamlets Environment Trust, due to complete in Autumn 2008. Claire became a partner of the practice in 2008.

Matthew Lloyd Architects LLP
1 the Hangar, Perseverance Works,
38 Kingsland Road
London E2 8DD,
United Kingdom
www.matthewlloyd.co.uk

Hans Weber, head of unit Climate Change, Impacts of Climate Change and Water Balance in the Bavarian Environment Agency. He is civil engineer specialised in water resources management and among other projects responsible for the german participation in the ESPACE project.

Bavarian Environment Agency
Climate Change
Impacts of Climate Change and Water Balance
Bayerisches Landesamt für Umwelt
Ref. 81: Klimawandel, Klimafolgen und Wasserhaushalt
Hans-Högn-Straße 12
D-95030 Hof,
Germany

Saskia Werners' main research focus is adaptation to climate change in water management. Based at the Centre for Water and Climate of Wageningen University and Research Centre, Saskia Werners' work is firmly rooted in the global change community.

Wageningen University and Research Centre
Box 47
6700 AA Wageningen
The Netherlands

Dr. *Sam Wong* is lecturer in the School of Earth and Environment at the University of Leeds, UK. His research interests lie in sustainable technology, water governance, natural resource management, collective action, social capital and community participation. He is currently a grant reviewer for the Natural Environmental Research Council and a contributing editor for the Journal of Technology. He is also a reviewer for more than 10 highly-regarded journals, including Journal of Environmental Planning and Management, of Environmental Values, Local of Environment, and Sustainable Development. He is an affiliated member of various professional institutions, includeing the Development Studies Association and the International Commission on Irrigation and Drainage.

School of Earth and Environment
University of Leeds
United Kingdom
s.wong@leeds.ac.uk

Karina Yager, Ph.D. Candidate, is in the Department of Anthropology at Yale University, New Haven. Her dissertation research is focused on pastoral practices in the Andes of Peru and Bolivia. Her research combines ethnography, remote sensing analysis and vegetation studies to consider the social and ecological dynamics of landscape change in the Andes. She has worked with interdisciplinary research teams focused on assessing climate change and its effects on local production in high Andean environments.

Department of Anthropology
Yale University
10 Sachem St.
New Haven 06511,
U.S.A.
karina.yager@yale.edu

Thematic Index

Umweltbildung, Umweltkommunikation und Nachhaltigkeit
Environmental Education, Communication and Sustainability

Herausgegeben von/Edited by Walter Leal Filho

Band 1 Walter Leal Filho (Hrsg.): Umweltschutz und Nachhaltigkeit an Hochschulen. Konzepte-Umsetzung. 1998.

Band 2 Walter Leal Filho / Farrukh Tahir (Eds.): Distance Education and Environmental Education. 1998.

Band 3 Walter Leal Filho (ed.): Environmental Engineering: International Perspectives. 1998.

Band 4 Walter Leal Filho / Mauri Ahlberg (Eds.): Environmental Education for Sustainability: Good Environment, Good Life. 1998.

Band 5 Walter Leal Filho (ed.): Sustainability and University Life. 1999. 2nd, revised ed. 2000.

Band 6 Wout van den Bor / Peter Holen / Arjen Wals / Walter Leal Filho (eds.): Integration Concepts of Sustainability into Education for Agriculture and Rural Development. 2000.

Band 7 Manfred Oepen / Winfried Hamacher (Eds.): Communicating the Environment. Environmental Communication for Sustainable Development. 2000.

Band 8 Walter Leal Filho (ed.): Communicating Sustainability. 2000.

Band 9 Walter Leal Filho (ed.): Environmental Careers, Environmental Employment and Environmental Training. International Approaches and Contexts. 2001.

Band 10 Rolf Jucker: Our Common Illiteracy. Education as if the Earth and People Mattered. 2002.

Band 11 Walter Leal Filho (ed.): Teaching Sustainability at Universities. Towards curriculum greening. 2002.

Band 12 Walter Leal Filho (ed.): International Experiences on Sustainability. 2002.

Band 13 Lars Wohlers (Hrsg.): Methoden informeller Umweltbildung. 2003.

Band 14 Ulisses Azeiteiro / Fernando Conçalves / Walter Leal Filho / Fernando Morgado / Mário Pereira (eds.): World Trends in Environmental Education. 2004.

Band 15 Walter Leal Filho / Arnolds Ubelis (eds.): Integrative approaches towards sustainability in the Baltic Sea Region. 2004.

Band 16 Walter Leal Filho / Michael Littledyke (eds.): International Perspectives in Environmental Education. 2004.

Band 17 Daniela Krumland: Beitrag der Medien zum politischen Erfolg. Forstwirtschaft und Naturschutz im Politikfeld Wald. 2004.

Band 18 Walter Leal Filho / Bernd Delakowitz (Hrsg.): Umweltmanagement an Hochschulen: Nachhaltigkeitsperspektiven. 2005.

Band 19 Walter Leal Filho / Volker Lüderitz / Gunther Geller (Hrsg.): Perspektiven der Ingenieurökologie in Forschung, Lehre und Praxis. 2006.

Band 20 Walter Leal Filho (ed.): Handbook of Sustainability Research. 2005.

Band 21 Walter Leal Filho / Dieter Greif / Bernd Delakowitz (eds.): Sustainable Chemistry and Biotechnology – A Contribution to Rivers Management. 2006.

Band 22 Walter Leal Filho / David Carpenter (eds.): Sustainability in the Australasian University Context. 2006.

Band 23 Walter Leal Filho / Arnolds Ubelis / Dina Berzina (eds.): Sustainable Development in the Baltic and Beyond. 2006.

Band 24 Walter Leal Filho (ed.): Innovation, Education and Communication for Sustainable Development. 2006.

Band 25 Walter Leal Filho / Mario Salomone (eds.): Innovative Approaches to Education for Sustainable Development. 2006.

Band 26 Walter Leal Filho / Franziska Mannke / Philipp Schmidt-Thomé (eds.): Information, Communication and Education on Climate Change – European Perspectives. 2007.

Band 27 Ulisses Miranda Azeiteiro, Fernando Gonçalves, Ruth Pereira, Mário Jorge Pereira, Walter Leal Filho, Fernando Morgado (eds.): Science and Environmental Education. Towards the Integration of Science Education, Experimental Science Activities and Environmental Education. 2008.

Band 28 Walter Leal Filho / Nils Brandt / Dörte Krahn / Ronald Wennersten (eds.): Conflict Resolution in Coastal Zone Management. 2008.

Band 29 Walter Leal Filho / Franziska Mannke (eds.): Interdisciplinary Aspects of Climate Change. 2009.

www.peterlang.de

Walter Leal Filho / Franziska Mannke / Philipp Schmidt-Thomé (eds.)

Information, Communication and Education on Climate Change – European Perspectives

Frankfurt am Main, Berlin, Bern, Bruxelles, New York, Oxford, Wien, 2007.
213 pp., num. fig., tab. and graph.
Environmental Education, Communication and Sustainability.
Edited by Walter Leal Filho. Vol. 26
ISBN 978-3-631-56682-4 · hardback € 36.90*

It is now beyond any doubt that climate change represents a major threat to the environmental, social and economic well-being of the planet. The United Nations Framework Convention on Climate Change and its Kyoto Protocol are examples of what is being done. A further means to work towards a reduction in greenhouse emissions is by raising awareness on the causes and consequences of climate change. The authors attempt to contribute to the global debate by outlining some educational methods, approaches and projects which have been focusing on climate issues in Europe. The examples illustrate that education, communication and training can play a key role in supporting efforts toward addressing climate change. This book is prepared as part of the INTERREG IIIB (Baltic Sea) Project "ASTRA" (Developing Policies and Adaptations Strategies to Climate Change in the Baltic Sea Region), a project involving Estonia, Finland, Germany, Latvia, Lithuania and Poland, and is also a further means of support to the UN Decade of Education for Sustainable Development, linking awareness, education and communication on climate change, one of the major challenges of modern times.

Contents: ESPACE – A New Transnational Approach to Adapting to the Impacts of Climate Change · Communication Climate Change impacts – an Approach Derived from Two Regional Development Projects · Baltex Assessment of Climate Change for the Baltic Sea Basin (BACC) · Promoting Climate Friendly Communities – Climate Change Mitigation and Sustainability: Educational Issues · and many more

Frankfurt am Main · Berlin · Bern · Bruxelles · New York · Oxford · Wien
Distribution: Verlag Peter Lang AG
Moosstr. 1, CH-2542 Pieterlen
Telefax 0041 (0) 32/376 17 27

*The €-price includes German tax rate
Prices are subject to change without notice

Homepage http://www.peterlang.de

www.ingramcontent.com/pod-product-compliance
Ingram Content Group UK Ltd.
Pitfield, Milton Keynes, MK11 3LW, UK
UKHW051846100926
8556IPUK00005B/474

9 783631 581537